AF789941

Handbook of Nuclear Medicine and Molecular Imaging for Physicists

Series in Medical Physics and Biomedical Engineering
Series Editors: Kwan Hoong Ng, E. Russell Ritenour, Slavik Tabakov

Recent books in the series:

Clinical Radiotherapy Physics with MATLAB: A Problem-Solving Approach
Pavel Dvorak

Advances in Particle Therapy: A Multidisciplinary Approach
Manjit Dosanjh and Jacques Bernier (Eds)

Radiotherapy and Clinical Radiobiology of Head and Neck Cancer
Loredana G. Marcu, Iuliana Toma-Dasu, Alexandru Dasu, and Claes Mercke

Problems and Solutions in Medical Physics: Diagnostic Imaging Physics
Kwan Hoong Ng, Jeannie Hsiu Ding Wong, and Geoffrey D. Clarke

Advanced and Emerging Technologies in Radiation Oncology Physics
Siyong Kim and John W. Wong (Eds)

A Guide to Outcome Modeling in Radiotherapy and Oncology: Listening to the Data
Issam El Naqa (Ed)

Advanced MR Neuroimaging: From Theory to Clinical Practice
Ioannis Tsougos

Quantitative MRI of the Brain: Principles of Physical Measurement, Second edition
Mara Cercignani, Nicholas G. Dowell, and Paul S. Tofts (Eds)

A Brief Survey of Quantitative EEG
Kaushik Majumdar

Handbook of X-ray Imaging: Physics and Technology
Paolo Russo (Ed)

Graphics Processing Unit-Based High Performance Computing in Radiation Therapy
Xun Jia and Steve B. Jiang (Eds)

Targeted Muscle Reinnervation: A Neural Interface for Artificial Limbs
Todd A. Kuiken, Aimee E. Schultz Feuser, and Ann K. Barlow (Eds)

Emerging Technologies in Brachytherapy
William Y. Song, Kari Tanderup, and Bradley Pieters (Eds)

Environmental Radioactivity and Emergency Preparedness
Mats Isaksson and Christopher L. Rääf

The Practice of Internal Dosimetry in Nuclear Medicine
Michael G. Stabin

For more information about this series, please visit: www.routledge.com/Series-in-Medical-Physics-and-Biomedical-Engineering/book-series/CHMEPHBIOENG

Handbook of Nuclear Medicine and Molecular Imaging for Physicists

Radiopharmaceuticals and Clinical Applications, Volume III

Edited by
Michael Ljungberg

CRC Press
Taylor & Francis Group
Boca Raton London New York

CRC Press is an imprint of the
Taylor & Francis Group, an informa business

First edition published 2022
by CRC Press
6000 Broken Sound Parkway NW, Suite 300, Boca Raton, FL 33487-2742

and by CRC Press
2 Park Square, Milton Park, Abingdon, Oxon, OX14 4RN

© 2022 Taylor & Francis Group, LLC

CRC Press is an imprint of Taylor & Francis Group, LLC

Reasonable efforts have been made to publish reliable data and information, but the author and publisher cannot assume responsibility for the validity of all materials or the consequences of their use. The authors and publishers have attempted to trace the copyright holders of all material reproduced in this publication and apologize to copyright holders if permission to publish in this form has not been obtained. If any copyright material has not been acknowledged please write and let us know so we may rectify in any future reprint.

Except as permitted under U.S. Copyright Law, no part of this book may be reprinted, reproduced, transmitted, or utilized in any form by any electronic, mechanical, or other means, now known or hereafter invented, including photocopying, microfilming, and recording, or in any information storage or retrieval system, without written permission from the publishers.

For permission to photocopy or use material electronically from this work, access www.copyright.com or contact the Copyright Clearance Center, Inc. (CCC), 222 Rosewood Drive, Danvers, MA 01923, 978-750-8400. For works that are not available on CCC please contact mpkbookspermissions@tandf.co.uk

Trademark notice: Product or corporate names may be trademarks or registered trademarks and are used only for identification and explanation without intent to infringe.

Library of Congress Cataloging-in-Publication Data
A catalog record has been requested for this book

ISBN: 978-1-138-59331-2 (hbk)
ISBN: 978-1-032-05956-3 (pbk)
ISBN: 978-0-429-48950-1 (ebk)

DOI: 10.1201/9780429489501

Typeset in Times
by Newgen Publishing UK

Access to and support material: www.routledge.com/9781138593312

Contents

Preface

During the spring of 2017, I was writing a review of a proposal for a book to potentially be published by CRC Press. Upon closing the discussion with CRC Press regarding the result of this review, I was asked to be an editor for a handbook of nuclear medicine, with focus on physicists of this field. After spending the summer thinking about a relevant table of contents and related potential authors, I formally accepted the offer. I soon realized that the field of nuclear medicine was too extensive to be covered in a single book. After consolidating with the publisher, it was decided that instead of one book it would be best to develop three volumes with the titles, (I) Instrumentation and Imaging Procedures, (II) Modelling, Dosimetry and Radiation Protection and (III) Radiopharmaceuticals and Clinical Applications

My vision was to create state-of-the-art handbooks, encompassing all major aspects relating to the field of Nuclear Medicine. The chapters should describe the theories in detail but also, when applicable, have a practical approach, focusing on procedures and equipment that are either in use today, or could be expected to be of importance in the future. I realized that the topic of each chapter would be broad enough, in principle, to lay the foundation for individual books of their own. As such, the chapters needed only cover the most relevant aspects of each topic. Therefore, this book series will, hopefully, serve as references for different aspects relating to both the academic and the clinical practice of a medical physicist.

I originally struggled with the definition of the word 'handbook'. I did not want the chapters to serve as point-by-point guidelines, but rather to function as independent chapters to be read more or less independently of one another. Consequently, there is some overlap in the content between chapters but, from a pedagogical point-of-view, I do not see this as a drawback, as repetition of key aspects may aid in the learning.

In this third and final volume, we highlight the process of production and application of radiopharmaceuticals. With this, we also describe the chemical composition of these compounds, as well as some of the main clinical applications where radiopharmaceuticals may be used. Following a brief introduction to the field of radiopharmacy, three chapters in this book are dedicated towards in-depth descriptions of common radionuclides and radiopharmaceuticals used during (i) diagnostic studies utilizing planar/Single Photon Emission Computed Tomography (SPECT) imaging; (ii) Positron Emission Tomography (PET) imaging; and (iii) radiotherapy. These chapters are followed by those describing procedures relating to quality control and manufacturing (good manufacturing practices (GMP)), also encompassing aspects like environmental compliance. Furthermore, we illustrate how facilities handling these chemicals should be designed to comply with set regulations.

Like many pharmaceuticals, the development of radiopharmaceuticals relies heavily on the use of mouse models. Thus, the translation of radiopharmaceuticals (i.e., the process undertaken to assure that the functionality and safety of a newly developed drug is maintained also in a human context), is covered in a later chapter. This is followed by a chapter emphasizing the importance of safe waste disposal and how to assure that these procedures meet the requirements set for the disposal of hazardous waste.

Eleven chapters have also been dedicated towards describing various medical procedures utilizing clinical nuclear medicine as a tool for diagnostics and therapeutics. As physicists may be involved in clinical trials, a chapter describing the procedures and regulations associated with these types of studies is also included. This is followed by a chapter focusing on patient safety and another on an imaging modality not based on ionizing radiation – ultrasound. Finally, the last chapter of this book discusses future perspectives of the field of nuclear medicine.

These three volumes are the result of the efforts of outstanding authors who, despite the exceptional circumstances related to the COVID-19 pandemic, have managed to keep to the deadline of the project – although, I must admit, there were times when I questioned the feasibility of doing this. As COVID-19 hit, many of us were faced with unexpected tasks to solve: Distance teaching, restrictions, and changes in administration, and sometimes also rapid modifications to local procedures at departments and hospitals. Naturally, the combined effect of these interruptions impacted the time available to dedicate to writing. However, despite these many challenges, we all did our utmost to complete the chapters according to the deadline.

I would like to thank all authors for their contributions, which made this book possible. You have all done a phenomenal job, especially considering the extraordinary circumstances we are currently faced with, but also considering the fact that you all have other obligations of high priority. I would especially like to thank Professor Philip H. Elsinga, who initially helped me define the content of the radiopharmaceutical section being prepared for Volume III. This subtopic of nuclear medicine is the one I have the least knowledge of, and I am therefore very grateful for the kind support I received during the initial planning of Volume III.

I would like to thank CRC Press officials for entrusting me with the position as editor of this series of books. I would also like to thank Kirsten Barr, Rebecca Davies and Francesca McGowan, who have been the points of contact for me during these years.

It is also important to acknowledge two authors who are sadly no longer with us: Anna Celler, University of British Columbia, Vancouver, Canada, and Lennart Johansson, Umeå University, Sweden. Both were dear friends and great scientists. Throughout the years, their work has made a huge impact in their respective fields of research.

Finally, I would like to dedicate this work to my wife, Karin, as well as to my beloved daughter Johanna, who lives in Brisbane, where she is pursuing her PhD at the University of Queensland. Karin – I am so grateful for your patience, especially during the intense period around Christmastime right before the submission of the manuscript for this volume. I love you both very much.

Michael Ljungberg, PhD
Professor Medical Radiation Physics, Lund
Lund University, Lund, Sweden

Access to colour images and support material: http://www.routledge.com/9781138593312

Editor

Michael Ljungberg is a Professor at Medical Radiation Physics, Lund, Lund University, Sweden. He started his research in the Monte Carlo field in 1983 through a project involving a simulation of whole-body counters but later changed the focus to more general applications in nuclear medicine imaging and SPECT. Parallel to his development of the Monte Carlo code, SIMIND, he began working in 1985 with quantitative SPECT and problems related to attenuation and scatter. After earning his PhD in 1990, he received a research assistant position that allowed him to continue developing SIMIND for quantitative SPECT applications and to establish successful collaborations with international research groups. At this time, the SIMIND program became used world-wide. Dr. Ljungberg became an associate professor in 1994 and, in 2005, after working clinically as a nuclear medicine medical physicist, received a full professorship in the Science Faculty at Lund University. He became the Head of the Department of Medical Radiation Physics at Lund in 2013 and a full professor in the Medical Faculty in 2015.

Aside from the development of SIMIND – including new camera systems such as CZT detectors – his research includes an extensive project in oncological nuclear medicine. In this project, he and colleagues developed dosimetry methods based on quantitative SPECT, Monte Carlo absorbed-dose calculations, and methods for accurate 3D dose planning for internal radionuclide therapy. Lately, his work has focused on implementing Monte Carlo–based image reconstruction in SIMIND. He is also involved in the undergraduate education of medical physicists and bio-medical engineers and supervises MSc and PhD students. In 2012, Professor Ljungberg became a member of the European Association of Nuclear Medicines task group on Dosimetry and served that association for six years. He has published over a hundred original papers, 18 conference proceedings, 18 books and book chapters, and 14 peer-reviewed papers.

Contributors

Erik H. J. G. Aarntzen
Department of Medical Imaging, Radboud University Medical Center, Nijmegen, The Netherlands

An Aerts
Radiobiology Unit, Interdisciplinary Biosciences, Institute for Environment, Health and Safety, SCK CEN, Belgian Nuclear Research Centre, Mol, Belgium

Kanhaiyalal Agrawal
Department of Nuclear Medicine, All India Institute of Medical Sciences (AIIMS), Bhubaneswar, India

Mohamed Altai
Department of Clinical Sciences, Lund, Lund University, Lund, Sweden

Rimma Axelsson
Division of Radiology, Department of Clinical Science, Intervention and Technology (CLINTEC), Karolinska Institute, and Medical Radiation Physics and Nuclear Medicine, Karolinska University Hospital, Huddinge, Stockholm, Sweden

Marika Bajc
Skåne University Hospital, Department of Clinical Sciences, University Hospital Lund, Lund, Sweden

Jim Ballinger
King's College London, London, United Kingdom

Lise Borgwardt
Department of Clinical Physiology, Nuclear Medicine and PET, Rigshospitalet and University of Copenhagen, Copenhagen, Denmark

Wen-Yi Chang
PET Center, Department of Nuclear Medicine, National Taiwan University Hospital, Taipei, Taiwan

Ching-Hung Chiu
Department of Nuclear Medicine, National Taiwan University Hospital, Taipei, Taiwan

Emre Demirci
Department of Nuclear Medicine, Faculty of Medicine, Yeditepe University, Istanbul, Turkey

Philip H. Elsinga
Department of Nuclear Medicine and Molecular Imaging, University of Groningen, University Medical Center Groningen, Groningen, The Netherlands

Ulrika Estenberg
Medical Radiation Physics and Nuclear Medicine, Karolinska University Hospital, Huddinge, Stockholm, Sweden

Pedro Fragoso Costa
Nuclear Medicine Department, University Clinics Essen, University Duisburg-Essen, Essen, Germany

Danique Giesen
Department of Nuclear Medicine and Molecular Imaging, University of Groningen, University Medical Center Groningen, The Netherlands

Nic Gillings
Department of Clinical Physiology and Nuclear Medicine, Copenhagen University Hospital Rigshospitalet, Denmark

Andor W. J. M. Glaudemans
University of Groningen, University Medical Center Groningen, Department of Nuclear Medicine and Molecular Imaging, Groningen, The Netherlands

Gopinath Gnanasegaran
Department of Nuclear Medicine, Royal Free London NHS Foundation Trust, London, United Kingdom

Sandra Heskamp
Department of Medical Imaging, Radboud University Medical Center, Nijmegen, The Netherlands

Liselotte Højgaard
Department of Clinical Physiology, Nuclear Medicine and PET, Rigshospitalet and University of Copenhagen and DTU, Copenhagen, Denmark

Søren Holm
Department of Clinical Physiology, Nuclear Medicine and PET, Rigshospitalet, Copenhagen, Denmark

Hanna Holstein
Medical Radiation Physics Lund, Lund University, Lund, Sweden

Maria Holstensson
Division of Medical Imaging Technology, Department of Clinical Science, Intervention and Technology (CLINTEC), Karolinska Institute, Stockholm, Sweden

Ya-Yao Huang
PET Center, Department of Nuclear Medicine, National Taiwan University Hospital, Institute of Medical Device and Imaging, National Taiwan University College of Medicine, Molecular Imaging Center, National Taiwan University, Taipei, Taiwan

Susanna Jakobson Mo
Department of Radiation Sciences, Diagnostic Radiology, Umeå University, Umeå, Sweden

Tomas Jansson
Division of Biomedical Engineering, Department of Clinical Sciences Lund, Lund University, Sweden

Lena Jönsson
Medical Radiation Physics Lund, Lund University, Lund, Sweden

Tomas Kirkhorn
Division of Biomedical Engineering, Department of Clinical Sciences Lund, Lund University, Sweden

Jacek Koziorowski
RadCad, Åtvidaberg, Sweden

Rebecca Krimins
Department of Radiology, Johns Hopkins University, Baltimore

Yvette Kruiter
Department of Nuclear Medicine and Molecular Imaging, University of Groningen, University Medical Center Groningen, The Netherlands

Anne Larsson Strömvall
Department of Radiation Sciences, Radiation Physics, Umeå University, Umeå, Sweden

Ari Lindqvist
Research Unit of Pulmonary Diseases, Clinical Research Institute HUCH, Helsinki University Hospital HUS and Helsinki University, Helsinki, Finland

Sofie Lindskov Hansen
Department of Clinical Physiology, Nuclear Medicine and PET, Rigshospitalet, Copenhagen, Denmark

Mark Lubberink
Nuclear Medicine & PET, Department of Surgical Sciences, Uppsala University, Uppsala, Sweden

Gert Luurtsema
Department of Nuclear Medicine and Molecular Imaging, University of Groningen, University Medical Center Groningen, Groningen, The Netherlands

David Minarik
Radiation Physics, Skåne University Hospital, Malmö and Nuclear Medicine, Malmö, Department of Translational Medicine, Lund University, Malmö, Sweden

Jessie R. Nedrow
Department of Radiology, University of Pittsburgh, Pittsburgh, USA

Oliver Neels
Institute of Radiopharmaceutical Cancer Research, Helmholtz-Zentrum Dresden-Rossendorf (HZDR), Dresden, Germany

Meltem Ocak
Department of Pharmaceutical Technology, Faculty of Pharmacy, Istanbul University, Istanbul, Turkey

László Pávics
Department. of Nuclear Medicine, University of Szeged, Hungary

Steffie M. B. Peters
Department of Medical Imaging, Radboud University Medical Center, Nijmegen, The Netherlands

Latifa Rbah-Vidal
CRCINA, INSERM, CNRS, Université d'Angers, Université de Nantes, Nantes, France

Martin Šámal
Institute of Nuclear Medicine, First Faculty of Medicine, Charles University and the General University Hospital in Prague, Prague, Czech Republic

Erik Samén
Department of Oncology and Pathology, Karolinska Institutet, and Department of Karolinska Radiopharmacy, Karolinska University Hospital, Stockholm, Sweden

Margret Schottelius
Laboratoire Translationnel des Sciences Radiopharmaceutiques, Centre Hospitalier Universitaire Vadois (CHUV), Lausanne, Switzerland

Katarina Sjögreen Gleisner
Medical Radiation Physics Lund, Lund University, Lund, Sweden

Joanna Strand
Medical Oncology, Department of Clinical Sciences Lund, Lund University, Lund, Sweden

Sven-Erik Strand
Medical Radiation Physics Lund, Lund University, Lund, Sweden

Anna Sundlöv
Medical Oncology, Department of Clinical Sciences Lund, Lund University, Lund, Sweden

Elin Trägårdh
Clinical Physiology and Nuclear Medicine, Skåne University Hospital, Malmö and Nuclear Medicine, Malmö, Department of Translational Medicine, Lund University, Malmö, Sweden

Thuy A. Tran
Department of Oncology and Pathology, Karolinska Institutet, and Department of Karolinska Radiopharmacy, Karolinska University Hospital, Stockholm, Sweden

Jiří Trnka
Institute of Nuclear Medicine, First Faculty of Medicine, Charles University and the General University Hospital in Prague, Czech Republic

David Ulmert
Department of Molecular and Medical Pharmacology David Geffen School of Medicine, UCLA, USA

Fijs W. B. van Leeuwen
Interventional Molecular Imaging laboratory, Department of Radiology, Leiden University Medical Centre, Leiden, The Netherlands

Rolf Zijlma
Department of Nuclear Medicine and Molecular Imaging, University of Groningen, University Medical Center Groningen, The Netherlands

1 Principles behind Radiopharmacy

Thuy A. Tran and Erik Samén

CONTENTS

1.1 HISTORY OF RADIOPHARMACY

The first radiotherapeutics: The history of the existence of the radiopharmacy field originated mainly from the United States sometime in the late 1930s, when the first use of sodium iodide-131 was introduced by Hamilton and Soley [1]. In 1951 Sodium iodide-131 became the first radiopharmaceutical approved by the US Food and Drug Administration (FDA) for use in thyroid patients [2].

The first Single Photon Emission Computed Tomography (SPECT) radiopharmaceutical: One of the most important developments in radiopharmacy was in the early 1960s with the clinical introduction of ^{99m}Tc generators for the preparation of radiopharmaceuticals by Powell Richards at Brookhaven National Laboratory [3]. Although more and more Positron Emission Tomography (PET) cameras are available, SPECT imaging using ^{99m}Tc-labelled radiopharmaceuticals accounts for at least 80 per cent of all nuclear medicine investigations worldwide due to its low cost and availability.

The first PET radiopharmaceutical: Historically, ^{11}C is the first positron-emitting radionuclide was discovered in 1934 by Lauritsen and colleagues [4]. It proved to be very useful and was first studied in humans in 1945 by Tobias, who investigated the interactions of [^{11}C]carbon monoxide with red blood cells [5]. ^{11}C-labelled compounds are today frequently used, both in clinical routine and in PET research studies.

1.2 BASIC REQUIREMENTS FOR A RADIOPHARMACY

The procedures and requirements for operating a radiopharmacy are largely dependent on the nature of the radiopharmaceuticals that are to be produced and their intended use. As general principles, the following basics requirements apply:

- **Good Manufacturing Practice (GMP):** Radiopharmaceuticals are strictly regulated as they are both medicinal products and radioactive materials. A radiopharmaceutical must be produced in accordance with GMP. The applicable legislation and guidelines and how to implement them are well-reviewed by Lange and colleagues [6]. The GMP guidelines for medicinal products are well described in EudraLex Vol. 4 (EU GMP) [7], and in supplementary guidelines applied specifically for the manufacture of radiopharmaceuticals according to EU GMP

DOI: 10.1201/9780429489501-1

Annex 3 [8]. More details on the GMP rules and recommendations are found in Chapter 8 in this volume, and further readings on the small-scale production of radiopharmaceuticals can be found in Gillings and colleagues [9].

- **Design considerations of a radiopharmacy:** One of the most important considerations, when constructing and designing a radiopharmacy, is to define early on the goals and scope of the radiopharmacy: What types of radiopharmaceuticals will it produce (manufacturing and/or compounding)? Will it be a centralized radiopharmacy, an industrial manufacturer and/or a PET centre? Will it produce established routine radiopharmaceuticals for nuclear medicine investigations, or will it also be a research centre? The radiopharmacy laboratories are then designed, constructed, and adapted to suit the purposes and operations to be carried out. These details are outlined in Chapters 5 and 7 in this volume. Useful reading on this topic can be found in Gillings and colleagues [9].
- **Instrumentation and equipment needed**: A large number of instruments and equipment are used in the production and quality control (QC) of radiopharmaceuticals. Essential equipment for the production of PET radiopharmaceuticals [such as Fludeoxyglucose F-18 (FDG)] includes cyclotrons (see further in Chapter 4 in Vol. I), hot cells, lead-shielded Laminar Air Flow (LAF) benches, synthesis modules, dose calibrators, and dispensers. A wide variety of equipment is also dedicated to the QC of radiopharmaceuticals before release for patient administration. These are, for example, high-performance liquid chromatography (HPLC) coupled with a radiation detector for analysing the radiochemical purity; a thin-layer chromatography (TLC) scanner for determining radiochemical impurities; gas chromatography (GC) for analysis of residual solvents such as ethanol, acetonitrile, or acetone; a pH-meter, equipment for checking endotoxin contents, bubble point tester for checking the filter integrity, and so forth. More about the methods and equipment for QC can be found in Chapter 6. Other critical instrumentation needed are, for example, radiation monitoring equipment, facility monitoring systems, and general laboratory equipment.
- **Permits from authorities**: In Europe, radiopharmaceuticals are considered a special classification of medicinal products. Manufacturers of radiopharmaceuticals must comply with different applicable requirements and regulations, and both local and EU directives. A manufacturer's authorization is needed before the radiopharmacy is allowed to manufacture medicinal products for human use. Before permission is granted to produce any radiopharmaceutical, the radiopharmacy or PET centre is inspected by the Medical Product Agency. When all requirements and conditions are fulfilled, authorization is issued (see example in Figure 1.1).

1.3 WORKFLOW FOR MANUFACTURING OF RADIOPHARMACEUTICALS

Radiopharmaceuticals differ from medicinal products in several aspects. One noteworthy difference is the short shelf-life, particularly for most of the current diagnostic radiopharmaceuticals. Due to the short half-lives of several radionuclides (for example ^{11}C, ^{68}Ga, ^{99m}Tc) used for radiolabelling, the radiopharmacy must be located close to the Nuclear Medicine Department, where injections and scanning of patients in the PET or SPECT cameras are performed. For therapeutic radiopharmaceuticals (such as ^{131}I, ^{177}Lu-based, ^{90}Y-based) and other long-lived radionuclides (for example ^{111}In), they can be supplied from external vendors.

The production of radiopharmaceuticals is a demanding process and involves several steps that are performed in a limited time frame of a few hours. These steps include radionuclide production (cyclotron bombardment), radiosynthesis, quality control, and release of the product for administration to patients. A schematic illustration of the workflow is shown in Figure 1.2. The various means to produce a radiopharmaceutical are outlined in section 1.4. The details of quality control (QC) in the production of radiopharmaceuticals can be found in the IAEA TECDOC series [10] and in Chapter 6. In general, QC procedures are always performed for manufactured products prior to patient administration. The main purpose of QC is to ensure that the radiopharmaceuticals are of acceptable quality in terms of purity, identity, and sterility and that they are safe before they are administered in humans.

The quality specifications that a product must meet are set in the *European Pharmacopoeia* (*Ph. Eur.*) or corresponding such as the *US Pharmacopoeia* (USP) or the *International Pharmacopoeia*. It is the responsibility of the manufacturer to use the applicable regulations and thereafter develop QC methods to adhere to. On another note, radiopharmaceuticals must comply with the QC attributes stated in the official monograph published in *Ph. Eur.*, but the monograph does not state that all QC tests have to be carried out before the release of a product. In practice, due to the short half-life of many radionuclides, it is not possible to perform all QC tests on every single batch that is produced. Therefore, a set of pre-release QC tests and a set of post-release QC tests are usually defined, and the establishment of release criteria should be based upon a risk-factor assessment [10]. A post-release QC test is, for example, a sterility analysis that requires 14 days before it can be finalized. Although there might be some tests and specifications that differ between radiopharmaceuticals, the following limits and methods, exemplified for [^{18}F]PSMA-1007, are commonly used for many radiopharmaceuticals (Table 1.1).

MANUFACTURER'S AUTHORISATION [1, 2]

1. Authorisation Number	X.X.X-YEAR-YYYYY
2. Name of authorisation holder	Karolinska Universitetssjukhuset FO Radiofarmaci
3. Address(es) of manufacturing site(s)	Karolinska Universitetssjukhuset FO Radiofarmaci, Akademiska stråket 1, Solna 171 64, Sweden
4. Legally registered address of authorisation holder	Akademiska stråket 1, Solna 171 64, Sweden
5. Scope of authorisation and dosage forms [2]	ANNEX 1 and/ or ANNEX 2
6. Legal Basis of authorisation	Art. 40 of Directive 2001/83/EC Art. 13 of Directive 2001/20/EC
7. Name of responsible officer of the competent authority of the member state granting the manufacturing authorisation	confidential
8. Signature	
9. Date	2019-06-15
10. Annexes attached	Annex 1 and/or Annex 2 Optional Annexes as required: Annex 3 (Addresses of Contract Manufacturing Site(s)) Annex 4 (Addresses of Contract laboratories) Annex 5 (Name of Qualified Person) Annex 6 (Name of responsible persons) Annex 7 (Date of inspection on which authorisation granted, scope of last inspection) Annex 8 (Manufactured/ imported products authorised) [3]

[1] *The authorisation referred to in paragraph 40(1) of Directive 2001/83/EC and 44(1) of Directive 2001/82/EC, as amended, shall also be required for imports coming from third countries into a Member State.*

[2] *Guidance on the interpretation of this template can be found in the Help menu of EudraGMDP database.*

[3] *The Competent Authority is responsible for appropriate linking of the authorisation with the manufacturer's application (Art. 42(3) of Directive 2001/83/EC and Art. 46(3) of Directive 2001/82/EC as amended).*

Online EudraGMDP. Ref key: 39817 MANUFACTURER'S AUTHORISATION: DE_RP_01_MIA_2019_0039 Page 1 of 3

FIGURE 1.1 Example of an authorization from the Swedish Medical Product Agency.

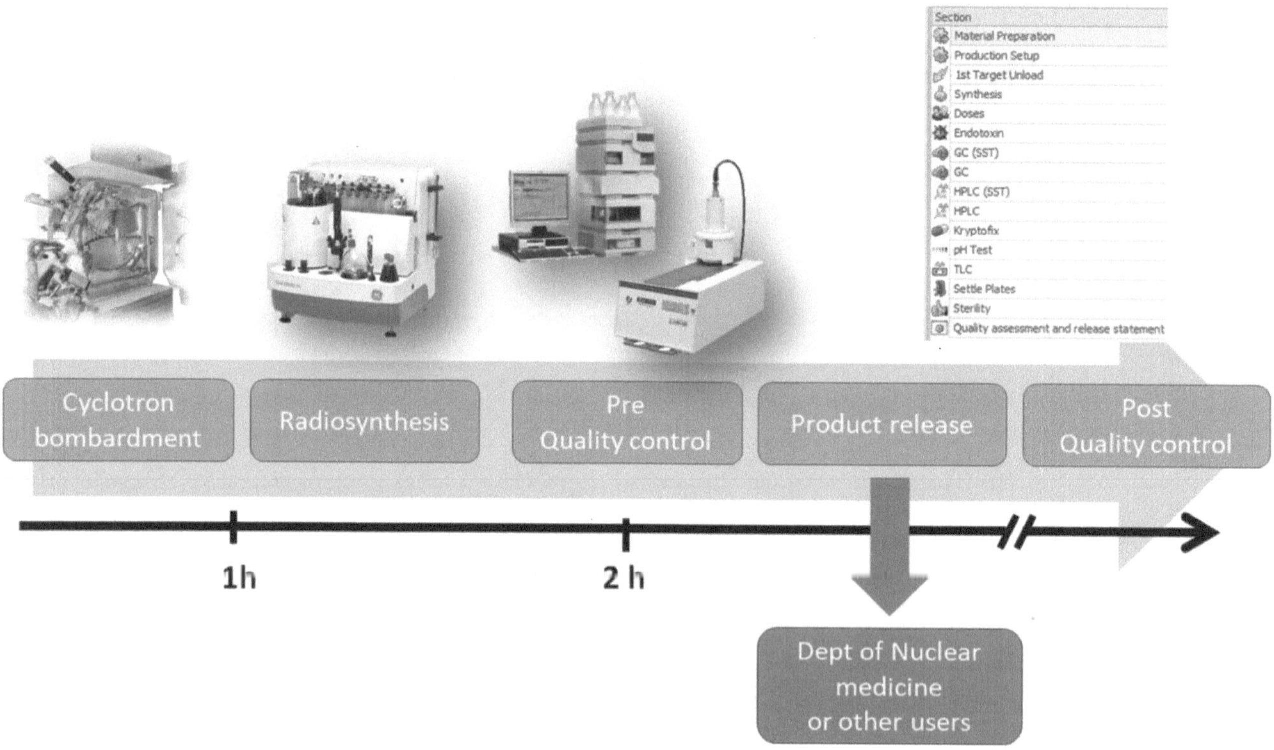

FIGURE 1.2 Illustration of a typical workflow for manufacturing a radiopharmaceutical for clinical use.

TABLE 1.1
Typical QC Tests, Specifications and Recommended Test Methods for the Most Commonly Manufactured Radiopharmaceuticals. The Example Below is Based on the Specifications for [^{18}F]PSMA-1007

QC parameter	Specifications	Test method
V_{max}	Maximum recommended injected volume, V_{max} is 10 mL	
Appearance	Clear and colourless	Visual inspection
pH	4.5 – 8.5	pH-meter or pH-paper
Radionuclidic identity	A. Half-life is 105 – 115 min B. Gamma photons of 511 keV and possible a summation peak of 1022 keV	A. Dose calibrator B. HPGe detector
Radionuclidic purity	≥ 99.9% of total radioactivity is due to ^{18}F	HPGe detector
Total radiochemical purity	≥ 91% of the total radioactivity is [^{18}F]PSMA-1007	HPLC and TLC
Chemical purity	PSMA-1007 ≤ 0.1 mg/ V_{max} Any other impurity ≤ 0.1 mg/V_{max} The sum of PSMA-1007 and other impurities ≤ 0.5 mg/V_{max}	HPLC
Chemical purity	Tetrabutylammonium ≤ 2.6 mg/V_{max}	Spot test
Residual solvents	DMSO ≤ 50 mg/V_{max} Ethanol ≤ 10 %	GC
Endotoxins	≤ 175 EU/V_{max}	Limulus amoebocyte lysate
Sterility	Sterile	Direct inoculation

Abbreviations: HPGe = High Purity Germanium; EU = Endotoxin Units

TABLE 1.2
Overview of the Principles and Steps of Preparation Versus Manufacturing.[1] Table Modified from Velikyan et al.[11]

Preparation: Kit-based synthesis	**Manufacturing: manual or automated synthesis**
Mostly generator-based products	Mostly cyclotron-produced based products
Example: ^{99m}Tc-based products, ^{68}Ga kit-based compounds	Examples: ^{11}C-, ^{18}F-based radiopharmaceuticals
• Radionuclide eluted from generator into the product vial containing the active product ingredient	• Radionuclide bombarded from cyclotron or eluted from generator, into reaction vial
• Labelling reaction occurs in the **product** vial	• Labelling reaction in the **reaction** vial
	• Purification of the product
• Formulation for example by adding physiological buffers or 0.9% NaCl	• Formulation for example by adding physiological buffers or 0.9% NaCl
	• Sterile filtration
	• Dispensing
• QC: very simple using iTLC (most products do not require QC)	• QC: **extensive QC**
• Approved for administration **by the production chemist or compounding staff**	• Released for administration **by QA/QP staff** (i.e. not production chemist)
• Cost: **Low**	• Cost: **Expensive**
• Time: < 1 h	• Time > usually 2 h

When a radiopharmaceutical fulfils all pre-release specifications, it can be released for use. Each batch of the final product is released by the Qualified Person (QP) or other staff holding a release delegation from the QP. The purpose of the batch release is to ensure that each individual batch has been manufactured in accordance with the principles and requirements of GMP and/or its marketing authorization or any other relevant legal obligations.

1.4 METHODS FOR SYNTHESIS OF RADIOPHARMACEUTICALS

There are different means to radiolabelling a molecule. The radiochemical reactions, or radiochemistry, differ between the types of radionuclides used. There is a vast list of literature describing the specific chemistry applied for each radionuclide, and several factors need to be considered. Common requirements for a radiopharmaceutical synthesis or preparation include that it should have a high labelling yield, high radiochemical purity, and high chemical purity as well as good microbiological and radiochemical stability. Moreover, the cost, the practical performance of the staff, and radiation-protection measures all need to be taken into account. In general, the synthesis and preparation of radiopharmaceuticals can largely be divided into, *kit-based* (also generator-based or compounding), *manual*, and *automated*, the last two of which are categorized as *manufacturing*, while kit-based procedures are considered as *compounding* or *preparation*. See Table 1.2 for the comparison of these two routes.

1.4.1 Kit-based Synthesis

A kit is supplied with sterile, pre-packed materials and ingredients, ready for the preparation of a specific radiopharmaceutical. Note, however, that the radionuclide is not a part of the kit, and the radionuclides are normally obtained from generators – for example, ^{99m}Tc-generators or ^{68}Ga/^{68}Ge-generators. Typical examples of radiopharmaceuticals that are prepared through a kit-based synthesis are the clinically approved ^{99m}Tc radiopharmaceuticals (see Chapter 2 on SPECT radiopharmaceuticals) and during the last three or four years also the ^{68}Ga radiopharmaceuticals (for example the ^{68}Ga-DOTATOC = SomaKIT TOC or ^{68}Ga-PSMA-11 kit). See Figure 1.3 for a schematic illustration of a kit-based labelling of SomaKIT TOC that is used for PET visualization of somatostatin receptor-expressing tumours [12].

As illustrated in Table 1.1, kit-based preparation of radiopharmaceuticals offers several advantages, including a shorter labelling process, shorter preparation time, and often the QC is simplified. The whole process normally requires less than one hour from preparation to approval for administration to humans.

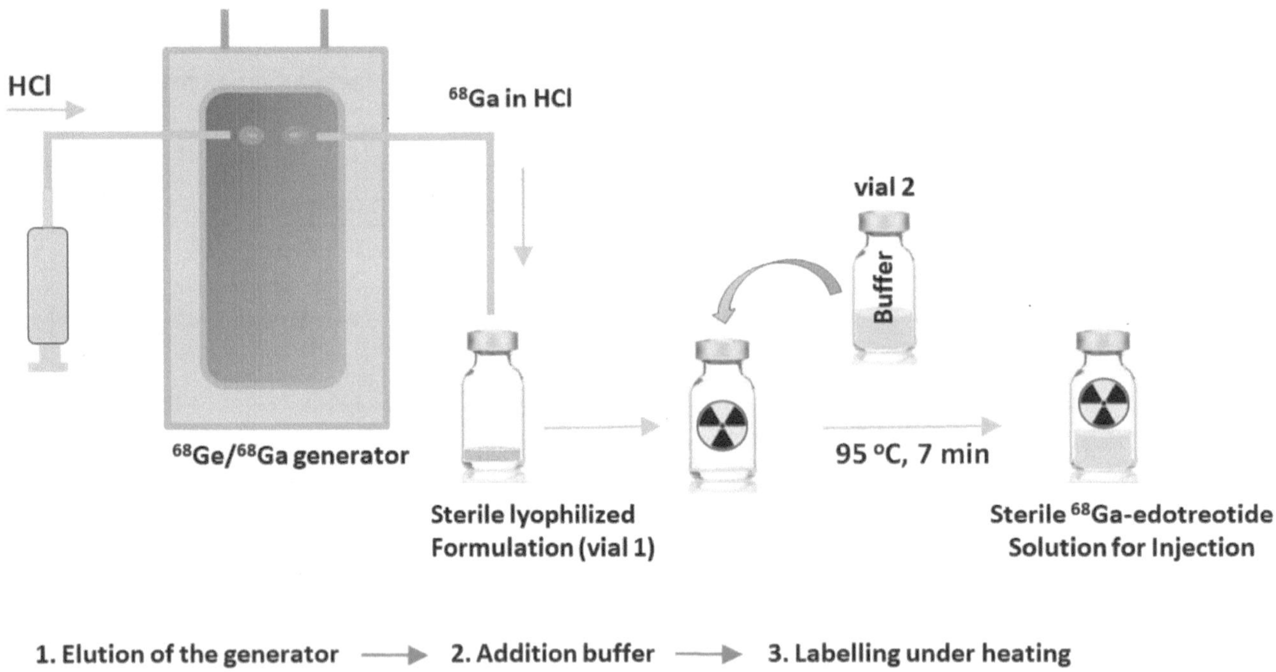

FIGURE 1.3 A schematic illustration of a kit-based preparation of ^{68}Ga-edotreotide (SomaKIT TOC).

1.4.2 Automated Synthesis

The manufacturing process of a radiopharmaceutical consists of several steps in order to yield a final sterile solution for injection, fulfilling all QC attributes required according to GMP regulations. The increasing clinical demand, radioprotection/safety, and regulatory requirements have driven the necessity for automation of the production process. Overall, the design and development of the automated synthesis of PET radiopharmaceuticals have played an important role in radiochemistry research, which today ranges from in-house built modules to sophisticated commercially available modules [13]. Using automated synthesis – particularly when automated dispensing of the bulk product to patient vials is also automatically integrated – would offer several advantages including high reproducibility, good radiation protection for the operators, and the feasibility of producing large batches (up to several hundred GBq). Automated synthesis is the recommended route for routine clinical production of radiopharmaceuticals, and the automation can be cassette-based or on reused tubings and semi-automated synthesis, as further described below. A clear trend in recent years is that the development of automation has expanded rapidly to include long-lived and large molecules, for examples ^{89}Zr- and ^{177}Lu-labelled antibodies [14, 15].

1.4.2.1 Cassette-based Automated Synthesis

A good example of a fully cassette-based automated synthesis is the synthesis of many ^{18}F-based radiopharmaceuticals, such as [^{18}F]FDG, [^{18}F]PSMA-1007, [^{18}F]FLT, and [^{18}F]FMISO. Briefly, in such a module or synthesizer, all chemical reactions/steps occur within a disposable cassette. The cassette is sterile, pre-packed, intended for one single-use and contains all reagent vials, syringes, tubings and valves that are assembled on a stationary manifold. Prior to the synthesis, the cassette is mounted on the module and is connected to the radioactivity inlet for receiving [^{18}F]fluoride from the cyclotron bombardment (Figure 1.4). The preparation steps, as well as the synthesis itself, can be started and operated from a programmed software adapted for each radiopharmaceutical outside the hot cell. Several automated systems/modules are today commercially available, giving a high multiscale production of a large range of radiopharmaceuticals.

1.4.2.2 Semi-automated Synthesis with Tubing Systems

Another type of synthesis module, also commonly used for the production of ^{18}F-labelled and ^{11}C-labelled tracers, comprise the semi-automated systems (see Figure 1.5). In these modules, the tubings, reagent vials, connectors, and reaction and collection vessels are permanent and are not changed or disposed of after each use. These components are

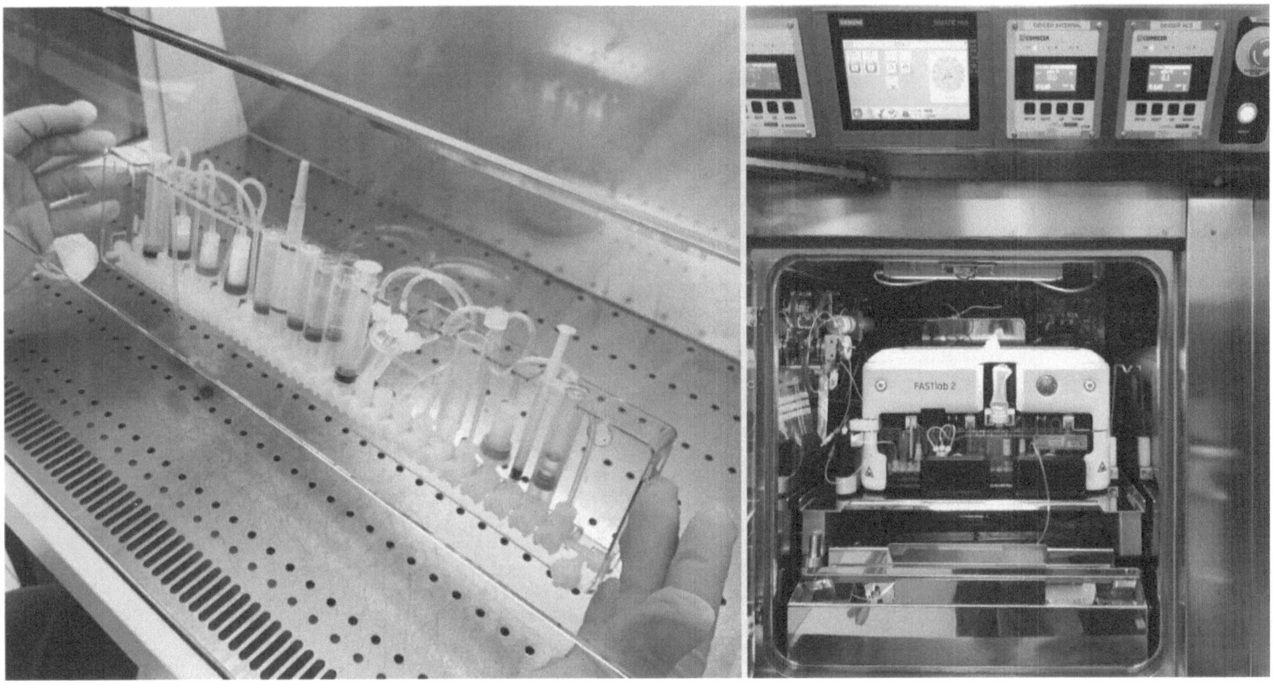

FIGURE 1.4 Example of an automated synthesis of radiopharmaceuticals. On the left is shown a single-use disposable sterile cassette for [^{18}F]FDG synthesis (GE Healthcare). The cassette is mounted on the Fastlab 2 synthesis module (GE Healthcare) that is placed inside a lead-shielded hotcell for radiation protection (image on the right). The synthesis is fully automated and operated through a software in a computer outside the hotcell.

only replaced after a certain defined number of uses, or when needed. As there is no use of disposable cassettes, the cost of each production is lower, but it requires an additional cleaning program, usually performed after the activity has decayed below a certain limit (usually the day after). Furthermore, before putting these systems into use, an extensive cleaning validation has to be performed to ensure that the method used is efficient in removing any chemicals, solvents, or products from a previous synthesis to avoid possible cross-contamination in the next batch. The cleaning validation has to show whether any present microorganism or bacterial endotoxin must also be removed by the cleaning procedure [16].

While a lot of effort is dedicated to developing and adapting the synthesis of many radiopharmaceuticals to be cassette-based, some radiotracers still require a semi-preparative HPLC purification to have a pure product. These semi-automated systems are therefore extended with an HPLC part.

1.4.3 Manual Synthesis

Early in the research and development of new radiopharmaceuticals, radiochemists usually start the synthesis manually, particularly when small batches (a low amount of radioactivity) are produced. For several reasons, manual synthesis would not, in the long term, be adequate for routine clinical applications. These reasons include the risk of a high radiation dose to the operator and the risk for high variability in the production. In routine clinical production for human use, the manual synthesis is often up-scaled, optimized, and adapted to automation to produce large batches that will suffice for administration in several patients [17]. Nonetheless, manual synthesis is still used and applied in several early research tracers.

1.5 SOME TRENDS AND DEVELOPMENTS IN RADIOPHARMACY

The fields of radiopharmacy and radiochemistry are evolving rapidly and have contributed to the nuclear medicine field with novel radiopharmaceuticals. Development of new radiopharmaceuticals is the driving force in molecular imaging. Some trends and developments in radiopharmacy are highlighted below.

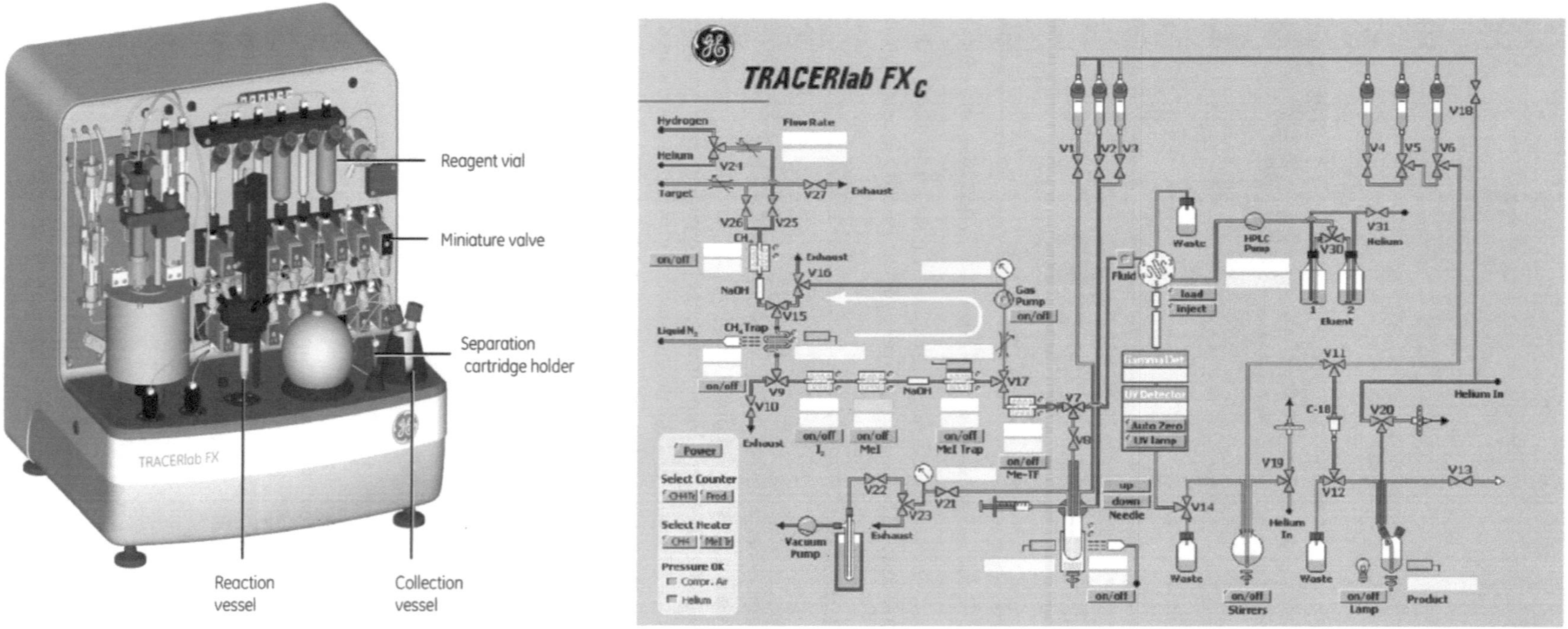

FIGURE 1.5 Example of a semi-automated synthesis module. On the left is the TRACERlab FX2 C module (GE Healthcare) that is used for the synthesis of for example [^{11}C] L-Methionine. Images taken from the supplier's manual. On the right is shown the flowchart of the components, reaction vessels and tubings. Most often all these parts are multi-use and therefore a cleaning method is required before synthesis. Colour image available at www.routledge.com/9781138593312.

TABLE 1.3
Examples of Radiotheranostics Pairs that are Clinically Used or in Clinical Studies

Diagnostics	Therapy	Biomarker	Cancer
^{68}Ga-DOTATATE	^{177}Lu-DOTATATE	Somatostatin receptor	Neuroendocrine tumours [12]
^{68}Ga-PSMA	^{177}Lu-PSMA	Prostate specific membrane antigen	Prostate cancer [19]
^{68}Ga-pentixafor	^{177}Lu/^{90}Y -pentixather	CXCR-4 receptors	Multiple myeloma and lymphoma [20]
^{68}Ga -or ^{18}F -FAPI	^{177}Lu-FAPI	Fibroblast activation protein	Several tumour types [21]

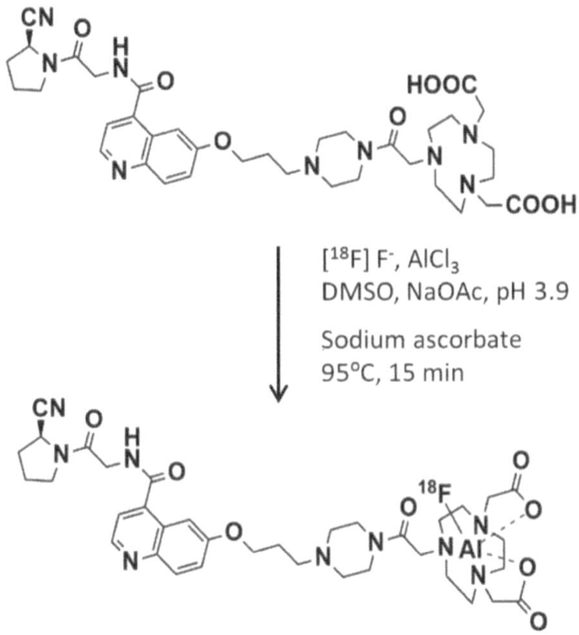

FIGURE 1.6 Labelling of FAPI molecule with conditions for [^{18}F] Aluminium fluoride.

1.5.1 Radiotheranostics Concept

Radiotheranostics is a concept in nuclear medicine that has emerged in the last decade and grown rapidly, offering many opportunities and developments for several cancers [18]. The radiotheranostics concept is based on the selection of patients for radionuclide targeted therapy after imaging of the same biomarker. The same targeting molecule can be used for diagnostic purposes when radiolabelled with a PET emitter (for example ^{68}Ga) or treatment when radiolabelled with a therapeutic radionuclide – for example, a beta-emitter such as ^{177}Lu. The chemical construct is often based on the ligand-linker-radionuclide design. Table 1.3 shows some successful examples that are in clinical use or currently in clinical studies. It is believed that the radiotheranostics concept will be even more expanded in the future, with new radionuclides – for example, alpha emitters of ^{225}Ac, ^{213}Bi, and ^{211}At. In combination with new targeting ligands and new biomarkers that are being explored, radiotheranostics will become one of the established clinical options for many cancers, and eventually for non-cancer diseases [18].

1.5.2 Aluminium [^{18}F]-fluoride radiolabelling of peptides

The radionuclide ^{18}F (β^+, 0.635 MeV, half-life of 110 min) is an ideal radionuclide for PET imaging, and ^{18}F-labelled radiopharmaceuticals are the most commonly used in PET. Fluorinations of small molecules are well-established, often through covalent radiolabellings, and generally require harsh conditions such as high temperatures and organic solvents in combination with inorganic bases to obtain good yields. ^{18}F-labelled peptides and small biomolecules would, however, represent an interesting class of radiopharmaceuticals. Conventional ^{18}F-radiolabelling methods to introduce ^{18}F directly into targeting compounds are not suitable for peptides. New methods have been developed, but they are lengthy and require several steps [22].

McBride and co-workers presented a new labelling method using aluminium [^{18}F]fluoride ([^{18}F]AlF), in which ^{18}F is first attached to aluminium as Al^{18}F, and this complex is then bound to a chelate (for example NOTA) present in a peptide. This stable Al^{18}F-chelate-peptide complex can be efficiently performed in a one-pot process [23–26]. The [^{18}F]AlF chemistry has since then been applied for radiolabelling of a number of peptides and molecules (see review by Fersing and colleagues [27]). Exploring this robust non-covalent approach, the GMP-compliant synthesis of the FAPI-74 has successfully been automatized [28]. FAPI-74 is a Fibroblast Activation Protein inhibitor (Figure 1.6) that is used for imaging the fibroblast activation protein biomarker highly expressed in several cancer types [29].

In general, [^{18}F]AlF provides a robust, convenient, simple, and fast way to label small molecules, peptides, and conjugates. Although no radiopharmaceutical is currently routinely established in clinics based on the [^{18}F]AlF, it is believed that it is only a matter of time before several PET radiotracers will find their way to the clinics.

REFERENCES

[1] J. G. Hamilton and M. H. Soley, "Studies in iodine metabolism by the use of a new radioactive isotope of iodine," *American Journal of Physiology-Legacy Content,* vol. 127, no. 3, pp. 557–72, 1939, doi: 10.1152/ajplegacy.1939.127.3.557.

[2] W. H. Beierwaltes, "The history of the use of radioactive iodine," *Seminars in Nuclear Medicine,* vol. 9, no. 3, pp. 151–55, 1979, doi: https://doi.org/10.1016/S0001-2998(79)80023-9.

[3] P. Richards, W. D. Tucker, and S. C. Srivastava, "Technetium-99m: an historical perspective," *Int J Appl Radiat Isot*, vol. 33, no. 10, pp. 793–99, 1982, doi: 10.1016/0020-708x(82)90120-x.

[4] C. C. Lauritsen, H. R. Crane, and W. W. Harper, "Artificial production of radio-active substances," *Science*, vol. 79, no. 2045, pp. 234–35, 1934, doi: 10.1126/science.79.2045.234.

[5] C. A. Tobias, J. H. Lawrence, F. J. W . Roughton, W. S. Root, and M. I. Gregersen, "The elimination of carbon monoxide from the human body with reference to the possible conversion of CO to CO2," *American Journal of Physiology-Legacy Content*, vol. 145, no. 2, pp. 253–63, 1945, doi: 10.1152/ajplegacy.1945.145.2.253.

[6] R. Lange *et al.*, "Untangling the web of European regulations for the preparation of unlicensed radiopharmaceuticals: A concise overview and practical guidance for a risk-based approach," *Nucl Med Commun*, vol. 36, no. 5, pp. 414–22, 2015, doi: 10.1097/mnm.0000000000000276.

[7] EudraLex (2010). *"The rules governing medicinal products in the European Union" – Good Manufacturing Practice (GMP) guidelines*. https://ec.europa.eu/health/documents/eudralex/vol-4_en

[8] *The European Commission, EudraLex. The Rules Governing Medicinal Products in the European Union Volume 4 EU Guidelines to Good Manufacturing Practice Medicinal Products for Human and Veterinary Use, Annex 3 Manufacture of Radiopharmaceuticals, 2017.*

[9] N. Gillings *et al.*, "Guideline on current good radiopharmacy practice (cGRPP) for the small-scale preparation of radiopharmaceuticals," *EJNMMI Radiopharmacy and Chemistry,* vol. 6, no. 1, p. 8, 2021, doi: 10.1186/s41181-021-00123-2.

[10] *Quality Control in the Production of Radiopharmaceuticals*. Vienna: International Atomic Energy Agency, 2018.

[11] I. Velikyan, "68Ga-based radiopharmaceuticals: Production and application relationship," *Molecules,* vol. 20, no. 7, pp. 12913–43, 2015, doi: 10.3390/molecules200712913.

[12] H. R. Kulkarni and R. P. Baum, "Patient selection for personalized peptide receptor radionuclide therapy using Ga-68 somatostatin receptor PET/CT," *PET Clin,* vol. 9, no. 1, pp. 83–90, 2014, doi: 10.1016/j.cpet.2013.08.015.

[13] P. Schubiger, L. Lehmann, M. Friebe, and D. Yang, "PET chemistry: The driving force in molecular imaging," *J Nucl Med*, vol. 48, pp. 1750–50, 2007, doi: 10.2967/jnumed.107.043810.

[14] A. J. Poot *et al.*, "Fully automated (89)Zr labeling and purification of antibodies," *J Nucl Med,* vol. 60, no. 5, pp. 691–95, 2019, doi: 10.2967/jnumed.118.217158.

[15] M. De Decker and J. H. Turner, "Automated module radiolabeling of peptides and antibodies with gallium-68, lutetium-177 and iodine-131," *Cancer Biother Radiopharm,* vol. 27, no. 1, pp. 72–76, 2012, doi: 10.1089/cbr.2011.1073.

[16] J. Aerts *et al.*, "Guidance on current good radiopharmacy practice for the small-scale preparation of radiopharmaceuticals using automated modules: A European perspective," *Journal of Labelled Compounds and Radiopharmaceuticals,* vol. 57, no. 10, pp. 615–20, 2014, doi: https://doi.org/10.1002/jlcr.3227.

[17] M. Meisenheimer, S. Kürpig, M. Essler, and E. Eppard, "Manual vs automated (68) Ga-radiolabelling-A comparison of optimized processes," *J Labelled Comp Radiopharm,* vol. 63, no. 4, pp. 162–73, 2020, doi: 10.1002/jlcr.3821.

[18] K. Herrmann *et al.*, "Radiotheranostics: A roadmap for future development," *Lancet Oncol*, vol. 21, no. 3, pp. e146–e156, 2020, doi: 10.1016/s1470-2045(19)30821-6.

[19] I. Virgolini, C. Decristoforo, A. Haug, S. Fanti, and C. Uprimny, "Current status of theranostics in prostate cancer," *European Journal of Nuclear Medicine and Molecular Imaging,* vol. 45, no. 3, pp. 471–95, 2018, doi: 10.1007/s00259-017-3882-2.

[20] M. Kircher *et al.*, "CXCR4-directed theranostics in oncology and inflammation," *Annals of Nuclear Medicine,* vol. 32, no. 8, pp. 503–11, 2018, doi: 10.1007/s12149-018-1290-8.
[21] F. Zhuravlev, "Theranostic radiopharmaceuticals targeting cancer-associated fibroblasts," *Current Radiopharmaceuticals,* 2020, pp. 374–393, doi: 10.2174/1874471013666201224114148.
[22] S. Richter and F. Wuest, "18F-Labeled Peptides: The Future Is Bright," *Molecules (Basel, Switzerland),* vol. 19, no. 12, pp. 20536–56, 2014, doi: 10.3390/molecules191220536.
[23] W. J. McBride, R. M. Sharkey, and D. M. Goldenberg, "Radiofluorination using aluminum-fluoride (Al18F)," *EJNMMI Res,* vol. 3, no. 1, p. 36, 2013, doi: 10.1186/2191-219X-3-36.
[24] W. J. McBride *et al.*, "Improved 18F labeling of peptides with a fluoride-aluminum-chelate complex," *Bioconjug Chem,* vol. 21, no. 7, pp. 1331–40, 2010, doi: 10.1021/bc100137x.
[25] W. J. McBride *et al.*, "A novel method of 18F radiolabeling for PET," *J Nucl Med,* vol. 50, no. 6, pp. 991–98, 2009, doi: 10.2967/jnumed.108.060418.
[26] L. Allott, C. Da Pieve, D. R. Turton, and G. Smith, "A general [18F]AlF radiochemistry procedure on two automated synthesis platforms," *Reaction Chemistry & Engineering,* 10.1039/C6RE00204H vol. 2, no. 1, pp. 68–74, 2017, doi: 10.1039/C6RE00204H.
[27] C. Fersing, A. Bouhlel, C. Cantelli, P. Garrigue, V. Lisowski, and B. Guillet, "A comprehensive review of non-covalent radiofluorination approaches using aluminum [(18)F]fluoride: Will [(18)F]AlF replace (68)Ga for metal chelate labeling?" *Molecules,* vol. 24, no. 16, 2019, doi: 10.3390/molecules24162866.
[28] F. L. Giesel *et al.*, "FAPI-74 PET/CT Using Either (18)F-AlF or Cold-Kit (68)Ga Labeling: Biodistribution, Radiation Dosimetry, and Tumor Delineation in Lung Cancer Patients," *J Nucl Med,* vol. 62, no. 2, pp. 201–07, 2021, doi: 10.2967/jnumed.120.245084.
[29] K. Dahl, E. Jussing, L. Bylund, M.M. Moein, E. Samén, T. Tran. Fully automated production of the fibroblast activation protein radiotracer [18F]FAPI-74. *J Label Compd Radiopharm*, vol. 64, pp. 346–352, https://doi.org/10.1002/jlcr.3926

2 Radiopharmaceuticals for Diagnostics

Planar/SPECT

Jim Ballinger and Jacek Koziorowski

CONTENTS

DOI: 10.1201/9780429489501-2

2.1 RADIONUCLIDES FOR GAMMA IMAGING

2.1.1 Technetium-99m

Technetium-99m (^{99m}Tc) is the most commonly used radionuclide for gamma imaging since its introduction ~1970, being used in ~ 85 per cent of nuclear medicine procedures worldwide. ^{99m}Tc (t½ 6 h; IT 99.99%; principal γ emission 140.5 keV; 89% abundance) has mainly been produced by decay of Molybdenum-99 (^{99}Mo) in a ^{99}Mo/^{99m}Tc generator system. ^{99}Mo (t½ 66 h) is a nuclear fission product of Uranium-235, the [^{235}U](n,f)[^{99}Mo] reaction having a yield of 6.1 per cent. ^{99m}Tc is conveniently available on-site from a generator system in which a chromatographic column (generally aluminium oxide) is loaded with ^{99}Mo which continually decays to ^{99m}Tc. The two radionuclides can be readily separated because ^{99}Mo remains on the column when it is eluted with 0.9 per cent sodium chloride solution (saline), while ^{99m}Tc emerges in the eluate. The ratio between the half-lives of the parent and daughter radionuclides is such that maximal yields of ^{99m}Tc are obtained at 24-h intervals, perfect for once-a-day elution. Because of the gradual decay of ^{99}Mo and concomitant reduced yield of ^{99m}Tc, the generator is generally replaced on a weekly basis. The generator is autoclaved by the manufacturer and, as long as the end-user maintains aseptic technique, the eluate remains sterile and suitable for direct injection into patients [1].

Historically, the world's supply of ^{99}Mo has come from a small number of nuclear research reactors, and government subsidy of these reactors had kept the price of ^{99m}Tc low [2]. However, the situation has changed since ~2008 as a series of prolonged outages at one or more of these reactors has led to severe shortages of ^{99}Mo for certain periods. Moreover, several of these aging reactors are due for decommissioning. Furthermore, there has been a shift from highly enriched uranium (HEU; >20% [^{235}U]), sometimes also called 'weapons grade uranium' (>85% [^{235}U]), to low enriched uranium (LEU: <20% [^{235}U]), resulting in lower yields of ^{99}Mo. In the last 10 years there has been greater emphasis on coordination of preventative maintenance schedules and recruitment of additional reactors to the international network.

In parallel with this, there has been work on development of alternative routes to production of ^{99}Mo, which does not rely on reactors. There is a non-fission nuclear route for ^{99}Mo, [^{98}Mo](n,γ)[^{99}Mo], but it still requires a nuclear reactor and results in low specific activity ^{99}Mo. For cyclotron production there are two routes, either the direct [^{100}Mo](p,n)[^{99m}Tc] nuclear reaction at <25 MeV proton energy or the [^{100}Mo](p,x)[^{99}Mo] at 20-40 MeV proton energy. The first, direct route gives good ^{99m}Tc yield but the radionuclidic purity is hampered by the co-production of other isotopes such as ^{97m}Tc (t½ 90.1 d) and ^{96}Tc (t½ 4.28 d), which makes it difficult to match the pharmacopoeia monograph on radionuclidic purity (Ph. Eur. states 99.9%). The second route gives good ^{99m}Tc radionuclidic purity, but the cross-sections are small and thus the yield is low [3]. There are alternative routes, such as the photonuclear reaction [^{100}Mo](γ,n)[^{99}Mo] and the accelerator-driven liquid reactor (an aqueous solution of LEU) where tritium is being irradiated with deuterium to produce neutrons which, in turn, fission [^{235}U]); d+t → n → [^{235}U](n,f)[^{99}Mo]. For the [^{100}Mo](γ,n)[^{99}Mo] photonuclear reaction, electrons are accelerated to >20 MeV (practical range 20-40 MeV) and directly hit either the ^{100}Mo target or a high-Z converter (e.g. Tantalum; in order to increase the conversion from electron energy into bremsstrahlung photons). Although none of these alternative routes has entered routine production, some may do so in the next few years. Finally, the direct production of ^{99m}Tc by cyclotron irradiation of ^{100}Mo targets has been explored; however, this creates an entirely different set of logistical problems.

2.1.2 Indium-111

Indium-111 [^{111}In] (t½ 67 h, EC 100%, principal γ emissions 171 keV (90.6%) and 245 keV (94.1%)) is a cyclotron-produced radiometal. The main commercial nuclear reaction is the [^{nat}Cd](p,xn)[^{111}In], which consists of the [^{111}Cd](p,n)[^{111}In] and [^{112}Cd](p,2n)[^{111}In] nuclear reactions and is performed at an incident proton energy of 25–35 MeV. The target is then purified by ion-exchange chromatography or solvent extraction to yield [^{111}In]-$InCl_3$ in dilute HCl [4]. Because of the requirement for cyclotron production and transport to the end-user, ^{111}In is considerably more expensive than ^{99m}Tc, but its moderately long half-life does not require daily production.

There are several applications of ^{111}In in nuclear medicine. ^{111}In-chloride can be used to label transferrin in whole blood as a blood pool marker or for diagnosis of protein-losing enteropathy. ^{111}In-oxine or -tropolonate have been used for in vitro labelling of white blood cells or platelets (see section 2.3.8). ^{111}In-pentetate (DTPA) is the preferred agent for cisternography, which requires injection into the cerebrospinal fluid via lumbar puncture. ^{111}In-pentetate is also used as a liquid phase marker in gastric emptying studies (see section 2.3.6.1). ^{111}In-pentetreotide is a ligand for the somatostatin receptor, which is overexpressed in neuroendocrine tumours (see section 2.3.10.3).

2.1.3 Iodine-123

Iodine-123 [^{123}I] (t½ 13.2 h, EC 100%, principal γ emission 159 keV (83%)) is a cyclotron-produced radiohalogen. The highest yielding nuclear reactions are the [^{124}Xe](p,x)[^{123}Xe] → [^{123}I] and [^{124}Xe](p,2n)[^{123}Cs] → [^{123}Xe] → [^{123}I]. The [^{124}Xe] is a gas target irradiated with 25–40 MeV protons, after the decay of [^{123}Cs] (t½ 6 min) and [^{123}Xe] (t½ 2 h), the ^{123}I is washed out from the target. These production routes thus give a high yield but have complicated post-irradiation processing. A simpler route is the [^{124}Te](p,2n)[^{123}I], which uses a solid target and dry distillation. The target is irradiated with 30 MeV protons and the target exit energy should be 12-15 MeV in order to minimize production of the contaminant ^{124}I by the [^{124}Te](p,n)[^{124}I] route [4].

^{123}I-iodide is used for imaging the thyroid gland (see section 2.1.3.7.1). A wide variety of ^{123}I-labelled molecules have been studied over the years, but the main ones – which are widely available commercially – are ^{123}I-iobenguane for imaging the adrenergic pool in a tumour or the myocardium (see section 2.3.10.2) and ^{123}I-ioflupane for imaging the dopamine transporter in the diagnosis of Parkinson's disease (see section 2.3.3.2). Because of its 13 h t½ and requirement for cyclotron production, ^{123}I is expensive to produce and distribute.

2.1.4 Thallium-201

Thallium-201 [^{201}Tl] (t½ 73 h, EC 100%, principal γ emissions 135 keV (2.6%) and 167 keV (10%); however, the low energy 69 keV (27%), 71 keV (46%) and 80.2 keV (10%) x-ray emissions from the decay product ^{201}Hg are used for imaging) is a cyclotron produced radiometal. The most common and versatile route is the [^{203}Tl](p,3n)[^{201}Pb] (t½ 9.3 h) → [^{201}Tl] nuclear reaction using protons of 25–35 MeV incident energy. The final form is [^{201}Tl]-thallous chloride in saline [4]. When ^{201}Tl was first used routinely ~1980 it was considered extremely expensive in comparison to ^{99m}Tc. Its half-life makes it convenient to supply.

The primary clinical use of ^{201}Tl is for myocardial perfusion imaging. Though largely replaced by the ^{99m}Tc agents sestamibi and tetrofosmin (see section 2.3.5), there is still some use of ^{201}Tl. For example, its half-life makes it convenient to supply for weekend use when ^{99m}Tc agents may not be available. Also, it can be used for dual-isotope myocardial perfusion imaging together with a ^{99m}Tc agent.

2.1.5 Krypton-81m

Krypton-81m [^{81m}Kr] (t½ 13 s, IT 100%, principal γ emission 190 keV (68%)) is a generator-produced radioactive noble gas that is obtained from Rubidium-81 using a [^{81}Rb]/[^{81m}Kr] generator. ^{81}Rb is a cyclotron produced radiometal with a half-life of 4.6 h, and which is produced either by the [^{82}Kr](p,2n)[^{81}Rb] reaction using an incident proton energy of 20–30 MeV or the [^{nat}Kr](p,x)[^{81}Rb] reaction using an incident proton energy of 40–70 MeV [4]. The half-life of ^{81}Rb means that the generator must be produced on a daily basis and is expensive to use. ^{81m}Kr is breathed in as a gas. The sole clinical use of ^{81m}Kr is for lung ventilation imaging, used in conjunction with lung perfusion imaging for diagnosis of pulmonary embolism (see section 2.3.4.2).

2.1.6 Gallium-67

Gallium-67 [^{67}Ga] (t½ 78 h, EC 100%, principal γ emissions 93 keV (39%), 185 keV (21.2%) and 300 keV (16,8%)) is a cyclotron-produced radiometal. The main production route is the [^{68}Zn](p,2n)[^{67}Ga] nuclear reaction using an incident proton energy of 20–40 MeV. As an alternative reaction for 'in house' production the [^{67}Zn](p,n)[^{67}Ga] nuclear reaction may be used by irradiating enriched ^{67}Zn with 10–20 MeV protons; however, there are 2 drawbacks to this reaction: (a) lower yields and (b) expensive starting material (the natural abundance of ^{67}Zn is 4.0 per cent whereas the natural abundance of ^{68}Zn is 18.5%) [4]. The physical characteristics and production of gamma-emitting radionuclides of clinical interest are summarized in Table 2.1.

2.2 PRODUCTION OF GAMMA-EMITTING RADIOPHARMACEUTICALS

2.2.1 Chelates

^{99m}Tc and ^{111}In are radiometals, which means they do not form covalent bonds but can be attached to a targeting molecule via coordinate bonds to a multidentate chelator. A chelator (from Latin; 'chela' – 'claw') is a polydentate ligand that coordinate bonds to a single atom. Figuratively, it is a claw that grasps the radiometal ion with a firm grip. Most radiometal labelling for production of radiopharmaceuticals uses so-called shake and bake chemistry. This means that

TABLE 2.1
Characteristics of Radionuclides for Gamma Camera Imaging

Radionuclide	Half-life	Gamma energies	Means of production
^{99m}Tc	6 h	140 keV	Generator
^{111}In	67 h	171, 245 keV	Cyclotron
^{123}I	13 h	159 keV	Cyclotron
^{201}Tl	73 h	135, 167 keV 69-80 keV (X-rays)	Cyclotron
^{81m}Kr	13 s	190 keV	Generator
^{67}Ga	78 h	93, 185, 300 keV	Cyclotron

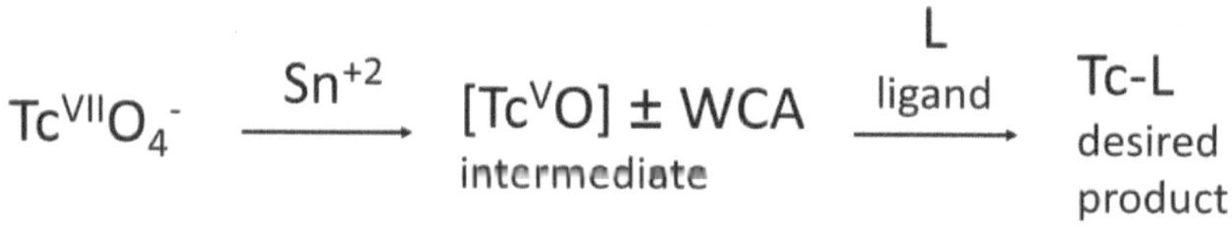

FIGURE 2.1 Schematic of formation of a ^{99m}Tc labelled radiopharmaceutical in a kit. WCA = weak chelating agent.

the radiometal is introduced into a sealed vial (a 'kit') that contains the precursor compound to be labelled to form the final radiopharmaceutical and all other necessary pharmaceutical-grade chemicals to perform the labelling and yield a ready-to-inject radiopharmaceutical [5]. ^{111}In has one oxidation state of +3, whereas ^{99m}Tc in its native form as pertechnetate, TcO_4, is in the +7 oxidation state and is not very reactive. In order to label a molecule via a chelator, the oxidation state of the Tc atom must be changed by a reducing agent, usually stannous chloride. The reduced species are unstable and must be trapped by rapid chelation or stabilized with an intermediate transfer ligand, or weak chelating agent (WCA), if the formation of the final complex is slow (see Figure 2.1). Ideally, these labelling reactions will take place rapidly at room temperature and quantitatively, that is >>95 per cent incorporation of label so that purification is not required. With most routinely used agents, this is the case, though some labelling reactions require heating at 100°C. ^{99m}Tc is eluted from the generator using a sterile isotonic saline, a 0.9 per cent sodium chloride (NaCl) solution. Some of the chelators, for example EDDA/HYNIC, also work with Rhenium, thus making [^{99m}Tc] and [^{186}Re] or [^{188}Re] a good theranostic pair.

2.2.2 Covalent Bonds

^{123}I-labelled molecules can be produced either by nucleophilic or electrophilic routes. ^{123}I-iobenguane is produced by heating [^{123}I]-NaI with cold iobenguane; by nucleophilic attack, most of the ^{123}I-iodide will exchange with cold iodide. This means that the final product is carrier-added. A no-carrier-added version of ^{131}I-iobenguane has recently been licensed in the United States for therapeutic use [6]. In contrast, ^{123}I-ioflupane is prepared by the electrophilic route in which an oxidant such as peracetic acid is used to convert ^{123}I-iodide to a formal [I^+], which reacts with a tri-alkyl-stannyl precursor to produce ^{123}I-ioflupane. After the reaction has been quenched, the mixture is purified by high-pressure liquid chromatography to yield a high-specific-activity product virtually devoid of cold iodine. For direct radioiodination of biomolecules such as antibodies and peptides, the iodination occurs at pH ~7, and the iodine is attached by a covalent bond at the ortho position on a tyrosine moiety. For small molecules the iodine displaces a leaving group, that is, a molecular fragment that is easily displaced by the electrophilic iodine (I^+).

2.2.2.1 Ions and Atoms

^{123}I-iodide is produced by cyclotron, as described in section 2.1.3, and formulated in dilute sodium hydroxide to minimize the release of elemental iodine as a vapour. ^{201}Tl is also produced by cyclotron, as described in section 2.1.4 ,and formulated as an isotonic solution of the chloride salt.

^{81m}Kr is formed by decay of ^{81}Rb in a generator system as described in section 2.1.5 and released as a gas. ^{67}Ga is produced by cyclotron as described in section 2.1.6 and formulated as the citrate salt.

2.3 RADIOPHARMACEUTICALS IN CLINICAL USE

2.3.1 Skeletal System

The bone scan is the most widely used nuclear medicine procedure in many countries. The tracers for this purpose are ^{99m}Tc-labelled complexes of the diphosphonates medronate (methylene diphosphonate, MDP) and oxidronate (hydroxymethylene diphosphonate, HDP). In both cases, the phosphonate groups perform the dual role of chelation of reduced ^{99m}Tc and binding to the biological target, hydroxyapatite crystals at the surface of bone. Neither of these agents is a single, chemically defined species but rather an ever-changing mixture of complexes with varying affinity for bone. Bone-scan kits contain an antioxidant to maintain the stability of the complex throughout its shelf-life.

Following intravenous injection, the tracer distributes throughout the body with about one-half of the activity localizing to bone while the remainder is excreted via glomerular filtration into the urine. Imaging is generally performed 2–3 hours after injection to allow for clearance of unbound activity through the kidneys, providing a clearer image of the skeleton [7]. Whole-body planar images are usually obtained, ideally with a dual-headed gamma camera. This may be followed by SPECT images of the affected area. SPECT/CT may be useful for precise anatomical localization of the lesion. The information produced by a radionuclide bone scan is quite different from an x-ray, which shows bone density. The bone scan reflects the rate of bone growth via osteoblastic activity, though vascularization also plays a role [8].

2.3.2 Renal System

2.3.2.1 Renal Function

Dynamic imaging of the kidneys with a tracer that is both passively filtered and actively secreted by the tubules allows assessment of the integrity of the kidney and collecting system, stenosis, and the function of a transplanted kidney. The original tracers, ^{131}I- or ^{123}I-orthoiodohippurate (hippuran), have largely been replaced by ^{99m}Tc-mertiatide (mercaptoacetyl triglycine, MAG3®, Nephromag®) and ^{99m}Tc-ethylenedicysteine (EDC or EC, Edicis®). Both are polar ^{99m}Tc complexes that are negatively charged at physiological pH. The kits contain stannous chloride and tartrate; the reduced ^{99m}Tc forms a weak complex with tartrate to stabilize it before the secondary transchelation reaction to the chelator. The original mertiatide formulation contained a precursor in which the S-atom of the N_3S chelator was protected with a benzoyl group. The labelling process required heating in a boiling water bath for 15 min to hydrolyse the protecting group and allow transchelation of reduced ^{99m}Tc from tartrate to mertiatide to take place. There is now a non-protected formulation of mertiatide in which labelling takes place at room temperature (since hydrolysis of the protecting group is not required) and high pH; a buffer is added after labelling to bring the formulation to physiological pH. Similarly, chelation of reduced ^{99m}Tc by the N_2S_2 chelator in EDC takes place at room temperature and high pH, followed by buffering.

A normal renogram shows rapid delivery of the tracer to both kidneys, peaking at ~1 min, followed by slower but continuous clearance over several minutes with a concomitant increase in activity in the bladder. The relative function of the two kidneys can be assessed [9].

The glomerular filtration rate (GFR) can be measured using ^{99m}Tc-pentetate (diethylenetriamine penta-acetic acid, DTPA), which is cleared from the circulation almost completely by glomerular filtration. The clearance of ^{99m}Tc-pentetate can be monitored either by gamma camera imaging or by blood sampling [10, 11].

2.3.2.2 Renal Structure

Static imaging of the structure of the kidney is performed using ^{99m}Tc-succimer (dimercaptosuccinic acid, DMSA), which forms a 2:1 complex with reduced ^{99m}Tc. Early formulations were extremely unstable, but the addition of an antioxidant such as ascorbic acid has solved that problem.

Following IV injection, the tracer binds extensively but reversibly to plasma proteins. Over several hours the tracer accumulates in the renal cortex. Imaging is generally performed 4 h after injection. This is useful for assessment of congenital malformations, trauma, or pyelonephritis following repeated urinary tract infections [12].

2.3.3 Neuro Imaging

2.3.3.1 Cerebral Blood Flow

Regional cerebral blood flow can be mapped using tracers that are rapidly extracted from the blood on first pass and trapped in that distribution long enough for SPECT imaging. The two ^{99m}Tc-labelled radiopharmaceuticals for this use are exametazime (hexamethyl propyleneamine oxime, HMPAO, Ceretec, Neuroscan, Brain-Spect, Cerebrotec) and bicisate (ethyl cysteinate dimer, ECD, Neurolite). Both are small, neutral, lipophilic complexes that are highly extracted by the brain on first pass by passive diffusion, then trapped in the brain following a chemical change that results in a charged moiety that cannot exit across the blood-brain barrier. With exametazime the change is believed to involve interaction with ubiquitous compounds such as glutathione, forming an unidentified product. With bicisate it involves sequential enzymatic hydrolysis of the two ethyl ester groups, leaving free carboxylic acids that are negatively charged at physiological pH. After initial clearance of background activity, the regional distribution of tracer remains stable over time, allowing flexibility in imaging time. SPECT images are generally obtained, except in the case of brain-death studies, where it may not be feasible to perform SPECT on a patient on life support.

From a radiopharmaceutical point of view, the two products differ in their ease of preparation and use. Exametazime requires a fresh eluate of the ^{99m}Tc generator and is stable for only 30 min after preparation (or 1 h after addition of a stabilizer); both of these requirements create logistical problems. The stabilizer in the North American formulation is methylene blue, while in the European formulation it is cobalt chloride. The inherent instability of ^{99m}Tc-exametazime is still not understood, and the identity of the alternative complexes, which do not enter the brain, has not been confirmed. In contrast, bicisate does not have these restrictions and is stable for 6 h [13].

The most common indications for use of cerebral blood flow imaging are cerebrovascular disease, presurgical localization of epileptogenic foci [14], early detection and differential diagnosis of dementia, evaluation of traumatic brain injury, and assessment of brain death [15]. The logistics of preparation make bicisate the preferred agent for ictal studies in epilepsy for localization of an active epileptogenic focus.

2.3.3.2 Dopamine Transporter

The function of the dopamine transporter is reuptake of dopamine released into the synapse into vesicles in the presynaptic nerve terminal. The dopamine transporter can be imaged with an ^{123}I-labelled analogue of cocaine, ioflupane (fluoropropyl carbomethoxy iodophenyl nortropane, FP-CIT, Datscan, Striascan). Ioflupane binds to the transporter to give an indication of the density of presynaptic nerve terminals. SPECT imaging is performed at a fixed time, 3-6 h after injection [16]. In normal individuals there is bilateral homogeneous accumulation of activity in the basal ganglia (striatum; caudate nucleus and putamen). Parkinson's disease involves degeneration of the nigrostriatal pathway with loss of dopaminergic nerve terminals. Thus, dopamine transporter imaging can be used to diagnose Parkinson's disease and to differentiate it from other causes of Parkinsonian tremor, such as essential tremor (i.e. non-degenerative) or that caused by certain neuroleptic drugs. In early Parkinson's disease there tends to be asymmetry, particularly in the putamen, while in advanced disease there is bilateral degeneration. Ioflupane is also useful in diagnosis of Lewy body disease [17].

2.3.4 Pulmonary System

2.3.4.1 Perfusion

The integrity of the arterial system in the lungs can be assessed using ^{99m}Tc-labelled macroaggregated human serum albumin particles (MAA). The diameter of capillaries is ~7 µm, so particles larger than this diameter will be trapped in capillaries. MAA particles are engineered to be 10–90 µm in diameter (90% within this range, none >150 µm; median typically 20–40 µm). The particles are formed by heat denaturation of human serum albumin in the presence of stannous chloride, which allows labelling of the particles by reduction of ^{99m}Tc-pertechnetate. When ^{99m}Tc-MAA is injected intravenously, it travels to the right ventricle, then to the pulmonary arterial system, where it encounters capillaries for the first time and is essentially completely trapped, outlining the pulmonary blood supply. This allows detection of pulmonary embolism as a cold spot. Lung perfusion imaging is routinely performed in conjunction with ventilation imaging to improve the differential diagnosis [18].

For a statistically valid scan, a minimum of 60×10^3 particles should be administered; however, for safety reasons (blockade of a significant fraction of the capillaries) the total number of particles should be no more than 700×10^3. It can be difficult to maintain acceptable particle numbers throughout the shelf-life of the product, since the number of particles administered increases as the amount of ^{99m}Tc decays. A further complication is administration of 200 MBq activities for SPECT rather than the usual 100 MBq for planar imaging. Additionally, the maximum particle number should be reduced to 100-200x10^3 in patients with pulmonary hypertension, right-to-left shunt, or after single lung transplantation [18].

2.3.4.2 Ventilation

The airways can be visualized with an inhaled gas or aerosol droplets. Historically ^{133}Xe or ^{127}Xe gas was used, but this has been largely replaced. Generator-produced ^{81m}Kr gas (t½ 13 s) is available in some countries (see section 2.1.5). ^{99m}Tc-pertechnetate burnt at high temperature generates submicron graphite particles, which behave like a gas (Technegas®). Most commonly, an aerosol of ^{99m}Tc-DTPA generated by wet nebulization is used. The disadvantage of an aerosol is that it only shows delivery, not washout, via the airways as a gas does [18].

Differential diagnosis is made by comparison of the ventilation and perfusion scans (V/Q) together with a recent chest x-ray to exclude other causes such as pneumothorax. There are four abnormal patterns that can be seen:

1. segmental perfusion defect with preserved ventilation = pulmonary embolism;
2. segmental matched perfusion and ventilation defect = pulmonary infraction or infection,
3. segmental or sub-segmental ventilation defect with preserved perfusion = infection, and
4. non-segmental, patchy, matched perfusion and ventilation defect = chronic obstructive pulmonary disease (COPD). The use of SPECT or SPECT/CT imaging can reduce the incidence of indeterminate scans [19].

2.3.4.3 Permeability

Altered alveolar permeability is seen in a number of conditions. This can be assessed with ^{99m}Tc-DTPA delivered as an aerosol (as for lung ventilation imaging) followed by repeated imaging over 90 min to allow generation of clearance curves. This is particularly useful for rapid screening of *Pneumocystis carinii* infection (PCP) in patients with a normal chest X-ray. Clearance is accelerated in these patients and resolves with successful treatment. However, clearance is also accelerated in smokers, limiting the applicability of screening [20].

2.3.5 Myocardial Perfusion

Myocardial perfusion imaging (MPI) became established in the 1980s using the tracer ^{201}Tl-thallous chloride (see section 2.1.4). ^{201}Tl is an analogue of potassium and is rapidly and efficiently transported into the myocardium by sodium-potassium ATPase. Thus, the distribution of ^{201}Tl in the myocardium reflects regional perfusion and when injected at rest allows assessment of infarcted areas as cold spots (permanent defects). When injected at peak stress (exercise or pharmacological), areas of coronary artery disease will be evident with reduced delivery of tracer. Imaging must begin immediately after stress in order to map the full extent of ischaemia, because redistribution of tracer begins immediately, and ischemic defects will gradually fill in. Imaging 4 h after stress is the equivalent of a rest injection, and this dual time point imaging allows differential diagnosis to be made from a single radiation dose [21].

The physical limitations of ^{201}Tl as a tracer (poor emission energies, long half-life limiting administered activity, poor radiation dosimetry) led to the search for a ^{99m}Tc-labelled equivalent. Two such agents came into clinical use around 1990: sestamibi (methoxy isobutyl isonitrile, MIBI, Cardiolite, Stamicis), and tetrofosmin (ethylene-bis[bis(2-ethoxyethyl)phosphine], TFOS, Myoview). Both are lipophilic cations designed to mimic the active transport of ^{201}Tl. However, it turns out that both enter cells by passive diffusion as lipophilic agents and are trapped in electronegative mitochondria due to their positive charge. Unlike ^{201}Tl, they do not undergo redistribution, so separate injections are required in order to image the rest and stress distributions. These are ideally performed on separate days, but sequential injections on the same day can be used, with a low activity first followed by a higher activity. Although planar imaging may be performed, SPECT provides better delineation of the affected territories of the coronary arteries. SPECT/CT allows attenuation correction and, with diagnostic quality CT detectors, coronary artery calcium scoring. Gating to the electrocardiogram may also be used [22].

From a radiopharmaceutical point of view there are differences between sestamibi and tetrofosmin. The chemical structure of sestamibi consists of six methoxy isobutyl isonitrile units coordinated around a central Tc atom with a delocalised net +1 charge. This complex forms by heating a vial containing methoxy isobutyl isonitrile (as a copper tetrafluoroborate salt for ease of handling), ^{99m}Tc-pertechnetate, and stannous chloride. After 10 min in a boiling water bath the incorporation of ^{99m}Tc into the sestamibi complex is >95 per cent. In contrast, tetrofosmin is a more traditional kit, one which labels at room temperature to form a complex in which two bidentate diphosphine ligands coordinate around a Tc dioxo core, again with a delocalized net +1 charge.

Sestamibi and tetrofosmin have generally similar biological properties but there are significant differences in their biodistribution. The initial extraction of tracer into the myocardium is slightly higher and more linear with flow for sestamibi compared to tetrofosmin, though both are lower than ^{201}Tl. Sestamibi is excreted almost exclusively via the gall bladder into the intestinal tract, whereas tetrofosmin also undergoes a degree of urinary excretion. In general, imaging begins later for sestamibi, requiring more time for clearance from the liver. With both radiotracers the high and changing activity in the intestine can create problems for reconstruction and interpretation of images.

The availability of sestamibi and tetrofosmin fuelled the expansion of MPI over the last 30 years to the extent that in some countries it comprises more than 50 per cent of nuclear medicine patient volume. The patents have expired on both agents and the availability of generic equivalents has caused prices to fall. The greatest value of MPI is the negative predictive value; a normal MPI study indicates a very low risk of cardiac events (<1% per year). Thus, it is used for risk stratification with asymptomatic airline pilots, truck drivers, and other high-risk professions, as well as in symptomatic patients with chest pain or as part of workup for cardiovascular surgery [23].

2.3.6 Gastrointestinal System

2.3.6.1 Gastric Emptying

Gastric emptying rates of solids and/or liquids can be determined by repeated imaging following ingestion of a radiolabeled meal. The requirement of the radiolabel, whether solid or liquid, is that it reflects the behaviour of food in the GI tract, and the radiolabel is not absorbed across the mucosa into the circulation. The solid phase is most often ^{99m}Tc-colloid labelled scrambled egg or omelet served with bread or toast. The liquid phase can be ^{99m}Tc-DTPA; however, if solid and liquid emptying is performed at the same time the liquid phase should be ^{111}In-DTPA so that dual-energy imaging can be used. Regions of interest are placed around the stomach in the anterior and posterior views at each time point (after correction for cross-talk if dual-energy imaging is used) and the geometric mean activity in the stomach is calculated (geometric mean is used to eliminate errors due to depth within the abdomen). These values are plotted versus time, and a gastric clearance curve is generated. Gastric emptying may be faster or slower than normal in different conditions, and repeated studies can be performed to monitor the effectiveness of drug therapies [24].

2.3.6.2 GI Transit

Imaging can be used to investigate disorders in transit in various parts of the gastrointestinal system. Rapid dynamic imaging as the patient swallows a small amount of ^{99m}Tc-labelled liquid or solid can be used to measure the esophageal transit time, which can be delayed in motility disorders, lower esophageal sphincter dysfunction, scleroderma, spasm, and achalasia. Gastroesophageal reflux with or without pulmonary aspiration can be particularly useful in children. Extension of a gastric emptying study can provide information about small-bowel transit time. However, studies of colon transit cannot be performed with ^{99m}Tc due to its half-life being too short; in this case ^{111}In-DTPA is used [25].

2.3.6.3 Liver

Although the liver scan was an important study in the early days of nuclear medicine, it is less commonly used since the introduction of CT and ultrasound. However, there are still some indications for the liver scan. The tracer is ^{99m}Tc-labelled colloidal particles that are cleared from the circulation by the Kupfer cells in the liver and spleen. ^{99m}Tc-sulphide colloid is prepared by heating ^{99m}Tc-pertechnetate with sodium thiosulphate at low pH; ^{99m}Tc is trapped in the particles as they grow. The reaction is terminated by cooling and addition of a buffer. Alternatively, there is a kit for preparation of ^{99m}Tc-tin colloid.

Clinical indications include functional morphology of the liver and/or spleen, detection of masses including hemangiomas, and evaluation of hepatic function in acute or chronic liver disease. Following an ^{111}In-leukocyte scan, liver/spleen imaging may be used to help identify abscesses and to differentiate infection from normal bone-marrow activity [26].

2.3.6.4 Hepatobiliary System

Hepatobiliary scintigraphy is performed with a lipophilic compound that is taken up by the liver, then rapidly clears through the gall bladder and is excreted into the small intestine. The most widely used tracer is ^{99m}Tc-mebrofenin (bromo-2,4,6-trimethylacetanilido iminodiacetic acid, BrIDA, Cholecis, Choletec, Cholediam), though ^{99m}Tc-disofenin (2,6-diisopropylacetanilido iminodiacetic acid, DISIDA, Hepatolite) is still used in some countries. These agents can be considered the first bifunctional chelators, that is, constructs designed to contain both a biological targeting group and a chelator for ^{99m}Tc.

Dynamic imaging is used at 1 frame/min for up to 60 min to monitor the extraction from the blood into the liver (normally 100%) then concentration in the gall bladder followed by excretion. Delayed imaging may be required if the gall bladder is not visualized within 60 min, as in acute cholecystitis. Likewise, delayed imaging may aid in detection of a bile leak [27].

2.3.7 Endocrine System

2.3.7.1 Thyroid Function

The role of the thyroid gland, located in the front of the neck, is to incorporate dietary iodine into the hormones triiodothyronine (T3) and thyroxine (T4) which, when secreted into the circulation, control a variety of metabolic, growth, and development processes. One of the first radiopharmaceuticals was ^{131}I-iodide, which is a perfect probe of thyroid function as it is taken up by what we now know is the sodium iodide symporter (NIS) then organified as hormones. NIS is a transporter located in the basolateral membrane of thyroid epithelial cells, and which simultaneously transports Na^+ and I^- ions from extracellular fluid into the thyroid epithelial cell via secondary active transport driven by the sodium gradient across the membrane. Because of the suboptimal radionuclidic properties of ^{131}I (long physical half-life, high gamma energy, beta particle emissions), the pure gamma emitter ^{123}I is now the preferred radionuclide. ^{99m}Tc-pertechnetate is an alternative radiotracer, which is also taken up by NIS but not organified [28]. ^{99m}Tc-pertechnetate is a substrate for NIS because it has some of the chemical properties of iodide (single negative charge, small size, hydrophilic), but it cannot be incorporated into thyroid hormones.

Hypothyroidism is diagnosed as reduced accumulation of ^{123}I-iodide and is associated with sensitivity to cold, little appetite, and low energy. In children it can result in cretinism with abnormal bone formation and mental retardation. Hypothyroidism can be treated with thyroxine tablets. In contrast, hyperthyroidism (increased accumulation of ^{123}I-iodide) is associated with sensitivity to heat, excessive appetite, and hyperactivity. It may be evident as a goiter or bulge in the neck. Hyperthyroidism can be treated with drugs, but radioiodine is also used. Hyperthyroidism can also result from thyroiditis, an inflammation that can eventually lead to hypothyroidism. The thyroid image in hypo- or hyperthyroidism is relatively uniform; however, in other conditions focal hot or cold spots may be seen. Solitary thyroid nodules are generally benign but should be investigated with a fine needle aspiration (FNA) biopsy to determine if it is cancerous.

2.3.7.2 Parathyroid Imaging

The parathyroid glands, located adjacent to the thyroid, regulate the body's use of calcium. Parathyroid adenoma is a benign condition causing hyperparathyroidism, usually resulting in elevated blood calcium levels, which is best treated by minimally invasive surgery. Imaging plays an important role in guiding surgery. Most commonly used now is the myocardial perfusion agent, ^{99m}Tc-sestamibi, which accumulates in both the thyroid and parathyroid glands but is only retained in the parathyroid glands. The mechanism of accumulation of ^{99m}Tc-sestamibi is not fully understood but is believed to be related to high mitochondrial activity. Two imaging approaches are taken. One uses early and late imaging (15 min and 2 h) following injection of ^{99m}Tc-sestamibi. Relative focal increase in the parathyroid glands (as the tracer washes out of the thyroid) is indicative of adenoma. An alternative approach uses a second, thyroid-specific tracer, such as ^{99m}Tc-pertechnetate or ^{123}I-iodide, to allow subtraction of thyroid activity for better definition of the parathyroid glands. In either case, SPECT or SPECT/CT imaging is useful to provide the surgeon with a roadmap [29].

2.3.8 Infection and Inflammation

Autologous leukocytes (white blood cells, WBC) labelled with ^{99m}Tc or ^{111}In are used to image sites of infection and inflammation. WBC cannot be labelled in whole blood and must first be separated by enhanced sedimentation, primarily to remove the ~1000-fold excess of erythrocytes (red blood cells, RBC). The isolated WBC are then incubated for ~15 min with ^{99m}Tc-HMPAO or ^{111}In-oxine or -tropolonate, then the excess labelling agent is washed off, and the cells are resuspended in plasma for reinjection into the patient. The entire process must be carried out in specialized, separate, aseptic facilities, both to protect the patient's blood from contamination and to protect the staff from blood-borne pathogens. The process is cumbersome, time-consuming (~2 h) and expensive, but has been in routine use since ~1980 [30].

The normal biodistribution of radiolabelled WBC includes the liver, spleen, and bone marrow on whole-body images 2–24 h after reinjection. There should be little retention of activity in the lungs unless the cells have been damaged during labelling. With ^{99m}Tc-labelled WBC there is also excretion of activity into the GI tract, which can interfere with interpretation of the image. Sites of infection or inflammation will be seen as hot spots. Among the most common indications for WBC imaging are osteomyelitis, prosthetic joints, inflammatory bowel disease, soft tissue infections, and fever of unknown origin.

Because of the complexity of WBC labelling techniques, alternative tracers have been investigated, largely without success. There are two ^{99m}Tc-labelled anti-granulocyte antibodies that have some limited use, particularly when blood cell labelling facilities are not available – sulesomab (Leukoscan) and besilesomab (Scintimun) [30]. ^{18}F-Fluorodeoxyglucose (FDG) is also useful for infection imaging with PET.

2.3.9 Lymphoscintigraphy

Although lymphoscintigraphy with ^{99m}Tc-labelled radiocolloids was used in the earliest days of nuclear medicine, it fell into decline. However, the technique has re-emerged in recent years to direct minimally invasive surgery based on the sentinel lymph node concept, which is the standard of care in an increasing range of solid tumours. The role of the radiocolloid is to act as a surrogate for cancer cells spreading from the tumour to adjacent lymph nodes. The radiocolloid is retained in lymph nodes due to phagocytosis by macrophages, a non-specific process. The radiocolloid thus allows identification of a lymph node but gives no information as to whether the node contains tumour cells – that can only be determined by histology of a biopsy specimen. The particular radiocolloid used depends on geography and commercial availability rather than optimal properties: Albumin nanocolloid (Nanocoll, Nanotop, NanoScan) in most of Europe; rhenium sulphide colloid (Nanocis) in France; sulphide colloid, filtered or unfiltered, in North America; antimony trisulphide colloid in Australia; and tin colloid in Japan. Further, there is a lack of agreement on optimal particle size – particles <5 nm in diameter are liable to penetrate capillaries and be carried away by blood, while particles >400 nm will not migrate in lymph and thus will remain at the injection depot. While it has been stated that a diameter of ~100 nm would be optimal, most of the available agents tend to be somewhat smaller than that (15-50 nm). Despite this, and the variety of agents used around the world, there are no obvious international differences in the sensitivity of the technique [31].

An alternative, non-colloidal agent was introduced in 2014. Tilmanocept (Lymphoseek) targets the CD206 mannose receptor on macrophages. Its soluble nature and moderate size allow it to leave the depot efficiently and migrate along lymphatic channels to lymph nodes. Its high affinity for the mannose receptor causes it to be tightly bound in first echelon nodes [32].

The first widespread implementation of the sentinel node concept was in melanoma and breast cancer. It is now established in oropharyngeal and gynaecological cancers and is being applied in other solid tumours [33, 34].

2.3.10 Oncology

2.3.10.1 Thyroid Cancer

Papillary thyroid cancer is the most common form (~80%). It generally occurs in younger patients (peak 30–50 years) and predominantly in women. The prognosis is very good for small, well-differentiated tumours in younger patients. Radioiodine (^{131}I-iodide) is the mainstay of treatment. Follicular thyroid cancer is the second most common form. Peak onset is slightly later (40–60 years) but again predominantly female. Prognosis is slightly poorer than for the papillary form.

Imaging, as with benign thyroid diseases, uses ^{123}I-iodide, although ^{131}I-iodide is still used since radiation exposure is less of a concern in these patients, and its longer half-life allows prolonged imaging of biodistribution. However, ^{99m}Tc-pertechnetate is not useful in diagnosis of thyroid cancer due to inadequate retention in primary and metastatic tissue since it is not organified [35].

2.3.10.2 Adrenergic Tumours

Imaging the adrenal glands is difficult because they are so close to the kidneys. In the early 1980s the tracer ^{123}I-iobenguane (metaiodobenzylguanidine, MIBG, Adreview) was developed and remains the agent of choice. Iobenguane is an analogue of noradrenaline and is taken up by the noradrenaline transporter and stored in presynaptic vesicles. Whole-body and SPECT images are obtained 24 h after injection [36]. Adrenergic tumours include pheochromacytoma, carcinoid tumour, and neuroblastoma (particularly paediatric). High activities of ^{131}I-iobenguane can be used to treat certain adrenergic tumours.

Iobenguane is also used in imaging of adrenergic innervation in the heart to determine prognosis in congestive heart failure. Early and delayed anterior planar images of the chest are obtained. On each set of images, regions of interest are placed around the heart (H) and the mediastinum (M). H/M ratios are calculated and also the fractional washout between time points. Low ratios and rapid washout are associated with poor prognosis in heart failure [37].

2.3.10.3 Neuroendocrine Tumours

Neuroendocrine tumours (NETs) arise from the neuroendocrine system and include gastrointestinal, lung, and pancreatic NETs. NETs overexpress the somatostatin receptor, which can be imaged with tracers that are ligands for the receptor, including ^{111}In-pentetreotide (DTPA-octreotide, Octreoscan) and ^{99m}Tc-hynic-TOC (EDDA/hynic-Tyr$_3$-octreotide, Tektrotyd). Planar and/or SPECT/CT images are acquired at 24 h with ^{111}In and at 1-2 and 4 h with ^{99m}Tc [38, 39]. Analogues of pentetreotide coupled to beta-emitting radionuclides are used for therapy of NETs, most notably ^{177}Lu-DOTATATE (DOTA-Tyr$_3$-octreotate, Lutathera) [40]. Diagnostic scans are important for selection of patients for therapy and for monitoring response to therapy. In particular, ^{68}Ga-DOTATATE (DOTA-Tyr$_3$-octreotate, NETSPOT) is becoming widely used. ^{68}Ga (t½ 68 min) is obtained on-site from a generator loaded with ^{68}Ge (t½ 271 days). ^{68}Ga-DOTATATE was originally prepared with an automated synthesis device followed by purification but now a 'kit' formulation is available that requires only heating in a vial with no further purification.

2.3.10.4 Prostate-specific Membrane Antigen

The prostate-specific membrane antigen (PSMA) is overexpressed in prostate cancer, especially in advanced-stage prostate carcinomas, and there is little expression in normal tissue. This makes it a good target for imaging and ^{68}Ga-DKFZ-PSMA-11 and analogues have been rapidly adopted for PET/CT imaging. These are small molecules with a peptide-urea recognition site linked to a chelator. The remarkable degree of targeting and high tissue contrast made this an ideal candidate for therapy, and a variety of ^{177}Lu-labelled analogues have been developed, notably ^{177}Lu-PSMA-617. Although most of the focus has been on PET, there are also SPECT agents under development. For example, ^{111}In-PSMA-617 can be used in evaluation of patients for therapy [41]. ^{99m}Tc-labelled analogues include MIP-1404 and PSMA-RGS (radioguided surgery) [42, 43].

REFERENCES

[1] A. Boschi, L. Uccelli, and P. Martini, "A picture of modern Tc-99m radiopharmaceuticals: Production, chemistry, and applications in molecular imaging," *Applied Sciences,* vol. 9, no. 12, p. 2526, 2019, doi: 10.3390/app9122526.

[2] J. R. Ballinger, "Short- and long-term responses to molybdenum-99 shortages in nuclear medicine," *Brit J Radiol,* vol. 83, no. 995, pp. 899–901, 2010, doi: 10.1259/bjr/17139152.

[3] IAEA, "Cyclotron-based production of Technetium-99m," IAEA, Vienna, 2017.

[4] IAEA, "Cyclotron-produced radionuclides: Physical characteristics and production methods," IAEA, Vienna, 2009.

[5] J. Ballinger, "Formulation of radiopharmaceuticals," in *Sampson's Textbook of Radiopharmacy*, T. Theobald, Ed., 4th ed. London: The Pharmaceutical Press, 2011, pp. 325–37.

[6] D. A. Pryma *et al.*, "Efficacy and safety of high-specific-activity ^{131}I-MIBG therapy in patients with advanced Pheochromocytoma or Paraganglioma," *J Nucl Med,* vol. 60, no. 5, pp. 623–30, 2019, doi: 10.2967/jnumed.118.217463.

[7] T. Van den Wyngaert *et al.*, "The EANM practice guidelines for bone scintigraphy," *Eur J Nucl Med Mol Imaging,* vol. 43, no. 9, pp. 1723–38, 2016, doi: 10.1007/s00259-016-3415-4.

[8] K. Agrawal, F. Marafi, G. Gnanasegaran, H. Van der Wall, and I. Fogelman, "Pitfalls and limitations of radionuclide planar and hybrid bone imaging," *Semin Nucl Med,* vol. 45, no. 5, pp. 347–72, 2015, doi: https://doi.org/10.1053/j.semnucl med.2015.002.

[9] M. D. Blaufox *et al.*, "The SNMMI and EANM practice guideline for renal scintigraphy in adults," *Eur J Nucl Med Mol Imaging,* vol. 45, no. 12, pp. 2218–28, 2018, doi: 10.1007/s00259-018-4129-6.

[10] A. T. Taylor, "Radionuclides in nephrourology, Part 1: Radiopharmaceuticals, quality control, and quantitative indices," *J Nucl Med,* vol. 55, no. 4, pp. 608–15, 2014, doi: 10.2967/jnumed.113.133447.

[11] J. S. Fleming, M. A. Zivanovic, G. M. Blake, M. Burniston, and P. S. Cosgriff, "Guidelines for the measurement of glomerular filtration rate using plasma sampling," *Nucl Med Commun,* vol. 25, no. 8, pp. 759–69, 2004, doi: 10.1097/01.mnm.0000136715.71820.4a.

[12] A. Piepsz *et al.*, "Guidelines for ^{99m}Tc-DMSA scintigraphy in children," *Eur J Nucl Med,* vol. 28, no. 3, pp. Bp37–41, 2001.

[13] Ö. L. Kapucu *et al.*, "EANM procedure guideline for brain perfusion SPECT using ^{99m}Tc-labelled radiopharmaceuticals, version 2," *Eur J Nucl Med Mol Imaging,* vol. 36, no. 12, p. 2093, 2009, doi: 10.1007/s00259-009-1266-y.
[14] W. Van Paesschen, "Ictal SPECT," *Epilepsia,* vol. 45, no. s4, pp. 35–40, 2004, doi: 10.1111/j.0013-9580.2004.04008.x.
[15] R. H. Reid, K. Y. Gulenchyn, and J. R. Ballinger, "Clinical use of technetium-99m HM-PAO for determination of brain death," *J Nucl Med,* vol. 30, no. 10, pp. 1621–26, 1989.
[16] J. Darcourt *et al.*, "EANM procedure guidelines for brain neurotransmission SPECT using ^{123}I-labelled dopamine transporter ligands, version 2," *Eur J Nucl Med Mol Imaging,* vol. 37, no. 2, pp. 443–50, 2010, doi: 10.1007/s00259-009-1267-x.
[17] J. L. Cummings, C. Henchcliffe, S. Schaier, T. Simuni, A. Waxman, and P. Kemp, "The role of dopaminergic imaging in patients with symptoms of dopaminergic system neurodegeneration," *Brain,* vol. 134, no. 11, pp. 3146–66, 2011, doi: 10.1093/brain/awr177.
[18] M. Bajc *et al.*, "EANM guidelines for ventilation/perfusion scintigraphy: Part 1. Pulmonary imaging with ventilation/perfusion single photon emission tomography," *Eur J Nucl Med Mol Imaging,* vol. 36, no. 8, pp. 1356–70, 2009, doi: 10.1007/s00259-009-1170-5.
[19] J. Mortensen and H. Gutte, "SPECT/CT and pulmonary embolism," *Eur J Nucl Med Mol Imaging,* vol. 41, no. 1, pp. 81–90, 2014, doi: 10.1007/s00259-013-2614-5.
[20] M. J. O'Doherty and A. M. Peters, "Pulmonary technetium-99m diethylene triamine penta-acetic acid aerosol clearance as an index of lung injury," *Eur J Nucl Med,* vol. 24, no. 1, pp. 81–87, 1997, doi: 10.1007/bf01728316.
[21] K. A. Brown, "The role of stress redistribution thallium-201 myocardial perfusion imaging in evaluating coronary artery disease and perioperative risk," *J Nucl Med,* vol. 35, no. 4, pp. 703–6, 1994.
[22] H. J. Verberne *et al.*, "EANM procedural guidelines for radionuclide myocardial perfusion imaging with SPECT and SPECT/CT: 2015 revision," *Eur J Nucl Med Mol Imaging,* vol. 42, no. 12, pp. 1929–40, 2015, doi: 10.1007/s00259-015-3139-x.
[23] G. A. Beller and R. C. Heede, "SPECT imaging for detecting coronary artery disease and determining prognosis by noninvasive assessment of myocardial perfusion and myocardial viability," *J Cardiovasc Transl Res,* vol. 4, no. 4, pp. 416–24, 2011, doi: 10.1007/s12265-011-9290-2.
[24] T. L. Abell *et al.*, "Consensus recommendations for gastric emptying scintigraphy: A joint report of the American Neurogastroenterology and Motility Society and the Society of Nuclear Medicine," *Am J Gastroenterol,* vol. 103, no. 3, pp. 753–63, 2008, doi: 10.1111/j.1572-0241.2007.01636.x.
[25] E. S. Bonapace, A. H. Maurer, S. Davidoff, B. Krevsky, R. S. Fisher, and H. P. Parkman, "Whole gut transit scintigraphy in the clinical evaluation of patients with upper and lower gastrointestinal symptoms," *Am J Gastroenterol,* vol. 95, no. 10, pp. 2838–47, 2000, doi: 10.1111/j.1572-0241.2000.03195.x.
[26] M. L. Middleton and M. D. Strober, "Planar Scintigraphic imaging of the gastrointestinal tract in clinical practice," *Semin Nucl Med,* vol. 42, no. 1, pp. 33–40, 2012, doi: https://doi.org/10.1053/j.semnuclmed.2011.07.006.
[27] M. Tulchinsky *et al.*, "SNM practice guideline for Hepatobiliary Scintigraphy 4.0," *J Nucl Med Technology,* vol. 38, no. 4, pp. 210–18, 2010, doi: 10.2967/jnmt.110.082289.
[28] S. D. Sarkar, "Benign Thyroid Disease: What is the role of Nuclear Medicine?" *Semin Nucl Med,* vol. 36, no. 3, pp. 185–93, 2006, doi: https://doi.org/10.1053/j.semnuclmed.2006.03.006.
[29] E. Hindié *et al.*, "2009 EANM parathyroid guidelines," *Eur J Nucl Med Mol Imaging,* vol. 36, no. 7, pp. 1201–16, 2009, doi: 10.1007/s00259-009-1131-z.
[30] A. Signore, F. Jamar, O. Israel, J. Buscombe, J. Martin-Comin, and E. Lazzeri, "Clinical indications, image acquisition and data interpretation for white blood cells and anti-granulocyte monoclonal antibody scintigraphy: An EANM procedural guideline," *Eur J Nucl Med Mol Imaging,* vol. 45, no. 10, pp. 1816–31, 2018, doi: 10.1007/s00259-018-4052-x.
[31] J. R. Ballinger, "The use of protein-based radiocolloids in sentinel node localisation," *Clin Transl Imaging,* vol. 3, no. 3, pp. 179–86, 2015, doi: 10.1007/s40336-014-0097-4.
[32] D. S. Surasi, J. O'Malley, and P. Bhambhvani, "^{99m}Tc-Tilmanocept: A novel molecular agent for lymphatic mapping and sentinel lymph node localization," *J Nucl Med Technology,* vol. 43, no. 2, pp. 87–91, 2015, doi: 10.2967/jnmt.115.155960.
[33] F. Giammarile *et al.*, "The EANM clinical and technical guidelines for lymphoscintigraphy and sentinel node localization in gynaecological cancers," *Eur J Nucl Med Mol Imaging,* vol. 41, no. 7, pp. 1463–77, 2014, doi: 10.1007/s00259-014-2732-8.
[34] F. Giammarile *et al.*, "The EANM and SNMMI practice guideline for lymphoscintigraphy and sentinel node localization in breast cancer," *Eur J Nucl Med Mol Imaging,* vol. 40, no. 12, pp. 1932–47, 2013, doi: 10.1007/s00259-013-2544-2.
[35] T. F. Heston and R. L. Wahl, "Molecular imaging in thyroid cancer," *Cancer Imaging,* vol. 10, no. 1, pp. 1–7, 2010, doi: 10.1102/1470-7330.2010.0002.
[36] E. Bombardieri *et al.*, "^{131}I/^{123}I-Metaiodobenzylguanidine (mIBG) scintigraphy: Procedure guidelines for tumour imaging," *Eur J Nucl Med Mol Imaging,* vol. 37, no. 12, pp. 2436–46, 2010, doi: 10.1007/s00259-010-1545-7.
[37] S. Kasama, T. Toyama, and M. Kurabayashi, "Usefulness of cardiac sympathetic nerve imaging using ^{123}Iodine-Metaiodobenzylguanidine scintigraphy for predicting sudden cardiac death in patients with heart failure," *Int Heart J,* vol. 57, no. 2, pp. 140–44, 2016, doi: 10.1536/ihj.15-508.
[38] E. Bombardieri *et al.*, "^{111}In-pentetreotide scintigraphy: procedure guidelines for tumour imaging," *Eur J Nucl Med Mol Imaging,* vol. 37, no. 7, pp. 1441–48, 2010, doi: 10.1007/s00259-010-1473-6.

[39] I. Garai, S. Barna, G. Nagy, and A. Forgacs, "Limitations and pitfalls of ^{99m}Tc-EDDA/HYNIC-TOC (Tektrotyd) scintigraphy," *Nucl Med Rev Cent East Eur,* vol. 19, no. 2, pp. 93–98, 2016, doi: 10.5603/NMR.2016.0019.

[40] J. Strosberg *et al.*, "Phase 3 Trial of ^{177}Lu-Dotatate for Midgut Neuroendocrine Tumors," *N Eng J Med,* vol. 376, no. 2, pp. 125–35, 2017, doi: 10.1056/NEJMoa1607427.

[41] M. Mix *et al.*, "Performance of ^{111}In-labelled PSMA ligand in patients with nodal metastatic prostate cancer: correlation between tracer uptake and histopathology from lymphadenectomy," *Eur J Nucl Med Mol Imaging,* vol. 45, no. 12, pp. 2062–70, 2018, doi: 10.1007/s00259-018-4094-0.

[42] T. Maurer *et al.*, "^{99m}Technetium-based prostate-specific membrane Antigen-radioguided surgery in recurrent prostate cancer," *Eur Urol,* vol. 75, no. 4, pp. 659–66, 2019, doi: 10.1016/j.eururo.2018.03.013.

[43] C. Schmidkonz *et al.*, "^{99m}Tc-MIP-1404-SPECT/CT for the detection of PSMA-positive lesions in 225 patients with biochemical recurrence of prostate cancer," *The Prostate,* vol. 78, no. 1, pp. 54–63, 2018, doi: 10.1002/pros.23444.

3 Radiopharmaceuticals for Diagnostics

PET

Philip H. Elsinga

CONTENTS

3.1 INTRODUCTION

In order to make PET scans, radiopharmaceuticals are needed to be administered to the patient. Depending on the biochemical or physiological process to be measured, the right radiopharmaceutical has to be selected that takes part in such a process. A radiopharmaceutical can be seen as a drug that is given in homeopathic amounts. Therefore, a PET-radiopharmaceutical is also called a PET-tracer [1].

A PET-radiopharmaceutical has two properties: (1) it participates in a biochemical / physiological process and (2) contains a scaffold that binds the radioactive PET-radionuclide. Ideally, the PET-radionuclide is completely integrated in the molecule so that the pharmacological properties of the radiopharmaceutical do not change compared to the original molecule; see Figure 3.1, where, in the left molecule, the radionuclide ^{11}C is fully integrated. This is only possible when using radionuclides derived from endogenous isotopes (^{11}C, ^{13}N, ^{15}O). In other cases, an additional chemical group is added to the molecule [1]. In case of ^{18}F or ^{124}I this is like ^{11}C performed using covalent bonds [2]. When using radiometals, chelators are used to catch the metal through electronic interactions [3, 4] (right molecule in Figure 3.1).

PET-radiopharmaceuticals (RPs) are usually produced on small scale at tracer dosages. A few exceptions exist, for example. in case of RPs based on proteins or peptides where cold compound is often added to prevent excessive metabolism. The molar or specific activity typically needs to be high (100-1000 GBq/umol) to ensure administration of tracer amount of compound. This is important to eliminate pharmacological and toxicological effects.

DOI: 10.1201/9780429489501-3

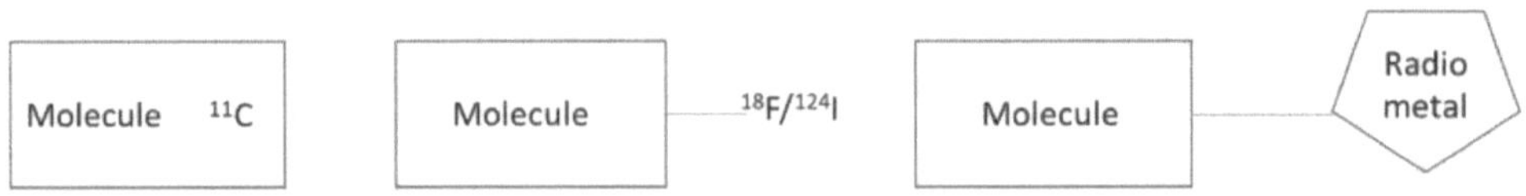

FIGURE 3.1 Several modes of incorporation of a PET-radionuclide in a molecule.

TABLE 3.1
Overview of Properties of Most Abundantly used PET-radionuclides

Radionuclide	Half-life	Positron energy (mean)	Positron abundance	On site Availability	Most important nuclear reaction
^{15}O	2 min	735 keV	>99%	cyclotron	$^{15}N(p,n)^{15}O$ $^{14}N(d,n)^{15}O$
^{13}N	10 min	491 keV	>99%	cyclotron	$^{16}O(p,\alpha)^{13}N$
^{11}C	20 min	385 keV	>99%	cyclotron	$^{14}N(p,\alpha)^{11}C$
^{18}F	110 min	250 keV	97%	cyclotron	$^{18}O(p,n)^{18}F$ $^{20}Ne(d,\alpha)^{18}F$
^{68}Ga	68 min	353 keV	1.19%	generator	$^{68}Ge/^{68}Ga$ generator
^{89}Zr	78 h	396 keV	23%	shipment	$^{89}Y(p,n)^{89}Zr$
^{124}I	100 h	687 keV	11 %	shipment	$^{124}Te(p,n)^{124}I$
^{64}Cu	12.7 h	278 keV	18%	cyclotron or shipment	$^{64}Ni(p,n)^{64}Cu$

Tissue uptake of RPs can be regulated by:

- Transporters
- Enzymes
- Receptors
- Antigen binding
- Ligand-protein interaction
- Blood flow
- Changes in pH or oxygenation

3.2 RADIONUCLIDES FOR PET

In order to prepare PET-radiopharmaceuticals, the starting PET-radionuclides have to be produced. They are either available through cyclotron irradiation or by generators that contain reactor-produced mother radionuclides. The short-lived radionuclides ^{15}O, ^{13}N, ^{11}C and ^{18}F can be produced relatively easily with all commercially available cyclotrons [5]. As ^{15}O and ^{13}N have very short half-lives it is not really possible to synthesize a wide range of PET-radiopharmaceuticals. ^{15}O-Based radiopharmaceuticals are limited to [^{15}O]water and ^{15}O-gasses such as [^{15}O]O_2, C[^{15}O]O and C[^{15}O]O_2. In the case of ^{13}N, usually [^{13}N]ammonia is produced as a perfusion tracer. Some groups perform some more sophisticated radiochemistry to prepare ^{13}N-based radiopharmaceuticals.

^{68}Ga is commercially available through (GMP-compliant) generators [6]. The main problem is their increasing costs and the relatively low yields of ^{68}Ga, which is in the order of magnitude of 1 GBq. Therefore, technology is being developed to produce higher amounts of ^{68}Ga using cyclotrons and solid-state or liquid targets. This technology is not yet widely available. ^{89}Zr, ^{124}I and ^{64}Cu are produced with solid-state targets. The number of production sites for these radionuclides is increasing. Because of their longer half-lives, these radionuclides and/or corresponding radiopharmaceuticals are suitable for shipment to other nuclear medicine sites.

Table 3.1 lists some properties of the most abundantly used PET-radionuclides [7]. Besides these, several new PET-radionuclides (isotopes of Scandium (Sc) and Terbium (Tb)) are upcoming; their production methods are optimized, and radiochemistry is in development.

3.3 PREPARATION OF PET-RADIOPHARMACEUTICALS

The preparation of RPs can be subdivided into 3 steps:

1. Radiochemistry: During the radiolabelling with PET-radionuclides, a reaction/complexation between a precursor molecule and the radiolabelled chemical entity occurs preferably at the latest stage possible. Radiolabelling can occur directly with the radionuclide obtained from the cyclotron or generator, or a small PET-labelled synthon is prepared first, followed by a second reaction step.
2. Purification: After the radiolabelling steps, which are considered to be the radiochemistry part, the radiolabelled product needs to be isolated from (radio)chemical impurities. This is mostly achieved by High Performance Liquid Chromatography (HPLC) or by using Solid Phase Extraction (SPE) cartridges.
3. Reformulation: Finally, a reformulation step is carried out to make sure that the RP is injected in a proper and safe form, and thereafter the solution is sterilized by passing through a 0.22 um sterile filter.

A suitable PET-radiopharmaceutical should comply with several criteria to ensure that PET-images represent the right physiological process: Amenability and reliability for radiolabelling, lipophilicity, affinity, selectivity, choice of radionuclide, and metabolic stability. In addition, in order to have a robust production method automation of the process is also crucial to reduce the radiation burden and to monitor and document in-process control parameters such as reaction times and temperature profiles and improve reproducibility of the whole production process. Documentation of production parameters is important for Good Manufacturing Practice (GMP) purposes, and can be of great use in case of troubleshooting and to perform trend analyses [8]. Automation can be performed by synthesis modules based on reaction vials, fixed tubings to transport chemicals and solutions and switchable valves. Fixed tubings can also be replaced by disposable (sterile) cassettes avoiding time-consuming cleaning procedures and cross-contaminations.

3.4 RADIOCHEMISTRY

In this chapter the general principles of radiolabelling procedures for the most used PET-radionuclides will be described, as follows:

- Fluorine-18 has a half-life of 110 min and is readily available in large quantities by cyclotron production and can be distributed in up to six hours transport time. The most used production method, which is the nucleophilic substitution reaction, will be described in more detail, but other methods are developed – electrophilic fluorinations (almost exclusively for [^{18}F]FDOPA production), late-stage ^{18}F-fluorinations with nucleophilic ^{18}F, but sparsely used. Some of them will be used more often in the future.
- Carbon-11, which is also available from cyclotrons, can only be used locally because of the short half-life of 20 min. A wide variety of ^{11}C-building blocks can be produced but most often require specialized equipment and expertise. Synthesis equipment to produce [^{11}C]methyl iodide and [^{11}C]methyl trifate is commercially available, and therefore the ^{11}C-methylation reactions that can be performed with these 2 building blocks are predominantly used.
- Gallium-68 has the advantage of a reasonable half-life of 68 min and can be eluted as [^{68}Ga]$GaCl_3$ from GMP-compliant generators. Nowadays many large and medium-sized hospitals have a Gallium generator to mainly produce PSMA and DOTATOC or DOTATATE.
- Zirconium-89 is an emerging long-lived PET-radionuclide that is almost exclusively used to radiolabel monoclonal antibodies. As the pharmacokinetics/tissue penetration of antibodies is relatively slow to achieve optimal contrast, the half-life of 78 hours perfectly suits this kinetic behaviour.

Radiochemistry with ^{18}F and ^{11}C involves formation through covalent bonds requiring expertise combining radiochemistry and organic chemistry. Radiochemistry with radiometals like ^{68}Ga, ^{89}Zr, but also ^{64}Cu or involves complexation, which is similar to ^{99m}Tc chemistry with chelator ligands that interact with the available metal electron orbitals. This type of radiochemistry combines radiochemistry knowledge with inorganic chemistry.

3.4.1 ^{18}F-RADIOLABELLING

^{18}F is the most used PET-radionuclide as it has favourable characteristics such as a half-life of 110 min, low positron energy, and many different radiolabelling opportunities [9]. A fluorine atom behaves as an electronegative atom with

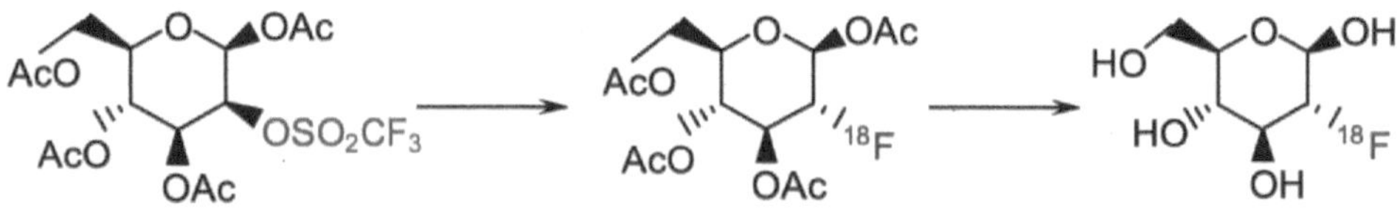

FIGURE 3.2 Reaction scheme of a nucleophilic substitution reaction with [^{18}F]fluoride.

the size of a hydrogen atom. This results on one hand in minor steric changes but on the other hand fluorine substitution has an effect on acidity, and interactions with proteins can change dramatically.

^{18}F can be produced by the cyclotron in two different chemical forms, that is, nucleophilic and electrophilic ^{18}F. Electrophilic ^{18}F is produced in the form of [^{18}F]F_2. During its production process relatively large amounts of 'cold' F_2 need to be added to be able get the highly reactive F_2 out of the target [10]. As a consequence, the molar activity of [^{18}F]F_2 is several orders of magnitude (100-1000) lower than of nucleophilic ^{18}F. Because of this, electrophilic [^{18}F]F_2 is hardly used (except for [^{18}F]FDOPA) despite the fact that most chemical reactions involving fluorine are electrophilic and include additions to double bonds and aromatic rings and electrophilic substitutions on aromatic rings with a trialkyl tin group often used as the leaving group.

Nucleophilic ^{18}F is produced in the form of [^{18}F]fluoride ([^{18}F]F^-). [^{18}F]F^- is produced from ^{18}O-enriched water. However, before [^{18}F]F^- can react with precursor molecules, it must be dried and a phase-transfer catalyst must be added to allow [^{18}F]F^- to dissolve in organic solvents that are required to perform nucleophilic substitution reactions. Dried [^{18}F]F^- acts as a nucleophilic particle to attack carbon atoms that are slightly positively charged because of a neighbouring electron-withdrawing group.

This electron-withdrawing group will be substituted by [^{18}F]F^- and is called the leaving group in this so-called nucleophilic substitution reaction (see Figure 3.2 for a typical example of a nucleophilic substitution reaction for the synthesis of [^{18}F]FDG) [11]. Typical reactions are performed with mg (micromoles) scale amounts of precursor, which is in large excess (100-1000-fold) compared to the amounts of [^{18}F]F^- expressed in moles and are usually performed at elevated temperatures ranging from 80–150 degrees and reaction times of 5–30 min. Currently used well-known RPs [^{18}F]FDG and [^{18}F]PSMA1007 [12] are synthesized through this radiochemical route.

3.4.2 ^{11}C-LABELLING

^{11}C is the radionuclide of choice to radiolabel drugs that are part of a drug development process as well as to label endogenous compounds. In addition, because of the short half-life of 20 min, multiple injections in the same subject on one day are possible while limiting the radiation dose for the subject. ^{11}C is produced with gas targets filled with N_2-gas containing either O_2 or H_2 yielding, respectively, [^{11}C]CO_2 or [^{11}C]CH_4. From these ^{11}C-target products, ^{11}C-labelling synthons can be produced. The most commonly used ^{11}C-synthons are ^{11}C-methyl iodide and ^{11}C-methyl triflate [13]. In by far the majority of the cases, ^{11}C-methylation (typically with a reaction time up to 10 min) is performed on amines, phenols, and thiols, with the corresponding examples [^{11}C]raclopride, [^{11}C]choline, and [^{11}C]methionine. Figure 3.3 shows an overview of ^{11}C-synthons that can be prepared. Most of them are used in specific laboratories using specialized equipment. A synthon that is emerging is [^{11}C]CO. Its synthesis has been optimized for routine use. Using [^{11}C]CO as labelling agent opens up new classes of molecules containing carbonyl groups that are present in many biological active molecules [14].

3.4.3 ^{68}GA-LABELLING

PET-radiopharmaceuticals labelled with the radiometal ^{68}Ga are very useful for Nuclear Medicine Departments without a cyclotron. ^{68}Ga is readily available from ^{68}Ge/^{68}Ga-generators that nowadays also can be obtained with GMP-quality. [^{68}Ga]$GaCl_3$ can be eluted a few times per day for radiopharmaceutical production. [^{68}Ga]Ga^{3+} can be complexed by a chelator, which is covalently connected to the radiopharmaceutical. Several chelators for ^{68}Ga are being used, including cyclic chelators DOTA, NOTA, NODAGA, or acyclic chelators like HBED (Figure 3.4). All these chelators have different characteristics regarding affinity for [^{68}Ga]Ga^{3+}, the stability of the complex, and the reaction conditions to perform the complexation. In the clinic, the most used ^{68}Ga-labelled radiopharmaceuticals are [^{68}Ga]-Ga-HBED-CC-PSMA and [^{68}Ga]Ga-DOTATOC or –TATE [3].

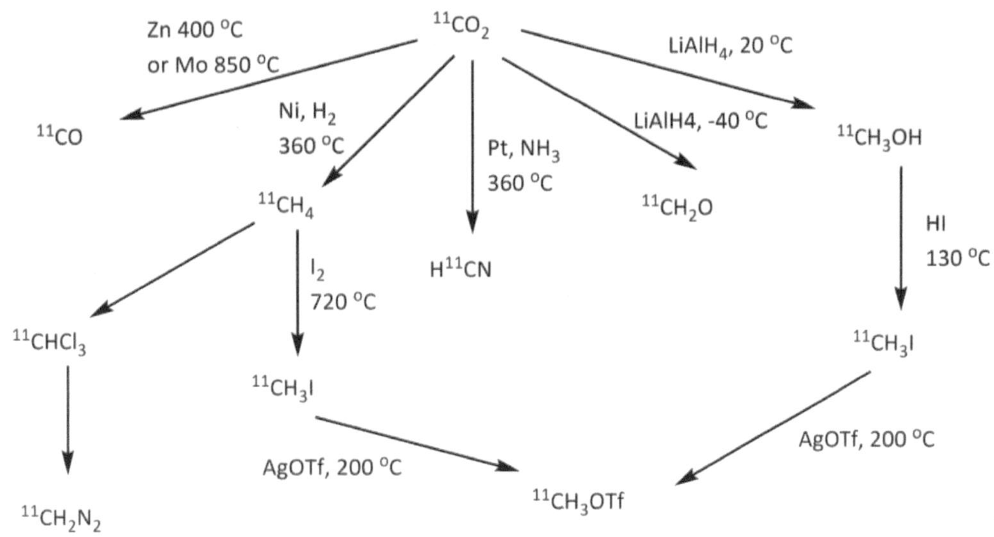

FIGURE 3.3 Overview of ^{11}C-building blocks.

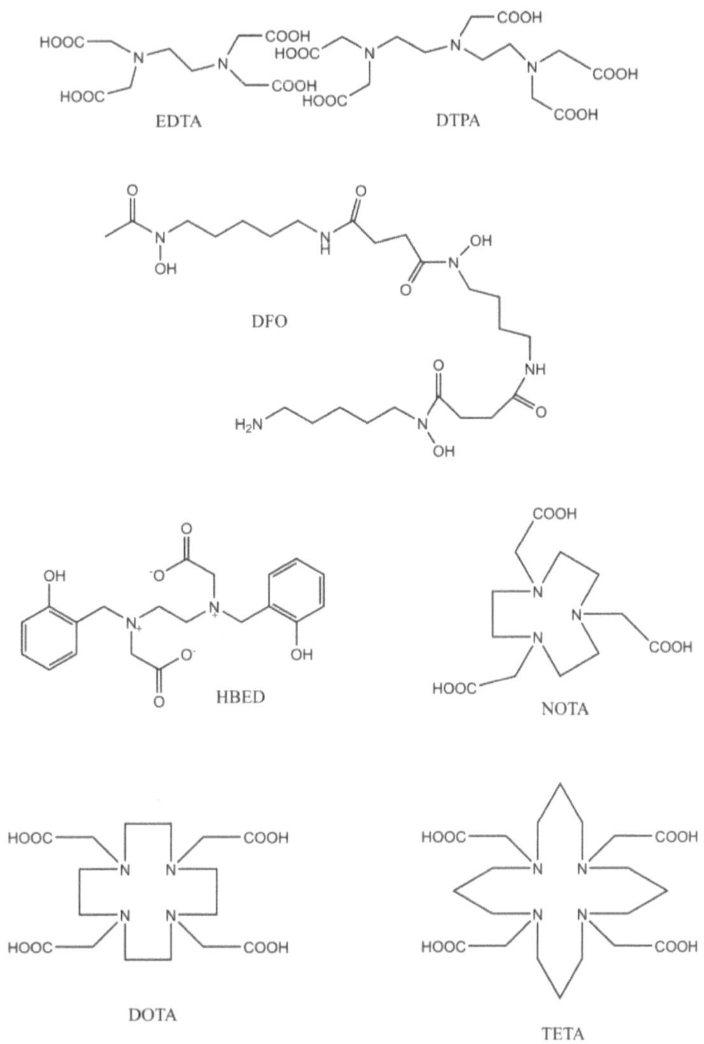

FIGURE 3.4 Examples of chelators used for ^{68}Ga-labelling.

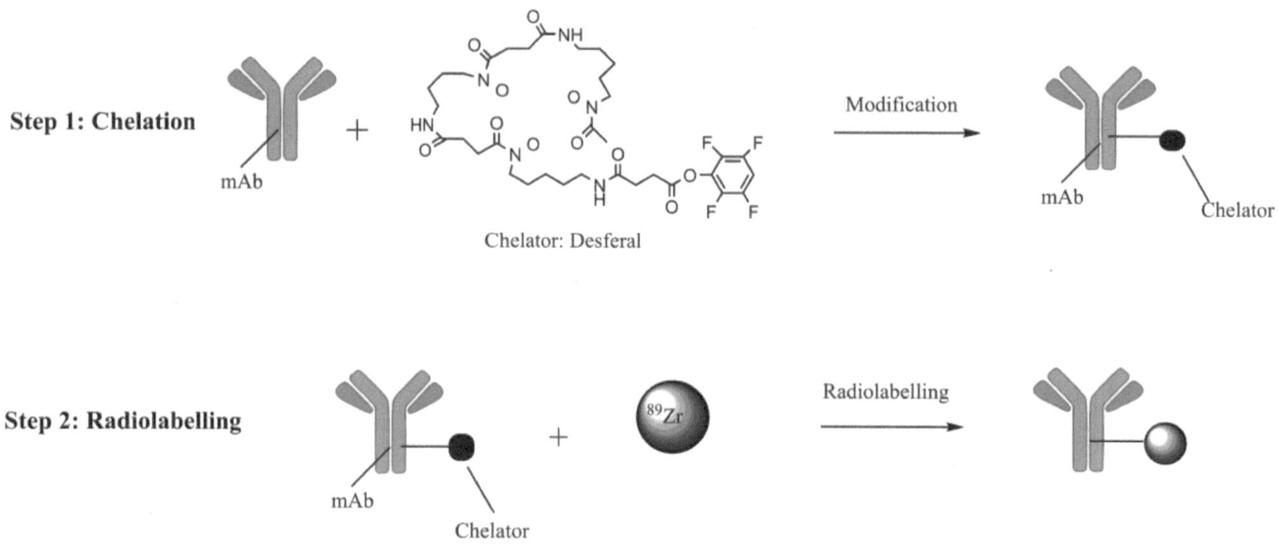

FIGURE 3.5 Procedure to prepare ^{89}Zr-labelled antibodies.

3.4.4 ^{89}ZR-LABELLING

The radionuclide ^{89}Zr with a half-life of 78 hours is the radionuclide of choice for large-molecule radiopharmaceuticals because of the slow pharmacokinetics. In practice this means that ^{89}Zr-labelling is applied for antibody-based PET-radiopharmaceuticals. The number of these ^{89}Zr-radiopharmaceuticals is exploding. As ^{89}Zr is also a radiometal, a chelator is required to attach it to the antibody. Suitable chelators are desferal, desferrioxamine (DFO), and its octadentate form, DFO* [15]. The general approach is that antibodies are conjugated to the chelator with standard coupling techniques. The chelators contain a reactive ester that can react with lysine residues in the antibody protein. After purification of the chelated antibody, the complexation with [^{89}Zr]Zr^{4+} usually obtained as [^{89}Zr]Zr-oxalate can be carried out followed by a purification step (Figure 3.5).

3.5 UPTAKE MECHANISMS OF COMMON PET-RADIOPHARMACEUTICALS

This paragraph describes the field of application and uptake mechanism of the most-used PET-radiopharmaceuticals.

The workhorse tracer of PET is FDG, 2-deoxy-2-[^{18}F]fluoro-D-glucose which is a sugar molecule. It strongly resembles D-glucose, the naturally occurring sugar that is one of the main sources for energy in cell metabolism. Like glucose, it is taken up in cells requiring an energy source, both by diffusion and active GLUT-1 transporters. Tissues with increased demand for energy are brain, heart, and tumour. Therefore, FDG is used for tumour and inflammation imaging and energy metabolism. After uptake in the cell, FDG is phosphorylated to FDG-6-phospate. Then, unlike glucose, FDG is trapped in the tissue and not further metabolized via the Krebs Cycle and oxidative phosphorylation to produce many ATP-molecules [16].

FLT, 3'-deoxy-3'-[^{18}F]fluoro-L-thymidine, is a thymidine analogue that is used to measure cell proliferation rates in tumour. Both thymidine and FLT are phosphorylated by thymidine kinase to the monophosphate derivative. Thymidine is subsequently further phosphorylated to its triphosphate and then incorporated into DNA. FLT gets stuck in the triphosphate state and is not incorporated into DNA but remains trapped in the cell. FLT is used for PET-imaging and shows benefit over FDG in the evaluation of anticancer treatment because its uptake is not affected by inflammatory responses [17].

There are several PET-labelled amino acids that are used in oncological PET. Initially, several natural ^{11}C-labelled amino acids (AAs) were produced for the measurement of protein synthesis rates in tumour. Currently, the most abundantly used are L-[methyl-^{11}C]methionine and [^{18}F]fluoroethyl tyrosine (FET). In addition, several artificial AAs have been labelled with ^{18}F, such as [^{18}F]FACBC ([^{18}F]fluciclovin). Many of these artificial AAs are not incorporated in proteins, but their uptake reflects AA transport [17]. Amino acid transporters are overexpressed on tumour cells and therefore AA uptake in tumour is increased. The use of AAs in brain tumours had great benefit as they give higher tumours to normal tissue ratio (T/N) than FDG, which (T/N) uptake is less differentiated because of high natural energy consumption in brain tissue.

Another molecular target on tumour cells (especially on prostate cancer cells and, more recently discovered, breast cancer) is Prostate Membrane Specific Antigen (PSMA). This protein with enzymatic activity has an intracellular domain that was previously targeted with ^{111}In-ProstaScint, and an extracellular domain with a binding site for Lys-urea-Glu to inhibit PSMA enzyme activity. This target has become a hot topic over the last few years as it can be used for radionuclide therapy, using ^{177}Lu-PSMA analogues. The PSMA-tracers all have the same binding motif, Lys-urea-Glu. This motif is then connected using a linker to the radionuclide binding motif which can be ^{68}Ga, ^{18}F or other radionuclides [18]. Upon binding of the PSMA PET-tracer, the formed complex internalizes into the cell. The PSMA tracer is trapped in the lysosome of the cancer cell.

An exploding field in oncology is the use of ^{89}Zr-labelled monoclonal antibodies (mAb), which bind very specifically to tumour antigens with high affinity. The production process of GMP-grade mAbs in large scale is still very expensive and time-consuming [19]. Therefore, its application in routine clinical PET is still pending. After binding of the radiolabelled antibody to the antigen, the complex is internalized into the cell by endocytosis. As the accumulation process of ^{89}Zr-mAbs in tumours tissue is relatively slow (takes several days), PET scans are typically performed 3–6 days after intravenous injection to both allow accumulation in tissue and wash-out of non-bound radioactivity.

Several myocardial flow tracers are available to measure cardiac perfusion. This is used to detect ischemia and infarction. Uptake mechanisms of several perfusion tracers are different [20].

- Myocardial uptake of [^{15}O]water is a result of passive diffusion and gives an excellent correlation between perfusion and tracer extraction from blood.
- [^{13}N]ammonia is diffusing across the cell membrane, where it forms an equilibrium with the ammonium ion NH_4^+, which subsequently is trapped in the cell as it is converted to [^{13}N]glutamine.
- [^{82}Rb]$RbCl_2$ uptake is determined by ^{82}Rb as being an analogue of potassium, which is actively transported by the Na^+/K^+ ATPase-pump.

Brain tracers consist of a variety of tracers with different uptake mechanisms [21]. The majority of the tracers is dedicated to several pathways within neurotransmission systems. Most studied systems are the dopaminergic system followed by the serotonergic, and cholinergic systems. Also, other receptor systems involved in the regulation of brain function are the subjects of PET-brain studies. As shown in Figure 3.6, several processes are involved in proper neurotransmission, which can all be studied with PET. These processes include biosynthesis of the neurotransmitter from their precursors, storage in presynaptic vesicles, metabolic breakdown of neurotransmitters, their release into the synaptic cleft and subsequent binding to postsynaptic receptors and/or reuptake into the presynaptic neuron. The corresponding tracers are usually inhibitors of enzymes involved in biosynthesis and break down of the neurotransmitters or ligands for the receptors (either antagonists or agonists).

Another important class of brain tracers are those targeting protein misfolding, like β-amyloid, tau-protein and α-synuclein [22]. Accumulation of these misfolded proteins are the end stage of neurodegenerative diseases like Parkinson's Disease (PD) and Alzheimers Disease (AD). Accumulation of Tau protein is predominantly related to AD, α-synuclein to PD, β-amyloid predominantly to Lewy Body Disease and AD, but the differences are not distinct, although there is a continuous spectrum. Misfolded proteins display a β-sheet tertiary conformation, which is the targeting domain for PET-tracers. Therefore, it is very challenging to develop PET-tracers for specific misfolded proteins as they all have a common binding domain.

3.6 PET-RADIOPHARMACEUTICALS APPLIED IN HUMANS

3.6.1 PET-Radiopharmaceuticals for Brain Studies

Regarding PET-radiopharmaceuticals for human brain studies, several new applications have emerged to image new targets. These include third-generation tracers for TSPO [22] (related to image activated microglia and neuroinflammation), ^{18}F-labelled tracers for the cholinergic system, tracers to image protein misfolding (beta-amyloid, tau and alpha-synuclein for AD, PD and other neurodegenerative diseases). Very recently, PET-radiopharmaceuticals also were published for SV2A receptors representing synaptic density, which is a factor that changes over time during neurodegeneration [23]. Table 3.2 gives an overview of PET-radiopharmaceuticals applied in humans and appeared in literature.

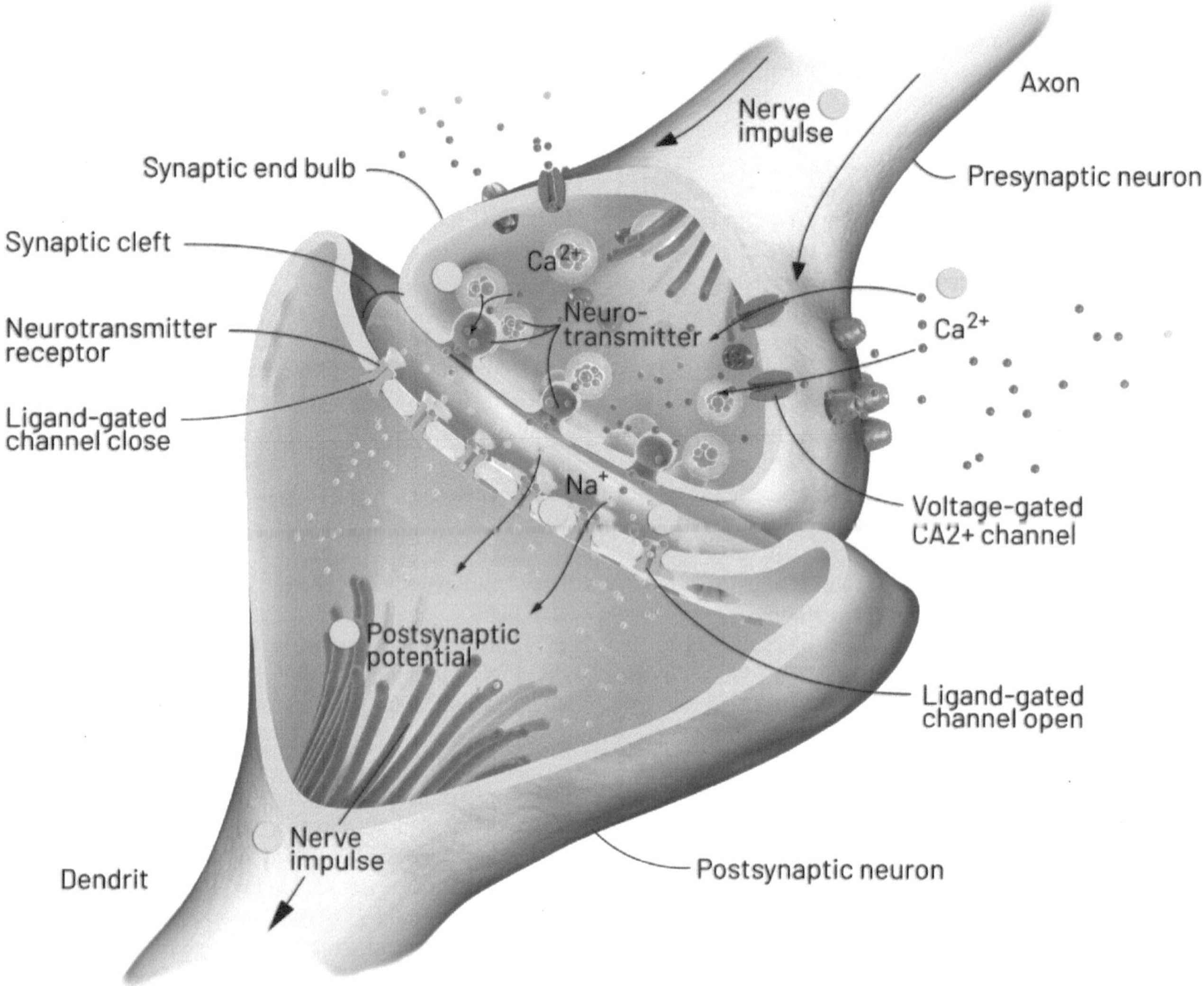

FIGURE 3.6 Overview of processes at the synapse with a neurotransmission system. Colour image available at www.routledge.com/9781138593312.

3.6.2 Oncology PET-Tracers

Table 3.3 represents published PET-radiopharmaceuticals that are used in oncological patients. The main development is the emerging of tracers for PSMA ([^{18}F]PSMA1007) to diagnose and treat prostate cancer [24]. Furthermore, tracers for EGFR-tk are developed for treatment follow-up. There is a striking explosion of using [^{89}Zr]antibodies for many different targets. In comparison to the neurology tracers there are a few new tracers labelled with ^{11}C or ^{18}F published in the last 5 years. Physicians still use established tracers such as [^{18}F]FDG, [^{18}F]FET, [^{18}F]FDOPA and [^{18}F]FLT.

With respect to cardiovascular PET-radiopharmaceuticals, most interest is in the PET-imaging of heart perfusion, which is affected in heart failure and infarction [20]. Traditionally [^{13}N]ammonia and [^{15}O]water are the most-used tracers but, because of the short half-life, only hospitals with a cyclotron are able to perform such PET studies. Other hospitals mostly still use [^{99m}Tc]Sestamibi for SPECT, but for those sites without a cyclotron a solution is to apply [^{82}Rb]$RbCl_2$ which is obtained from a generator. Another option is [^{18}F]flurpiridaz, however its clinical value is not proven yet. Recently mini-cyclotrons in combination with user-friendly equipment and software have been developed by industry to boost the use of [^{13}N]ammonia and [^{15}O]water.

Another application is re-use of PET-radiopharmaceuticals for atherosclerosis imaging. Several tracers that are already in use for oncology or neuroinflammation have found their way to probe molecular processes in atherosclerotic plaques.

TABLE 3.2
Neuro PET-tracers Applied in Humans

Target	Tracer	Physiological process
TSPO	[^{11}C]PK11195, DAA and PBR derivatives, [^{18}F]GE180	Antagonist
GABA	[^{11}C]/[^{18}F]Flumazenil	Antagonist
Dopaminergic system	[^{18}F]FDOPA, [^{11}C]SCH23390	Vesicular storage
		D_1 antagonist
	[^{11}C]Raclopride	D_2 antagonist
	[^{11}C]PHNO	D_2 agonist
	[^{18}F]FP-CIT, [^{18}F]FE-PE2I	Dopamine transporter
ß-Amyloid	[^{11}C]PIB	Staining agent
	[^{18}F]Florbetaben/florbetapir	
NMDA	[^{11}C]GSK-931145	Antagonist
	[^{18}F]GE179	
P-Glycoprotein	[^{11}C]Verapamil, [^{11}C]dLop,	Substrate
	[^{11}C]Metoclopramide	
Cholinergic system	[^{11}C]MP4A	Acetylcholinesterase inhibitor
	[^{18}F]FEOBV	VAChT ligand
	[^{18}F]FP-TZTP	M_2 antagonist
	[^{18}F]ASEM, [^{11}C]CHIBA-1001	α_7-nAChR ligands
	[^{18}F]A-85360, [^{18}F]flubatine	$\alpha_4\beta_2$-nAChR ligands
mGlu-5	[^{11}C]ABP688, [^{18}F]PSS232	Antagonist
	[^{18}F]FPEB	
VMAT2	[^{11}C]DTPZ, [^{18}F]AV-133	Inhibitor
	[^{18}F]FP-DTBZ	
Adenosine receptor	[^{11}C]MPDX, [^{11}C]preladenant	A_1 antagonist
		A_{2a} antagonist
	[^{11}C]TSMX	A_{2a} antagonist
Serotonergic system	[^{11}C]DASB	Serotonin transporter ligand
	[^{11}C]WAY100635, [^{18}F]MPPF	5-HT_{1A} antagonist
	[^{11}C]Cimbi-36	5-HT_{2A} agonist
Norephedrine	[^{11}C]Methylreboxetine	NE transporter ligand
P2X7	[^{18}F]JNJ-64413739	Antagonist
	[^{11}C]JNJ-54173717	
Opioid receptors	[^{11}C]Carfentanil	Mu opioid receptor antagonist
Monoamine oxidase	[^{11}C]Deprenyl	MAO inhibitors
Energy	[^{18}F]FDG	Glucose consumption
Tau protein	[^{18}F]THK523, [^{18}F]AV1541	Protein misfolding in AD
Phosphodiesterase-PDE-4	[^{11}C]rolipram,[^{11}C]IMA107, [^{11}C]LuAE92680, [^{18}F]MNI589	Breakdown of cAMP
Myeline	[^{11}C]MeDAS	Staining of demyelination
Cannabinoid-2 Receptor	[^{11}C]NE40	antagonist
Neuronal activity	[^{15}O]Water	Blood flow
SV2A	[^{11}C]/[^{18}F]UCB-J	Synaptic density Ligands
Sigma	[^{11}C]SA4503, [^{18}F]fluspidine	Antagonists

3.6.3 Infection/Inflammation Tracers (Excluding Neuroinflammation and Atherosclerosis)

PET-radiopharmaceuticals to investigate inflammatory and infectious processes are relatively new [25]. Traditionally, radiolabelled white blood cells (WBC) are used to image these processes. Furthermore, [^{18}F]FDG is applied as a non-specific tracer to measure metabolic flair as a result of infection/inflammation. For bacterial infection imaging, first PET-tracers have become available based on ubiquicidine peptides.

TABLE 3.3
Oncology PET-tracers Applied in Humans

Target	Tracer	Physiological process
GLUT-transporters and Hexokinase	[^{18}F]FDG	Glucose consumption
Thymidine kinase type 1	[^{18}F]FLT	DNA-synthesis
Choline synthase	[^{18}F]/[^{11}C]choline	Membrane synthesis
Dopamine storage	[^{18}F]fluoro-DOPA	Dopaminergic system
Amino acid transporter	[^{18}F]fluoroethyl tyrosine, [^{11}C]methionine	Amino acid transports
Hydroxyapatite matrix	[^{18}F]sodium fluoride	Bone metastases
Oxidative metabolism	[^{11}C]acetate	Energy metabolism
Androgen receptor	[^{18}F]FDHT	Prostate cancer
Estrogen receptor	[^{18}F]FES	Breast cancer
Hypoxia	[^{18}F]FMISO, [^{18}F]FAZA, [^{64}Cu]ATSM	Oxygenation levels in tissue
PSMA	[^{68}Ga]HBED-CC-PSMA, [^{18}F]PSMA1007	Prostate cancer
Somatostatin receptor	[^{68}Ga]DOTATOC/TATE	Neuroendocrine tumour
Amino acid transporter	[^{18}F]fluciclovine	Prostate cancer
Integrins, alphaV-beta3	[^{18}F]galacto-RGD	Angiogenesis
HER2	[^{89}Zr]trastuzumab, pertuzumab	Tumour antigen, Induction of downstream signalling pathways in regulation of cancer
HER3	[^{89}Zr]GSK2849330, [^{89}Zr]lumretuzumab	Tumour antigen, Induction of downstream signalling pathways in regulation of cancer
EGFR	[^{89}Zr]cetuximab	Tumour antigen, Induction of downstream signalling pathways in regulation of cancer
VEGF	[^{89}Zr]bevacizumab	Tumour angiogenesis
EGFR tk	[^{11}C]erlotinib	Tyrosine kinase involved in signalling pathways in regulation of cancer
Folate receptor	[^{89}Zr]M9346A	Regulation of tumour progression
CD20 (B-cells)	[^{89}Zr]rituximab	Immunotherapy of B-cell lymphoma
PDL1	[^{89}Zr]girentuximab	Immune checkpoint inhibition, regulation of cancer
PD1	[^{89}Zr]nivolumab	Immune checkpoint inhibition, regulation of cancer
TNFα	[^{89}Zr]certolizumab	Tumour necrosis
CD3	[^{89}Zr]AMG211	Activated T-cells in tumourimmunology
IL2 receptor	[^{89}Zr]cytokine	Activated T-cells in tumourimmunology
CD38	[^{89}Zr]daratumumab	Activated T-cells in tumourimmunology
PSMA	[^{89}Zr]IAB2M	Enzyme on cell membrane of prostate cancer
CD44	[^{89}Zr]RG7356	Aggregation and migration of cancer cells

3.7 CONCLUSIONS

An increasing number of PET-radiopharmaceuticals is becoming available every year. This is a result of increased availability of PET-radionuclides and an expanding toolbox of radiochemical production methods. Production methods become more robust and can be carried out on large scale in combination with GMP-compliant synthesis modules. This increased availability goes hand in hand with increasing interest in using PET-imaging in the clinic.

TABLE 3.4
Myocardial PET-tracers Applied in Humans

Target	Tracer	Physiological process
Passive diffusion followed by trapping	[^{13}N]ammonia	Blood flow
Mitochondrial complex	[^{18}F]flurpiridaz	Blood flow
Passive diffusion	[^{15}O]water	Blood flow
Na^+-K^+ transporter	[^{82}Rb]$RbCl_2$	Blood flow
Adrenergic system	[^{11}C]mHED	Innervation
Fatty acid-CoA ligase	[^{11}C]palmitate	Fatty acid metabolism
Muscarinic receptor	[^{11}C]MQNB	Innervation
Beta-adrenergic receptor	[^{11}C]CGP12388/[^{11}C]CGP12177	Innervation
PDE-10	[^{11}C]rolipram	Second messenger in innervation
Several targets see table above	[^{11}C]PK11195, Na[^{18}F]F, [^{18}F]/[^{11}C]choline, [^{18}F] galacto-RGD, [^{68}Ga]DOTATOC	Atherosclerotic plaques

TABLE 3.5
Infection/Inflammation PET-tracers Applied in Humans

Target	Tracer	Physiological process
Iron	[^{68}Ga]Gallium citrate	Host defence
Energy consumption	[^{18}F]FDG	Activated immune cells
Antimicrobial peptides	[^{68}Ga]/[^{18}F]Ubiquicidin peptides	Immune response
T-cells	[^{18}F]Interleukin-2	Immune response
CD8	[^{89}Zr]IAB22M2C	Pancreatic Beta-cells, diabetes

REFERENCES

[1] P. H. Elsinga and R. A. Dierckx, "Small molecule PET-radiopharmaceuticals," *Curr Pharm Des,* vol. 20, no. 14, pp. 2268–74, 2014, doi: 10.2174/13816128113196660661.

[2] V. Pichler *et al.*, "An Overview of PET Radiochemistry, Part 1: The Covalent Labels (18)F, (11)C, and (13)N," *J Nucl Med,* vol. 59, no. 9, pp. 1350–54, 2018, doi: 10.2967/jnumed.117.190793.

[3] I. Velikyan, "68Ga-based radiopharmaceuticals: production and application relationship," *Molecules,* vol. 20, no. 7, pp. 12913–43, 2015, doi: 10.3390/molecules200712913.

[4] M. A. Deri, B. M. Zeglis, L. C. Francesconi, and J. S. Lewis, "PET imaging with (8)(9)Zr: From radiochemistry to the clinic," *Nucl Med Biol,* vol. 40, no. 1, pp. 3–14, 2013, doi: 10.1016/j.nucmedbio.2012.08.004.

[5] S. M. Qaim, "Nuclear data for production and medical application of radionuclides: Present status and future needs," *Nucl Med Biol,* vol. 44, pp. 31–49, 2017, doi: 10.1016/j.nucmedbio.2016.08.016.

[6] L. Martiniova, L. Palatis, E. Etchebehere, and G. Ravizzini, "Gallium-68 in medical imaging," *Current Radiopharmaceuticals,* vol. 9, no. 3, pp. 187–207, 2016, doi: 10.2174/1874471009666161028150654.

[7] L. Radford, "Methods for the production of radionuclides for medicine," in *Radiopharmaceutical Chemistry*, J. Lewis, A. Windhorst, and B. Zeglis, Eds.: Springer International Publishing, 2019, pp. 63–83.

[8] J. Aerts *et al.*, "Guidance on current good radiopharmacy practice for the small-scale preparation of radiopharmaceuticals using automated modules: A European perspective," *J Labelled Comp Radiopharm,* vol. 57, no. 10, pp. 615–20, 2014, doi: 10.1002/jlcr.3227.

[9] H. H. Coenen, "Fluorine-18 labeling methods: Features and possibilities of basic reactions," *Ernst Schering Res Found Workshop,* no. 62, pp. 15–50, 2007, doi: 10.1007/978-3-540-49527-7_2.

[10] F. Fuchtner and J. Steinbach, "Efficient synthesis of the 18F-labelled 3-O-methyl-6-[18F]fluoro-L-DOPA," *Appl Radiat Isot,* vol. 58, no. 5, pp. 575–78, 2003, doi: 10.1016/s0969-8043(03)00062-9.

[11] K. Hamacher, H. H. Coenen, and G. Stocklin, "Efficient stereospecific synthesis of no-carrier-added 2-[18F]-fluoro-2-deoxy-D-glucose using aminopolyether supported nucleophilic substitution," *J Nucl Med,* vol. 27, no. 2, pp. 235–38, 1986.

[12] J. Cardinale *et al.*, "Procedures for the GMP-compliant production and quality control of [(18)F]PSMA-1007: A next generation radiofluorinated tracer for the detection of prostate cancer," *Pharmaceuticals (Basel),* vol. 10, no. 4, p. 77, 2017, doi: 10.3390/ph10040077.

[13] F. Wuest, M. Berndt, and T. Kniess, "Carbon-11 labeling chemistry based upon [11C]methyl iodide," *Ernst Schering Res Found Workshop,* no. 62, pp. 183–213, 2007, doi: 10.1007/978-3-540-49527-7_7.

[14] C. Taddei and V. W. Pike, "[(11)C]Carbon monoxide: advances in production and application to PET radiotracer development over the past 15 years," *EJNMMI Radiopharm Chem,* vol. 4, no. 1, p. 25, 2019, doi: 10.1186/s41181-019-0073-4.

[15] D. J. Vugts *et al.*, "Comparison of the octadentate bifunctional chelator DFO*-pPhe-NCS and the clinically used hexadentate bifunctional chelator DFO-pPhe-NCS for (89)Zr-immuno-PET," *Eur J Nucl Med Mol Imaging,* vol. 44, no. 2, pp. 286–95, 2017, doi: 10.1007/s00259-016-3499-x.

[16] K. Izuishi *et al.*, "Molecular mechanism underlying the detection of colorectal cancer by 18F-2-fluoro-2-deoxy-D-glucose positron emission tomography," *J Gastrointest Surg,* vol. 16, no. 2, pp. 394–400, 2012, doi: 10.1007/s11605-011-1727-z.

[17] R. Sala *et al.*, "Phosphorylation status of thymidine kinase 1 following antiproliferative drug treatment mediates 3'-deoxy-3'-[18F]-fluorothymidine cellular retention," *PLoS One,* vol. 9, no. 7, p. e101366, 2014, doi: 10.1371/journal.pone.0101366.

[18] S. M. Schwarzenboeck *et al.*, "PSMA ligands for PET imaging of prostate cancer," *J Nucl Med,* vol. 58, no. 10, pp. 1545–52, 2017, doi: 10.2967/jnumed.117.191031.

[19] S. Heskamp, R. Raave, O. Boerman, M. Rijpkema, V. Goncalves, and F. Denat, "(89)Zr-Immuno-Positron emission Tomography in oncology: State-of-the-art (89)Zr radiochemistry," *Bioconjug Chem*, vol. 28, no. 9, pp. 2211–23, 2017, doi: 10.1021/acs.bioconjchem.7b00325.

[20] R. A. deKemp, J. M. Renaud, R. Klein, and R. S. Beanlands, "Radionuclide tracers for myocardial perfusion imaging and blood flow quantification," *Cardiol Clin,* vol. 34, no. 1, pp. 37–46, 2016, doi: 10.1016/j.ccl.2015.08.001.

[21] C. Y. Sander and S. Hesse, "News and views on in-vivo imaging of neurotransmission using PET and MRI," *Q J Nucl Med Mol Imaging,* vol. 61, no. 4, pp. 414–28, 2017, doi: 10.23736/s1824-4785.17.03019-9.

[22] A. Jovalekic, N. Koglin, A. Mueller, and A. W. Stephens, "New protein deposition tracers in the pipeline," *EJNMMI Radiopharm Chem,* vol. 1, no. 1, p. 11, 2017, doi: 10.1186/s41181-016-0015-3.

[23] Z. Cai, S. Li, D. Matuskey, N. Nabulsi, and Y. Huang, "PET imaging of synaptic density: A new tool for investigation of neuropsychiatric diseases," *Neurosci Lett,* vol. 691, pp. 44–50, 2019, doi: 10.1016/j.neulet.2018.07.038.

[24] I. Rauscher *et al.*, "Matched-pair comparison of (68)Ga-PSMA-11 and (18)F-PSMA-1007 PET/CT: frequency of pitfalls and detection efficacy in biochemical recurrence after radical prostatectomy," *J Nucl Med,* 2019, doi: 10.2967/jnumed.119.229187.

[25] M. Sollini *et al.*, "PET/MRI in infection and inflammation," *Semin Nucl Med,* vol. 48, no. 3, pp. 225–41, 2018, doi: 10.1053/j.semnuclmed.2018.02.003.

4 Radiopharmaceuticals for Radionuclide Therapy

Meltem Ocak, Emre Demirci, Jessie R. Nedrow and Rebecca Krimins

CONTENTS

4.1 THE HISTORY OF RADIONUCLIDE THERAPY

The idea of using radium in medicine to treat tumours dates back to the early 1900s, a significant discovery for advancing the idea of radionuclide therapy. Contrary to common knowledge, the first self-exposure experiments were made by two Germans, Friedrich Walkoff and Friedrich Giesel [1], and not Becquerel self-exposing himself to a tube of radium carried in his waistcoat pocket and resulting in severe inflammation of his skin. In October 1900, Walkoff and Giesel reported that an exposure of the arm to radium salt for a set period of time (Giesel applied to his arm for two hours; Walkoff applied to his arm for two 20-minute sessions) produced an inflammation of the skin with the same aspects as that obtained after a long exposure to X-rays. Following the announcement of this discovery by Walkoff and Giesel, Pierre Curie voluntarily underwent self-exposure for 10 hours to test the reported findings and concluded the experiment with a radium burn on his arm. This discovery led Curie to loan radium to dermatologist Henri Alexandre Danlos and Eugene Bloch, his physicist, for the first local application of radium to treat tuberculous skin lesions at the Saint Louis Hospital in Paris in 1901. Not long after this discovery, Alexander Graham Bell suggested placing sources containing radium in or near tumours for treatment in 1903. In 1913, Frederick Proescher published the first study on the intravenous injection of soluble radium salts for therapy of various types of diseases [2]. These key experiments lead the way for the first radionuclide treatment in March 1936, with artificially produced radionuclides, starting with the administration of Sodium-24 Chloride in three patients with leukemia and allied diseases [3]. A few months later Phorphorus-32,

DOI: 10.1201/9780429489501-4

TABLE 4.1
History of Radionuclide Therapy

Year	Application
Dates back to 1941	^{32}P was the first radioisotope to be evaluated for palliative treatment of bone metastases [4]
In 1941	Saul Hertz first used ^{131}I (^{130}I + ^{131}I) to treat a patient with Graves' Disease [5]
In 1951	First treatment by radioimmunotherapy was performed by William H. Beierwaltes to treat patient in a metastatic melanoma [6]
In 1980	meta-[^{131}I] iodobenzyl-guanidine (^{131}I-MIBG) was introduced for treatment of neuroendocrine tumours [7]
1980s	Trans-arterial radiotherapy for liver malignancies began with iodine-131 labeled lipiodol for hepatocellular carcinoma [8]
1994	First Peptide Receptor Radionuclide Therapy (PRRT) with ^{111}In-pentetreotide for NET patients [9]
1996	First PRRT with ^{90}Y-DOTA-TOC for NET patients [9]
2000	First PRRT with ^{177}Lu-DOTA-TATE for NET patients [9]
2002	^{90}Y Ibritumomab Tiuxetan was introduced to treat CD20 positive non-Hodgkin's Lymphoma
2003	^{131}I tositumomab was introduced to treat CD20 positive non-Hodgkin's Lymphoma
2005	^{177}Lu-labeled anti-PSMA antibody in patients with castration-resistant prostate cancer published on *pubmed* [10]
2013	FDA approves the first α—emitting radiotherapy, Xofigo (Radium-223 dichloride) for patients with castration-resistant prostate cancer
2015	^{177}Lu-PSMA Ligand in patients with castration-resistant prostate cancer first time published on *pubmed* [11]
2016	^{225}Ac-PSMA-ligand in patients with castration-resistant prostate cancer first time published on *pubmed* [12]
2019	^{225}Ac-DOTATATE in patients with NET first time published on *pubmed* [13]

as orthophosphate, was utilized for the treatment of leukemia by John Lawrence; and shortly thereafter by Hamilton in 1936 for treatment of polycythemia rubra vera. While radionuclide therapy has evolved over the years (Table 4.1), the basic theory of using ionizing radiation to kill or shrink abnormal cells and tumours by damaging the cells' DNA has stayed the same. In the last 30–40 years, radionuclide therapy has been widely used in various clinical malignant and pain-management applications and is continually expanded to develop new treatments.

4.2 THERAPEUTICAL RADIOPHARMACEUTICALS IN CLINICAL USE

4.2.1 Radiopharmaceuticals for Local Treatments

4.2.1.1 Transarterial Radioembolization

In accordance with the American Cancer Society, hepatocellular carcinoma (HCC) is the most common form of primary liver cancer, with rates of incidences tripling since 1980 and the rate of death doubling. Liver cancer, worldwide, is the leading cause of cancer deaths. Patients with unresectable disease have limited options for treatment. Furthermore, underlying diseases (hepatitis B or C, cirrhosis, fatty liver disease, etc.) and the role of liver in the metabolism of drugs increases the risk of adverse effect of chemotherapy [14–15], leading to the exploration of alternative therapeutic approaches, such as Transarterial Radioembolization (TARE), for primary liver cancer as well as metastatic tumours located in the liver. TARE selectively targets embolic agents loaded with therapeutic radionuclides (Iodine-131, Yttrium-90, etc.) to tumours through intra-arterial injections. TARE is primarily used for HCC as well as metastatic liver disease, most notably for liver metastases of colorectal cancer. Interventional radiologists are able to use intra-arterial injections for targeted delivery to hepatic tumours by exploiting their preference to derive their blood supply primarily from the hepatic artery while the normal liver is mainly fed by the portal vein [16]. The embolic properties of TARE agents allow them to become trapped in the vasculature of the targeted tumours, embolizing the tumour with a radiotherapeutic. Microspheres and Lipiodol are the two main types of embolic agents that are modified to be used in TARE. The TARE approach allows for a greater dose to be delivered to hepatic tumours as compared to external beam irradiation, where the dose is more limited due to radiation induced toxicities [17, 18].

4.2.1.2 Yttrium-90 Glass Spheres/Resin Microspheres

Yttrium-90 ($T_{1/2}$=64.2 h) is a therapeutic radionuclide that emits a β-particle with a mean energy of 0.94 MeV and an average tissue penetration range of 2.5 mm. The selective targeting of Yttrium-90-TARE was first reported by Dr. Irving Ariel. In his study, patients with primary pancreatic and liver cancers were treated with ^{90}Y-labeled ceramic microspheres injected intra-arterial at either the celiac axis or hepatic artery. The study concluded that the complications were minimal, and a significant number of patients experienced palliative effects as well as a few patients appearing to have increased survival [19]. Furthermore, the results were supportive of the continuation of exploring the use of Yttrium-90 for TARE. Yttrium-90 TARE agents are today still utilizing microspheres, which are either glass-based (Thrashers) or resin-based (SIR-Spheres) [20, 21]. Both TARE agents are FDA approved as devices under the humanitarian device exemption. Clinical trials mainly throughout the 1990s demonstrated that these agents were safe and effective in HCC (Thrashers) and metastatic colorectal cancer (SIR-Spheres) leading to their FDA approval in the late 1990s and early 2000s.

TheraSpheres are insoluble glass microspheres (20-30 μM) that have Yttrium-90 embedded within a glass matrix following neutron bombardment, helping to minimize the loss of Yttrium-90 from the microsphere [22]. Initial dose-escalation trials of TheraSpheres in patients with primary hepatobiliary tumours, or liver metastatic tumours from colorectal or neuroendocrine cancers found that the use of the microspheres was well-tolerated even at whole-liver doses much greater than external beam irradiation. The lung was the only other organ that was significantly irradiated due to shunting of the microspheres, but the dose to the lungs could be monitored by SPECT imaging of ^{99m}Tc-MAA [23]. Phase I/II trials of TheraSpheres continued to demonstrate the therapy was well-tolerated. In addition, as the studies progress, improvements to the interventional procedure were implemented to help reduce side effects [24]. The comparison of TheraSpheres to transarterial chemoembolization (TACE) did not demonstrate any significance in the survival benefits in HCC patients. The comparisons did find that TheraSpheres were better tolerated, and required less hospitalization and treatments than TACE [25, 26]. However, a recent clinical trial (NCT02582034) of patients with advanced HCC investigated the dosing of TheraSpheres based on a standard dose of 120 ± 20 Gy to the whole liver as compared to a personalized dose of ≥205 Gy to the tumour with no more than 120 Gy to normal tissue. The study demonstrated a significant increase in overall survival in patients with personalized dose (26.7 months) as compared to a standard dose (10.7 months) [27]. These results challenge the results of TheraSpheres clinical trials without the incorporation of a personalized dosimetry approach, including the comparison of TheraSpheres to TACE.

Selective Internal Radiation (SIR)-Spheres are biocompatible resin micropolymers ranging in size between 20–60 μM, averaging 32 ± 10 μM, with Yttrium-90 permanently embedded. Gray and colleagues demonstrated in early studies that patients with liver metastases from primary tumours of the large bowel had a high rate of tumour regression following SIR therapy [28]. A Phase III study further investigated the potential of SIR-Spheres in combination with chemotherapy administered via the hepatic artery to provide increased benefit to patients with liver metastases from primary adenocarcinoma of the large bowel. Phase III demonstrated significant improvement tumour area and volumes, carcino-embryonic antigen (CEA), and median time to disease progress in patients receiving SIR-Spheres and intra-arterial chemotherapy as compared to chemotherapy alone with limited toxicity and low impact on patients' quality of life [29]. The SIRvsNIB (NCT01135056) clinical study found HCC patients treated with yttrium-90 resin microspheres did not see an improvement in overall survival as compared to patients treated with sorafenib, a multi-kinase inhibitor; however, SIR-Spheres–treated patients did have fewer adverse effects [30]. As subsequent therapy, SIR-Spheres have demonstrated positive clinical benefits; however, Phase 3 trials investigating SIR-Spheres in combination with chemotherapy as a first-line therapy in metastatic colorectal cancer with liver metastases did not improve overall survival or adverse effects [31].

4.2.1.3 Holmium-166 PLLA Microspheres

Holmium-166 is an attractive radioisotope for TARE due to its theranostic properties, leading to the development and clinical evaluation of holmium-166 poly(L-lactic acid) microspheres (^{166}Ho-PLLA-MS). Holmium-166 emits therapeutic β-particles (1774 and 1855 keV) for therapy, and a γ-ray (81 keV) that can be detected through SPECT imaging. Furthermore, Holmium-166 is highly paramagnetic and is detected by MRI [32–34]. The phase I HEPAR trial's primary objective was to investigate the max tolerated radiation dose (MTRD) as well as compare the SPECT imaging of a tracer dose of ^{166}Ho-PLLA-MS as compared to the ^{99m}Tc-MAA SPECT scans to accurately predict the microsphere distribution and monitor lung shunting. The HEPAR study determined the MTRD to be 60 Gy to the whole liver and demonstrated that the ^{166}Ho-PLLA-MS had a similar pattern of distribution as ^{99m}Tc-MAA, providing an image-guided platform for TARE in patients with hepatic tumours [35]. Phase II studies demonstrated that the median survival of colorectal cancer patients with liver metastases treated with 166Ho-PLLA-MS was similar to Ytrrium-90 microspheres. In the phase II studies ^{166}Ho-PLLA-MS was associated with several grade 3 or higher adverse effects with abdominal pain and nausea being the most prevalent, but the adverse effects were manageable [36].

4.2.1.4 Iodine-131 Lipiodol

Lipiodol, an FDA-approved radio-opaque agent, serves as an embolic agent for delivery of chemotherapy as well as therapeutic radionuclides. Initially discovered in 1901, Lipiodol is an ethiodized oil composed of a variety of ethyl ester fatty esters from poppy seed oil. The mechanism of action has not been fully elucidated; however, it has been proposed that Lipiodol droplets achieve transient dual (arterial and portal) embolization based on tendency of oil droplets to select larger tumour vasculature due to their size [37]. Lipiodol accumulates in liver tumours while clearing out of normal liver tissue, helping to reduce post-embolization syndrome, the main side effect associated with intra-arterial therapies [38]. As compared to Yttrium-90 microspheres that accumulate into arterioles, ^{131}I-Lipiodol seems to be directly up taken by tumour cells, resulting in a higher tumour radiation dose [39]. Clinical trials of ^{131}I-iodized oil (^{131}I-lipiodol) started in the late 1980s [40–42] and continued through 2013 (NCT00116454, NCT00870558, NCT00027768). Initial treatment with ^{131}I-Lipiodol following resection or radiofrequency ablation in patients with HCC demonstrated a 6-month improvement in recurrence-free survival and a 24-month increase in median overall survival [43]. Furthermore, as adjuvant treatment 131I-lipiodol led to a significant increase in overall survival at 3 years of 86.4 per cent in the treated group versus 46.3 per cent in the control group [44]. Similar to Yttrium-90 microspheres, the survival benefits of ^{131}I-Lipiodol were similar to standard TACE treatments; However, patients who had a portal vein thrombosis or more advanced disease demonstrated a significantly higher mean survival [39]. The high cost of incorporating Iodine-131 into Lipiodol led to the exploration of alternate β-emitting radionuclides, including Rhenium-188 [45–47]. Initial clinical studies of ^{188}Re-lipiodol demonstrated that Rhenium-188 provided a radionuclide for incorporation into Lipiodol that was cost-efficient and safe and demonstrated a 25 per cent objective response rate [47]. In Phase I, trails ^{188}Re-SSS Lipiodol was shown to deliver a mean absorbed dose of 45.6±36.3 Gy to the tumour while only 8.42 ± 3.87 Gy to the whole liver, helping to limit the overall toxicity [48].

4.2.1.5 Radionuclide Synovectomy in Humans

Radiation synovectomy (RS) is a minimally invasive technique to treat persistent joint inflammation. The fundamental thought of the technique is to destroy the hypertrophic synovium with ionizing radiation. Over the long haul, its impact is equivalent to arthroscopic or open synovectomy without all the side effects of surgery. The principal idea of joint radiotherapy goes back to 1924, however the first large clinical trial utilizing intra-articular radioactive agents was done in 1963, when colloidal ^{198}Au was infused to treat knee joint effusions. Since the late 1960s and until now, ^{90}Y-citrate has been a state-of-the-art joint therapy [49].

The joint is penetrated using ultrasonographic or fluoroscopic direction, and a solution of colloidal radioactive agent is intra-articularly infused. The diameter of the colloid molecule is somewhere in the range of 2 and 5 μm, which is small enough to be phagocytised, yet large enough not to enter the circulatory system by capillary fenestrations. Following the infusion, a large portion of the radiocolloid is phagocytised by type 2 synoviocytes (synovial macrophages) [49, 50] With a half-life going from 3 to 10 days, radionuclides utilized for RS continuously emit radiation for several weeks. As expected, this leads to necrosis and resulting fibrosis of the synovial layer, a decrease in synovial liquid creation and, clinically, decrease of inflammation. β-radiation has restricted tissue penetration, depositing over 90 per cent of the energy inside 10 mm from the point of origin, consequently affecting only the joint cavity [49, 51]

Rheumatoid joint pain (RA) is the most common indication, followed by seronegative spondyloarthropathies and adolescent idiopathic joint pain (JIA). More uncommon indication includes reactive arthritis, enteropathy-associated arthritis, other systemic diseases with joint involvement (e.g. Sjögren's syndrome, Behçet's disease), calcium pyrophosphate dehydrate deposition disease (CPPD), and pigmented villonodular synovitis (PVNS). Imaging of choice before RS to confirm the presence of aggravation are ultrasonography (US) and bone scintigraphy [49, 52].

According to the guidelines of the EANM, the following radionuclides are in use at present: Yttrium (^{90}Y-silicate/citrate), Rhenium (^{186}Re-sulfide) and Erbium (^{169}Er-citrate) [52]. Their use in various localizations depends on their tissue-infiltrating distance. ^{90}Y, with a soft tissue range of 3.6 mm, is utilized in knees; ^{186}Re, with a soft tissue range of 1.1 mm, is utilized in medium-sized joints (hips, shoulders, elbows, wrists, lower legs, and the subtalar joint), and 169Er, a soft tissue range of 0.3 mm, is utilized in little joints (metacarpophalangeal, metatarsophalangeal and proximal interphalangeal) [53]. ^{90}Y is a pure β emitter with a particle size of 10–20 μm, a penetration rate of 4–10 mm and a half-life of 2.4 days. It is used for knees, ankles, elbows and shoulder. Rhenium-186 sulphide is a consolidated β and γ producer with a molecule size of 0.1 μm, an infiltration pace of 1–4 mm and a half-life of 3.8 days. The EANM suggests

TABLE 4.2
Radioisotope Specifications [53]

Radioisotope	Emitted Radiation	Particle Size (μm)	Penetration Range	Half Life
Au-198	β and γ	3	1-4	2.7
P-32	β	6-20	3-5	14
Sm-153	β and γ	1-15 (*macroagraggregate*) 40-60nm (*hydroxyapatite nonaparticles*)	0.5-3.0	1.9
Y-90	β	10-20	4-10	2.4
Ho-166	β and γ	1-8	2.2-8.7	1.1
Re-186	β and γ	0.1	1-4	3.8

^{186}Re-sulfide for treatment of medium-sized joints such as elbows and ankles [52, 53]. The other radiopharmaceuticals were listed in Table 4.2.

4.2.2 Radiopharmaceuticals for Local Treatments in Pets

4.2.2.1 Radionuclide Synovectomy in Pets

Approximately 20 per cent of the canine pet population suffers from osteoarthritis [54]. A safety study published in 2019 discussed the use of intra-articular injection of ^{117m}Sn as a radiosynoviorthesis agent in normal canine elbows and concluded there were no adverse effects of the injections, including no changes in joint fluid cytology, no dog exhibited lameness after injection, there was absence of joint damage, and there was a lack of systemic effects after injection [55]. The authors concluded that the agent could be safely used for radiosynoviorthesis in dogs with osteoarthritis [55]. A more recent publication reviews the safety considerations in the treatment of canine skeletal conditions using ^{153}Sm, ^{90}Y, and ^{117m}Sn. The authors found that radiopharmaceutical treatments of canine skeletal conditions using ^{153}Sm, ^{90}Y, or ^{117m}Sn as the radionuclide did not require isolation of the treated animal from human beings, but the ^{90}Y and ^{117m}Sn treatments did require restrictions on prolonged, close human interaction with the animals for 3–4 weeks after treatment [56].

Synovetin OA is a homogeneous tin (^{117m}Sn) stannic colloid in ammonium salt that was brought to the veterinary market in 2020 as a veterinary device for use in dogs. Synovetin OA is intended to reduce synovitis and associated pain of canine elbow joints afflicted with osteoarthritis. In the veterinary clinic, Synovetin OA is administered as a single intra-articular injection into the canine elbow. Sedation or anesthesia of the canine patient is recommended for treatment. If bilateral elbow osteoarthritis with synovitis exists, the individual dog may have both elbows treated. Indications for use include one injection per elbow per twelve-month period; if warranted, a single repeat dose to the affected elbow(s) may be offered twelve months or more after the initial dose. The pet is discharged on the same day as treatment once recovered from sedation.

4.2.2.2 Other Radionuclide Therapy in Pets

The most frequently diagnosed endocrinopathy in feline geriatric patients is hyperthyroidism associated with benign thyroid adenoma [57]. It is estimated that hyperthyroidism is diagnosed in 1.5–11.4 per cent of geriatric cats worldwide [58]. Feline hyperthyroidism is similar to toxic nodular goitre in humans [59]. Common historical signs, physical examination signs, and laboratory findings in hyperthyroid cats may include a palpable thyroid nodule, muscle wasting, thin body condition, dental disease, tachycardia, cardiac murmur, high alanine aminotransferase, high alkaline phosphatase, high aspartate aminotransferase, elevated hematocrit, high serum T_4, high serum fT_4, high serum T_3, and low serum TSH [60]. Treatment of feline hyperthyroidism can include any or a combination of the following: medical management with methimazole or carbimazole, nutritional management (low-iodine diet), surgical thyroidectomy, and radioactive iodine treatment (^{131}I) [61]. Radioactive iodine is generally considered the optimal treatment for hyperthyroidism in cats because of its safety and efficacy [62]. During the 1990s, multiple studies were published lauding the benefits of ^{131}I treatment in cats. Radioactive iodine treatment has also been used to treat feline thyroid carcinomas and canine thyroid tumours [63, 64]. Radioactive iodine treatment in cats is currently offered throughout the United States in local veterinary clinics (i.e., general practitioner offices), specialty referral veterinary settings, and some veterinary academic institutions. Veterinary hospital requirements for offering ^{131}I treatment to pets include an active radioactive materials license, proper evaluation of each veterinary case, facilities built for conducting treatments with radioactive agents (most facilities will hospitalize each patient for 5–8 days before sending the animal home), review of safety discussions with the pet owners, maintaining paperwork and more.

Quadramet is a therapeutic agent consisting of radioactive samarium and a tetraphosphonate chelator, ethylenediami netetramethylenephosphonic acid (EDTMP). In 1998, a team from South Africa published results from nine individual dogs with bone tumours that were treated with Sm-153-EDTMP [65]. This therapy has subsequently been evaluated in dogs with metastatic bone lesions including osteosarcoma [66, 67].

Over the past decade numerous studies have looked at various radionuclide treatments for diseases commonly seen in the veterinary population. For example, in 2017 a prospective cohort study in thirteen cats with oral squamous cell carcinoma looked at radioactive Holmium-166 (^{166}Ho) microspheres injected directly into the tumour [68]. They concluded that intratumoral injection of ^{166}Ho has potential as a minimally invasive, single procedure radio-ablation treatment of unresectable tumours with minimal morbidity [68]. A 2016 report evaluated the pharmacokinetics, dosimetry and therapeutic efficacy of ^{177}Lu-EDTMP in dogs with different types of primary and metastatic bone lesions [69]. In this study of four dogs, they reported that one dog suffering from skeletal metastases showed remarkable improvement in disease conditions, two dogs with primary bone cancer showed stable disease initially with palliative effect of ^{177}Lu-EDTMP and the fourth dog with metal-induced osteosarcoma did not show any response to treatment [69].

4.3 RADIOPHARMACEUTICALS FOR SYSTEMIC TREATMENTS

4.3.1 Bone-seeking Radiopharmaceuticals for Therapy

Bone metastasis is a typical complication of advanced-stage disease. Especially, prostate and breast cancers frequently metastasize to the bone. Patients with bone metastases frequently experience a reduced quality of life because of severe pain, and an expanded occurrence of skeletal related functions, including pathological fractures, spinal cord compression, and hypercalcemia. Therapy of pain related to bone metastases includes regular analgesics, opiates, and external beam radiation. However, external radiation is improper if osseous metastases are present in numerous regions or may be constricted by the cumulative absorbed dose limits on repeated treatments. In this situations, systemic therapies using bone-seeking radionuclides or radioisotope labelled bone-seeking tracers are an alternative treatment modality. Several tracers have been evaluated in clinical trials, including $^{89}SrCl_2$, ^{153}Sm–EDTMP and the α-emitter $^{223}RaCl_2$ [70].

For the treatment of metastatic bone disease, an ideal tracer would be a radiopharmaceutical that predominantly accumulates in bone lesions. For this purpose, calcimimetic and phosphonate-based radiopharmaceuticals are used (Table 4.3). Phosphonates have a particular affinity for calcium and are adsorbed by attachment to the calcium atoms in hydroxyapatite. The hydroxyapatite affinity is strongly dependent on the molecular structures of these compounds [71, 72].

4.3.1.1 Samarium-153-EDTMP

[153Sm] Samarium–Ethylene Diamine Tetramethylene Phosphonate (^{153}Sm–EDTMP) is a well-studied tracer for bone-palliation and available since 1997 under the brand name 'Quadramet'. It is FDA-approved for various osteoblastic metastasis, especially in prostate and breast cancer. It rapidly binds to hydroxyapatite crystals, leading to less than 1 per cent availability in the blood 5 h after injection. Samarium-153's half-life is 46.3 hours. The excretion occurs mainly through the kidneys [73]. The standard dose of the treatment is 37 MBq/kg of body weight, but the maximum tolerable dose is reported to be 111 MBq/kg in initial studies [74]. However, Tian and colleagues also reported similar success for low dose administration of Sm-153 (18.5 MBq/kg) [75].

TABLE 4.3
Calcimimetic-based Bone-seeking Radiopharmaceuticals

Radionuclide	$t_{1/2}$	max range in soft tissue	Radiopharmaceuticals/ Clinical Status
P-32	14.3 d	8.5 mm	^{32}P-orthophosphate/ administered to humans
Sr-89	50.5 d	7 mm	$^{89}SrCl_2$/ administered to humans
Ra-223	11.4 d	<10 µm	$^{223}RaCl_2$/ administered to humans

In numerous studies, a response rate of 40–97 per cent with a mean of 70 per cent has been reported. The quality of life of the patients also improved, and a significant decrease in analgesic consumption has been noted. A mild bone marrow toxicity is noted in most the patients. The platelet and white blood cell (WBC) counts have been decreased to the lowest levels in 3–5 weeks but recover in 6–8 weeks after therapy. For the repeated treatments, the main dose-limiting toxicity is bone marrow suppression [73]. Overall, ^{153}Sm–EDTMP has been successfully used for pain control for more than three decades.

4.3.1.2 Radium-223 Dichloride

Radium-223 (^{223}Ra) is a calcium mimetic isotope and is deposited on hydroxyapatite crystal [73]. 223Ra-dichloride ($^{223}RaCl_2$) is developed to target metastatic bone disease, accumulating in areas of increased bone turnover. The availability of a long-lived ^{227}Ac/^{227}Th generator and a physical half-life of 11.4 d, which enables fast delivery to end users due to the long time between development and expiry, are among its advantages. Ra-223 has a complex decay mechanism in which, during each decay, four α-particles are emitted, resulting in high energy deposition (28.2 MeV), with 95 per cent of the emission energy. A radiation's high linear energy transfer results in greater biological effectiveness than β-radiation, as well as the generation of double-strand DNA breaks, resulting in cytotoxicity that is independent of dose rate, phase of cell cycle formation, and concentration of oxygen. Due to the range of the alfa particles, less hematologic toxicity would be expected compared to β-radiation [76].

The approved injection dose is 50 KBq/kg body weight for a total of 6 cycles. The excretion is mainly by gastrointestinal tract. In contrast to ^{153}Sm–EDTMP, urinary excretion is minimal (around 5%) [73]. Pharmacokinetics testing of ^{223}Ra showed rapid clearance from the blood, with less than 1 per cent remaining at 24 h [76]. Mild and reversible myelosuppression occurs after intravenous injection, with a nadir in 2–4 weeks, resolving in 6 weeks after administration. Furthermore, diarrhoea, nausea and vomiting occur in around 10 per cent of cases [73].

In the phase III ALSYMPCA trial, overall survival was 14.9 months for men treated with 223Ra and 11.3 months for those treated with placebo [77]. Ra-223 is also effective in the palliation of pain. In the initial studies, pain relief was observed at all dose levels, ranging from 46 to 250 KBq/kg without a definite dose–response relationship [76, 78].

Gȧmma ray imaging from Ra-223 and its daughters is, while feasible, of poor quality, but allows biodistribution to be understood and lesion targeting to be identified [76]. Lassman and colleagues estimated the absorbed dose to organs from Ra-223 by using an International Commission on Radiological Protection biokinetic model for alkaline earth elements. 3.8×10^{-6} Gy/Bq to bone endosteum and 3.7×10^{-7} Gy/Bq to bone marrow were the highest calculated doses [76, 79]

4.3.1.3 Lutetium-177 Labelled Bone-seeking Complexes

[177Lu]Lutetium ethylenediaminetetramethylene phosphonate (^{177}Lu-EDTMP) is another radiopharmaceutical recommended for systemic radionuclide therapy in patients with metastatic bone involvement, often in patients with breast or prostate cancer but not yet approved by the Food and Drug Administration (FDA). The radiochemistry of Lu-177 is similar to that of Sm-153 [72]. Even though ^{177}Lu-DOTMP showed encouraging outcomes, ^{177}Lu-EDTMP was developed for human use, since EDTMP was already in use in the approved ^{153}Sm–EDTMP treatment [80]. With favourable β- and γ-characteristics and favourable synthesis process, ^{177}Lu-EDTMP is a good candidate for bone palliation purposes. Also, the formulation of a freeze-dried kit identical to the product ^{153}Sm–EDTMP (Quadramet) was reported by Das and colleagues [81]. According to a review of Askari and colleagues, an average of 4.84-point drop (of 10) was observed in various metastatic cancers. The meta-analysis revealed a significant effect on the frequencies of summed overall palliative pain response (84 per cent, 95 per cent CI: 75 per cent–90 per cent; $p < 0.001$). However, a high level of grade III/IV transient anaemia (19 per cent) was observed in the patients [72].

4.4 PROGRESS IN TARGETED ALPHA-PARTICLE THERAPY

Targeted α-therapy (TAT) has emerged as a highly potent treatment modality against a variety of metastatic cancers. The high linear-energy transfer (LET) of α-particles allows them to deposit large amounts of energy over a short range (50-100 μm in tissue), minimizing damage to nearby untreated cells. The high LET results in cell death due to the high prevalence of irreparable double-stranded DNA breaks [82]. The potential of TAT was evident when it was demonstrated that castrate-resistant prostate cancer patients with bone metastases had an increase in overall survival when treated with Xofigo with minimal toxicity and was not just palliative treatment [83].The success of Xofigo and more recently, the remarkable imaging responses seen in patients with metastatic prostate cancer following treatment

TABLE 4.4
Phosphonate-based Bone-seeking Radiopharmaceutical

Radionuclide	$T_{1/2}$	max range in soft tissue	Radiopharmaceuticals/ Clinical Status
Ho-166	27 h	9 mm	^{166}Ho-DOTMP/ administered to humans
			^{166}Ho-EDTMP/R&D
			^{166}Ho-APD/R&D
			^{166}Ho-APDDMP, ^{166}Ho-PDTMP/ R&D
			^{166}Ho-Zoledronate / R&D
I-131	8.0 d	2.3 mm	^{131}I-BDP3 /administered to humans
			^{131}I-HPEB/R&D
Lu-177	6.7 d	2 mm	^{177}Lu-EDTMP/administered to humans
			^{177}Lu-DTPMP, ^{177}Lu-TTHMP/R&D
			^{177}Lu-DOTMP, ^{177}Lu-CTMP/R&D
			^{177}Lu-MDP,^{177}Lu-HEDP,^{177}Lu-TTHMP/R&D
			^{177}Lu-PYP/R&D
			^{177}Lu-zoledronate/R&D
			^{177}Lu-pamidronate, ^{177}Lu-alendronate/R&D
			^{177}Lu-BPAMD/R&D
Re-186	89 h	5 mm	^{186}Re-HEDP /administered to humans
			^{186}Re-MDP/R&D
			^{186}Re-EDTMP, ^{186}Re-PDTMP, ^{186}Re-DMPDTMP/R&D
			^{186}Re-CTMP/R&D
			^{186}Re-MAMA-BP/R&D
			^{186}Re-MAMA-HBP/R&D
			^{186}Re-MAG3-HBP/R&D
			^{186}Re-CpTR-Gly-APD/R&D
Re-188	17 h	10 mm	^{188}Re-HEDP/administered to humans
			^{188}Re-MDP,^{188}Re-HDP,^{188}Re-SEDP/R&D
			^{188}Re-risedronate,^{188}Re-aledronate/R&D
			^{188}Re-AEDP/R&D
			^{188}Re-EDTMP, ^{188}Re-EDBMP and ^{188}Re-NTMP/R&D
			^{188}Re-DTPMP,^{188}Re-TTHMP/R&D
			^{188}Re-DTPA-BP,^{188}Re-5FU-BP,^{188}Re-Gem-BP/R&D
Rh-105	35 h	2 mm	^{105}Rh-EDTMP/R&D
Sm-153	46 h	4 mm	^{153}Sm-EDTMP/administered to humans
			^{153}Sm-pamidronate, ^{153}Sm-alendronate, ^{153}Sm-neridronate/R&D
			^{153}Sm-APDDMP,^{153}Sm-BPAMD/R&D
Tm-170	128 d	5 mm	^{170}Tm-EDTMP/R&D
			^{170}Tm-DTPMP, ^{170}Tm-TTHMP, ^{170}Tm-DOTMP, ^{170}Tm-CTMP/R&D
Yb-175	4.2 d	2 mm	^{175}Yb-EDTMP, ^{175}Yb-PDTMP, ^{175}Yb-DTPMP, ^{175}Yb-TTHMP/R&D
			^{175}Yb-DOTMP,^{175}Yb-TTHMP,^{175}Yb-pamidronate, ^{175}Yb-alendronate/R&D
Y-90	64 h	11 mm	^{90}Y-DPD,^{90}Y-DOTA-HBP/R&D
			^{90}Y-EDTMP/R&D
Sn-117m	13.6 d	300 μm	^{117m}Sn(II)–APDDMP,^{117m}Sn(IV)-PEI-MP/R&D
Ac-225	10.0 d	< 100 μm	^{225}Ac-EDTMP,^{225}Ac-DOTMP/R&D
			^{225}Ac-DOTA^{ZOL}
At-211	7.2 h	70 μm	^{211}At-BPB and ^{211}At-PPB/R&D
Bi-212	1.1 h	90 μm	^{212}Pb/^{212}Bi-EDTMP
			^{212}Bi-DOTMP, ^{212}Pb/^{212}Bi-DOTMP.
Th-227	18.7 d	< 100 μm	^{227}Th-DOTMP, ^{227}Th-DTMP
			^{227}Th-EDTMP

with an Actinium-225 (^{225}Ac) labelled anti-PSMA small molecule has cemented TAT as a highly potent therapy for metastatic cancer [84]. Subsequently, the success of TAT has highlighted limitations including the supply/availability of α-emitting radioisotopes, the complexities of the decay chains of radioisotopes, the high potency of α-particles within normal tissues, and the possible development of resistant to TAT [85].

There are a handful of alpha-particle–emitting isotopes that are utilized for TAT, including the following: Actinium-225, Astatine-211, Bismuth-212, Bismuth-213, Radium-223, Terbium-149 and Thorium-227.

4.4.1 Actinium-225/Bismuth-213

The use of ^{225}Ac ($T_{1/2}$ = 10.0 d) for TAT was described in the early 1990s [86]. Ac-225 is a potent α-particle emitter and emits four α-particles from its daughters (Francium-221, Astatine-217, Bismuth-213 and Polonium-213) with energies ranging between 60–240 keV/μm over α-tracks with a range of 50–80 μm in tissue [87]. A variety of targeting agents (e.g. antibodies, peptides, small-molecules, etc.) have been radiolabelled with ^{225}Ac for investigation in the clinic with the most prevalent being ^{225}Ac-PSMA-617 [13, 88–90]. As Morgenstern and colleagues state, the development of ^{225}Ac-PSMA617 for therapy of prostate cancer can be considered a milestone in the evolution of TAT [91]. The success of ^{225}Ac-PSMA-617 has led to an increase in demand for Ac-225 for the development of novel TAT agents, resulting in a bottleneck due to limited supply.

4.4.2 Progress in Auger Electron-based Radionuclide Therapy

Auger electrons are low-energy orbital electrons emitted from the atomic shells via post-electron capture (EC), internal conversion (IC), or incident x-rays excitation. Auger electron emitters can be divided in two major groups: halogens (^{125}I; ^{123}I; ^{77}Br; ^{88m}Br) and metals (^{201}Tl; ^{195m}Pt; ^{193m}Pt; ^{111}In; ^{114m}In, ^{99m}Tc; ^{67}Ga; ^{55}Fe and ^{51}Cr). ^{201}Tl has the greatest yield of electrons with 36.9 electrons/decay and average energy of 15.27 keV. Other Auger electron-emitting radioisotopes with high yields include, ^{195m}Pt and ^{125}I with 33 electrons/decay (average energy 22.53 keV) and 25.8 electrons/decay (average energy 19.4 keV) respectively. ^{111}In and ^{123}I have intermediate electron yield (14.7 and 14.9 electrons/decay, respectively) associated with average energies of 32.7 and 27.6 keV, respectively [92]. The amount of energy deposited per decay in a five nanometer sphere is much greater for ^{195m}Pt (2000 eV) than for either ^{125}I (1000 eV), ^{123}I (550 eV), or ^{111}In (450 eV) [93].

There are several factors. including (1) the number of electrons emitted per decay or electron yield (2), the ratio of penetrating (x- and γ-rays) to non-penetrating (electron or α-particle) forms of radiation (3), the isotope's half-life, and (4) suitable chemistry for the radiolabelling process should be taken into consideration when selecting an Auger electron-emitting radionuclide for the development of Auger-based radiopharmaceuticals [94, 95].

Auger electron emitters have shorter range than α-particle emitters, making them well suited for treatment of micrometastases with higher anti-tumour efficacy and avoiding normal tissue toxicity [96–98].The first requirement for the delivery of Auger electron-emitting nuclides is the efficient incorporation into tumour-selective carrier molecule. Once inside the target tissue, the selective carrier molecule should be able to associate with the DNA complex for a time corresponding to the radionuclide half-life, they become highly toxic in the vicinity of DNA within the cell nucleus [99, 100].

Auger radiation therapy remains mostly pre-clinical. Antitumor efficacy has been demonstrated in pre-clinical models, including metastatic melanoma [101], glioblastoma [102], micrometastatic model of prostate cancer [103], bladder cancer [104]. Recently PSMA-targeted radiopharmaceutical therapy with the Auger emitter ^{125}I-DCIBzL showed significantly delayed development of detectable metastatic disease and improved survival in a micrometastatic model of PC, with no long-term toxicities noted at 12 months [103]. And in another recent study with ^{111}In attached to epidermal growth factor receptor (EGFR)-targeted MNTs (Modular nanotransporters) therapy on a human bladder cancer (overexpressing EGFR) animal model showed significantly enhanced cytotoxicity and tumour growth inhibition up to complete tumour resorption, prolonged tumour retention of radioactivity [104]. Several phase I clinical trials have also been done with Auger electron-emitting radiopharmaceuticals. In one of the trials, patients with neuroendocrine tumours treated with ^{111}In-DTPA-octreotide at doses of 20–160 GBq. Therapeutic effects were seen in about 50 per cent of patients; however, a maximum tolareble dose (MTD) of 100 GBq was established due to development of myelodysplastic syndrome or leukaemia [105]. In another trial using ^{111}In-DTPA-octreotide, patients received two treatments of 6.6 GBq each, which were well-tolerated but only resulted in partial response in a small per cent of patients [106], possibly related to only modest uptake of this compound into cell nuclei [107]. ^{111}In-DTPA-hEGF has been evaluated in a Phase I trial of patients with epidermal growth factor receptor-positive metastatic breast cancer at activities up to 2.3 GBq, which were well tolerated (MTD not reached), but did not result in objective antitumor responses [108].

4.5 PROGRESS IN THERAPEUTIC RADIOPHARMACEUTICALS BASED ON NANOTECHNOLOGY

In recent years, there is increasing interest on developing therapeutic radioisotope-incorporated nanosized materials (nanocarriers) [109]. Radioisotope-incorporated nanosized materials show several advantages to bare radioisotopes: They have the ability to contain multivalent radioactive elements in a single carrier that enables transporting numerous α-and beta-emitters to cancer cells.

In general, a wide range of materials ranging from inorganic particles to proteins and organic structures such as polymers and liposomes can be used as nanocarriers, and they can accumulate via passive targeting into tumour (enhanced permeability and retention effect is responsible for it) [110]. The size, surface properties, shape, and charge of nanocarriers can alter accumulation [111]. To prevent renal excretion, nanocarriers should typically be larger than ~6 nm. On the other hand, nanocarriers larger than ~200 nm is rapidly recognized and sequestered by the reticuloendothelial system (RES). A common strategy to prevent RES recognition of nanocarriers is shielding them from RES facilitating proteins (opsonins) by decorating the surface with neutralizing molecules such as poly(ethylene glycol) (PEG) [112]. Preclinical development of passively targeted liposome and others as a carrier for targeted radionuclide therapy of cancers was started before 2000 with β-emitter radionuclides (^{131}I, ^{90}Y, ^{188}Re, and ^{67}Cu). Further clinical trial studies are still required to translate those advanced radiolabelled nanocarriers to the health care of cancer patients.

Besides β-emitter radionuclides, longer-lived α-emitters (^{225}Ac, ^{223}Ra, ^{227}Th and ^{212}Pb) exhibit complex decay chains and release of daughter radionuclides from radioligands. This is main problem in off-target toxicity. There are three different approaches to deal with this problem: Cell internalisation, local administration, or encapsulation of α-emitters in nanocarriers [113]. Encapsulation of α-emitters (mother radionuclide and its daughters) in a nanocarriers approach prevents escape of daughter radionuclides from the target site. In this approach the size, shape and type of material needed to fully encapsulate ^{225}Ac decay products [114]. Holzwarth and colleagues found that if the mother of the decay chains localized at the centre of the spherical nanoparticles a 12 nm gold layer or 39 nm graphite layer may prevent the release of ^{221}Fr, the first decay product of ^{225}Ac. And they also found that to sequester all radionuclides up to the third daughter, a layer of 35 nm gold or 143 nm of graphite is necessary [114]. Radionuclide α- therapy using nanoparticles have great potential in nuclear medicine applications. Nowadays immobilization of ^{223}Ra in inorganic nanoparticles and targeted nanobrachytherapy using nanoparticles labelled with α-emitters seems to have great potential for the treatment of small tumours and tumour metastases. However, using nanoparticles to immobilize, ^{225}Ac and its daughter radionuclides still need to be deeply investigated due to usual behaviour of the nanoparticles after injection. As is known, nanoparticles, before reaching the tumour, usually accumulate in spleen, liver, or lungs. This issue is very important when you apply ^{225}Ac-labelled nanoparticles to the body. Studies are still going on for ^{225}Ac-labelled nanoparticles. Recently, the study done with [^{225}Ac]Ac-DOTA-TDA-Lipiodol showed us the combination of a TAT agent labelled with a long-lived α-emitting radionuclide emulsified with Lipiodol was capable of being selectively delivered to hepatic tumours, delivering a highly potent therapeutic dose to hepatic tumours over an extended period of time [115].

REFERENCES

[1] R. F. Mould, "Pierre Curie, 1859–1906," *Curr. Oncol.*, vol. 14, no. 2, pp. 74–82, Apr. 2007, doi: 10.3747/co.2007.110.

[2] C.-H. Yeong, M. Cheng, and K.-H. Ng, "Therapeutic radionuclides in nuclear medicine: current and future prospects," *J. Zhejiang Univ. Sci. B*, vol. 15, no. 10, pp. 845–63, Oct. 2014, doi: 10.1631/jzus.B1400131.

[3] J. G. Hamilton and R. S. Stone, "The intravenous and intraduodenal administration of Radio-Sodium," *Radiology*, vol. 28, no. 2, pp. 178–88, 1937, doi: 10.1148/28.2.178.

[4] F. D. C. Guerra Liberal, A. A. S. Tavares, and J. M. R. S. Tavares, "Palliative treatment of metastatic bone pain with radiopharmaceuticals: A perspective beyond Strontium-89 and Samarium-153," *Appl. Radiat. Isot. Incl. data, Instrum. methods use Agric. Ind. Med.*, vol. 110, pp. 87–99, Apr. 2016, doi: 10.1016/j.apradiso.2016.01.003.

[5] B. Hertz, "A tribute to Dr. Saul Hertz: The discovery of the medical uses of radioiodine," *World J. Nucl. Med.*, vol. 18, no. 1, pp. 8–12, 2019, doi: 10.4103/wjnm.WJNM_107_18.

[6] J. Barbet, J.-F. Chatal, and F. Kraeber-Bodéré, "Radiolabeled antibodies for cancer treatment," *Med. Sci. (Paris).*, vol. 25, no. 12, pp. 1039–45, Dec. 2009, doi: 10.1051/medsci/200925121039.

[7] A. R. Wafelman, C. A. Hoefnagel, R. A. A. Maes, and J. H. Beijnen, "Radioiodinated metaiodobenzylguanidine: A review of its biodistribution and pharmacokinetics, drug interactions, cytotoxicity and dosimetry," *Eur. J. Nucl. Med.*, vol. 21, no. 6, pp. 545–59, 1994, doi: 10.1007/BF00173043.

[8] D. K. Leung and C. Divgi, "Trans-arterial I-131 lipiodol therapy of liver tumors BT – Nuclear medicine therapy: Principles and clinical applications," C. Aktolun and S. J. Goldsmith, Eds. New York: Springer, 2013, pp. 199–205.

[9] R. Levine and E. P. Krenning, "Clinical History of the Theranostic Radionuclide Approach to Neuroendocrine Tumors and Other Types of Cancer: Historical review based on an interview of Eric P. Krenning by Rachel Levine," *J. Nucl. Med.*, vol. 58, no. Suppl 2, pp. 3S–9S, Sep. 2017, doi: 10.2967/jnumed.116.186502.

[10] N. H. Bander, M. I. Milowsky, D. M. Nanus, L. Kostakoglu, S. Vallabhajosula, and S. J. Goldsmith, "Phase I trial of 177lutetium-labeled J591, a monoclonal antibody to prostate-specific membrane antigen, in patients with androgen-independent prostate cancer," *J. Clin. Oncol. Off. J. Am. Soc. Clin. Oncol.*, vol. 23, no. 21, pp. 4591–4601, Jul. 2005, doi: 10.1200/JCO.2005.05.160.

[11] C. Kratochwil *et al.*, "[^{177}Lu]Lutetium-labelled PSMA ligand-induced remission in a patient with metastatic prostate cancer," *Eur. J. Nucl. Med. Mol. Imaging*, vol. 42, no. 6, pp. 987–88, May 2015, doi: 10.1007/s00259-014-2978-1.

[12] C. Kratochwil *et al.*, "225Ac-PSMA-617 for PSMA-targeted radiation therapy of metastatic castration-resistant prostate cancer," *J. Nucl. Med.*, vol. 57, no. 12, pp. 1941–44, 2016, doi: 10.2967/jnumed.116.178673.

[13] S. Ballal, M. P. Yadav, C. Bal, R. K. Sahoo, and M. Tripathi, "Broadening horizons with (225)Ac-DOTATATE targeted alpha therapy for gastroenteropancreatic neuroendocrine tumour patients stable or refractory to (177)Lu-DOTATATE PRRT: First clinical experience on the efficacy and safety," *Eur. J. Nucl. Med. Mol. Imaging*, vol. 47, no. 4, pp. 934–46, Apr. 2020, doi: 10.1007/s00259-019-04567-2.

[14] C. F. Gonsalves, D. B. Brown, and B. I. Carr, "Regional radioactive treatments for hepatocellular carcinoma," *Expert Review of Gastroenterology and Hepatology*. 2008, doi: 10.1586/17474124.2.4.453.

[15] J. R. Senior, "Unintended hepatic adverse events associated with cancer chemotherapy," *Toxicologic Pathology*. 2010, doi: 10.1177/0192623309351719.

[16] H. R. Bierman, R. L. Byron, K. H. Kelley, and A. Grady, "Studies on the blood supply of tumors in man. iii. Vascular patterns of the liver by hepatic arteriography in vivo," *J. Natl. Cancer Inst.*, 1951, doi: 10.1093/jnci/12.1.107.

[17] B. Emami *et al.*, "Tolerance of normal tissue to therapeutic irradiation," *Int. J. Radiat. Oncol. Biol. Phys.*, 1991, doi: 10.1016/0360-3016(91)90171-Y.

[18] C. C. Pan *et al.*, "Radiation-associated liver injury," *Int. J. Radiat. Oncol. Biol. Phys.*, 2010, doi: 10.1016/j.ijrobp.2009.06.092.

[19] I. M. ARIEL, "Treatment of inoperable primary pancreatic and liver cancer by the intra-arterial administration of radioactive isotopes (Y90 radiating microspheres)," *Ann. Surg.*, 1965, doi: 10.1097/00000658-196508000-00018.

[20] A. Riaz, R. Awais, and R. Salem, "Side effects of yttrium-90 radioembolization," *Front. Oncol.*, 2014, doi: 10.3389/fonc.2014.00198.

[21] A. Van Der Gucht *et al.*, "Resin versus glass microspheres for 90Y transarterial raDioembolization: Comparing survival in unresectable hepatocellular carcinoma using pretreatment partition model dosimetry," *J. Nucl. Med.*, 2017, doi: 10.2967/jnumed.116.184713.

[22] G. J. Ehrhardt and D. E. Day, "Therapeutic use of 90Y microspheres," *Int. J. Radiat. Appl. Instrumentation. Part B. Nucl. Med. Biol.*, vol. 14, no. 3, pp. 233–42, 1987, doi: https://doi.org/10.1016/0883-2897(87)90047-X.

[23] J. C. Andrews *et al.*, "Hepatic radioembolization with yttrium-90 containing glass microspheres: Preliminary results and clinical follow-up," *J. Nucl. Med.*, 1994.

[24] M. J. Herba and M. P. Thirlwell, "Radioembolization for hepatic metastases," *Semin. Oncol.*, 2002, doi: 10.1053/sonc.2002.31672.

[25] A. El Fouly *et al.*, "In intermediate stage hepatocellular carcinoma: Radioembolization with yttrium 90 or chemoembolization?" *Liver Int.*, 2015, doi: 10.1111/liv.12637.

[26] L. E. Moreno-Luna *et al.*, "Efficacy and safety of transarterial radioembolization versus chemoembolization in patients with hepatocellular carcinoma," *Cardiovasc. Intervent. Radiol.*, 2013, doi: 10.1007/s00270-012-0481-2.

[27] E. Garin *et al.*, "Major impact of personalized dosimetry using 90Y loaded glass microspheres SIRT in HCC: Final overall survival analysis of a multicenter randomized phase II study (DOSISPHERE-01)," *J. Clin. Oncol.*, 2020, doi: 10.1200/jco.2020.38.4_suppl.516.

[28] B. N. Gray *et al.*, "Regression of liver metastases following treatment with Yttrium microspheres," *Aust. N. Z. J. Surg.*, 1992, doi: 10.1111/j.1445-2197.1992.tb00006.x.

[29] B. Gray *et al.*, "Randomised trial of SIR-Spheres plus chemotherapy vs. chemotherapy alone for treating patients with liver metastases from primary large bowel cancer," *Ann. Oncol.*, 2001, doi: 10.1023/A:1013569329846.

[30] P. K. H. Chow *et al.*, "SIRveNIB: Selective internal radiation therapy versus sorafenib in Asia-Pacific patients with hepatocellular carcinoma," *J. Clin. Oncol.*, 2018, doi: 10.1200/JCO.2017.76.0892.

[31] H. S. Wasan *et al.*, "First-line selective internal radiotherapy plus chemotherapy versus chemotherapy alone in patients with liver metastases from colorectal cancer (FOXFIRE, SIRFLOX, and FOXFIRE-Global): A combined analysis of three multicentre, randomised, phase 3 trials," *Lancet Oncol.*, 2017, doi: 10.1016/S1470-2045(17)30457-6.

[32] N. J. M. Klaassen, M. J. Arntz, A. Gil Arranja, J. Roosen, and J. F. W. Nijsen, "The various therapeutic applications of the medical isotope holmium-166: A narrative review," *EJNMMI Radiopharm. Chem.*, 2019, doi: 10.1186/s41181-019-0066-3.

[33] G. H. Van De Maat *et al.*, "MRI-based biodistribution assessment of holmium-166 poly(L-lactic acid) microspheres after radioembolisation," *Eur. Radiol.*, 2013, doi: 10.1007/s00330-012-2648-2.

[34] M. L. J. Smits *et al.*, "Holmium-166 radioembolization for the treatment of patients with liver metastases: Design of the phase i HEPAR trial," *J. Exp. Clin. Cancer Res.*, 2010, doi: 10.1186/1756-9966-29-70.

[35] M. L. J. Smits *et al.*, "Holmium-166 radioembolisation in patients with unresectable, chemorefractory liver metastases (HEPAR trial): a phase 1, dose-escalation study," *Lancet. Oncol.*, vol. 13, no. 10, pp. 1025–34, Oct. 2012, doi: 10.1016/S1470-2045(12)70334-0.

[36] J. F. Prince *et al.*, "Efficacy of radioembolization with 166Ho-Microspheres in salvage patients with liver metastases: A phase 2 study," *J. Nucl. Med.*, 2018, doi: 10.2967/jnumed.117.197194.

[37] R. C. Gaba, R. M. Schwind, and S. Ballet, "Mechanism of Action, Pharmacokinetics, Efficacy, and Safety of Transarterial Therapies Using Ethiodized Oil: Preclinical review in liver cancer models," *J. Vasc. Interv. Radiol.*, 2018, doi: 10.1016/j.jvir.2017.09.025.

[38] T. J. Vogl, S. Zangos, K. Eichler, D. Yakoub, and M. Nabil,"Colorectal liver metastases: Regional chemotherapy via transarterial chemoembolization (TACE) and hepatic chemoperfusion: An update," *Eur. Radiol.*, vol. 17, no. 4, pp. 1025–1034, Apr. 2007, doi:10.1007/s00330-006-0372-5.

[39] L. Marelli *et al.*, "Transarterial injection of 131I-lipiodol, compared with chemoembolization, in the treatment of unresectable hepatocellular cancer," *J. Nucl. Med.*, 2009, doi: 10.2967/jnumed.108.060338.

[40] J. L. Raoul *et al.*, "Hepatic artery injection of I-131-labeled lipiodol. Part I. Biodistribution study results in patients with hepatocellular carcinoma and liver metastases," *Radiology*, vol. 168, no. 2, pp. 541–45, Aug. 1988, doi: 10.1148/radiology.168.2.2839866.

[41] J. I. Raoul *et al.*, "Internal radiation therapy for hepatocellular carcinoma. Results of a French multicenter phase II trial of transarterial injection of iodine 131-labeled lipiodol," *Cancer*, 1992, doi: 10.1002/1097-0142(19920115)69:2<346::AID-CNCR2820690212>3.0.CO;2-E.

[42] J. L. Raoul *et al.*, "Randomized controlled trial for hepatocellular carcinoma with portal vein thrombosis: Intra-arterial iodine-131-iodized oil versus medical support," *J. Nucl. Med.*, 1994.

[43] L. Schwarz *et al.*, "Adjuvant I-131 Lipiodol after resection or radiofrequency ablation for hepatocellular carcinoma," *World J. Surg.*, 2016, doi: 10.1007/s00268-016-3502-5.

[44] W. Y. Lau *et al.*, "Adjuvant intra-arterial iodine-131-labelled Lipiodol for resectable hepatocellular carcinoma: A prospective randomised trial," *Lancet*, 1999, doi: 10.1016/S0140-6736(98)06475-7.

[45] P. Bernal *et al.*, "Intra-Arterial Rhenium-188 Lipiodol in the Treatment of Inoperable Hepatocellular Carcinoma: Results of an IAEA-sponsored multination study," *Int. J. Radiat. Oncol. Biol. Phys.*, 2007, doi: 10.1016/j.ijrobp.2007.05.009.

[46] P. B. Zanzonico and C. Divgi, "Patient-specific radiation dosimetry for radionuclide therapy of liver tumors with Intrahepatic Artery Rhenium-188 Lipiodol," *Semin. Nucl. Med.*, 2008, doi: 10.1053/j.semnuclmed.2007.10.005.

[47] P. Bernal *et al.*, "International Atomic Energy Agency-Sponsored Multination Study of Intra-Arterial Rhenium-188-Labeled Lipiodol in the Treatment of Inoperable Hepatocellular Carcinoma: Results with special emphasis on prognostic value of dosimetric study," *Semin. Nucl. Med.*, 2008, doi: 10.1053/j.semnuclmed.2007.10.006.

[48] K. Delaunay *et al.*, "Preliminary results of the Phase 1 Lip-Re I clinical trial: biodistribution and dosimetry assessments in hepatocellular carcinoma patients treated with 188Re-SSS Lipiodol radioembolization," *Eur. J. Nucl. Med. Mol. Imaging*, 2019, doi: 10.1007/s00259-019-04277-9.

[49] M. M. Chojnowski, A. Felis-Giemza, and M. Kobylecka, "Radionuclide synovectomy – essentials for rheumatologists," *Reumatologia/Rheumatology*, vol. 3, no. 3, pp. 108–16, 2016, doi: 10.5114/reum.2016.61210.

[50] E. Savio *et al.*, "188Re radiopharmaceuticals for radiosynovectomy: Evaluation and comparison of tin colloid, hydroxyapatite and tin-ferric hydroxide macroaggregates," *BMC Nucl. Med.*, 2004, doi: 10.1186/1471-2385-4-1.

[51] C. Turkmen, "Safety of radiosynovectomy in hemophilic synovitis: It is time to re-evaluate," *J Coagul Disord*, 2009.

[52] G. Clunie and M. Fischer, "EANM procedure guidelines for radiosynovectomy," *Eur. J. Nucl. Med. Mol. Imaging*, 2003, doi: 10.1007/s00259-002-1058-0.

[53] P. Teyssler, K. Kolostova, and V. Bobek, "Radionuclide synovectomy in haemophilic joints," *Nuclear Medicine Communications*. 2013, doi: 10.1097/MNM.0b013e32835ed50c.

[54] D. Cimino Brown, "What can we learn from osteoarthritis pain in companion animals?" *Clin. Exp. Rheumatol.*, vol. 35 Suppl 1, no. 5, pp. 53–58, 2017.

[55] J. C. Lattimer *et al.*, "Intraarticular injection of a Tin-117 m radiosynoviorthesis agent in normal canine elbows causes no adverse effects," *Vet. Radiol. ultrasound Off. J. Am. Coll. Vet. Radiol. Int. Vet. Radiol. Assoc.*, vol. 60, no. 5, pp. 567–74, Sep. 2019, doi: 10.1111/vru.12757.

[56] R. E. I. I. Wendt *et al.*, "Radiation safety considerations in the treatment of canine skeletal conditions using 153Sm, 90Y, and 117mSn," *Health Phys.*, vol. 118, no. 6, 2020, [Online]. Available: https://journals.lww.com/health-physics/Fulltext/2020/06000/Radiation_Safety_Considerations_in_the_Treatment.16.aspx.

[57] L. A. Boland, J. K. Murray, C. P. Bovens, and A. Hibbert, "A survey of owners' perceptions and experiences of radioiodine treatment of feline hyperthyroidism in the UK," *J. Feline Med. Surg.*, vol. 16, no. 8, pp. 663–70, Aug. 2014, doi: 10.1177/1098612X13518939.

[58] H. C. Carney *et al.*, "2016 AAFP guidelines for the management of feline hyperthyroidism," *J. Feline Med. Surg.*, vol. 18, no. 5, pp. 400–16, May 2016, doi: 10.1177/1098612X16643252.

[59] M. E. Peterson, "Animal models of disease: Feline hyperthyroidism: An animal model for toxic nodular goiter," *J. Endocrinol.*, vol. 223, no. 2, pp. T97–114, Nov. 2014, doi: 10.1530/JOE-14-0461.

[60] M. E. Peterson, C. A. Castellano, and M. Rishniw, "Evaluation of body weight, body condition, and muscle condition in cats with hyperthyroidism," *J. Vet. Intern. Med.*, vol. 30, no. 6, pp. 1780–89, Nov. 2016, doi: 10.1111/jvim.14591.

[61] M. E. Peterson, "Hyperthyroidism in Cats: Considering the impact of treatment modality on quality of life for cats and their owners," *Vet. Clin. North Am. Small Anim. Pract.*, vol. 50, no. 5, pp. 1065–84, Sep. 2020, doi: 10.1016/j.cvsm.2020.06.004.

[62] R. J. Milner, C. D. Channell, J. K. Levy, and M. Schaer, "Survival times for cats with hyperthyroidism treated with iodine 131, methimazole, or both: 167 cases (1996–2003)," *J. Am. Vet. Med. Assoc.*, vol. 228, no. 4, pp. 559–63, Feb. 2006, doi: 10.2460/javma.228.4.559.

[63] J. M. Liptak, "Canine thyroid carcinoma," *Clin. Tech. Small Anim. Pract.*, vol. 22, no. 2, pp. 75–81, 2007, doi: https://doi.org/10.1053/j.ctsap.2007.03.007.

[64] A. Hibbert, T. Gruffydd-Jones, E. L. Barrett, M. J. Day, and A. M. Harvey, "Feline thyroid carcinoma: diagnosis and response to high-dose radioactive iodine treatment," *J. Feline Med. Surg.*, vol. 11, no. 2, pp. 116–24, Feb. 2009, doi: 10.1016/j.jfms.2008.02.010.

[65] R. J. Milner, I. Dormehl, W. K. Louw, and S. Croft, "Targeted radiotherapy with Sm-153-EDTMP in nine cases of canine primary bone tumours," *J. S. Afr. Vet. Assoc.*, vol. 69, no. 1, pp. 12–17, Mar. 1998, doi: 10.4102/jsava.v69i1.802.

[66] J. M. Vancil *et al.*, "Use of samarium Sm 153 lexidronam for the treatment of dogs with primary tumors of the skull: 20 cases (1986-2006)," *J. Am. Vet. Med. Assoc.*, vol. 240, no. 11, pp. 1310–15, Jun. 2012, doi: 10.2460/javma.240.11.1310.

[67] S. M. Barnard, R. M. Zuber, and A. S. Moore, "Samarium Sm 153 lexidronam for the palliative treatment of dogs with primary bone tumors: 35 cases (1999–2005)," *J. Am. Vet. Med. Assoc.*, vol. 230, no. 12, pp. 1877–81, Jun. 2007, doi: 10.2460/javma.230.12.1877.

[68] S. A. van Nimwegen *et al.*, "Intratumoral injection of radioactive holmium ((166) Ho) microspheres for treatment of oral squamous cell carcinoma in cats," *Vet. Comp. Oncol.*, vol. 16, no. 1, pp. 114–24, Mar. 2018, doi: 10.1111/vco.12319.

[69] S. Chakraborty *et al.*, "Evaluation of ^{177}Lu-EDTMP in dogs with spontaneous tumor involving bone: Pharmacokinetics, dosimetry and therapeutic efficacy," *Curr. Radiopharm.*, vol. 9, no. 1, pp. 64–70, 2016, doi: 10.2174/1874471008666150312164255.

[70] H. D. Zacho, N. N. Karthigaseu, R. F. Fonager, and L. J. Petersen, "Treatment with bone-seeking radionuclides for painful bone metastases in patients with lung cancer: A systematic review," *BMJ Supportive and Palliative Care*. 2017, doi: 10.1136/bmjspcare-2015-000957.

[71] G. Henriksen, D. R. Fisher, J. C. Roeske, O. S. Bruland, and R. H. Larsen, "Targeting of osseous sites with alpha-emitting 223Ra: Comparison with the beta-emitter 89Sr in mice," *J. Nucl. Med.*, vol. 44, no. 2, pp. 252–59, Feb. 2003.

[72] E. Askari, S. Harsini, N. Vahidfar, G. Divband, and R. Sadeghi, "177 Lu-EDTMP for Metastatic Bone Pain Palliation: A systematic review and meta-analysis," *Cancer Biother. Radiopharm. Cancer Biother Radiopharm.* vol. 36, no. 5, pp. 383–390, Jun. 2021, doi: 10.1089/cbr.2020.4323.

[73] R. Manafi-Farid *et al.*, "Targeted palliative radionuclide therapy for metastatic bone pain," *J. Clin. Med.*, vol. 9, no. 8, Aug. 2020, doi: 10.3390/jcm9082622.

[74] S. Petersdorf *et al.*, "Samarium-153-EDTMP biodistribution and estimation," pp. 1031–36, 2015.

[75] J. H. Tian *et al.*, "Multicentre trial on the efficacy and toxicity of single-dose samarium-153-ethylene diamine tetramethylene phosphonate as a palliative treatment for painful skeletal metastases in China," *Eur. J. Nucl. Med.*, vol. 26, no. 1, pp. 2–7, 1999, doi: 10.1007/s002590050351.

[76] N. Pandit-Taskar, S. M. Larson, and J. A. Carrasquillo, "Bone-seeking radiopharmaceuticals for treatment of osseous metastases, Part 1: α therapy with 223Ra-dichloride," *J. Nucl. Med.*, vol. 55, no. 2, pp. 268–74, 2014, doi: 10.2967/jnumed.112.112482.

[77] C. Parker *et al.*, "Alpha Emitter Radium-223 and survival in metastatic prostate cancer," *N. Engl. J. Med.*, vol. 369, no. 3, pp. 213–23, Jul. 2013, doi: 10.1056/NEJMoa1213755.

[78] C. C. Parker *et al.*, "A randomized, double-blind, dose-finding, multicenter, phase 2 study of radium chloride (Ra 223) in patients with bone metastases and castration-resistant prostate cancer," *Eur. Urol.*, vol. 63, no. 2, pp. 189–97, 2013, doi: 10.1016/j.eururo.2012.09.008.

[79] L. M. and N. D., "Dosimetry of 223Ra-chloride: Dose to normal organs and tissues," *Eur. J. Nucl. Med. Mol. Imaging*, 2013.

[80] R. Lange, R. Ter Heine, R. F. Knapp, J. M. H. de Klerk, H. J. Bloemendal, and N. H. Hendrikse, "Pharmaceutical and clinical development of phosphonate-based radiopharmaceuticals for the targeted treatment of bone metastases," *Bone*, vol. 91, pp. 159–79, Oct. 2016, doi: 10.1016/j.bone.2016.08.002.

[81] T. Das, H. D. Sarma, A. Shinto, K. K. Kamaleshwaran, and S. Banerjee, "Formulation, preclinical evaluation, and preliminary clinical investigation of an in-house freeze-dried EDTMP kit suitable for the preparation of 177Lu-EDTMP," *Cancer Biother. Radiopharm.*, 2014, doi: 10.1089/cbr.2014.1664.

[82] G. Sgouros *et al.*, MIRD Pamphlet No. 22 (abridged): *Radiobiology and dosimetry of alpha-particle emitters for targeted radionuclide therapy. J. Nucl. Med.*, vol. 51, no. 2, pp. 311–28, Feb. 2010, doi: 10.2967/jnumed.108.058651.

[83] "Radium – 223 (Xofigo) for prostate cancer," *Med. Lett. Drugs Ther.*, vol. 55, no. 1426, pp. 79–80, Sep. 2013.

[84] C. Kratochwil *et al.*, "Targeted Alpha Therapy of mCRPC with ^{225} Actinium-PSMA-617: Dosimetry estimate and empirical dose finding," *J. Nucl. Med.*, vol. 58, no. 10, p. jnumed.117.191395, 2017, doi: 10.2967/jnumed.117.191395.

[85] C. Kratochwil *et al.*, "Patients resistant against PSMA-targeting α-radiation therapy often harbor mutations in DNA damage-repair-associated genes," *J. Nucl. Med.*, vol. 61, no. 5, pp. 683–88, May 2020, doi: 10.2967/jnumed.119.234559.

[86] M. W. Geerlings, F. M. Kaspersen, C. Apostolidis, and R. van der Hout, "The feasibility of 225Ac as a source of alpha-particles in radioimmunotherapy," *Nucl. Med. Commun.*, vol. 14, no. 2, pp. 121–25, Feb. 1993.

[87] J. F. Ziegler, "Comments on ICRU report no. 49: Stopping powers and ranges for protons and alpha particles," *Radiat. Res.*, vol. 152, no. 2, pp. 219–22, Aug. 1999.

[88] S. T. Tagawa *et al.*, "Phase I dose-escalation study of 225Ac-J591 for progressive metastatic castration resistant prostate cancer (mCRPC)," *J. Clin. Oncol.*, vol. 36, no. 6_suppl, pp. TPS399–TPS399, Feb. 2018, doi: 10.1200/JCO.2018.36.6_suppl.TPS399.

[89] A. Bandekar, C. Zhu, R. Jindal, F. Bruchertseifer, A. Morgenstern, and S. Sofou, "Anti-prostate-specific membrane antigen liposomes loaded with 225Ac for potential targeted antivascular α-particle therapy of cancer," *J. Nucl. Med.*, vol. 55, no. 1, pp. 107–14, Jan. 2014, doi: 10.2967/jnumed.113.125476.

[90] M. Sathekge *et al.*, "(225)Ac-PSMA-617 in chemotherapy-naive patients with advanced prostate cancer: A pilot study," *Eur. J. Nucl. Med. Mol. Imaging*, vol. 46, no. 1, pp. 129–38, Jan. 2019, doi: 10.1007/s00259-018-4167-0.

[91] A. Morgenstern, C. Apostolidis, and F. Bruchertseifer, "Supply and clinical application of Actinium-225 and Bismuth-213," *Semin. Nucl. Med.*, vol. 50, no. 2, pp. 119–23, Mar. 2020, doi: 10.1053/j.semnuclmed.2020.02.003.

[92] N. Falzone, B. Cornelissen, and K. A. Vallis, "Auger emitting radiopharmaceuticals for cancer therapy BT – Radiation damage in biomolecular systems," G. García Gómez-Tejedor and M. C. Fuss, Eds. Dordrecht: Springer, 2012, pp. 461–78.

[93] K. S. R. Sastry *et al.*, "Dosimetry of Auger emitters: Physical and phenomenological approaches," United States, 1987. [Online]. Available: http://inis.iaea.org/search/search.aspx?orig_q=RN:19076402.

[94] N. Falzone, B. Cornelissen, and K. A. Vallis, "Auger emitting radiopharmaceuticals for cancer therapy BT – Radiation damage in biomolecular systems," G. García Gómez-Tejedor and M. C. Fuss, Eds. Dordrecht: Springer, 2012, pp. 461–78.

[95] G. Pirovano, T. C. Wilson, and T. Reiner, "Auger: The future of precision medicine," *Nucl. Med. Biol.*, vol. 96–97, pp. 50–53, Mar. 2021, doi: 10.1016/j.nucmedbio.2021.03.002.

[96] J. C. Roeske, B. Aydogan, M. Bardies, and J. L. Humm, "Small-scale dosimetry: Challenges and future directions," *Semin. Nucl. Med.*, vol. 38, no. 5, pp. 367–83, Sep. 2008, doi: 10.1053/j.semnuclmed.2008.05.003.

[97] A. I. Kassis, "Molecular and cellular radiobiological effects of Auger emitting radionuclides," *Radiat. Prot. Dosimetry*, vol. 143, no. 2–4, pp. 241–47, Feb. 2011, doi: 10.1093/rpd/ncq385.

[98] F. Buchegger, F. Perillo-Adamer, Y. M. Dupertuis, and A. Bischof Delaloye, "Auger radiation targeted into DNA: A therapy perspective," *Eur. J. Nucl. Med. Mol. Imaging*, vol. 33, no. 11, pp. 1352–63, 2006, doi: 10.1007/s00259-006-0187-2.

[99] F. Buchegger, F. Perillo-Adamer, Y. M. Dupertuis, and A. Bischof Delaloye, "Auger radiation targeted into DNA: A therapy perspective," *Eur. J. Nucl. Med. Mol. Imaging*, vol. 33, no. 11, pp. 1352–63, 2006, doi: 10.1007/s00259-006-0187-2.

[100] P. Unak, "Targeted tumor radiotherapy," *Brazilian Arch. Biol. Technol.*, vol. 45, pp. 97–110, 2002, [Online]. Available: www.scielo.br/scielo.php?script=sci_arttext&pid=S1516-89132002000500014&nrm=iso.

[101] M. Gardette *et al.*, "Evaluation of two (125)I-radiolabeled acridine derivatives for Auger-electron radionuclide therapy of melanoma," *Invest. New Drugs*, vol. 32, no. 4, pp. 587–97, Aug. 2014, doi: 10.1007/s10637-014-0086-5.

[102] H. Thisgaard *et al.*, "Highly effective Auger-electron therapy in an orthotopic glioblastoma Xenograft Model using convection-enhanced delivery," *Theranostics*, vol. 6, no. 12, pp. 2278–91, 2016, doi: 10.7150/thno.15898.

[103] C. J. Shen *et al.*, "Auger radiopharmaceutical therapy targeting prostate-specific membrane antigen in a micrometastatic model of prostate cancer," *Theranostics*, vol. 10, no. 7, pp. 2888–96, 2020, doi: 10.7150/thno.38882.

[104] A. A. Rosenkranz *et al.*, "Antitumor activity of Auger electron emitter 111In delivered by modular nanotransporter for treatment of bladder cancer with EGFR overexpression," *Front. Pharmacol.*, vol. 9, p. 1331, 2018, doi: 10.3389/fphar.2018.01331.

[105] R. Valkema *et al.*, "Phase I study of peptide receptor radionuclide therapy with [In-DTPA]octreotide: The Rotterdam experience," *Semin. Nucl. Med.*, vol. 32, no. 2, pp. 110–22, Apr. 2002, doi: 10.1053/snuc/2002.31025.

[106] L. B. Anthony, E. A. Woltering, G. D. Espenan, M. D. Cronin, T. J. Maloney, and K. E. McCarthy, "Indium-111-pentetreotide prolongs survival in gastroenteropancreatic malignancies," *Semin. Nucl. Med.*, vol. 32, no. 2, pp. 123–32, Apr. 2002, doi: 10.1053/snuc.2002.31769.

[107] E. T. Janson, J. E. Westlin, U. Ohrvall, K. Oberg, and A. Lukinius, "Nuclear localization of 111In after intravenous injection of [111In-DTPA-D-Phe1]-octreotide in patients with neuroendocrine tumors," *J. Nucl. Med.*, vol. 41, no. 9, pp. 1514–18, Sep. 2000.

[108] M. R. Jackson, N. Falzone, and K. A. Vallis, "Advances in anticancer radiopharmaceuticals," *Clin. Oncol. (R. Coll. Radiol).*, vol. 25, no. 10, pp. 604–9, Oct. 2013, doi: 10.1016/j.clon.2013.06.004.

[109] L. Farzin, S. Sheibani, M. E. Moassesi, and M. Shamsipur, "An overview of nanoscale radionuclides and radiolabeled nanomaterials commonly used for nuclear molecular imaging and therapeutic functions," *J. Biomed. Mater. Res. A*, vol. 107, no. 1, pp. 251–85, Jan. 2019, doi: 10.1002/jbm.a.36550.

[110] E. J. L. Stéen *et al.*, "Pretargeting in nuclear imaging and radionuclide therapy: Improving efficacy of theranostics and nanomedicines," *Biomaterials*, vol. 179, pp. 209–45, Oct. 2018, doi: 10.1016/j.biomaterials.2018.06.021.

[111] Y. Matsumura and H. Maeda, "A new concept for macromolecular therapeutics in cancer chemotherapy: mechanism of tumoritropic accumulation of proteins and the antitumor agent smancs," *Cancer Res.*, vol. 46, no. 12 Pt 1, pp. 6387–92, Dec. 1986.

[112] T. Stylianopoulos and R. K. Jain, "Design considerations for nanotherapeutics in oncology," *Nanomedicine*, vol. 11, no. 8, pp. 1893–1907, Nov. 2015, doi: 10.1016/j.nano.2015.07.015.

[113] R. M. de Kruijff, H. T. Wolterbeek, and A. G. Denkova, "A critical review of Alpha Radionuclide Therapy: How to deal with recoiling daughters?," *Pharmaceuticals (Basel).*, vol. 8, no. 2, pp. 321–36, Jun. 2015, doi: 10.3390/ph8020321.

[114] U. Holzwarth, I. Ojea Jimenez, and L. Calzolai, "A random walk approach to estimate the confinement of α-particle emitters in nanoparticles for targeted radionuclide therapy," *EJNMMI Radiopharm. Chem.*, vol. 3, no. 1, p. 9, 2018, doi: 10.1186/s41181-018-0042-3.

[115] Y. Du and colleagues *Development of Alpha-emitting radioembolization for Hepatocellular Carcinoma: Longitudinal Monitoring of Actinium-225's Daughters Through SPECT Imaging*. 2020.

5 Design Considerations for a Radiopharmaceutical Production Facility

Nic Gillings

CONTENTS

5.1 INTRODUCTION

This chapter focuses on the design of a production facility for radiopharmaceuticals for human administration (i.e. sterile medicinal products). For diagnostic (short-lived) radiopharmaceuticals, such facilities are most often hospital-based. For radiopharmaceuticals for PET, an on-site cyclotron is often required. The design considerations presented here do not cover full details of cyclotron facilities but will include details of solutions for transfer of cyclotron-produced radionuclides to the radiopharmaceutical production facility. A more extensive guideline has been published by the International Atomic Energy Agency [1].

Facilities for production of radiopharmaceuticals for human administration need to be designed to comply with relevant rules and guidelines related to aseptic pharmaceutical manufacture. Since the vast majority of radiopharmaceuticals are in the form of solutions for in vivo injection (parenterals), this chapter will focus on the design of facilities for this type of product. Along with aseptic manufacturing considerations, protection of personnel from exposure to ionizing radiation is also a fundamental requirement.

5.2 CLEANROOM CONSIDERATIONS

Generally, the production of radiopharmaceuticals for human administration is classified as aseptic manufacture. Since a terminal sterilization of products is rarely possible (e.g. autoclaving), special precautions must apply to ensure the quality of the radiopharmaceutical products. Guidance for the requirements for such activities can be found in, for example, *The rules governing medicinal products in the European Union* [2]. To protect products from contamination, cleanroom areas are required. The classification of cleanrooms is well established and applies for both pharmaceutical manufacture and in the electronic industry, where extremely clean air is also important. Cleanroom classification is based on the measurement of airborne particles. Table 5.1 gives the EU limits for the different grades of cleanrooms.

For production of radiopharmaceuticals, where closed systems are used (e.g. automated radiosynthesis modules), a grade C air quality is sufficient. For critical processes (e.g. preparation of final product vials), a grade A environment is required. This can be achieved using either a laminar air flow work bench (LAF-bench, Class II) or a pharmaceutical isolator (Class III). It is generally considered acceptable that LAF-benches may be placed in grade C cleanrooms (grade A in C). For pharmaceutical isolators, a grade B airlock ensures the correct cascade of air quality (grade C to B to A).

DOI: 10.1201/9780429489501-5

TABLE 5.1
EU Limits for Airborne Particles Showing the Maximum Permitted Number of Particles Per m³ Equal To or Greater Than the Tabulated Size

	At rest		In operation	
Grade	0.5 µm	5.0 µm	0.5 µm	5.0 µm
A	3520	20	3520	20
B	3520	29	352000	29
C	352000	2900	3520000	29000
D	3520000	29000	Not defined	Not defined

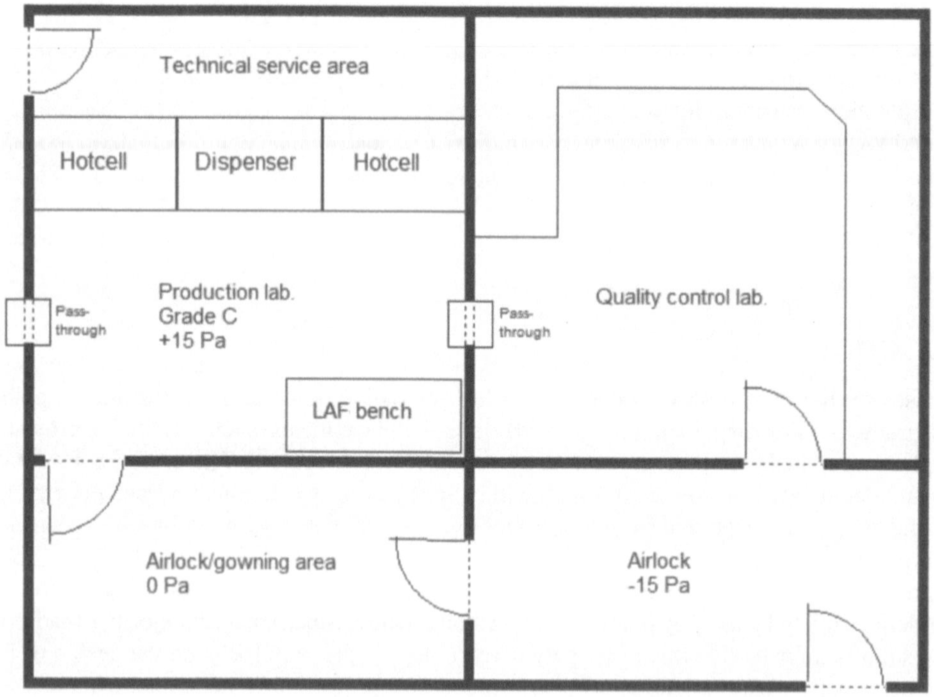

FIGURE 5.1 A possible layout of a small PET radiopharmaceutical production facility.

To ensure the air quality of cleanrooms, the design of the facility is crucial. Cleanrooms should be supplied with filtered air, with sufficient air exchange to maintain the air quality. Generally, a 20-fold air exchange per hour in grade C cleanrooms is recommended. To prevent ingress of particulates from the surround areas, cleanrooms should maintain an over-pressure compared with the surrounding rooms. Furthermore, cleanrooms should be constructed of suitable materials to avoid unnecessary particles. Construction materials such as wood and plaster board should be avoided and, if used, they should be treated in such a way as to prevent them shedding particles (several layers of paint and sealant between any joints). All materials should be easy to clean and there should be no areas that are inaccessible to cleaning. From a radiation protection point of view, all surfaces should be easy to decontaminate. and a liquid-proof/tight floor (normally made of vinyl) should be used.

An example of a small radiopharmaceutical production facility is given in Figure 5.1. Cyclotron-produced radionuclides in the form of liquids or gases (e.g. [^{18}F]fluoride or [^{11}C]carbon dioxide) are transferred from cyclotron targets to hot cells in suitable capillary tubing (target lines) in larger plastic conduits. These conduits are normally placed under the floor and, here, shielding considerations are important. The most practical solution is to set target line transfer conduits in concrete and enter from the back of the hot cells in a technical area normally free from personnel. If this is not possible, concrete shielding may not be sufficient, and a layer of lead shielding may be necessary under the floor and up into the hot cell. Ventilated pass-throughs are often used for transfer of samples to the quality-control laboratory and for transfer of materials to the production laboratory and finished products out. Pass-throughs should be equipped with a mechanical interlock system to ensure that both doors cannot be opened simultaneously. There should be a high air-extraction rate to ensure that contaminated air cannot enter the production area. There should be an air-pressure cascade to protect the production area from airborne contamination. The negative pressure (-15 Pa) in the final airlock will also protect surrounding areas in the case of a gaseous radioactivity leak.

5.3 RADIATION PROTECTION CONSIDERATIONS

Exposure to ionizing radiation should be strictly controlled. There is a general principle that this should be as low a reasonably achievable, the so-called ALARA principle. Full details on requirements in the EU can be found in *European Council Directive 2013/59/Euratom* [3].

For production of PET-radiopharmaceuticals, shielding of personnel against exposure to ionizing radiation is of utmost importance. Due to the high-energy gamma emissions (511 keV), significant amounts of lead or tungsten shielding are required. Lead-shielded cabinets (hot cells) or isolators are available in a large variety of designs for such purposes. The amount of shielding depends on the activity and nature of the radionuclides used. For PET radionuclides, 75 mm of lead shielding is standard for most hot cells. The amount of shielding required can be estimated based on the half-value layer (HVL) for a given radionuclide in a given material. For example, for fluorine-18 (which emits only 511 keV gamma energy), HVL in lead is approximately 4 mm. A standard PET hot cell with 75 mm shielding will thus reduce the radiation level by a factor of more than 300,000.

For non-PET radionuclides, for example technetium-99m, shielding requirements are much reduced. For technetium-99m, the HVL in lead is approximately 0.3 mm, thus only a few millimetres of lead shielding will provide considerable protection.

Radiopharmaceutical production with PET isotopes normally requires fully automated systems to avoid excessive exposure of personnel to ionizing radiation. Any manual procedures should be limited, and personnel should be monitored for radiation exposure to extremities (finger-doses). On the other hand, most procedures using technetium-99m can be carried out manually with limited radiation doses.

5.4 PERSONNEL AND WORKFLOW

For aseptic production, the environment in the production facility should be controlled. One of the main sources of contamination is people. To limit this, all personnel entering the facility should change to dedicated clothing. For entry into the production cleanroom, additional clothing should be worn. This can be either a one or two piece cleanroom suit, clean footwear, and hairnets. For hand-washing and disinfection, a sink should be available in the changing area or airlock. To avoid contamination during entry, a mechanical door interlock system should be in place to ensure that only one door can be opened at a time. There should be enough time for the ventilation system to re-establish the air pressure before the next door can be opened. Due to the potential risk of radioactive contamination, there should be monitoring equipment available in the airlock/changing areas.

5.5 EQUIPMENT

For automated preparation of radiopharmaceuticals, a large range of commercial systems are available. Most modern systems are cassette-based and thus utilize single-use disposable plastic synthesis cassettes consisting of manifold stopcocks connected to tubing, syringes, and reagent vials. Synthesis cassettes are often provided by the manufacturer of the system and can be either sterile or certified as low bioburden. The use of these system avoids any cleaning considerations and facilitates fully automated preparations with minimal radiation exposure to personnel. Finished products should, whenever possible, be transferred automatically from the production hot cell to avoid the necessity to open the hot cell to retrieve the product, with the associated radiation exposure to personnel. This can be achieved by automated removal of the product vial to a lead container or by transfer of the product in a capillary tube to a separate hot cell, which will be the case if automated dispensing of the product is required.

5.6 QUALITY CONTROL

For radiopharmaceuticals with short half-lives (e.g. carbon-11), a quality-control laboratory should ideally be adjacent to the production area, thus allowing samples to be transferred directly via a pass-through. For quality controls a variety of equipment and techniques are used. Only small aliquots (5-100 μl) of products are normally needed for each test, and thus radiation exposure to personnel can be minimized. For simple manual tests, such as pH measurement, a small lead-shielded area (lead castle) should be available. For HPLC analysis, liquid waste containers should be shielded to avoid unnecessary radiation doses. Shielding of HPLC columns may also be required.

REFERENCES

[1] IAEA. "Cyclotron produced radionuclides: Guidance on facility design and production of Fluorodeoxyglucose (FDG)," IAEA. Vienna, 2012.

[2] EudraLex (2010). *The rules governing medicinal products in the European Union – Good Manufacturing Practice (GMP) guidelines*. https://ec.europa.eu/health/documents/eudralex/vol-4_en

[3] (2013). *European Council Directive 2013/59/Euratom of 5 December 2013*. http://data.europa.eu/eli/dir/2013/59/2014-01-17

6 Methods and Equipment for Quality Control of Radiopharmaceuticals

Rolf Zijlma, Danique Giesen, Yvette Kruiter, Philip H. Elsinga and Gert Luurtsema

CONTENTS

6.1 INTRODUCTION

After production of a radiopharmaceutical, a quality control procedure (QC) is required in order to assess whether the prepared radiopharmaceutical meets the predefined quality standards. The purpose of performing quality control analysis is (1) to assure that the radiopharmaceutical can safely be administered to a patient, and (2) to assure that the intended PET scan will supply the correct information to the physician. In this chapter quality control methods will be described, including the equipment used and the rationale for the QC methods. To show practical examples, the radiopharmaceutical [$_{18}$F]fluorodeoxyglucose is taken as a representative example. FDG is the most-often used radiopharmaceutical; this "workhorse" of PET, is used for molecular imaging of patients in the field of cardiology, neurology, inflammation, and oncology [1, 2]. In addition, some additional methods that are not applicable for FDG, but which are crucial for other radiopharmaceuticals will be described briefly.

FDG is produced by several pharmaceutical companies as well as at production sites in the academic world equipped with an onsite cyclotron and dedicated to good manufacturing production (GMP) facilities. After the production, FDG is sterilized and diluted in clinically practical doses. QC of the final product is performed, and if the product meets the criteria, it will be released by a qualified person (QP) for human intravenous administration [3].

The release criteria for commonly used clinical radiopharmaceuticals are described in the *Eur. Pharmacopeia*. In each *Pharmacopeia* "monograph," a detailed description of the product specifications, release criteria, and the recommended analytical methods are given. (See website for more information [4].)

Quality control equipment can be purchased from various vendors and be installed independently of each other. There is an increasing trend to link QC-equipment into one Laboratory Information Management System (LIMS). The advantage is that the analytical data from the different types of equipment are combined in one software system that facilitates the overall judgement of the combined data and release of the radiopharmaceutical for clinical use. Some vendors even combine the different equipment to reduce the overall laboratory space required (Figure 6.1).

DOI: 10.1201/9780429489501-6

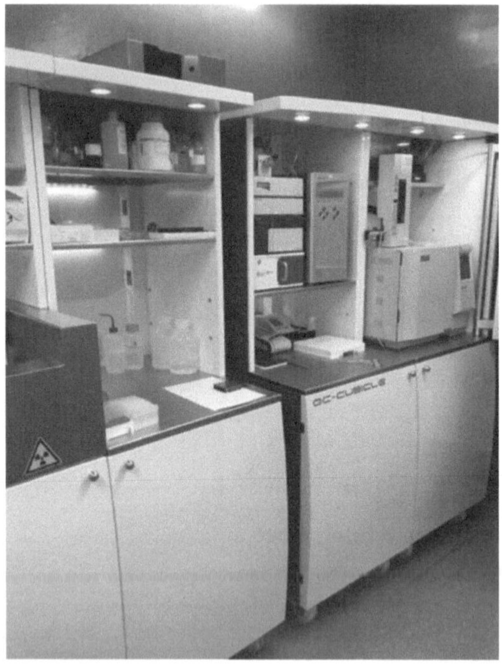

FIGURE 6.1 Example of a QC-setup for FDG-quality control (All in one QC cubicle).

6.2 RATIONALE FOR USING SPECIFIC QUALITY CONTROL METHODS

For several radiopharmaceuticals, the quality requirements are described in the *Pharmacopeia* (*Eur*, *USP*). For other radiopharmaceuticals these requirements need to be specified based on knowledge of the production process (presence of specific impurities) and on general requirements obtained by complemented methods (e.g. pH, sterility, osmolality, organic solvents). To understand some of the quality control methods for release of FDG, one must understand the production method and any potential deviations that may occur during this process.

6.3 GENERAL DESCRIPTION OF THE FDG SYNTHESIS

The worldwide method used to produce FDG is based on the radio-synthesis according to the literature described by Hamacher and colleagues [5]. Briefly, the radionuclide ^{18}F is produced using a medical 18 MeV cyclotron via the ^{18}O (p,n)^{18}F nuclear reaction. Dedicated synthesis modules are placed in a class C hot cell to protect the laboratory personnel against radiation. The synthesis is remotely performed and based on nucleophilic substitution reaction of $^{18}F/K_{222}$ complex with mannose triflate precursor in acetonitrile (ACN) (Figure 6.2).

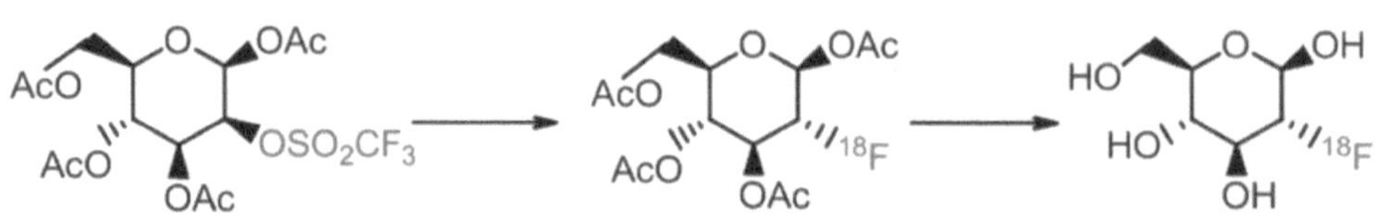

FIGURE 6.2 Nucleophilic substitution reaction of $^{18}F/K_{222}$ complex with mannose triflate precursor to produce FDG.

After the reaction, ^{18}F-labelled glucose precursor with protecting acetyl groups on the 1,3,4,6 carbon positions dissolved in ACN, is hydrolyzed under basic conditions using sodium hydroxide (NaOH). In the next step, the complete reaction mixture is purified using dedicated cartridges. The cartridges are activated using ethanol (ETOH) and are then washed with water. The FDG solution is transferred to a class A hot cell and diluted with physiological saline and sterilized using a sterile filter. See Figure 6.3 for a schematic overview of the remote FDG synthesis.

An assessment was done to define which possible side products or impurities can be formed during the procedure briefly described above. The formed side products or impurities could contaminate the final end product if a technical

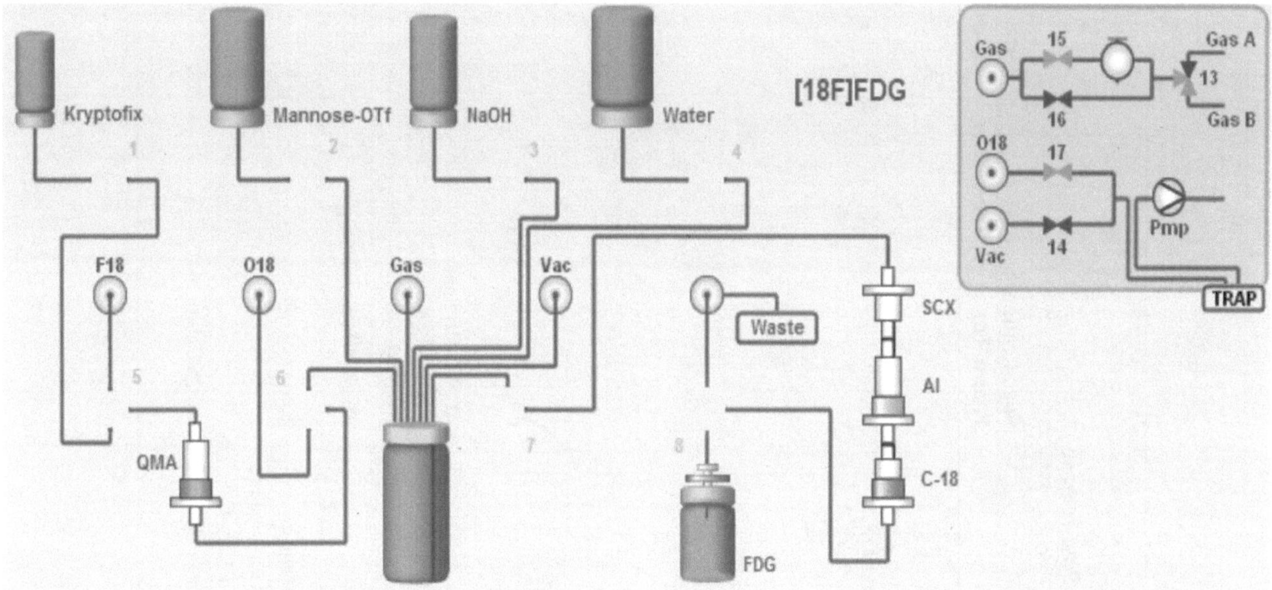

FIGURE 6.3 A schematic overview of FDG production.

failure occurs during the production process. During a successful production these contaminants are avoided or properly removed using a purification step. Information on potential impurities is essential to define the required analyses (Table 6.1). Besides this, a regular check on identity, residual solvents, and sterility must be performed. Because it not possible to perform all these tests before release due to the short half-life of ^{18}F, post-release and pre-release analyses are defined and performed. Below is a list of all release specifications that will be analyzed, the rational for their analysis, and whether they are suitable for pre-or post-release (see Table 6.1).

6.4 DESCRIPTION OF THE QC-METHODOLOGY AND EQUIPMENT

The following QC equipment is mostly used: High-performance liquid chromatograph (HPLC) in combination with electrochemical detector (ECD) or ultraviolet light (UV) detector and radioactivity detector (RA); a TLC scanner; gas chromatograph (GC) in combination with a flame ionization detector (FID) pH meter, dose calibrator, osmometer and an endotoxin test module (Endosafe). These QC-methods are also applied for analysis of FDG prior to administration to patients.

Before using the equipment to release FDG for human use, it is required to validate the equipment according to the IQ/OQ/PQ concept (see also section validation of the process).

6.4.1 Endotoxin Test Module

When a product is not produced under sterile conditions, it could happen that the end solution contains endotoxins. The Endosafe testing module from Charles River is designed for bacterial endotoxin test (Figure 6.4). Endotoxins can be described as a toxic substance released from the membrane of gram-negative bacteria also known as lipopolysaccharide (LPS). The Limulus amebocyte lysta (LAL) test reflects the endotoxin substance from the membrane of gram-negative bacteria. The specification for release criteria of radiopharmaceuticals is <2.50 Endotoxin Units (EU)/ml.

The endotoxin quantification method used by the Endosafe testing module is based on the reaction cascade of the following proteins; Factor C (FC), Factor B (FB), clotting pro-enzyme and a chromogenic substrate (Figure 6.5). The cascade of endotoxin detection in the LAL assay starts when the endotoxin LPS reacts with FC, which activates FB. FB turns the clotting pro-enzyme into an enzyme. The resulting clotting enzyme is responsible for cleaving the chromogenic substrate, which produces a yellow colouring. The yellow colouring is detected by the Endosafe device by measuring the solution at 280 nano meters (nm). The resulting data is compared to a pre-programmed standard curve and a concentration is calculated.

The Endosafe system uses microchips that contain four channels, to each of which 25μl of sample is added (Figure 6.6). The sample travels through the channels within the chip using pumps. Two of the four channels contain a reference

TABLE 6.1
Topics That Are Subject to QC, the Rational for the Analysis, and Whether It Is Suitable for Pre-or Post-release

Release specification	Rational	Equipment	Specifications Pharm. Eur	Pre-release	Post-release
% radionuclide purity	During radionuclide production, other radionuclides than ^{18}F may be formed (^{13}N, ^{58}Co, ^{3}H).	GE- detector			X
% radiochemical purity	The conversion to the desired product may not be fully quantitative. Therefore, free ^{18}F can be a part of the reaction mixture. The hydrolysis step may not be quantitative so analysis of the FDG-intermediate is required See Figure 6.2.1, the molecule in the middle). Radiochemical impurities disturb the quality of the PET-image as their radioactivity biodistribution profile does not behave similar to FDG.	HPLC-RA and TLC reader	>95%	X	
Chemical purity	During synthesis also non-radioactive glucose, mannose, fluoro-deoxy mannose can be formed.	HPLC-ECD		X	
K_{222}/ Kryptofix	K_{222} is used to activate ^{18}F in organic-ACN solution and is ideally trapped on the used purification columns. K_{222} is toxic, therefore its presence in the final product is undesirable.	Spot test	< 2.2 mg/V	X	
pH	Basic conditions are used in the radiosynthesis to hydrolyze the protecting acetyl-groups. pH should be within a specific range to prevent adverse events.	pH-meter	4-8	X	
Solvent	ACN and EtOH are both used during the radiosynthesis and are toxic when administered to patients.	GC-FID	ACN < 410 ppm ETOH < 3.5 g/L		X
Isotonic solution/ mosmolarity	Final FDG product is dissolved in water. If the mosmolarity of the final product is lower than the specifications, it is not allowed for i.v. injections. Therefore FDG is diluted with saline in the final step of the production process.	Osmometer	200-800 Mosmol/kg		
Sterility	To check sterility of the final product. Non-sterile products may cause infections when administered.	TSB medium and at incubation temperature of 22.5 and 32.5 °C.	Sterile		X
Microbacterial growth	Endotoxin reflects the endotoxin substance from the membrane of gram-negative bacteria. Bacteria can cause infectious disease if present in de final product.	Endosafe	< 0.25 EU/ml		X

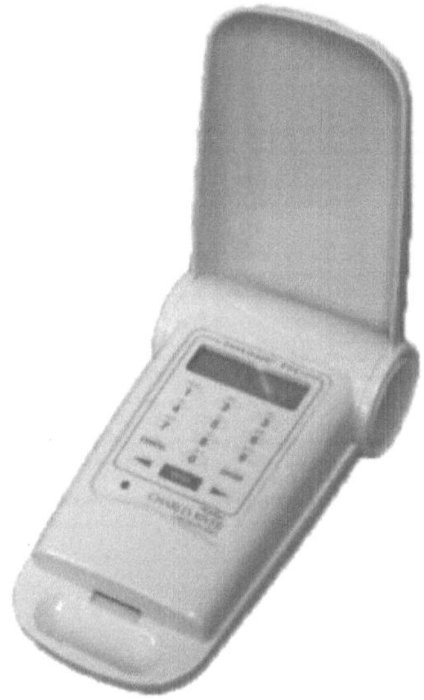

FIGURE 6.4 Picture of the Endosafe to check for bacterial endotoxin.

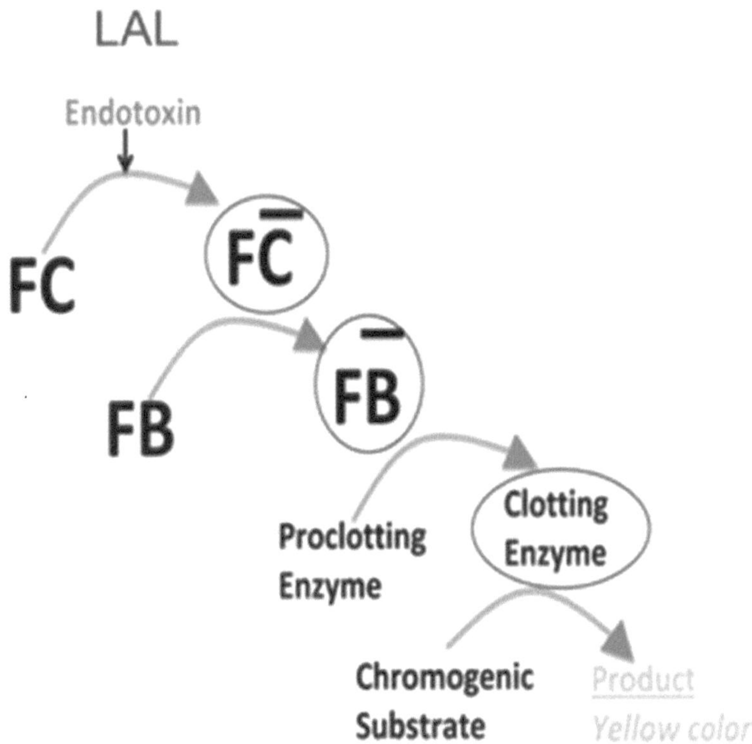

FIGURE 6.5 Example of protein cascade activated by endotoxin.

solution, "a spike sample" with a known amount of endotoxin. This reference solution functions as a positive control to test the analyzer. After the analyzer is tested, a sample travels through the LAL reagent, containing FC, FB and the clotting pro-enzyme. The protein cascade reaction has started when the sample moves through the chromogenic substrate and lastly enters the optical cells, where the yellow colouring is measured. This process takes about 15 minutes.

See for more detailed information the website www.sigmaaldrich.com/technical-documents/articles/biology/what-is-endotoxin.html.

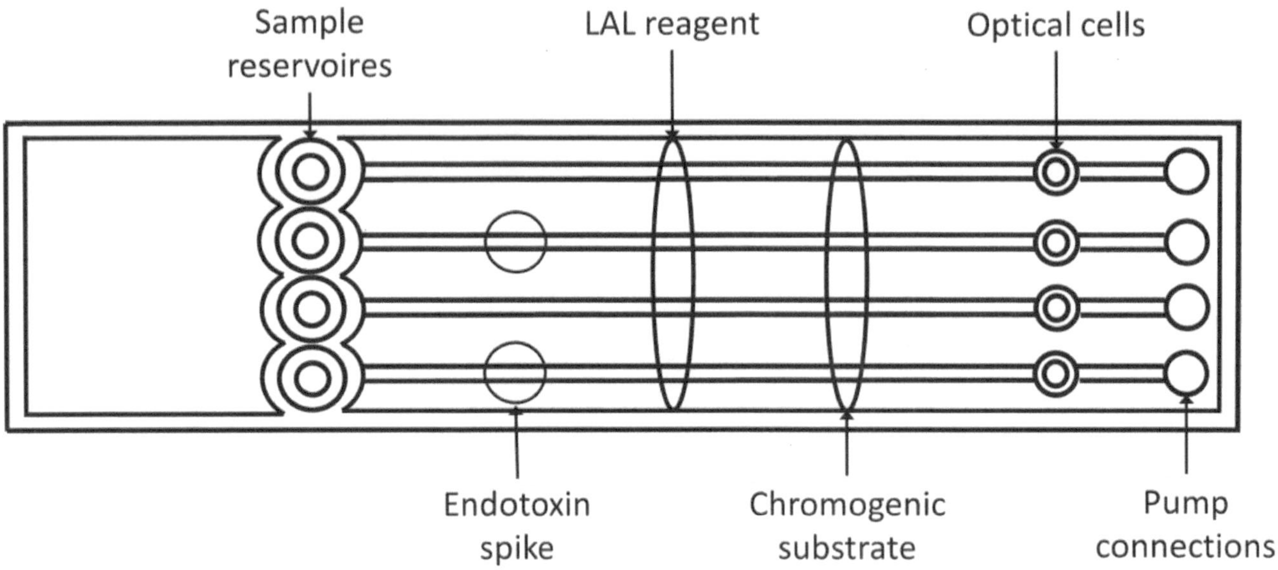

FIGURE 6.6 Example of chip used by the Endosafe system.

FIGURE 6.7 A typical example of pH paper with a range from 1–10 and the QC result for FDG. Colour image available at www.routledge.com/9781138593312.

6.4.2 pH Analysis

Radiopharmaceuticals are administrated mostly via intravenous injection. To avoid damage to the blood vessels it is important to use a neutral pH solution. A range between 4-8 is acceptable. Due to the radioactive content, a minimal exposure time and quantity are preferred for the quality control employee. Therefore, pH paper is used where only a drop (< 10 μL) of product can determine the pH with an error of approximately ± 0.5 (Figure 6.7).

With pH paper there is a colour change when the acid or base accepts or donates a proton; when this happens depends on the characteristics of the indicator molecule used. Some indicators can have more than one colour change.

6.4.3 TLC Reader

Thin Layer Chromatography (TLC) is a chromatography technique based on a solid phase, like paper or silica-coated plates (stationary phase) and a mobile phase (liquid). The analysis is run by putting the TLC plate with the sample on

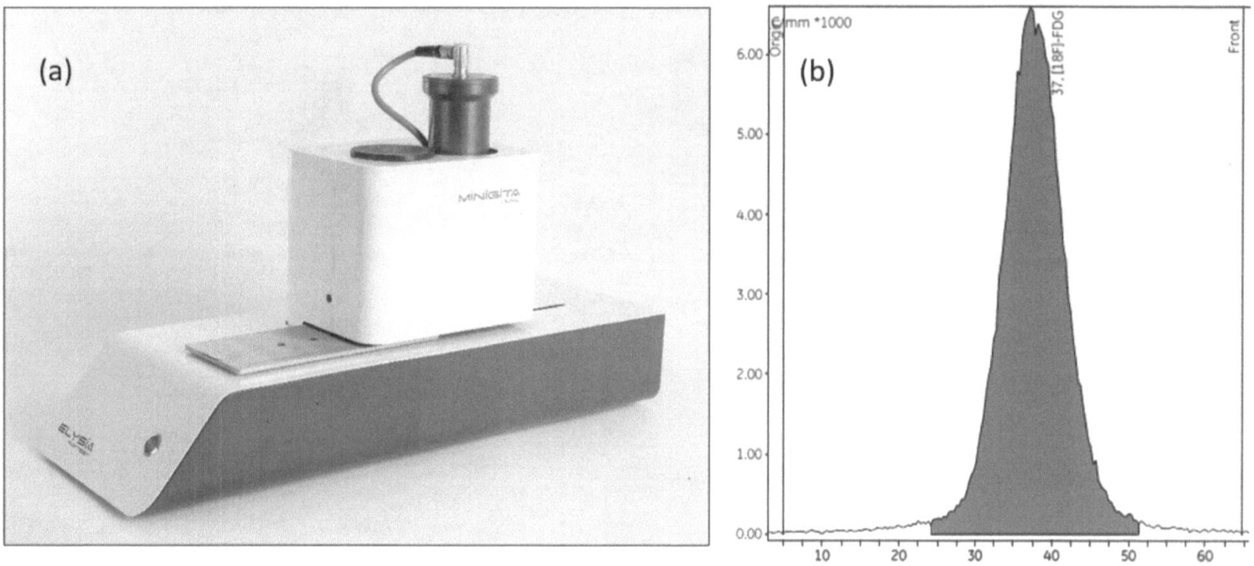

FIGURE 6.8 Picture of the TLC reader (a) and a typical chromatogram of a TLC readout (b).

the origin spot in a glass container containing a small layer of mobile phase. The mobile phase is moving upwards along the TLC plate until the liquid has reached the top. The compound and its impurities have different affinities and absorption to the stationary and the mobile phases. The technique is easy to perform and does not require dedicated equipment and, if combined with a reader or scanner (Figure 6.8a), the readout of the plate of paper is simple to determine the per cent of radiochemical purity and the radionuclide purity of the radiopharmaceutical. Typically, TLC takes only a few minutes. (See Figure 6.8b for a typical chromatogram.)

6.4.4 Osmometer

It is important that a radiopharmaceutical used for i.v. injection must be physiological neutral. Therefore, the injection solution must be isotone. This means that the ion concentration is similar to that in plasma. To check the osmolality, a dedicated osmometer is needed.

Osmolality is defined as osmoles per liter. For instance, 1 mol NaCl gives 2 osmol. The criterium for human plasma osmolarity is about 285 mosmol/kg. To check if a radiopharmaceutical meets the criteria for osmolality, a freezing point measurement can be performed using an osmometer (Figure 6.9).

The measurement principle is built on the colligative property of freezing-point depression (Figure 6.10). For example, the freezing point of the solution will decrease upon adding a solution to a liquid. The depression is 1.858 K per 1 mole of ideally solved compound in one litre of water. This effect depends solely on the number of particles in the liquid and not on the physical or chemical properties of the solutes. Due to this linear correlation, the osmolality of a sample can be determined by precisely measuring its freezing point.

A microprocessor-controlled peltier element is used to cool the sample at the beginning of the measurement. During this procedure, the solution is super-cooled to below 0 °C but remains in a liquid state. At a specific temperature, the freezing process is triggered by rotating the stirring wire. When ice crystals are formed, thermal energy is released, raising the sample's temperature. After a short time, an equilibrium is reached when melting and thawing of ice crystals are balanced and the temperature of the sample remains constant. This plateau represents the sample's actual freezing point. During the entire procedure, a high-precision thermistor is used to measure the solution's temperature. A resolution of 1/1000 K allows for accurate determination of the freezing point temperature and also enables the measurement of slight variations in the osmolality of two samples.

Typically, 140 µl is pipetted into an Eppendorf cup and subsequently the probe of the osmometer is submersed into the fluid. The freezing of the fluid and measurement takes about 3 minutes. Afterwards, the sample needs to thaw to avoid damage to the probe. For more information, check the website [6].

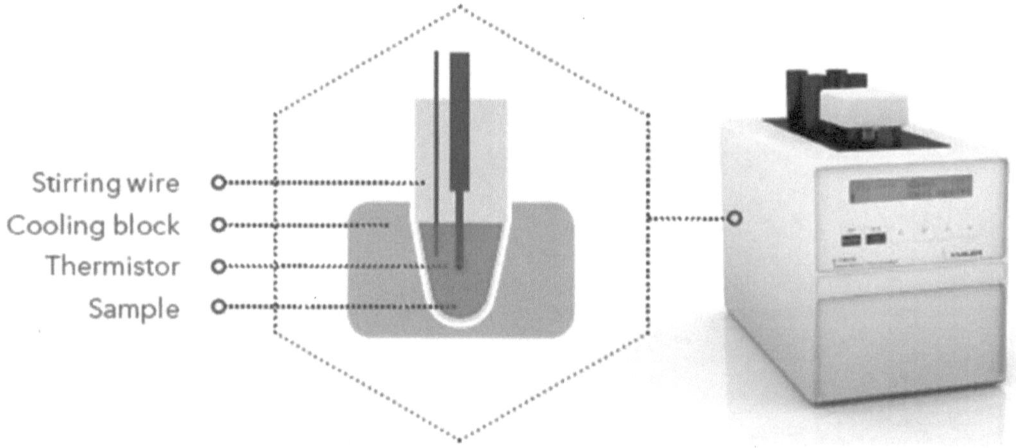

FIGURE 6.9 Schematic illustration of the measurement unit.

6.4.5 High Performance Liquid Chromatography with Online Radioactivity and UV Detector

Radiochemical purity is defined as a proportion of the total desired radioactive compound in the solution. The radiochemical purity is important because only the desired compound reflects the right distribution and kinetics and not the radiolabelled impurities.

To determine the radiochemical purity, the chemical purity, and to calculate the molar activity, a High-Performance Liquid Chromatography (HPLC) system combined with an online radioactivity (RA) and-UV detector are needed. HPLC is standard lab equipment, and its performance depends on the selection of the columns and choice of the detector. A high pressure HPLC pump pushes the mobile phase through an HPLC-column. As described for TLC, the analysis of the radiopharmaceutical is based on different affinities of the components in the sample for either the stationary or mobile phase. The higher the affinity for the stationary phase, the longer the component will remain on the column. The time it takes for a component to elute from the HPLC-column is called retention time. Typically, an HPLC-analysis takes 5–20 min. In terms of radiochemical and chemical purity, two different detectors are required: One to measure the mass, most often with a UV detector, and a second online radioactivity detector (Figure 6.11). The ratio between the measured radioactivity and the calculated mass reflects the molar activity of the desired compound. Although UV detection is quite standard, electrochemical detection (ECD) is also used in radiopharmaceutical analysis. The ECD is more sensitive than UV detection for molecules with electroactive functional groups such as amine or phenols [7].

The online radioactivity detector is generally a scintillation detector combined with a suitable flow cell installed close to the scintillator. The nature of the scintillator depends on the type of emission to be detected. For gamma detection a solid scintillator like NaI or BGO crystal will be used while, for high-energy beta detection, a plastic scintillator is used. For low beta energy or alpha detection, a liquid scintillator has to be used. For quality control of PET radiotracers, either a gamma probe is used to detect the 511 keV gammas or a beta probe to monitor the positron emission. Beta probes are less cumbersome as they are easy to shield from external radiations present in the laboratory. On the other hand, gamma detectors are suitable for detecting non-beta emitters like ^{99m}Tc or ^{125}I.

In the online radioactivity detector, the emission is indirectly measured by transformation of its energy into a proportional amount of light using a scintillation material. The emission of light is measured through a photocounting device like a photomultiplier tube (PMT).

The PMT tube consists of a photocathode to produce one electron from light emission and a series of dynodes to produce an electron cascade for signal amplification and an anode to create the signal output. Since PMTs are sensitive to light, the system has to be shielded from other sources of light.

The current generated by this measurement is processed electronically to determine the number of counts per second. Ideally, this signal is directly shared with the PC software and chromatography software (fully digital radiodetector). When this solution is not applicable, the analog output of the detector can be used to transfer the signal to the chromatography software, but this solution may reduce the dynamic range of the detector, and the detector parameters are not recorded by the main chromatography software.

The limit of detection achievable using online radiodetectors is determined by the type and geometry of the scintillator (efficiency) but also by the implementation of the liquid cell by example system with coil of HPLC tubings or

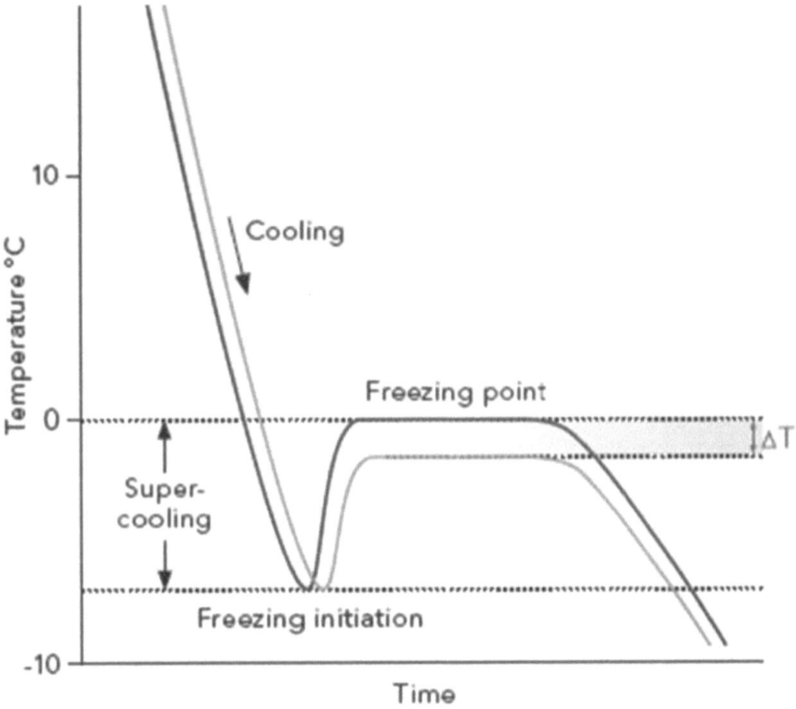

FIGURE 6.10 Principle graph illustration for the determination of the freezing point temperature. Colour image available at www.routledge.com/9781138593312.

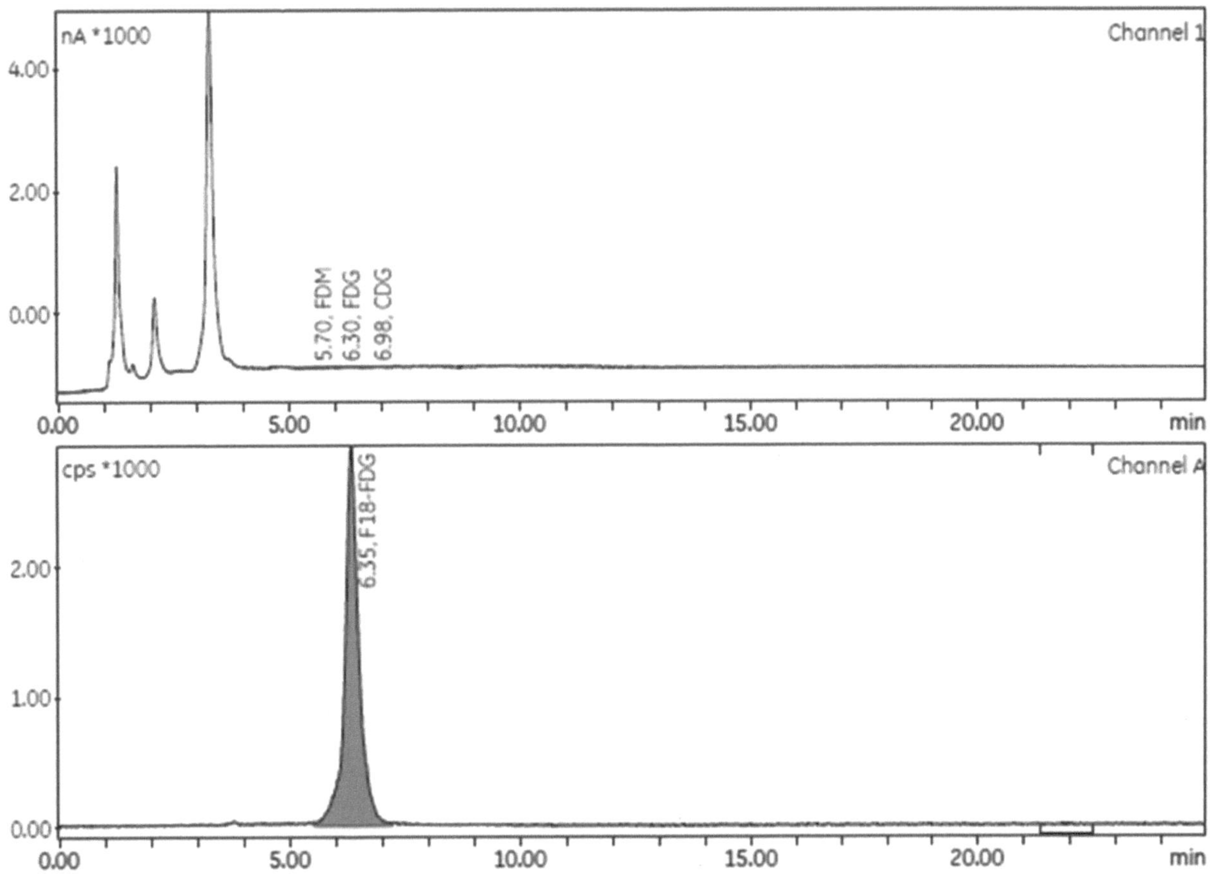

FIGURE 6.11 A typical chromatogram of the FDG end product with online RA and UV detection.

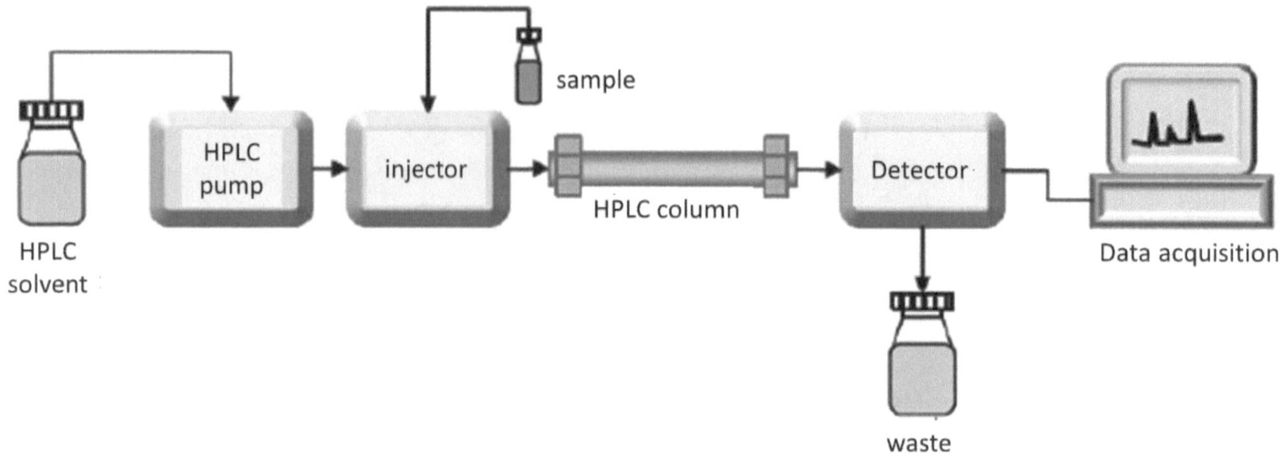

FIGURE 6.12 A schematic overview of the HPLC.

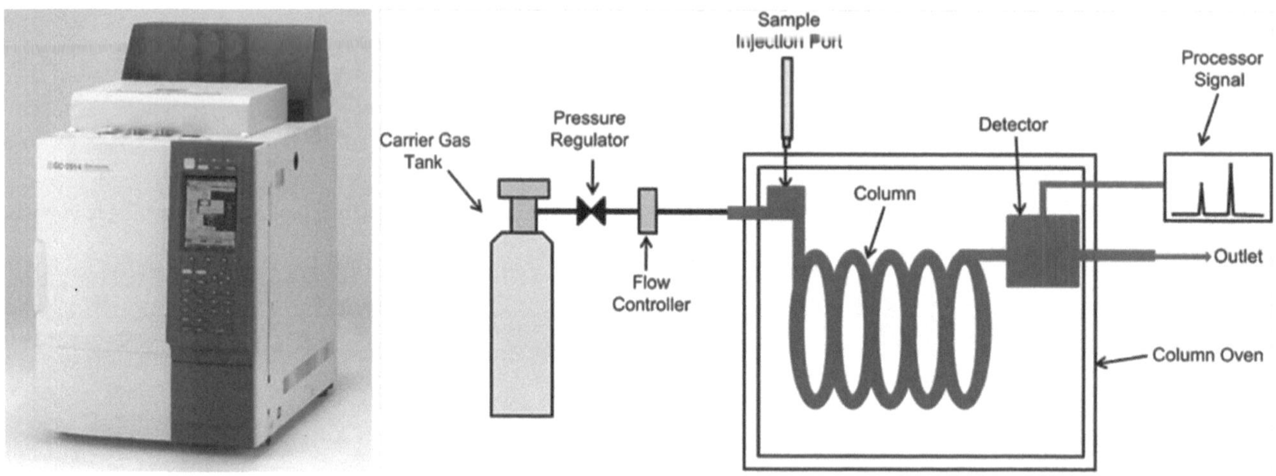

FIGURE 6.13 Picture of the GC-FID.

pinhole system allowing 4 π detection. The radiodetector loop is generally made of 1/16" HPLC tubing (inert plastic or stainless steel). Its size will impact the maximum limit of detection while a too-large cell will create peak widening. By consequence, it is recommended to select the cell volume in function of the HPLC peak width.

6.4.6 Gas Chromatography

Because the radiopharmaceutical process is performed using several organic solvents, it is required to test the final product on residual solvents. Residual solvents are defined as trace amounts of volatile organic compounds present in the final product. Analysis of residual solvents is performed using gas chromatography (GC) combined with a detector suitable for detection of trace amounts of organic solvents [8] (Figure 6.13). GC uses a gas as the mobile phase, in contrast to using a liquid, as described for TLC and HPLC. GC is particularly suitable for the analysis of volatile components. The most commonly used detector is a flame ionization detector (FID), sometimes in combination with a mass spectrometer (MS). Special attention must be given to the sample extraction procedure. The most common method to extract organic solvent from the final product solution is using a headspace GC. Volatile material or solvents are extracted from the matrix and moved into the gas phase or headspace and injected into the GC. Another possibility is the direct-injection approach, which requires a special column suitable for water-based injections.

6.4.7 Sterility

To check for sterility, 100 µL end product is used and added to two bottles filled with TSB and FTM medium. The two bottles are incubated at temperature of 22.5°C and 32.5°C and used for the sterility testing of cultivation of anaerobes,

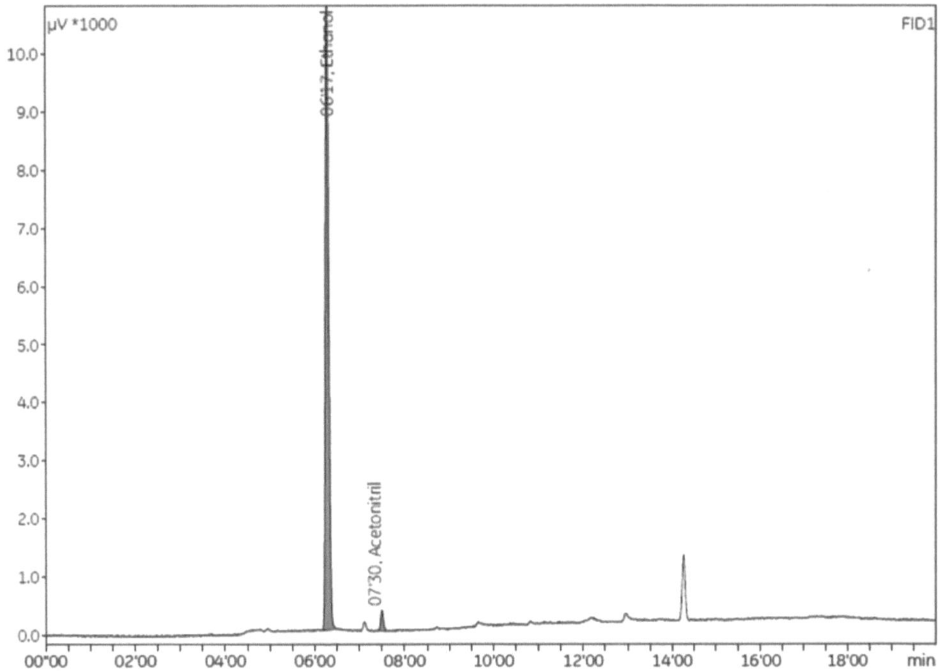

Sample description

Study:	FDG		
Measurement:	FDG-21-53., injection :		24/03/2021 10:21
Method:	FDG	Date:	26/10/2020 10:48
Instrument:	GC		
Injection volume:	0.50 µL (Method volume)		
Batch:	FDG-21-52		
Sample Type:	QC Sample		
Software Version: 10.4, Build 6482			

Integration FID1

Substance	R/T s	Type	Area µV*s	%Area %	Conc mg/l
Ethanol	06'17	BB(M)	54021.31	97.39	441.57
Acetonitril	07'30	BB(M)	1448.00	2.61	11.44
Sum in ROI	-	-	55469.31	100.00	-

FIGURE 6.14 A typical example of a GC chromatogram reflecting the analysis on residual solvents in FDG end product. Using this method, the retention time of ethanol is 6.17 min and 7.30 for acetonitrile.

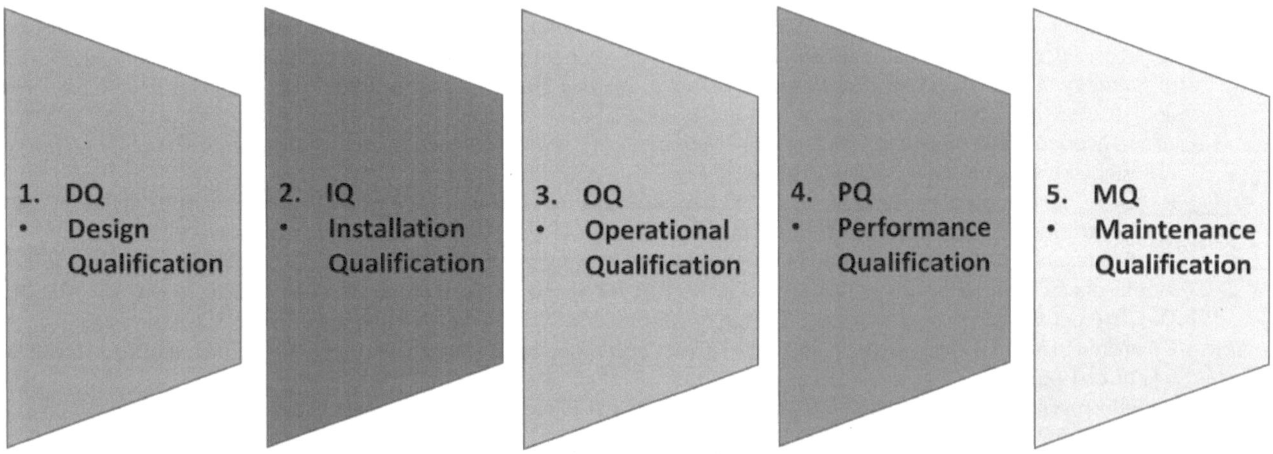

FIGURE 6.15 Scheme of the validation process, including IQOQ and PQ.

aerobes and microaerophiles. When the medium changes from a clear to a cloudy solution, this suggests that microorganism were present in the end product.

Next to the commonly used QC-methods as described above, some other methods are being used for specific classes of radiopharmaceuticals which are mainly based on proteins. Such methods include determination of radiochemical purity by trichloroacetic acid (TCA) precipitation and immunoreactivity in case of conjugation of chelators to monoclonal antibodies [9]. The advantage of the TCA method to determine the %RCP of labelled peptides and antibodies is the easy separation between labelled free compounds in the supernatant and the radiolabelled peptide in the residue.

6.5 VALIDATION OF THE PROCESS

Before the QC equipment can be used to perform QC analysis to release radiopharmaceuticals, all equipment must be validated. This means that the whole process is in a controlled state, and the qualification of the process is continuously verified. To do this, validation protocols on different levels are written and performed. The validation protocol consists of three topics: installation qualification, operation qualification, and performance qualification – also named IQOQ and PQ. If the process is validated and approved, then the equipment can be used for QC analysis of radiopharmaceuticals (Figure 6.15).

See, for more background information about validation processes of analytical methods, the "EANM guideline on the validation of analytical methods for radiopharmaceuticals" [10].

6.6 CONCLUSION

In this chapter, an overview and description of the analysis equipment required for release of radiopharmaceuticals for human use was described. Furthermore, rationale for the required quality controls and analysis was given. Finally, we provided an overview of the validation processes.

ACKNOWLEDGEMENT

The authors would like to thank David Thonon and Paul Bulters from Elysia Raytest for their contribution.

REFERENCES

[1] S. Basu, S. Hess, P. E. Nielsen Braad, B. B. Olsen, S. Inglev, and P. F. Høilund-Carlsen, "The Basic Principles of FDG-PET/CT Imaging," *PET Clin,* vol. 9, no. 4, pp. 355–70, v, 2014, doi: 10.1016/j.cpet.2014.07.006.

[2] Y.-S. Lee, "Radiopharmaceutical Chemistry," in *Handbook of Nuclear Medicine and Molecular Imaging*: World Scientific Publishing Co., pp. 21–51.

[3] S. Yu, "Review of F-FDG Synthesis and Quality Control," *Biomed Imaging Interv J,* vol. 2, no. 4, pp. e57–e57, 2006, doi: 10.2349/biij.2.4.e57.

[4] "General Notices, and Ph Eur. 2021. "Fludeoxyglucose (18-F) Injection." 1–8." www.pharmacopoeia.com/bp-2019/radiopharmaceutical/fludeoxyglucose--18f--injection.html

[5] K. Hamacher, G. Blessing, and B. Nebeling, "Computer-aided synthesis (CAS) of no-carrier-added 2-[18F]Fluoro-2-deoxy-d-glucose: an efficient automated system for the aminopolyether-supported nucleophilic fluorination," *International Journal of Radiation Applications and Instrumentation. Part A. Applied Radiation and Isotopes,* vol. 41, no. 1, pp. 49–55, 1990, doi: 10.1016/0883-2889(90)90129-5.

[6] "The Theory of Osmolality and Freezing Point Measurement." www.news-medical.net/whitepaper/20171026/The-Theory-of-Osmolality-and-Freezing-Point-Measurement.aspx

[7] R. Nakao, K. Furutuka, M. Yamaguchi, and K. Suzuki, "Quality control of PET radiopharmaceuticals using HPLC with electrochemical detection," *Nucl Med Biol,* vol. 33, no. 3, pp. 441–7, 2006, doi: 10.1016/j.nucmedbio.2005.12.008.

[8] T. Nowak *et al.*, "GC-FID method for high-throughput analysis of residual solvents in pharmaceutical drugs and intermediates," *Green Chemistry,* 10.1039/C6GC01210H vol. 18, no. 13, pp. 3732–3739, 2016, doi: 10.1039/C6GC01210H.

[9] D. Giesen *et al.*, "Probody Therapeutic Design of (89)Zr-CX-072 Promotes Accumulation in PD-L1-Expressing Tumors Compared to Normal Murine Lymphoid Tissue," *Clin Cancer Res,* vol. 26, no. 15, pp. 3999–4009, 2020, doi: 10.1158/1078-0432.Ccr-19-3137.

[10] N. Gillings *et al.*, "EANM guideline on the validation of analytical methods for radiopharmaceuticals," *EJNMMI Radiopharm Chem,* vol. 5, no. 1, p. 7, 2020, doi: 10.1186/s41181-019-0086-z.

7 Environmental Compliance and Control for Radiopharmaceutical Production

Commercial Manufacturing and Extemporaneous Preparation

Ching-Hung Chiu, Ya-Yao Huang, Wen-Yi Chang and Jacek Koziorowski

CONTENTS

7.1 INTRODUCTION

Radiopharmaceuticals have become a powerful tool of increasing importance, not only for diagnosis, but also for therapy. Particularly, the rise of theragnostics (also commonly known as theranostics) since the early 2000s, has pushed radiopharmaceuticals to evolve towards precision medicine [1]. However, the radioactive and time-related decay properties of radiopharmaceuticals attached to various radionuclide with specific half-life has significantly increased the operational differences and practical difficulties compared to conventional drugs, even if this has not been well recognized until now within the global regulatory framework [2].

Radiopharmaceuticals for positron-emission tomography (PET) have attracted growing interest since 1976. They are characterized by a short-half-life, sample lots consisting of only few units, and thus magistral (extemporaneous or in-hospital) preparation is generally adopted, especially for the clinical use with newly developed PET radiopharmaceuticals [3]. In the view of legislational development for PET radiopharmaceuticals, the governance of radiopharmaceuticals in most countries is still based on the regulatory framework for conventional drugs [4] and this has a significant negative impact on patient benefit and radiopharmaceutical development, although clinical administration of PET radiopharmaceuticals at micro-dose level has been considered as low risk.

DOI: 10.1201/9780429489501-7

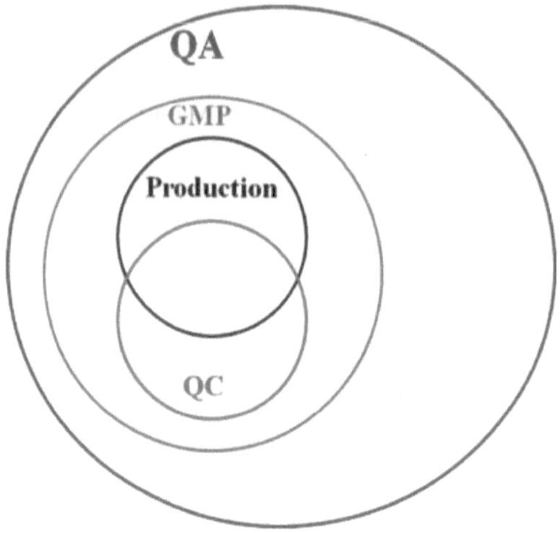

FIGURE 7.1 The interrelationship for whole quality system in drug manufacturing.

Generally, good manufacturing practice (GMP) regulations are applied to the large-scale and centralized manufacturing of medicinal products. The concept of Quality by Design (QbD), based on guidelines of the International Conference on Harmonization (ICH) (ICH Q8 [5], ICH Q9 [6], and ICH Q10 [7]) has been the fundamental topic for large-scale and centralized manufacturing of medical products (Figure 7.1) [8].

For PIC/S guidelines, Annex 3 (Manufacture of Radiopharmaceuticals) and Annex 13 (Investigational Medicinal Products) of the PIC/S GMP guide (PE 009-13) [9] have, respectively, been the requirement for radiopharmaceuticals with marketing authorization (MA) [10] and been used as investigational medicinal products (IMPs) [11], although there are several important specific exceptions for radiopharmaceuticals used in clinical trials [2, 12]. In order to set up an appropriate quality system and provide practice-oriented guidance for hospitals supplying medicinal products to their own patients in line with national legislation, a corresponding PIC/S GMP guide for healthcare establishments (PE 010-4) [13] has been added, effective March 2014. In this important annex, several issues for good practices for in-hospital preparation of radiopharmaceuticals are fully elaborated.

Furthermore, PET radiopharmaceuticals are commonly administered parenterally as a sterile injection. The aseptic production process in high-maintenance environmentally controlled cleanrooms usually results in high drug costs in commercial large-scale manufacture. The in-hospital, small-scale preparation of PET radiopharmaceuticals provides a potential solution to delivering such special medicines at affordable cost. However, in recent years, the GMP-compliant quality system has been requested by the competent authorities (medicines agencies) to be implemented in hospital radiopharmacies, and the appropriateness of this GMP standard implementation [14, 15] is still being debated. This chapter aims to provide an overview of environmental control in a cleanroom and further comparison between different PIC/S GMP guidelines for use in commercial manufacture or in a hospital radiopharmacy.

7.2 OVERVIEW OF ENVIRONMENTAL CONTROL IN ASEPTIC RADIOPHARMACEUTICAL MANUFACTURING AND IN-HOSPITAL PREPARATION

Generally, radiopharmaceuticals used in a clinical situation need to be sterile, and the production process involves two parts: aseptic manufacture or on-site preparation, and terminal sterilization. In other words, radiopharmaceutical manufacturing and in-hospital radiopharmaceutical preparation are carried out either in laminar flow cabinets (LFC), lead-shielded workstations (hot cells), or in closed radiological protection workstations in a well-controlled cleanroom, and the aseptic operational process conducted at some or all stages, and terminal sterile filtration of final products [13]. In Europe, all aseptic manufacturing of sterile radiopharmaceuticals has to comply with the requirements described in the current [16] version (noted as EU GMP 2020) [17] of Annex 1 of the Eudralex Volume 4 guide. However, the basic requirements presented in PIC/S Guide PE 010-4 (noted as PE 010-4) [13] cover on-site preparation of sterile radiopharmaceuticals for direct supply to patients in hospital radiopharmacies of the PIC/S member country.

Compared to the commonly used GMP guides in pharmaceutical manufacturing, such as the current version of Annex 1 of EU GMP guide [16] and Annex 3 (Manufacture of Radiopharmaceuticals) of PIC/S GMP guide (PE 009-13) [18], some new concepts and technologies were first introduced in EU GMP 2020 [17]. First, cleanroom qualification is an important process for assessing the level of compliance of a classified cleanroom or clean air equipment with their intended use, and is well defined in EU GMP 2020 [17] including

a) installed filter leakage testing
b) installed filter integrity testing
c) airflow volume measurement
d) airflow velocity measurement
e) air pressure difference measurement
f) airflow direction and visualization
g) microbial airborne contamination
h) microbial surface contamination
i) temperature and relative humidity measurement
j) recovery testing
k) containment leak testing, and
l) particulate concentration measurement at rest and in operation states.

Moreover, the maintenance of cleanrooms and clean air equipment plays a key role in ascetic operation. Therefore, the periodic requalification of cleanrooms and clean air equipment is further included with minimum test requirements [17] and the suggested maximum time interval for requalification is 6 and 12 months for Grades A & B areas and Grades C & D areas, respectively, in normal situations. However, when out-of-compliance equipment or facilities need to be rectified, or after changes to equipment, facility, or processes, appropriate requalification is necessary [17]. Secondly, the concept of a contamination control strategy (CCS) was first introduced in EU GMP 2020 [17], where the CCS should define all critical control points, implementation across the facility, and assessment of the effectiveness of all the controls and monitoring measures during risk management associated with contamination. Finally, several new technologies, such as restricted access barrier systems (RABS), robotic systems, rapid microbial testing, and monitoring systems, are considered in the latest version of Annex 1 of EU GMP guide [17] in order to increase the protection of the product from potential contamination and to assist in the rapid detection of potential contaminants in the environment and product. In particular, RABS has been known to be beneficial in minimizing microbial contamination during aseptic operation processes in the critical zone with a positive airflow from the critical zone to the supporting background environment.

For sterile radiopharmaceutical production, four-grade clean areas (A, B, C and D) of a cleanroom with different required environmental characteristics are included in a well-controlled cleanroom, and each clean area with different cleanliness grade must first be well defined according to specific operation activities [13, 17, 19] (Table 7.1) to minimize the contamination risk within this area. The unusual microbiological risks for manufacturing and in-hospital preparation of sterile radiopharmaceuticals have been cautiously taken into consideration, because they are both generally processed using automated synthesizers instead of some simple closed vessels. However, certain critical operations still have to been performed in higher-grade clean areas [13, 19]. For example, for general sterile products, all aseptic activities at critical stages of sterile radiopharmaceutical production must be carried out in a Grade A area – activities such as handling and filling the final products in an LFC or a positive pressure pharmaceutical isolator [13] – with an appropriate background area. The recommended minimal grades of working environment and background areas listed in EU GMP 2020 and PE 010-4 guides are shown in Table 7.2 [13, 17].

7.3 COMPARISON FOR EU GMP AND PIC/S GMP GUIDES FOR MANUFACTURING AND IN-HOSPITAL PREPARATION OF PET RADIOPHARMACEUTICALS

In order to ensure that the process, operators, and facility are operating under appropriate control, two monitoring states are involved in the EU GMP 2020 and PE 010-4 guides, and are respectively classified as 'in operation' and 'at rest' (Table 7.3) [13, 17]. In general, the 'in operation' and 'at rest' states should be defined for each cleanroom or suite of cleanrooms. Also, in order to perform particle monitoring (see next section) in the 'at rest' state, a short, unmanned period of 15–20 minutes (defined as a 'clean up' period) is suggested after full completion of the operation. Moreover, cleanroom qualification should be included for both at rest and in operation states.

7.3.1 Particle Monitoring in a Cleanroom

Cleanrooms and clean air devices should be classified in accordance with EN ISO 14644-1. Classification should be clearly differentiated from operational process environmental monitoring. However, there is no difference between EU

TABLE 7.1
Comparative Definitions and Operational Descriptions Listed in EU GMP 2020 and PE 010-4 Guides

	Definitions and Operational Descriptions	
	EU GMP 2020	**PE 010-4**
Grade A	1. Critical zone for high risk operations or for making aseptic connections by ensuring protection by first air (e.g. aseptic processing line, filling zone, stopper bowl, open ampoules and vials) 2. Critical zone of the RABS or open isolator used for aseptic processes with unidirectional airflow. 3. Critical zone of the closed isolator systems where airflow may not be unidirectional and be demonstrated to provide adequate protection for exposed products during processing. 4. Minimal operational area for filling of products for terminal sterilization at a high or unusual risk of microbial. 5. Critical zone for aseptic assembly of filling equipment, the final sterilizing filter during aseptic operation. 6. Critical zone for aseptic filling, compounding, mixing, refilling, dispensing and sealing.	1. Operational area for filling of terminal-sterilized products at unusual risk. 2. Operational area for handling of sterile starting materials and components used in aseptic process unless those will be sterilization later in process. 3. Operational area for aseptic preparation of unfiltered solutions. 4. Operational area for handling and filling of aseptically prepared products such as an LFC or a positive pressure pharmaceutical isolator. 5. Operational area for any open method of preparation and handling of aseptic radiopharmaceuticals products.
Grade B	1. Background area for aseptic preparation and filling. 2. Background area for RABS used for aseptic processing. 3. Background area for aseptic assembly of filling equipment, the final sterilizing filter during aseptic operation. 4. Background area for aseptic filling, compounding, mixing, dispensing, and sealing. 5. Background area for transport of equipment, components, and ancillary items for introduction into the Grade A zone.	1. Minimal background area for LFCs. 2. Minimal background zone for handling of open-aseptic-prepared radiopharmaceuticals in LFC.
Grade C	1. Areas used for carrying out less critical stages in the manufacture of aseptically filled sterile products but can be used for the preparation/filling of terminally sterilized products. 2. Background area for open isolators based on a risk assessment. 3. Minimal operational area for preparation of components and materials for terminally sterilized products at a high or unusual risk of microbial. 4. Minimal operational area for filling of products for terminal sterilization. 5. Minimal background area for filling of products for terminal sterilization at a high or unusual risk of microbial. 6. Operational area of preparation of solutions to be filtered, including weighing for aseptic-prepared radiopharmaceuticals.	1. Minimal operational area for preparation of terminal-sterilized products with high or unusual risk of microbial contamination. 2. Minimal background zone of the fill process in Grade A area for those terminal-sterilized products at unusual contamination risk from the environment. 3. Operational area for preparation of solutions, when unusually at risk for terminal-sterilized products. 4. Minimal operational area for filling of terminal-sterilized products. 5. Operational area for aseptic preparation of solutions that are to be sterile filtered during the process. 6. Potential operational area for the filling of immediate-use radiopharmaceutical products prepared via closed method if an adequate risk assessment is carried out. 7. Operational area for the filling of terminal-sterilized radiopharmaceutical products. 8. Operational area of radiochemical synthesis of radiopharmaceuticals.

TABLE 7.1 (Continued)
Comparative Definitions and Operational Descriptions Listed in EU GMP 2020 and PE 010-4 Guides

	Definitions and Operational Descriptions	
	EU GMP 2020	**PE 010-4**
	7. Minimal operational area for use of closed and automated systems (chemical synthesis, purification, online sterile filtration) inside the hot cell that should meet a high degree of air cleanliness with filtered feed air, when closed.*	9. Minimal background zone for handling of open-aseptic-prepared and immediate-use radiopharmaceuticals in LFC after an adequate risk assessment. 10. Minimal background zone for handling of close-aseptic-prepared radiopharmaceuticals in isolators or hot cells.
Grade D	1. Operational areas for carrying out less critical stages in the manufacture of aseptically filled sterile products but can be used for the preparation/filling of terminally sterilized products. 2. Background area for open isolators based on a risk assessment. 3. Minimal background area for closed isolators based on a risk assessment. 4. Minimal operational area for preparation of components and materials for terminally sterilized products. 5. Operational area for cleaning of equipment for aseptic preparation. 6. Operational area for handling and assembly of cleaned components, equipment and accessories prior to sterilization.	1. Minimal operational area for filling of terminal-sterilized products. 2. Operational areas for preparation of solutions and components for subsequent filling of terminal-sterilized products. 3. Minimal operational area for handling all components for aseptic preparation process, after washing. 4. Minimal background environment for pharmaceutical isolators. 5. Operational area of radiochemical synthesis of radiopharmaceuticals under an adequate risk assessment. 6. Minimal background zone for handling of aseptic-prepared radiopharmaceuticals in isolators. 7. Minimal background zone for handling of terminal-sterilized radiopharmaceuticals in isolators, LFC, or hot cells.

* Described in the current version of Annex 3 of EU GMP guide [19].

TABLE 7.2
An Overview of the Recommended Minimal Grades Listed in EU GMP 2020 and PE 010-4 Guides

	Working Environment	Background environment
EU GMP 2020		
RABs	A*	B
Open Isolators	A*	C or D
Closed Isolator	A	D
PE 010-4		
LFC	A	B
Isolator	A	D

* Need to be with unidirectional airflow and positive airflow from the critical zones to the supporting background environment.

TABLE 7.3
Comparative Definition of Two Monitoring States in EU GMP 2020 and PE 010-4 Guides

	Definitions and Operational Descriptions	
	EU GMP 2020*	**PE010-4**
In operation	The condition where the installation of the cleanroom is complete, the HVAC system fully operational, equipment installed and functioning in the manufacturer's defined operating mode with the maximum number of personnel present performing or simulating routine operational work.	The condition where the installation is functioning in the defined operating mode with the specified number of personnel working.
At rest	The condition whereby the installation of all the utilities is complete, including any functioning HVAC, with the main manufacturing equipment installed as specified and standing by for operation without personnel in the room.	The condition where the installation is installed and operating, complete with production equipment but with no operating personnel present.

* Heating, ventilation, and air conditioning (HVAC) system.

TABLE 7.4
Limits and Recommended Frequencies for Particle Monitoring in Cleanroom

	At Rest		In Operation	
	0.5 μm	**5.0 μm**	**0.5 μm**	**5.0 μm**
A	3,520	20	3,520	20
B	3,520	29	352,000	2,900
C	352,000	2,9000	3,520,000	29,000
D	3,520,000	29,000	N/D	N/D
Monitoring Frequency				
Direct working area (Grade A)	Full duration of critical processing	N/D	N/D	N/D
Other areas	justified by quality risk assessment	N/D	N/D	N/D

N/D: not defined

GMP 2020 and PE 010-4 guides [13, 17], and the maximum permitted airborne particle concentration for each grade is given in Table 7.4.

For Grade A zones, particle monitoring should be undertaken for the full duration of critical processing, including equipment assembly, except where justified by contaminants in the process – contaminants that would damage the particle counter or present a hazard, for example live organisms and radiological hazards. In such cases, monitoring during routine equipment set-up operations should be undertaken prior to exposure to the risk. Monitoring during simulated operations should also be performed. The Grade A zone should be monitored at such a frequency, and with suitable sample size, that all interventions, transient events, and any system deterioration would be captured, and alarms would be triggered if alert limits were exceeded [17].

7.3.2 Environmental Monitoring in a Cleanroom: Ventilation and Pressure

Cleanrooms and clean air devices should be routinely monitored in operation, and the monitoring locations based on a formal risk analysis study and the results obtained during the classification of cleanrooms and/or clean air devices. For both commercial manufacturing and in-hospital preparation of sterile radiopharmaceuticals, recommended limits of ventilation and pressure in all clean areas in EU GMP 2020 and PE 010-4 guides are shown in Table 7.5 [13, 17].

TABLE 7.5
Limits for Monitoring of Ventilation and Pressure in a Cleanroom Listed in EU GMP 2020 and PE 010-4 Guides

Grade	Air changes (number per hour)		Air-flow velocity (m/s ± 20%)		Pressure differential to adjacent low-classroom (Pa)	
	EU GMP 2020	PE 010-4	EU GMP 2020	PE 010-4	EU GMP 2020	PE010-4
A	No specific recommendation	N/D	0.54 HLF 0.36 VLF	0.45 HLF 0.30 VLF	N/D	N/D LFC >15 Isolator
B	No specific recommendation	>20	N/D	N/D	≥10	>10
C	No specific recommendation	>20	N/D	N/D	≥10	>10
D	No specific recommendation	>10	N/D	N/D	≥10	>10

Monitoring Frequency for LFC

	Classification test			Physical monitoring	
	EU GMP 2020	PE 010-4		EU GMP 2020	PE 010-4
Particle counts	6 months	Yearly	Pressure differentials between rooms	At beginning and end of each batch	Before beginning of work, usually daily
Room air changes per hour	6 months	Yearly	Pressure differentials across HEPA filters	At beginning and end of each batch	Before beginning of work, usually daily
Air velocities on workstations	6 months	Yearly	Particle counts	At beginning and end of each batch	Quarterly in the operational state
HEPA filter integrity checks	6 months	Yearly			

Monitoring Frequency for Isolator

	Classification test			Physical monitoring	
	EU GMP 2020	PE 010-4		EU GMP 2020	PE 010-4
Isolator alarm functional tests	6 months	Yearly	Pressure differentials across HEPA filters	At beginning and end of each batch	Before beginning of work, usually daily
Isolator leak test	6 months	Yearly	Isolator glove integrity	At beginning and end of each batch	Visual checks every session
HEPA filter integrity checks	6 months	Yearly	Isolator pressure hold test (with glove attached)	At beginning and end of each batch	weekly

N/D: not defined; HLF = horizontal laminar flow; VLF = vertical laminar flow

TABLE 7.6
Limits and Recommended Frequencies for Microbiological Monitoring Listed in EU GMP 2020 and PE 010-4 Guides

Grade	Air sample (cfu/m3)	Settle plates (diam. 90mm) (cfu/4 hrs)	Contact plates (diam. 55 mm) (cfu/plate)	Glove print (5 fingers) (cfu/glove)
A*	<1	<1	<1	<1
B	10	5	5	5
C	100	50	25	-
D	200	100	50	-
Recommended frequency in EU GMP 2020				
Direct working rea (Grade A)	Full duration of critical processing		At the end of each working session and on each exit from the Grade B area	
Background area	Full duration of critical processing			
Recommended frequency in PE 010-4				
Direct working area (Grade A)	Quarterly	Every working session	Weekly	At the end of each working session
Background area	Quarterly	Weekly	Monthly	At the end of each working session

* Any growth should result in an investigation for Grade A area.

Monitoring frequencies for classification test and physical monitoring of LFC and isolators are shown in Table 7.6 [13, 17].

7.3.3 Environmental Monitoring: Microbiology

Microbiological monitoring plays a vital role in confirming that the product is unlikely to be contaminated because radiopharmaceuticals are always used before end-product testing. Consequently, frequent microbiological monitoring of whole environment cleanness and final product and prompt reporting of results to the responsible person should help to reduce the possibility of patient pyrexia or septicaemia caused by microbiological contamination. Where aseptic operations are performed, various microbiological monitoring methods could be frequently performed, such as settle plates and volumetric air sampling for environmental microbiological monitoring, and surface sampling (e.g. contact plates and glove prints) for personnel microbiological monitoring. A combination of the above microbiological monitoring methods is suggested in EU GMP 2020 guide [17].

For both commercial manufacturing and in-hospital preparation of sterile radiopharmaceuticals, recommended limits (average values) and frequencies for microbiological monitoring of clean areas during operation are shown in Table 7.4 [13, 17]. It was noted that each settle plate is proposed to be exposed for less than 4 hours and a minimum sample volume of 1 m^3 is necessary per sample location in the direct working area (Grade A area). Moreover, the frequency of environmental microbiological monitoring could be justified based on the individual unit and the activities undertaken [13]. However, surfaces and personnel have to be monitored after critical operations. Besides general production operations, an additional monitoring for environmental microbiology is also required for other situations, such as after validation of systems, cleaning, and sanitization.

7.4 OVERVIEW FOR MICROBIOLOGY MONITORING FOR RADIOPHARMACEUTICAL PRODUCTS

Radiopharmaceutical microbiology is the application of microbiology to radiopharmaceutical and compounding environments. The scope of radiopharmaceutical microbiology is wide-ranging, but its overriding function is the safe manufacture of radiopharmaceutical and healthcare preparations and medical devices. This involves risk assessment (both proactive and reactive), together with testing materials and monitoring environments and utilities.

Some of the essential tests [20] are described below. which include:

- sterility test
- bacterial endotoxin test
- microbial limits test
- disinfectant efficacy
- pyrogen and abnormal toxicity tests
- environmental monitoring
- biological indicators.

It would be a mistake to think of radiopharmaceutical microbiology as confined to a range of laboratory tests. The concept of 'testing to compliance' is outdated. To address contamination risks, radiopharmaceutical microbiology places an emphasis upon contamination control. For this reason, quality assurance staff are included as microbiologists in several aspects of the production process, utility supply, and cleanroom environments. This generally involves (a) testing and the assessment of data; (b) conducting risk assessments, either proactively or in response to a problem; and (c) helping to design systems as part of a contamination control strategy. The twin areas of testing and control are intermixed throughout this book.

In this chapter, the major relevant points of radiopharmaceutical microbiology will be introduced relating to endotoxin/pyrogen testing, sterilization, and bioburden determination for radiopharmaceutical products and compounding environments.

7.4.1 Endotoxin and Pyrogen Testing

Bacterial endotoxin is the lipopolysaccharide (LPS) component of the cell wall of gram-negative bacteria. It is pyrogenic, and it is a risk to patients who are administered intravenous and intramuscular preparations [21]. The pathological effects of endotoxin, when injected, are a rapid increase in body temperature followed by extremely rapid and severe shock, often followed by death, before the cause is even diagnosed. The limulus amebocyte lysate (LAL) test is the most widely method used for endotoxin tests [22]. The pharmacopoeia monographs for the LAL test are long-established and relatively comprehensive and have been applied to the testing of parenteral products for bacterial endotoxin since the 1980s. Pyrogens are a concern for pharmaceutical drug products and for many of the ingredients used to formulate them. This is especially for radiopharmaceuticals that have direct contact with human blood. Here, by far the most concerning pyrogen is bacterial endotoxin. In relation to this, the risks of endotoxin to radiopharmaceutical processing and some of the control measures in place to reduce the risk of endotoxin contamination should be considered.

7.4.2 Sterilization

Pharmaceutical products are required to be sterile, which includes injections, infusions, and pharmaceutical forms for application on eyes and on mucous membranes. Because of the route of administration such; parenteral (*i.e.* intravenous injection) medicines are required to be sterile. Importantly, if such medicines are not sterile, then this could lead to patient harm or even death. Manufacturing sterile products requires more than just that the product solution be sterile [23].

Protecting the patient not only requires manufacturing a sterile product, but it also needs to include the use of sterile tools for administering the drug (sterile syringe or needle) and administering it under aseptic conditions.

Thus, sterile manufacturing is a continuum that stretches from development to manufacturing, to the finished product, to marketing and distribution, and to utilization of the drugs and biologicals in hospitals, as well as in patients' homes. There is no generic approach to the manufacturing of sterile products. Each plant or process will differ in relation to the technologies, products, and process steps. The common aim is that a product is produced that is sterile and where there is no risk of contamination until the contents of the outer packaging are breached (such as through the injection of a needle through a bung on a product vial).

Methods and conditions of sterilization include (European pharmacopoeia):

1. steam sterilization
2. dry heat sterilization
3. ionizing radiation sterilization
4. gas sterilization (vapor phase sterilization)
5. membrane filtration.

Because of the physical and chemical characteristics of radiopharmaceuticals, membrane filtration (aseptic filling) is a common method of radiopharmaceutical sterilization. Aseptic manufacturing is used in cases where the drug substance is unstable when subjected to heat (sterilization in the final container closure system is not possible). Aseptic filling is the most difficult type of sterile operations because the product cannot be terminally sterilized and, therefore, there are greater contamination risks during filling. With aseptic processing, there is always a degree of uncertainty, particularly because of the risk posed to the environment by personnel. Quality assurance staff should follow the SOP of aseptic assembly for evaluating the sterility of radiopharmaceutical production. Aseptic assembly includes the environment, personnel, critical surface, container/closure sterilization and transfer procedures and the maximum holding period of the product before filling the final container.

7.4.3 Bioburden Determination

Bioburden [24] is used to describe the microbial numbers on a surface or inside a device or from a portion of liquid. In some literature, bioburden testing relates to raw materials testing, environmental monitoring, or in-process sample testing. These areas will have a given, even 'natural' bioburden. When this bioburden rises above typical levels or ends up in the wrong place, it becomes biocontamination [25].

For sterile products, bioburden assessment is a key requisite prior to sterilization. This is necessary for radiopharmaceutical products that are to be aseptically filled. With aseptic processing, one of the most important samples is taken from the bulk material prior to transfer through a sterilizing grade filter in preparation for aseptic filling. The filters used are generally of a pore size of 0.22 µm. The 'sterilizing filter' was defined in 1987 by the US Food and Drug Administration (FDA). Within Europe, there is a requirement for the challenge liquid to contain no more than 10 CFU/100 mL. While bioburden assessment is important, aseptic processing carries continued risks. The sterile filtered liquid must subsequently be dispensed into sterile containers under a protective airflow. At this stage, contamination may occur if controls are not properly maintained

7.5 COMPROMISES BETWEEN GMP-COMPLIANCE AND RADIATION PROTECTION DURING RADIOPHARMACEUTICAL PREPARATION.

Radiopharmaceuticals are known for their dual properties. That is, they are regarded as medicinal products and as radioactive substances. Therefore, radiopharmaceuticals are simultaneously regulated by two sources of legislation, drug quality regulations and radiation safety regulations. Moreover, the design of the radiopharmacy facility – including the air pressure regime – needs to be approved by both drug and radiation safety authorities. However, two regulatory systems sometimes result in conflicting situations. The most notorious example is the air pressure regime in a radiopharmacy department, and the compromise between the GMP and the radiation protection regulations often results in a complicated regime [26].

First, the hot lab including the cyclotron room is generally required to be under pressure compared with the outer world, and such an air pressure regime favours both the microbiological quality of the drug product and the radiation safety of the workers. However, an earlier study [27] showed that individuals are the most common contamination source of cytotoxic agents, and pressure differences of manufacturing (or preparation) environment still cannot control this potential risk. On the other hand, a regular air pressure cascade with only overpressurized rooms is preferable for a GMP-compliant regime with overpressurized and underpressurized rooms. Generally, an acceptable and economically compromised solution between GMP- and radiation-based thinking is that the hot lab is operated at atmospheric pressure appropriately ventilated with HEPA filtered air, with overpressurized airlocks, but the cyclotron room is maintained underpressurized. However, when volatile hazardous radioactive substances are involved in radiopharmaceutical preparation (e.g. radioactive iodine-labelled radiopharmaceuticals), an above air pressure regime may not be applicable [26].

As one can see from Table 7.7, in a pharmacy, there is a pressure increase from low to high grade cleanliness, but in a radiopharmacy there is a need to have pressure sinks in all locations adjacent to the non-controlled area in order to avoid a radioactive release escaping from the radiopharmacy. In this example, the pass-through chambers have the air intake from the cleanest adjacent room and the exhaust connected to the X/D sluice exhaust. Also, the doors should open against the higher area pressure, which is why there are two drawings that are identical, with the exception of how the doors are opened. As a consequence, it is not easy to turn a radiopharmacy into a pharmacy or vice versa.

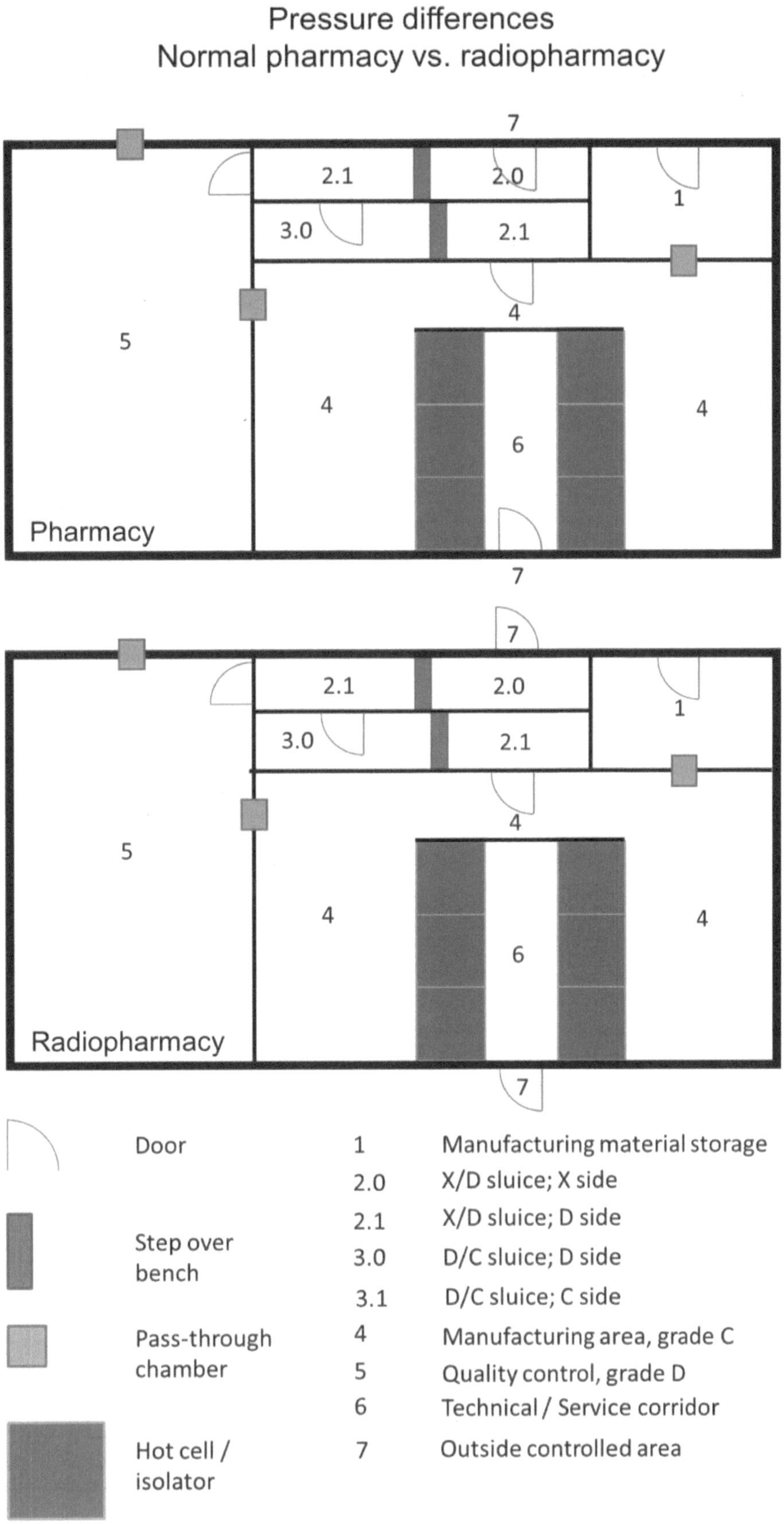

FIGURE 7.2 Pressure differences: normal pharmacy vs. radiopharmacy. Colour image available at www.routledge.com/9781138593312.

TABLE 7.7
Room Pressures in Pharmacy vs. Radiopharmacy

Location	#	Pharmacy (Pa)*	Radiopharmacy (Pa)*
Manufacturing material storage	1	0	0
X/D sluice	2.0/2.1	10	-10
D/C sluice	3.0/3.1	25	5
Manufacturing area, grade C	4	40	20
Hot cell / isolator		55	-30
Pass-through chamber°		10	-10
Quality control, grade D	5	25	5
Technical / service corridor	6	10	-10
Outside controlled area	7	0	0

* N.B. All pressures are relative to ambient pressure (0 Pa) and are only indicative.
° Pass-through chambers are fed air from the cleaner side.

Second, from the review of radiation protection, syringe shielding is the common approach for dispensing, distribution, and administration of radiopharmaceuticals. Generally, sterility assurance level (SAL) is required to be 10^{-6} and 10^{-3}, for drug products with terminal sterilization and aseptic manufacturing, respectively. However, such strict SAL requirements for intravenously administered radiopharmaceuticals that are usually administered within hours of preparation is debatable and thus results in troublesome decontamination of syringe shielding. Recently, a SAL of 10^{-2} for radiopharmaceuticals has been proposed [26] and has been adopted in Dutch hospital pharmacies for extemporaneous aseptic preparations of parenterals [28]. This proposed SAL value has proved to be consistent with the proposed contamination recovery rate in a class A (ISO 5) aseptic environment of the USP without huge numbers of media fills [26, 29].

Third, the radiation level can impede fast microbiological examination of the preparation when microbiological quality control tests are performed with adding some eluate to a concentrated growth-promoting medium. Hence, such tests are generally used as part of a trend analysis for microbiological quality monitoring of radiopharmaceuticals [26].

Fourth, in addition to aseptically performed operations, membrane filtration is a commonly used approach for radiopharmaceutical preparation using a bacterial retention filter to abate the microbiological risk derived from the above compromised solutions. Accordingly, an integrity test usually has to be performed immediately after the preparation. However, it adds an additional risk for radiation exposure, even if rinsing the filter before the integrity test is performed. Thus, an alternative filter-based sterilization method of radiopharmaceuticals has been suggested, with the use of two filters in series and waiving the integrity testing before release [26].

Finally, because endotoxin testing is limited for microbiological quality monitoring and is also a risk source for radiation safety, such a test has been suggested to be performed periodically except for the validation of new or changed processes, products, or systems [26].

For the small-scale preparation of radiopharmaceuticals that are not intended for commercial purposes, EANM has published two guidance documents with detailed consideration of both GMP guidelines and radiation protection [30, 31].

7.6 CONCLUSION AND OUTLOOK

For special characteristics of PET radiopharmaceuticals compared to conventional drugs, this chapter provides an overview of environmental control in a cleanroom and further compares the guidelines listed in respective PIC/S GMP guides for a commercial manufacturer and a hospital pharmacy [9, 13].

The definition of PET radiopharmaceuticals used in a clinic is prerequisite for selection of appropriate PIC/S GMP references. Based on EU guidance [32], PET radiopharmaceuticals are usually used to assess end points in the clinical trial and thus mostly belong to the category of non-investigational medicinal products (NIMP), unless they are being administrated to humans for the first time.

Fortunately, in addition to the United States [33], the authorities in Europe have gradually realized and recognized that radiopharmaceuticals are indeed a special group of drugs. Thus, for the patients' benefit, a specific framework for the safe and efficient preparation of radiopharmaceuticals is under construction in order to avoid disproportionate regulations that hamper the development of novel diagnostic tools for the benefit of the patient [2, 13, 34]. For example, in order to facilitate and harmonize clinical research in Europe, the preparation of radiopharmaceuticals from GMP requirements has been exempted from EU regulation (No 536/2014) [35] and Article 3 of Directive 2001/83 [36]. However, notwithstanding PIC/S GMP guidance documents, these annexes for MA or IMPs are not applicable to extemporaneous preparation of PET radiopharmaceuticals in hospital radiopharmacies with limited facilities and resources [2]. Actually, there are indeed no legally binding documents for this special case in Europe. On the contrary, the rules for extemporaneous preparation of PET radiopharmaceuticals are interpreted by the individual member states of the EU, and this has resulted in a huge variation and many different situations within Europe [37].

In order to set up an appropriate quality control system and provide practice-oriented guidance for those radiopharmaceuticals that are not intended for commercial purposes, in 2007, the European Association of Nuclear Medicine (EANM) issued Guidance on Current Good Radiopharmacy Practice for the small-scale preparation of radiopharmaceuticals [30, 31]. On the other hand, the new version of the PIC/S GMP guide to good practices for preparation of medicinal products in healthcare establishments (PE 010-4) [13] has been expanded; it became effective in March 2014. In this important annex, several issues for good practice for in-hospital preparation of radiopharmaceuticals are fully elaborated. More recently, a new General Chapter for the Ph. Eur. has been published: '5.19 Extemporaneous preparation of radiopharmaceutical preparations' [34]. It provides minimal requirements for kit-based preparations, PET radiopharmaceuticals, and radiolabelled blood cells for in-house preparation in radiopharmacies, and it aims to bridge the differences between European countries and standardize radiopharmacy practice [26]. In these three documents a clear distinction is made between industrial manufacturing and extemporaneous preparation of radiopharmaceuticals (on a small scale, locally), providing guidance on 'Good Practice', as distinct from industrial standards [3], and it further fills the gap in regulatory requirements between conventional drugs and PET radiopharmaceuticals. However, in order to promote sustainable development of personalized medicine and to move in-hospital production of radiopharmaceuticals forward in a reasonable and cost-effective manner, legislation and guidance with more flexibility – for the implementation of appropriate GMP-based quality assurance standards for the preparation of unlicensed radiopharmaceuticals – still need to be worked for in most countries in the world.

REFERENCES

[1] T. Langbein, W. Weber, and M. Eiber, "Future of Theranostics: An outlook on precision oncology in nuclear medicine," *J Nucl Med,* vol. 60, no. Suppl 2, pp. 13S–19S, 2019.

[2] C. Decristoforo *et al.*, "Radiopharmaceuticals are special, but is this recognized? The possible impact of the new Clinical Trials Regulation on the preparation of radiopharmaceuticals," *Eur J Nucl Med Mol Imaging,* vol. 41, no. 11, pp. 2005–7, 2014.

[3] C. Decristoforo and M. Patt, "Are we 'preparing' radiopharmaceuticals?" *EJNMMI Radiopharm Chem,* vol. 1, no. 1, p. 12, 2017.

[4] European Commission, 2015. *EudraLex – Volume 4 – Good Manufacturing Practice (GMP) Guidelines.*

[5] International Conference on Harmonisation. "ICH Harmonised Tripartite Guideline: Pharmaceutical Development Q8 (R2)." www.ich.org/fileadmin/Public_Web_Site/ICH_Products/Guidelines/Quality/Q8_R1/Step4/Q8_R2_Guideline.pdf.

[6] International Conference on Harmonisation. "ICH Harmonised Tripartite Guideline, Quality Risk Management, Q9." www.ich.org/fileadmin/Public_Web_Site/ICH_Products/Guidelines/Quality/Q9/Step4/Q9_Guideline.pdf.

[7] International Conference on Harmonisation. "ICH Harmonised Tripartite Guideline, Pharmaceutical Quality System, Q10." www.ich.org/fileadmin/Public_Web_Site/ICH_Products/Guidelines/Quality/Q10/Step4/Q10_Guideline.pdf.

[8] Y.-Y. Huang, "An overview of PET radiopharmaceuticals in clinical use: Regulatory, quality and pharmacopeia monographs of the United States and Europe," in *Nuclear Medicine Physics*: IntechOpen, 2018.

[9] Pharmaceutical Inspection Co-operation Scheme, 2017. *Document PE 009–13 (Annexes), Guide to Good Manufacturing Practice for Medicinal Products Annexes.*

[10] European Commission, 2008. *Eudralex Volume 4. EU guidelines to good manufacturing practice medicinal products for human and veterinary use –Annex 3. Manufacture of radiopharmaceuticals.*

[11] European Commission, 2009. *Eudralex Volume 4. EU guidelines to good manufacturing practice medicinal products for human and veterinary use –Annex 13. Investigational medicinal products.*

[12] I. Peñuelas, D. J. Vugts, C. Decristoforo, and P. H. Elsinga, "The new regulation on clinical trials in relation to radiopharmaceuticals: When and how will it be implemented?" *EJNMMI Radiopharm Chem,* vol. 4, p. 2, 2019.

[13] Pharmaceutical Inspection Co-operation Scheme, 2014. *Document PE 010–4, Annex 3: Good practices for the preparation of radiopharmaceuticals in healthcare establishments.*

[14] B. Långström and P. Hartvig, "GMP – three letters with many interpretations. Protection of patients or killing the clinical and research applications of PET?" *Eur J Nucl Med Mol Imaging,* vol. 35, no. 4, pp. 693–94, 2008.

[15] C. Decristoforo *et al.*, "The specific case of radiopharmaceuticals and GMP – activities of the Radiopharmacy Committee," *Eur J Nucl Med Mol Imaging,* vol. 35, no. 7, pp. 1400–1, 2008.

[16] EU EudraLex. *The Rules Governing Medicinal Products in the European Union, Volume 4 EU Guidelines for Good Manufacturing Practice for Medicinal Products for Human and Veterinary Use – Annex 1: Manufacture of Sterile Medicinal Products.* [Online] Available: https://ec.europa.eu/health/sites/health/files/files/eudralex/vol-4/2008_11_25_gmp-an1_en.pdf.

[17] *EU EudraLex. The Rules Governing Medicinal Products in the European Union, Volume 4 EU Guidelines for Good Manufacturing Practice for Medicinal Products for Human and Veterinary Use – Annex 1: Manufacture of Sterile Medicinal Products.* [Online] Available: https://ec.europa.eu/health/sites/health/files/files/gmp/2020_annex1ps_sterile_medicinal_products_en.pdf.

[18] Pharmaceutical Inspection Co-operation Scheme, 2018. *Document PE 009–14 (Annexes), Guide to Good Manufacturing Practice for Medicinal Products Annexes.*

[19] *EU EudraLex. The Rules Governing Medicinal Products in the European Union, Volume 4 EU Guidelines for Good Manufacturing Practice for Medicinal Products for Human and Veterinary Use – Annex 3. Manufacture of Radiopharmaceuticals.* [Online] Available: https://ec.europa.eu/health/sites/health/files/files/eudralex/vol-4/2008_09_annex3_en.pdf.

[20] T. Sandle, "1 – Introduction to pharmaceutical microbiology," in *Pharmaceutical Microbiology*, T. Sandle, Ed. Oxford: Woodhead Publishing, 2016, pp. 1–14.

[21] T. Sandle, "11 – Endotoxin and pyrogen testing," in *Pharmaceutical Microbiology*, T. Sandle, Ed. Oxford: Woodhead Publishing, 2016, pp. 131–145.

[22] D. Guy, "Endotoxins and depyrogenation," in *Hodges, N. and Hanlon, G., Industrial Pharmaceutical Microbiology: Standards and Controls.* Euromed, pp 12, pp. 1–12.15, 2003.

[23] T. Sandle, "12 – Sterilization and sterility assurance," in *Pharmaceutical Microbiology*, T. Sandle, Ed. Oxford: Woodhead Publishing, 2016, pp. 147–60.

[24] T. Sandle, "7 – Bioburden determination," in *Pharmaceutical Microbiology*, T. Sandle, Ed. Oxford: Woodhead Publishing, 2016, pp. 81–91.

[25] A. M. Cundell, "Managing the microbiological quality of pharmaceutical excipients," *PDA Journal of Pharmaceutical Science and Technology,* vol. 59, no. 6, pp. 381–95, 2005.

[26] R. Lange *et al.*, "Untangling the web of European regulations for the preparation of unlicensed radiopharmaceuticals: a concise overview and practical guidance for a risk-based approach," *Nucl Med Commun*, vol. 36, no. 5, pp. 414–422, 2015.

[27] S. Spaan, B. Blankvoort, and W. Fransman, "Risico's tijdens bereiden steriele cytostatica-oplossingen in apotheken (Risks during preparation of sterile solutions of cytostatics in pharmacies.)", Zeist, Netherlands, 2010.

[28] Dutch Association of Hospital Pharmacists (NVZA) and Royal Dutch Association for the Advancement of Pharmacy (KNMP), *Good Manufacturing Practice Hospital Pharmacy*. The Hague: NVZA, KNMP, 1998.

[29] United States Pharmacopeial Convention, "USP General Chapter, 1116. Microbiological control and monitoring of aseptic processing environments," in *Pharmacopeial Forum*, vol. 36, no. 6. Rockville, MD: United States Pharmacopeial Convention, 2013.

[30] EANM Radiopharmacy Committee, "Guidelines on current good radiopharmacy in the preparation of practice radiopharmaceuticals." 2007.

[31] P. Elsinga *et al.*, "Guidance on current good radiopharmacy practice (cGRPP) for the small-scale preparation of radiopharmaceuticals," *Eur J Nucl Med Mol Imaging,* vol. 37, no. 5, pp. 1049–62, 2010.

[32] European Commission, 2011. "The rules governing medicinal products in the European Union Volume 10 – Guidance documents applying to clinical trials; guidance on investigational medicinal products (IMPS) and non-investigational medicinal products" (NIMPS).

[33] U.S. Department of Health and Human Services Food and Drug Administration and Center for Drug Evaluation and Research (CDER), "PET Drugs – Current Good Manufacturing Practice (CGMP) (Small Entity Compliance Guide)," 2011.

[34] EDQM. 5.19, 2019. *Extemporaneous preparation of radiopharmaceutical preparations.* Pharmeuropa.

[35] European Parliament and Council of the European Union, "Regulation (EU) No 536/2014 of the European Parliament and of the Council of 16 April 2014 on clinical trials on medicinal products for human use, and repealing Directive 2001/20/EC," *Off J Eur Union*, vol. 1, 2014.

[36] European Parliament and Council of the European Union, "Directive 2001/83/EC of the European Parliament and the Council of 6 November 2001 on the Community code relating to medicinal products for human use," *Off J Eur Union,* vol. 31, pp. 67–128, 2001.
[37] A. Decristoforo and I. Peñuelas, "Towards a harmonized radiopharmaceutical regulatory framework in Europe?" *Q J Nucl Med Mol Imaging,* vol. 53, no. 4, pp. 394–401, 2009.

8 GMP

Rules and Recommendations

Oliver Neels

CONTENTS

8.1 INTRODUCTION

This chapter gives an overview on organizations, rules, and guidelines in good manufacturing practice (GMP) of medicinal products. The focus is on rules and guidelines for GMP of radiopharmaceuticals in the field of Nuclear Medicine in Europe; international guidelines are taken into consideration where applicable. In terms of the current legislation, radiopharmaceuticals can be divided into three different categories, depending on the status of the radiopharmaceutical:

- Radiopharmaceuticals with a marketing authorization
- Radiopharmaceuticals to be used in clinical trials
- Unlicensed radiopharmaceuticals extemporaneously (just before use) prepared, not for use in clinical trials

A brief overview on binding and non-binding documents is presented in this chapter, and more detailed information, in particular for country-specific situations, is available [1, 2]. Quality and safety of medicinal products are essential parts of GMP. Radiation safety regulations, which are often contradictive to GMP regulations, are not part of this chapter, but specific safety aspects for personnel or premises, for example, radiation protection, are explained.

8.2 ORGANIZATIONS

This section describes important organizations and stakeholders (in alphabetic order) that are actively involved in the legislation of (radio)pharmaceuticals and/or publication of corresponding GMP guidelines / guidance documents.

- **EANM** is the European Association of Nuclear Medicine. The aim is to advance science and education in nuclear medicine for the benefit of public health.
- **EC** is the European Commission, the executive branch of the European Union (EU), responsible for proposing legislation, implementing decisions, upholding EU treaties, and managing day-to-day business of the EU. The body of European Union legislation (EudraLex) in the pharmaceutical sector is compiled in Volume 1 (*Medicinal Products for human use*) and Volume 5 (*Medicinal Products for veterinary use*) of the publication, *The rules governing medicinal products in the European Union.* A series of EudraLex guidelines support the basic legislation, for example, *Volume 4 – Guidelines for good manufacturing practice for medicinal products for human and veterinary use.*
- **EDQM** is the European Directorate for the Quality of Medicines and Healthcare and is a directorate of the Council of Europe. The aim is to contribute to the basic human right of access to good quality medicines and healthcare and to promote and protect human and animal health. EDQM publishes the *European Pharmacopoeia*

DOI: 10.1201/9780429489501-8

(*Ph. Eur.*) and collaborates with national authorities as well as with the European Commission and European Medicines Agency.

- **EMA** is the European Medicines Agency, a decentralized agency of the European Union responsible for monitoring the scientific evaluation, supervision, and safety of medicines in the EU.
- **IAEA** is the International Atomic Energy Agency, the world's central intergovernmental forum for scientific and technical cooperation in the nuclear field. It works to assure the safe, secure, and peaceful use of nuclear science and technology, contributing to international peace and security and to the United Nations' Sustainable Development Goals.
- **ICH** is the International Council for Harmonisation of Technical Requirements for Pharmaceuticals for Human Use. The aim is to bring together regulatory authorities and the pharmaceutical industry to discuss scientific and technical aspects of drug registration.
- **PIC/S** is the Pharmaceutical Inspection Co-Operation Scheme, with a non-binding and informal co-operative arrangement between regulatory authorities in the field of GMP of medicinal products for human or veterinary use. The aim is to harmonize inspection procedures by using common standards in the field of GMP. The PIC/S GMP guide has been derived from the WHO GMP guide and is today almost identical with the EC GMP guide. The contact with the EC is mainly governed by a memorandum of understanding with the EMA.
- **WHO** is the World Health Organisation and is a specialized agency of the United Nations responsible for international public health. WHO publishes the *International Pharmacopoeia* (*Ph. Int.*) and jointly works with the IAEA on monographs for radiopharmaceuticals.

8.3 RULES AND GUIDELINES

As mentioned in the introduction, radiopharmaceuticals can, legally, be divided into three groups. Each group is individually affected by different documents, some that are legally binding and others that are non-binding.

- Radiopharmaceuticals with a marketing authorization
  - Binding documents
    - *Ph. Eur.: General Monograph 0125* "Radiopharmaceutical Preparations" and specific monographs [3]
    - Directive 2001/83/EC (community code on medicinal products for human use) [4]
    - Directive 2003/94/EC (GMP directive) [5]
    - Directive 2004/27/EC (amendment to directive 2001/83/EC) [6]
  - Non-binding documents
    - EudraLex Volume 4 EU *Guidelines to GMP*, Annex 3 [7]
    - *EMA Guideline on Radiopharmaceuticals* [8]
- Radiopharmaceuticals to be used in clinical trials
  - Binding documents
    - *Ph. Eur.: General Monograph 0125* "Radiopharmaceutical Preparations" and specific monographs [3]
    - Directive 2001/20/EC (Clinical trial directive) [9]
    - Directive 2003/94/EC (GMP directive) [5]
    - Directive 2005/28/EC (GCP directive) [10]
    - Regulation 536/2014 [11]
  - Non-binding documents
    - EudraLex Volume 4 EU *Guidelines to GMP*, Annex 13 [12]
    - EC Guidance on Investigational Medicinal Products (IMPs) and 'Non Investigational Medicinal Products'(NIMPs) [13]
    - *EMA Guideline on the requirements for the chemical and pharmaceutical quality documentation concerning investigational medicinal products in clinical trials* [14]
    - *EMA Guideline on strategies to identify and mitigate risks for first-in -human and early clinical trials with investigational medicinal products* [15]
- Radiopharmaceuticals prepared extemporaneously
  - Binding documents
    - *Ph. Eur.: General Monograph 0125* "Radiopharmaceutical Preparations" and Specific Monographs [3]
    - No further binding documents; National competence

- Non-binding documents
  - Ph. Eur. General Chapter 5.19 "Extemporaneous preparation of radiopharmaceutical preparations" [16]
  - PIC/S Guide to Good Manufacturing Practice for Medicinal Products in Healthcare Establishments, Annex 3 [17]
  - EANM guidelines on current good Radiopharmacy practice (cGRPP) [18]
  - National guidance documents

It is noticeable that for radiopharmaceuticals that hold a marketing authorization or are used in clinical trials, the applicable rules concerning GMP are given in directives and regulations dealing with large-scale, centralized production of medicinal products. Regulation 536/2014, when coming into force, replaces the clinical trial directive 2001/20/EC and will soften the GMP requirements for diagnostic radiopharmaceuticals in clinical trials [19, 20]. Nevertheless, radiopharmaceuticals are often small-scale preparations for extemporaneously prepared radiopharmaceuticals, no binding documents being applicable. For these radiopharmaceuticals, the competence is with national authorities, which results in a large diversity within Europe. Overall, GMP, when applicable, is defined and described more or less harmonized in several documents, which is reflected in

(a) The definition of GMP according to a number of organizations (in alphabetical order):

Good Manufacturing Practice means the part of quality assurance which ensures that products are consistently produced and controlled in accordance with the quality standards appropriate to their intended use. (EC)

Good Manufacturing Practice is that part of quality assurance which ensures that products are consistently produced and controlled to the quality standards appropriate to their intended use and as required by the marketing authorisation or product specification. (PIC/S)

Good Manufacturing Practices (GMP, also referred to as "cGMP" or "current Good Manufacturing Practice") is the aspect of quality assurance that ensures that medicinal products are consistently produced and controlled to the quality standards appropriate to their intended use and as required by the product specification. (WHO)

and

(b) Categorization into different chapters in guidance documents, the basic requirements:
1. Pharmaceutical Quality System (PQS) / Quality Assurance System
2. Personnel
3. Premises and Equipment
4. Documentation
5. Production
6. Quality Control
7. Outsourced Activities /Work Contracted Out
8. Complaints and Product Recall
9. Self-Inspection
10. Labelling

The specific characteristics of radiopharmaceuticals and its coherent requirements are often not taken into consideration in general documents. Specific GMP guidelines for radiopharmaceuticals from the below organizations (in alphabetical order) recognize the special properties of radiopharmaceuticals and give valuable additional information where necessary:

EANM: Guidelines [21] and guidance [22] on current good Radiopharmacy practice (cGRPP) are published regularly by the EANM Radiopharmacy Committee. More recently a guideline on the validation of analytical methods for radiopharmaceuticals was published [23].

EC: Annex 3 Manufacture of Radiopharmaceuticals [7] complementing the EU *Guidelines to Good Manufacturing Practice Medicinal Products for Human and Veterinary Use* [24], which contains guidance for good manufacturing practice laid down in Directive 2003/94/EC [5].

EDQM: EDQM has published revised guidance for elaborating monographs on radiopharmaceutical preparations [25], including a new section on validation of methods. General chapters that are relevant for the quality of medicinal products, for example, on extemporaneous preparation of radiopharmaceuticals [16], are part of the *European Pharmacopoeia.*

EMA: No specific GMP guideline document for radiopharmaceuticals has been published by EMA, but a guideline on radiopharmaceuticals [8] providing additional information in the context of applications for marketing authorization or variations to authorized medicinal products is available.

IAEA: A guideline for quality control in the production of radiopharmaceuticals [2], and a guideline for the design of a cyclotron-based radionuclide production facility [26] have been published by IAEA.

ICH: No specific GMP guideline document for radiopharmaceuticals has been published by ICH; the *Good Manufacturing Practice Guide for Active Pharmaceutical Ingredients* explicitly excludes radiopharmaceuticals.

PIC/S: Annex 3 *Manufacture of Radiopharmaceuticals* [27] complementing the PIC/S *Guide to Good Manufacturing Practice for Medicinal Products* [28] and the PIC/S *Guide to Good Practices for Medicinal Products in Healthcare Establishments* [17].

WHO: Annex 3 *Guidelines on Good Manufacturing Practices for Radiopharmaceutical Products* [29] complementing the good manufacturing practices for pharmaceutical products [30, 31] and sterile pharmaceutical products [32].

In most of the documents, radioactive products are classified into four categories.

1. Radiopharmaceuticals or ready-for-use radioactive products.
2. Radionuclide generators.
3. Radionuclide kits or non-radioactive components ("kits") for the preparation of labelled compounds with a radioactive component (usually the eluate from a radionuclide generator).
4. Radionuclide precursors or precursors used for radiolabelling other substances before administration.

It should be noticed that for some of these categories the definitions in the respective documents are potentially outdated, namely "radionuclide precursor", and should be reconsidered for updating [33].

Annex 3 of the EU *Guidelines to GMP* addresses the relevant additional aspects for the manufacture of radiopharmaceuticals. Most parts of the manufacturing process have to be in line with GMP, while the target and transfer system of the radionuclide produced from a cyclotron may be considered as GMP (Table 8.1). For the latter part, the manufacturer is reluctant upon the decision of the responsible authority.

Because of the characteristics of radiopharmaceuticals (e.g. short half-life, smaller volumes compared to other medicinal products) and their release for administration before completion of all tests (e.g. sterility), a quality assurance system is of great importance. All products must be protected against (cross-)contamination. In addition, staff and environment have to be protected from radiation hazards. The extent of qualification and validation processes of radiopharmaceuticals can be determined using a risk analysis / risk management approach combining GMP and radiation protection.

All staff involved in the production, quality control, and release should be trained in specific aspects of radiopharmaceutical manufacture; the same rules apply for cleaning and maintenance staff. Key personnel include the

TABLE 8.1
GMP and Non-GMP in the Manufacture of Radiopharmaceuticals. Adopted from [7]

Type of manufacture	Non-GMP*	GMP (Part I & II)**			
Radiopharmaceuticals PET Radiopharmaceuticals Radioactive Precursors	Reactor/Cyclotron Production	Chemical Synthesis	Purification steps	Processing, formulation and dispensing	Aseptic or final sterilization
Radionuclide Generators	Reactor/Cyclotron Production	Processing			

* Target and transfer system from cyclotron to synthesis rig may be considered as the first step of active substance manufacture. **Part I: Basic Requirements for Medicinal Products, Part II: Basic Requirements for Active Substances used as Starting Materials

TABLE 8.2
Cleanroom Classification for Maximum Permitted Airborne Particle Concentration

	Maximum permitted number of particles per m^3 equal to or greater than the tabulated size			
	At rest		In operation	
Grade	0.5 μm	5.0 μm	0.5 μm	5.0μm
A	3520	20	3520	20
B	3520	29	352000	2900
C	352000	2900	3520000	29000
D	3520000	29000	Not defined	Not defined

TABLE 8.3
Recommended Limits for Microbiological Monitoring of Clean Areas During Operation. Adopted from [35]

	Recommend limits for microbial contamination*			
Grade	Air sample cfu/m^3	Settle plates (diameter 90 mm) cfu/4 hours**	Contact plates (diameter 55 mm) cfu/plate	Glove print 5 fingers cfu/glove
A	<1	<1	<1	<1
B	10	5	5	5
C	100	50	25	-
D	200	100	50	-

* Average values, **individual settle plates might be exposed for less than 4 hours

Head of Production and the Head of Quality Control who should be independent from each other and engaged full-time. The duties and responsibilities of key personnel should be laid down in written form. At least one person should be appointed as Qualified Person (QP) [34], responsible for batch release. Depending on the size of the organization a Head of Quality Assurance should be appointed. The hierarchy and relationship of Head of Production, Head of Quality Control and QP and, if necessary, Head of Quality Assurance, are laid down in an organizational chart. If production facilities are shared with research institutions, the research staff must be adequately trained in GMP.

Manufacturing of radiopharmaceuticals must take place in controlled areas. Radioactive and environmental (particulate and microbiology quality) monitoring (Tables 8.2 and 8.3) and precautions to avoid contaminations should be applied. Access to the manufacturing areas should be via gowning areas with sluice/interlock systems and be restricted to authorized personnel. To protect the radiopharmaceuticals from environmental contamination and at the same time contain radioactive particles inside the controlled areas, it is recommended to establish a system using pressure sinks, airlocks, and barrier technology [26]. As most radiopharmaceuticals are manufactured aseptically, Annex 1 [35] of the *EU Guidelines to GMP* applies giving detailed information on monitoring. It should be noted that Annex 1 is currently undergoing revision, and major changes might be made in the future. Where closed systems (e.g. automated radiosynthesizers) are used, a grade C environment is suitable. All aseptic activities (e.g. assembling of tubings, sterile filters, and vials) have to be carried out in a grade A area. Table 8.4 shows some examples for the method of preparation and the required surrounding area, which should be additionally evaluated based on a risk-assessment approach.

All documents related to the manufacture of radiopharmaceuticals should be prepared, reviewed, approved, and distributed according to written procedures. Records should be retained for at least 3 years or according to national requirements. Specifications should be established, not only for the radiopharmaceutical, but also for raw materials, labelling material, and packing material. The quality and, in particular, identity and purity of the non-radioactive precursor

TABLE 8.4
Grade of Surrounding Areas. Adopted from [17]

	Open workstation	Closed workstation*
Method of preparation		
Open aseptic	B	D
Closed aseptic	C	D
Open terminally sterilized	D	D
Closed terminally sterilized	D	D

* Isolators or hot cells which can be leak tested

should be verified as the precursor is often linked to the pharmacokinetic behaviour of the radiopharmaceutical. Specific acceptance criteria for radiopharmaceuticals should be established and may include,

- Radionuclidic purity
- Radioactivity concentration
- Specific activity
- Radiochemical purity
- Appropriate half-life

Production of more than one radioactive product in the same hot cell or laminar flow unit should be avoided to minimize cross-contamination. A dedicated area and equipment should be provided if a radiopharmaceutical contains human blood or plasma. The membrane filter used for sterile filtration for aseptically filled radiopharmaceuticals should be tested for integrity prior to release, taking into account radiation protection. The labelling of the primary vial or direct container can be done prior to manufacture and, in any circumstances sterility and visual control of the filled vial should not be compromised.

Radiopharmaceuticals are special because of their short half-life. If distributed, batch release might occur in up to three stages:

- Before transportation (under quarantine) a designated person should check all batch processing records including production and quality-control testings that have been performed so far.
- The QP does the final assessment of all deviations and out-of-specifications, the results are documented and justified by the QP prior to release. The radiopharmaceutical is not administered to patients before batch release through the QP. A system should be established to prevent administration to patients prior to release.
- Some quality-control tests may not be performed before clinical use of the radiopharmaceutical (e.g. sterility, radionuclidic purity). Nevertheless, sterility testing should be performed as soon as possible (when the analytical sample has decayed sufficiently). It is worth mentioning that, in certain cases, the sterility testing of the radioactive probe has been required by the responsible authority. When all test results are obtained, the batch is finally certified by the QP. In case of out-of-specification results, corrective and preventive actions (CAPA) are taken for investigation to prevent future events.

Sufficient amounts of retain samples of the radiopharmaceutical and starting material should be stored for at least 6 months or 2 years, respectively. Exceptions from the storage durations are possible if the shelf-life of a starting material is shorter or other conditions apply, defined by agreement with the responsible authority.

8.4 SUMMARY

Within Europe we see a large variety of regulations with regard to GMP for radiopharmaceuticals and – in particular for small-scale, in-house preparations – the diversity in Europe is confusing. This chapter gives an overview on existing binding and non-binding regulatory documents issued by important organizations and stakeholders and present guidelines that take the specific character and requirements of radiopharmaceuticals into account.

REFERENCES

[1] R. Lange *et al.*, "Untangling the web of European regulations for the preparation of unlicensed radiopharmaceuticals: A concise overview and practical guidance for a risk-based approach," *Nucl Med Commun,* vol. 36, no. 5, pp. 414–22, 2015, doi: 10.1097/mnm.0000000000000276.

[2] *Quality Control in the Production of Radiopharmaceuticals.* Vienna: International Atomic Energy Agency, 2018.

[3] *EDQM, European Pharmacopoeia,* Vol. 10.2, Council of Europe, Strasbourg (2020).

[4] The European Parliament and the Council of the European Union. "Directive 2001/83/EC of the European Parliament and of the Council of 6 November 2001 on the community code relating to medicinal products for human use," *Off J Eur Communities,* vol. L311, pp. 67–128, 2001.

[5] The European Commission, "Commission Directive 2003/94/EC of 8 October 2003 laying down the principles and guidelines of good manufacturing practice in respect of medicinal products for human use and investigational medicinal products for human use."
https://eur-lex.europa.eu/legal-content/EN/TXT/PDF/?uri=CELEX:32003L0094&from=EN.

[6] *Official Journal of the European Union.* "Directive 2004/27/EC of the European Parliament and of the Council of 31 March 2004 amending Directive 2001/83/EC on the Community code relating to medicinal products for human use." L136/34 https://ec.europa.eu/health/sites/health/files/files/eudralex/vol-1/dir_2004_27/dir_2004_27_en.pdf.

[7] The European Commission, EudraLex, "The Rules Governing Medicinal Products in the European Union Volume 4 EU Guidelines to Good Manufacturing Practice Medicinal Products for Human and Veterinary Use, Annex 3 Manufacture of Radiopharmaceuticals." https://ec.europa.eu/health/sites/health/files/files/eudralex/vol-4/2008_09_annex3_en.pdf.

[8] *EMEA/CHMP/QWP/306970/2007.* www.ema.europa.eu/en/documents/scientific-guideline/guideline-radio pharmaceuticals-revision-1_en.pdf.

[9] *Official Journal of the European Communities:* "Directive 2001/20/EC of the European Parliament and of the Council of 4 April 2001 on the approximation of the laws, regulations and administrative provisions of the Member States relating to the implementation of good clinical practice in the conduct of clinical trials on medicinal products for human use." L121/34. https://eur-lex.europa.eu/legal-content/EN/TXT/PDF/?uri=CELEX:32001L0020&from=EN

[10] *Official Journal of the European Union.* "Commission Directive 2005/28/EC of 8 April 2005 laying down principles and detailed guidelines for good clinical practice as regards investigational medicinal products for human use, as well as the requirements for authorization of the manufacturing or importation of such products." L91/13.
https://eur-lex.europa.eu/LexUriServ/LexUriServ.do?uri=OJ:L:2005:091:0013:0019:en:PDF.

[11] The European Parliament and the Council of the European Union. "Regulation 536/2014 of the European Parliament and of the Council of 16 April 2014 on clinical trials on medicinal products for human use." *Off J Eur Union.* 2014: L158/1–76. https://eur-lex.europa.eu/legal-content/EN/TXT/PDF/?uri=CELEX:02014R0536-20140527&from=EN

[12] The European Commission, EudraLex, *The Rules Governing Medicinal Products in the European Union* Volume 4; EU Guidelines to Good Manufacturing Practice Medicinal Products for Human and Veterinary Use, Annex 13 Investigational Medicinal Products. https://ec.europa.eu/health/sites/health/files/files/eudralex/vol-4/2009_06_annex13.pdf.

[13] The European Commission. "Guidance on Investigational Medicinal Products (IMPs) and 'Non-investigational Medicinal Products' (NIMPs)." 18 March 2011."

[14] European Medicines Agency. *Guideline on the requirements for the chemical and pharmaceutical quality documentation concerning investigational medicinal products in clinical trials.* www.ema.europa.eu/en/documents/scientific-guideline/ guideline-requirements-chemical-pharmaceutical-quality-documentation-concerning-investigational_en.pdf.

[15] European Medicines Agency. *Guideline on strategies to identify and mitigate risks for first-in -human and early clinical trials with investigational medicinal products.*

[16] (2020). *EDQM,* Chapter 5.19. "Extemporaneous preparation of radiopharmaceutical preparations."

[17] *PE 010-4, Annex 3.* https://picscheme.org/layout/document.php?id=156

[18] "EANM guidelines on current good Radiopharmacy practice," www.eanm.org/publications/guidelines/radiopharmacy.

[19] I. Peñuelas, D. J. Vugts, C. Decristoforo, and P. H. Elsinga, "The new Regulation on clinical trials in relation to radiopharmaceuticals: When and how will it be implemented?" *EJNMMI Radiopharmacy and Chemistry,* vol. 4, no. 1, p. 2, 2019, doi: 10.1186/s41181-019-0055-6.

[20] C. Decristoforo *et al.*, "Radiopharmaceuticals are special, but is this recognized? The possible impact of the new Clinical Trials Regulation on the preparation of radiopharmaceuticals," *Eur J Nucl Med Mol Imaging,* vol. 41, no. 11, pp. 2005–07, 2014, doi: 10.1007/s00259-014-2838-z.

[21] P. Elsinga *et al.*, "Guidance on current good radiopharmacy practice (cGRPP) for the small-scale preparation of radiopharmaceuticals," *Eur J Nucl Med Mol Imaging,* vol. 37, no. 5, pp. 1049–62, 2010, doi: 10.1007/s00259-010-1407-3.

[22] J. Aerts *et al.*, "Guidance on current good radiopharmacy practice for the small-scale preparation of radiopharmaceuticals using automated modules: A European perspective," *J Labelled Comp Radiopharm,* vol. 57, no. 10, pp. 615–20, 2014, doi: 10.1002/jlcr.3227.

[23] N. Gillings *et al.*, "EANM guideline on the validation of analytical methods for radiopharmaceuticals," *EJNMMI Radiopharmacy and Chemistry,* vol. 5, no. 1, p. 7, 2020, doi: 10.1186/s41181-019-0086-z.

[24] The European Commission, EudraLex. The Rules Governing Medicinal Products in the European Union Volume 4 EU Guidelines to Good Manufacturing Practice Medicinal Products for Human and Veterinary Use https://eur-lex.europa.eu/legal-content/EN/TXT/PDF/?uri=CELEX:32003L0094&from=EN.

[25] *Guide for the elaboration of monographs on radiopharmaceutical preparations.* www.edqm.eu/en/news/revised-guidance-elaborating-monographs-radiopharmaceutical-preparations-new-section-validation.

[26] *Cyclotron Produced Radionuclides: Guidelines for Setting Up a Facility.* Vienna: International Atomic Energy Agency, 2009.

[27] *PE 009-14 (Annexes)* https://picscheme.org/layout/document.php?id=1407.

[28] *PE 009-14 (Part I)* https://picscheme.org/layout/document.php?id=1408.

[29] *Annex 3, WHO Technical Report Series 908, 2003.* www.who.int/medicines/areas/quality_safety/quality_assurance/GMPRadiopharmaceuticalProductsTRS908Annex3.pdf?ua=1.

[30] *Good manufacturing practices for pharmaceutical products. In: WHO Expert Committee on Specifications for Pharmaceutical Preparations. Thirty-second report.* Geneva, World Health Organization, 1992, Annex 1 (WHO Technical Report Series, No. 823).

[31] *Quality assurance of pharmaceuticals. A compendium of guidelines and related materials. Vol. 2. Good manufacturing practices and inspection.* Geneva, World Health Organization, 1999. www.who.int/medicines/areas/quality_safety/quality_assurance/QualityAssurancePharmVol2.pdf.

[32] *Good manufacturing practices for sterile pharmaceutical products. In: WHO Expert Committee on Specifications for Pharmaceutical Preparations.* Thirty-sixth report. Geneva, World Health Organization, 2002, Annex 6 (WHO Technical Report Series, No. 902).

[33] O. Neels, M. Patt, and C. Decristoforo, "Radionuclides: Medicinal products or rather starting materials?" *EJNMMI Radiopharmacy and Chemistry,* vol. 4, no. 1, p. 22, 2019, doi: 10.1186/s41181-019-0074-3.

[34] The European Commission, EudraLex. "The Rules Governing Medicinal Products in the European Union Volume 4 EU Guidelines to Good Manufacturing Practice Medicinal Products for Human and Veterinary Use, Annex 16 Certification by a Qualified Person and Batch Release." https://ec.europa.eu/health/sites/health/files/files/eudralex/vol-4/v4_an16_201510_en.pdf

[35] The European Commission, EudraLex "The Rules Governing Medicinal Products in the European Union Volume 4 EU Guidelines to Good Manufacturing Practice Medicinal Products for Human and Veterinary Use, Annex 1 Manufacture of Sterile Medicinal Products." https://ec.europa.eu/health/sites/health/files/files/eudralex/vol-4/2008_11_25_gmp-an1_en.pdf

9 Management of Radioactive Waste in Nuclear Medicine

Lena Jönsson and Hanna Holstein

CONTENTS

9.1 INTRODUCTION

The use of radionuclides in medicine will inevitably result in various forms of radioactive waste. This waste emerges from the production of radionuclides and radiopharmaceuticals, from diagnostic and therapeutic use, and from biomedical research. Radioactive waste can also include spent sealed sources used for external radiation therapy, brachytherapy, calibration, or quality control of different kinds of medical equipment. The waste can lie within a wide range of activities and half-lives and be in different states: solid, liquid, or airborne. In nuclear medical applications the main portion of the radioactive waste consists of radionuclides with a short half-life and low radiotoxicity, but other risks associated with the waste must also be considered.

The philosophy of all work with radioactive material is to minimize any hazards to human health and any impact on the environment, both in the short and the long term. To comply with this, the basic principles of radiation safety must be applied – that is, justification, optimization, and the use of dose limits. This also includes radioactive waste management, and therefore the radioactive waste generated must be kept to a minimum as well as adapted to the work situation.

The management of radioactive waste from medical applications is guided by international recommendations and regulated by regional and national authorities. The organization and regulations may vary in different countries due to the national legal framework, but the purpose is the same – to minimize a negative impact of the waste in all aspects. The local organization, management, and regulations must state that the procedures of the waste management comply with national and international standards and regulations. Safe management of radioactive waste is based on an organization with a clear allocation of responsibilities and provision of independent regulatory functions.

9.2 ORGANIZATION AND MANAGEMENT SYSTEM

A nuclear medicine department needs to have permission for all its work with ionizing radiation. The permission should also include the management of radioactive waste and transportation of radioactive material. The permission means, among other things, that there must be a well-functioning and well-documented organization in order to meet the basic principles of work with ionizing radiation. To ensure this, the responsibilities for the work environment and the work

DOI: 10.1201/9780429489501-9

with ionizing radiation must be clearly laid out. The organization must ensure that the entire process from planning use, purchase, use and applications, and waste handling right up to the decommissioning of the facility is optimized regarding radiation protection. This is the most important step towards a practical, economical, and safe handling of the waste. To ensure correct and sustainable handling of waste, the organization must be stable and robust in accordance with national and international standards and regulations. The organization must develop and implement a strategy for waste handling that is integrated in a way that secures correct handling in all steps where waste arises and throughout its whole life cycle. This should include appropriate evidence that protection and safety is optimized. Staff with correct, good competence in radiation safety are required, which often means that a radiation protection physicist should be involved. The waste management should be subject to periodic review and continuous development to ensure optimization of waste management practice concerning requirements, practical, economical, and safe handling of the waste.

A good way to fulfil all requirements in a structured and organized way, is to integrate the handling of the waste into a management system. The waste management system should include a comprehensive description of how to minimize waste hazards, including minimization of waste, waste storage, safety and risk assessments of facilities and transports, and the practical handling of the waste. The system should also contain regulations for training of staff, documentation and record keeping, and quality assurance.

The planning of a facility should include the choice of radionuclides with regard to the half-life, type of radiation and the activity. The work must also be optimized regarding the workflow, the number of operations and materials involved in the procedures, the risk of contamination, as well as management to minimize waste production.

Control of the amount of radioactive waste should be carried out through measurement and documentation of stored waste. Radiation dose assessment is a very important tool in all steps of waste management and risk assessment. However, in the daily routine the handling of the waste is most often based on direct measurements of the activity or the dose rate.

Waste minimization is fundamental to reducing the activity, activity concentration and volume of waste, waste storage, treatment, and disposal, to minimize the radiation doses to the staff and the public as well as the environmental impact. In many countries it is not allowed to reduce the concentration of activity by dilution, but in some countries, under certain well-defined circumstances, it may be accepted to dilute low-level waste to comply with specific activity limits laid down by the regulatory authority.

Waste categorized as non-radioactive waste, and waste with activity below the exemption levels given by the regulatory authorities, can be sent to a waste facility. The waste must be monitored to confirm that it is below these levels, and the measurements should be performed using instruments appropriate with regard to the kind of radiation emitted and the activity.

9.2.1 Produced Radioactive Waste

The radioactive waste in nuclear medicine is generated in a variety of applications, for instance, cyclotron production of radionuclides, synthesis of PET-radiopharmaceuticals at PET-centra, labelling procedures of ^{99m}Tc-pharmaceuticals, and preparation and administration of radiopharmaceuticals to patients. Waste may also arise as contaminated medical supplies and equipment during diagnostics and therapy with radiopharmaceuticals. Furthermore, the waste can consist of biological substances such as blood or tissue samples but also as excreta for example, after radiation therapy with ^{131}I.

The radioactive waste generated in medicine can be of high or low activity and with short or long half-life and can be in solid, liquid, or gaseous form (airborne). It can also consist of sealed sources used for radiation therapy or for calibration or quality control of different kinds of medical equipment. In nuclear medicine most of the radioactive waste consists of short-lived radionuclides that could be handled as non-radioactive waste after sufficient time of decay. Waste generated in radionuclide therapies consists of radionuclides with a longer half-life and higher activity and requires special precautions.

All waste must be segregated according to the type of radionuclide and emitted radiation, half-life, toxicity, and other pertinent characteristics. The waste can be in different physical and chemical forms and must also be managed based on associated properties and any chemical or biological risk associated with the waste.

9.3 STORAGE OF WASTE

Storage rooms for on-site interim storage of radioactive waste may be needed for different reasons: Storage for radioactive decay, storage before handling, or storage before return to the supplier. Each facility must define a policy for the storage of radioactive waste, and the design of the storage facility shall follow government guidelines and regulations.

Storage for decay is particularly useful in the medical use of radionuclides since the half-life is usually short and the activity used is known. A room intended for temporary storage of waste must be locked, properly marked, and ventilated in accordance with the type of waste. The dose rate, activity, and external contamination of the waste package should be measured at collection and further handling of the waste in order to minimize the exposure. After sufficient storage for decay and measurement to check that the activity is below activity limits for release, the waste can be handled according to the local regulatory requirements. Interim storage capacity should be available for spent sealed radiation sources prior to their return to the original supplier or to an appropriate waste facility.

Radioactive waste including chemical or biological hazards may require separate handling and arrangements such as metal cabinets for storage of flammable organic liquids and a freezer / refrigerator cabinet for biological waste.

9.3.1 Solid Waste

The use of radiopharmaceuticals gives rise to solid waste in nuclear medicine, such as used syringes, needles, vials with residual activity, cotton swabs, gloves and cover paper, and other potentially contaminated materials. This waste should be collected in appropriate shielded containers and segregated according to the half-life of the radionuclide. Containers for separation of different types of waste should be in place where the waste is generated. All containers should be fit for purpose, lined with plastic bags to prevent leakage, appropriately shielded, and labelled with the radiation symbol. When the container is full it should be sealed and placed in a room intended for temporary storage and decay of the waste.

Contaminated sharp objects such as needles, blood lancets, and glass ampoules should be collected in special containers. Contaminated clothes and bed linen from radionuclide therapy patients at hospital wards should be stored for decay in the interim storage room before being sent to the laundry.

Sealed radiation sources used for calibration and control measurement of nuclear medicine equipment and anatomical markers should be stored until the activity has decayed and the source can be handled according to the local regulations. Decommissioning of high-activity sealed sources (HASS) is strictly regulated and must be specifically considered.

9.3.2 Liquid Waste

Liquid waste can, for instance, include residuals of radiopharmaceuticals, ^{3}H or ^{14}C in organic solutions, contaminated water and solutions, blood, body fluids, or excreta from patients after radionuclide therapy procedures.

Liquid waste has to be segregated, according not only to radiological properties, but also to physical, chemical, and biological properties. Chemical, organic, toxic, carcinogenic, or infectious waste has to be treated in an appropriate way according to the regulations for the same. Organic solutions containing very low activity of ^{3}H or ^{14}C can be considered as non-radioactive but should be handled with regard to the chemical toxicity of the solvent.

Aqueous-based liquids should be separated, taking into account the radionuclide, activity, and half-life, and suitably stored for decay until the waste meets the clearance levels. All containers should be suitable for the intended use, lined with plastic bags to prevent leakage, appropriately shielded and labelled with the radiation symbol.

Low-activity liquid waste that meets the exempt levels, that is, if the activity is below the discharge limits, may be discharged directly to an approved drainage/sewage system such as a municipal sewer.

Excreta is regulated very differently in various countries. The waste can be collected in tanks at the hospital which, after a decay time of a couple of months, are emptied into the local sewage system. In some countries discharges of excreta into the normal sewage system is allowed without delay.

9.3.3 Airborne Waste

The release of small amounts of airborne activity may be inevitable in nuclear medicine, for example when radioactive gases, aerosols, or particulate matter such as ^{99m}Tc-aerosols, ^{99m}Tc-technegas, ^{133}Xe, and ^{81m}Kr are used for lung ventilation imaging. Cyclotron production of PET radionuclides can also result in the release of airborne radioactivity. This airborne activity can leak into the environment. However, during normal operation the doses to the staff, the general public, or the environment will be small, or even insignificant. Even in the event of an accident the consequences will probably be small. Special arrangements for ventilation, air extraction, and exhaust systems may be needed, depending on operations and the risk of airborne activity being released during handling.

The release of airborne activity to the environment is normally monitored or evaluated to ensure that the exempt levels are not exceeded. Noble gases, such as xenon and krypton, are difficult to handle and often ventilated out of the administration room through an exhaust system. It is important to ensure that there is no re-entry for airborne activity back into the building through the ventilation system.

Volatile substances such as radioiodine should be handled in fume cupboards or extraction cabinets, where released vapours get trapped in charcoal or HEPA filters. The filters are treated as solid radioactive waste when changed.

9.3.4 Radionuclide Generators

The most commonly used radionuclide generator is the ^{99}Mo-^{99m}Tc generator. After its use the technetium generator can be sent back to the supplier, if this is agreed upon. Otherwise, the generator can be stored for decay for a couple of months, and then the molybdenum column can be dismounted, and the rest of the generator can be handled as non-radioactive material after all radiation labels have been removed. Applicable conditions for other radionuclide generators should be checked with the manufacturer.

9.4 TRANSPORTATION

The transportation of radioactive waste should take place in such a way that the safety of everyone involved is ensured, under normal conditions as well as in the case of an accident during the transportation. Detailed regulations and guidelines that describe transportation of the waste are required – both internal transportation within the facility and external transportation to other waste facilities. These procedures must be in accordance with local, national, regional, and international regulations and recommendations.

The packaging of the waste should be appropriate for the type of transportation in accordance with applicable national legal requirements. International guidelines of the packaging requirements for transportation of radioactive material are described in IAEA Safety Standards Series No. ST-1 [15]. The transportation and transfer of radioactive waste within the facility should be carried out in accordance with a local framework and rules specifying requirements for handling, shielding, marking of the waste, handling of spills, or other accident scenarios.

9.4.1 Internal Transportation

Internal transportation of radioactive waste shall, as far as reasonably possible, comply with current regulations concerning transportation of dangerous goods by road (ADR-S). The transport routes must be as short and safe as possible and reasonable. These routines must include a plan for how unwanted events are managed.

All packaging must be well planned and properly radiation shielded to avoid contamination and unnecessary irradiation of staff.

9.4.2 External Transportation

Most waste in nuclear medicine has a short half-life, which enables radioactive decay before transportation to a disposal facility. Transportation of radioactive waste or other radioactive sources is often classified as transportation of dangerous goods and must be performed in accordance with international recommendations and regional and national regulations.

9.5 FURTHER READING

- Council Directive 2013/59/Euratom on basic safety standards for protection against the dangers arising from exposure to ionizing radiation and repealing Directives 89/618/Euratom, 90/641/Euratom, 96/29/Euratom, 97/43/Euratom and 2003/122/Euratom
- International Atomic Energy Agency. 2000. *Management of radioactive waste from the use of radionuclides in medicine. IAEA-TECDOC-1183.* International Atomic Energy Agency.
- International Atomic Energy Agency. 2005. *Applying radiation safety standards in nuclear medicine.* IAEA safety report series No. 40. International Atomic Energy Agency.
- International Atomic Energy Agency. 2005. *Management of waste from the use of radioactive material in medicine, industry, agriculture, research and education.* IAEA safety standards series no. WS-G-2.7. International Atomic Energy Agency.

- International Atomic Energy Agency. 2009. *Release of patients after radionuclide therapy.* IAEA safety standards series no. IAEA safety report series No.63. International Atomic Energy Agency.
- International Atomic Energy Agency. 2011. *Disposal of radioactive waste. Specific safety requirements.* IAEA safety standards series no. IAEA safety standards series no. SSR-5. International Atomic Energy Agency.
- International Atomic Energy Agency. 2012. *Regulations for the safe transport of radioactive material.* IAEA safety standards series no. IAEA safety standards series no. SSR-6. International Atomic Energy Agency.
- International Atomic Energy Agency. 2013 *Management of discharge of low level liquid radioactive waste generated in medical, educational, research and industrial facilities.* IAEA-tecdoc-1714. International Atomic Energy Agency.
- International Atomic Energy Agency. 2014. *Radiation protection and safety of radiation sources: International basic safety standards.* IAEA safety standards series no. GSR Part 3. International Atomic Energy Agency.
- International Atomic Energy Agency. 2014. *Nuclear Medicine Physics: A Handbook for Teachers and Students.* Technical Editors: D. L. Bailey, J. L. Humm, A. Todd-Pokropek, and A. Van Aswegen. Chapters 3, 9 20.
- International Atomic Energy Agency. 2014. *Regulations for the safe transport of radioactive material. 2012 Edition.* IAEA safety standards series no. SSR-6. International Atomic Energy Agency.
- International Atomic Energy Agency. 2019. *Predisposal management of radioactive waste from the use of radioactive material in medicine, industry, agriculture, research and education.* IAEA safety standards series No. SSG-45. International Atomic Energy Agency.

10 Translation of Radiopharmaceuticals

Mouse to Man

Pedro Fragoso Costa, Latifa Rbah-Vidal, An Aerts, Fijs W.B. van Leeuwen and Margret Schottelius

CONTENTS

10.1 INTRODUCTION

Perhaps the most widely known example of translational research in medicine, where basic experimental and theoretical discoveries are implemented into benefiting humanity, comes from the physics discipline [1]. Particle acceleration as a means to create x rays, Röntgen's discovery in 1895, has had a profound impact on modern medicine and continues to benefit millions of patients worldwide. Since then, many other basic research breakthroughs have entered the healthcare domain.

DOI: 10.1201/9780429489501-10

All translational research efforts in medicine start with a clinical question. This question will be progressively examined into its fundamental constituents until it reaches the basic research domain. This allows basic research scientists to interact with clinical scientists to understand the pathophysiological and biochemical basis of disease. In that sense, nuclear medicine is a uniquely translational discipline, since it unites the fields of medicine, biology, chemistry, and physics. Using specifically designed, targeted molecules labelled with radioisotopes with diverse nuclear characteristics for *in vivo* molecular imaging (and, in some cases, subsequent therapy within the framework of a theranostic concept) provides important and oftentimes decisive biochemical information on the presence, state, and dynamics of a given disease.

The particular precision of nuclear medicine, increasingly allowing for personalized precision diagnosis and therapy, is due on the one hand to the fact that including engineering principles and approaches, merging into highly advanced and sensitive instrumentation, renders this discipline a computable, measurable, and systems-based discipline [2]. On the other hand, the success of nuclear medicine also relies on the availability of increasingly sophisticated targeted radiopharmaceuticals, which allow in vivo targeting of specific molecular processes with oftentimes high selectivity.

Reaching this goal, however – ultimately leading to clinical translation a radiopharmaceutical candidate – requires a long and strenuous cascade of chemical and radiochemical optimizations as well as extensive in vitro and in vivo testing to ensure both targeting efficiency and therapeutic efficacy, and to characterize potential adverse effects. There are many diverse and highly interdependent factors that determine the success of such a candidate; these include not only the capability of a tracer to successfully act as a diagnostic or therapeutic agent, but also the ease to comply with regulatory procedures as well as questions concerning the availability of radionuclide and precursor. This chapter will guide the reader through the steps of radiopharmaceutical translation, with a section reserved for a description of the typical tasks in which physicists (specifically medical physicists) might be involved, throughout the process.

10.2 TARGET SELECTION

By nature, the identification and characterization of a relevant molecular target is the first and pivotal point in all tracer developments and in virtually all nuclear medicine applications. Generally, all nuclear medicine techniques aim at distinguishing healthy from diseased tissue by visualizing and ideally quantifying a pathological *change* in a given physiological process. Of course, this "change" may cover a vast range of structures and processes, ranging from solid structures (e.g., bone, plaques) over general dynamic processes (e.g., perfusion) down to deregulation and upregulation on the molecular level (e.g., receptor expression, transporter expression, enzyme activity, antigen presentation).

However, if the principle of "molecular imaging" is to provide relevant and valuable information to the clinician and, for targeted therapies to be precise and efficient, the molecular target generally needs to fulfil some, if not all, of the following prerequisites:

- The change in the targeted physiological process should correlate with
  - the localization,
  - the extent and,
  - (ideally) the kinetics of the disease;
- the targeted change has sufficient specificity for the disease;
- the molecular target is accessible for a given radiopharmaceutical (see section 10.3), and
- there is a sufficient accumulation mechanism for a targeted radiopharmaceutical (transport, binding, internalization, metabolic trapping).

Of course, there are almost no examples in nuclear medicine, where a molecular target fulfils *all* requirements. For example, [^{18}F]FDG uptake in tumours is clearly based on a pathologically changed glucose utilization that correlates with the localization and the extent of the disease. The target (GLUT1 and Hexokinase) is readily accessible, and there is an efficient accumulation mechanism (metabolic trapping). However, increased [^{18}F]FDG uptake is *not specific* for the disease – pathologically enhanced glucose metabolism is also observed in inflammation. Nevertheless, most clinically addressed targets in nuclear oncology, neurology and cardiology owe their success to adhering to (almost all) of the above requirements.

Thus, successful target selection and, ultimately, correct interpretation of the imaging results will always require a maximum of in-depth understanding of the observed molecular process and its regulation in health and disease. It is then the task of radiopharmaceutical developers to address all "downstream" factors leading to more or less successful clinical translation of an imaging approach or therapeutic tracer (Figure 10.1).

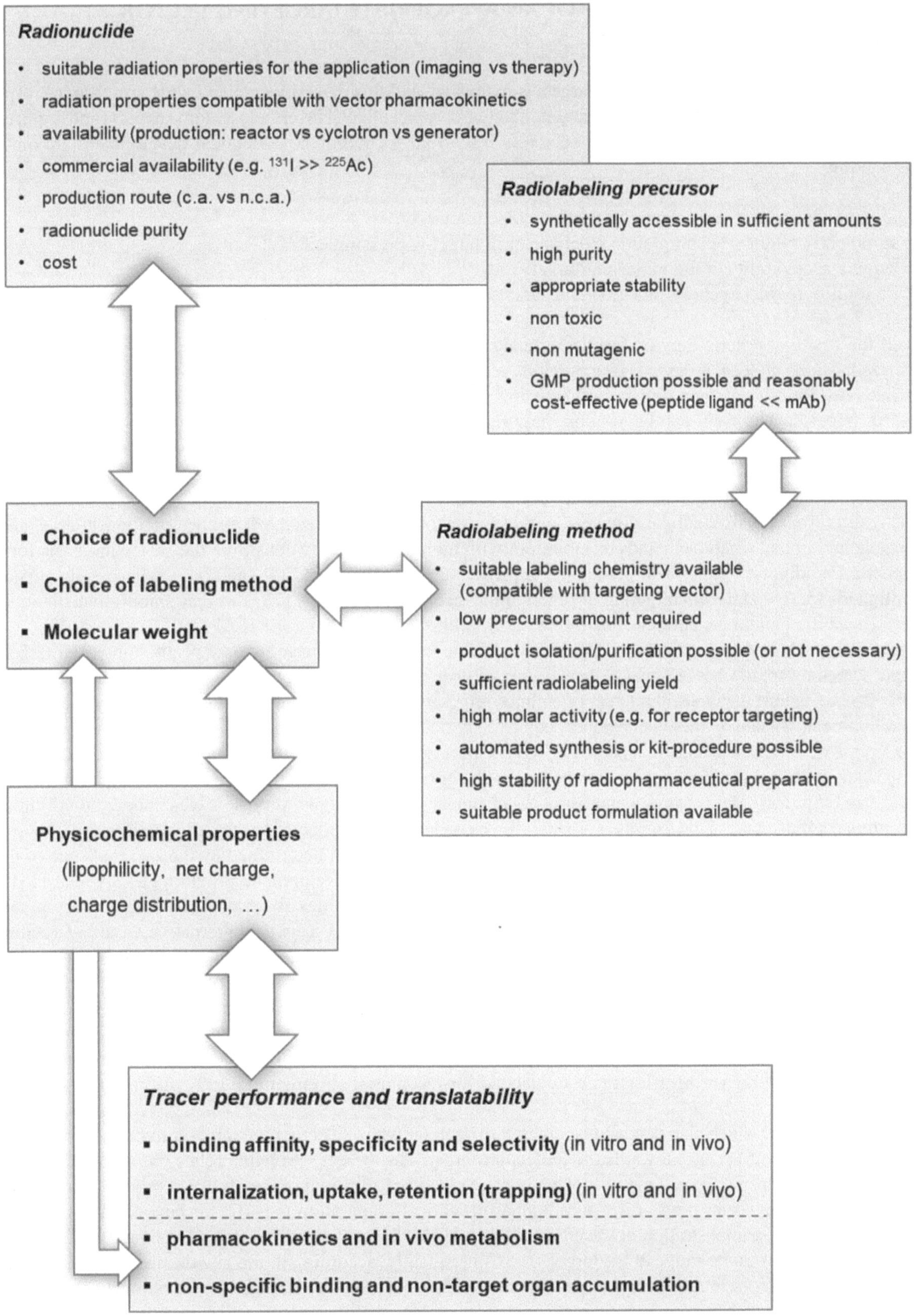

FIGURE 10.1 Synopsis of aspects determining tracer performance and translatability.

10.3 DESIGN AND DEVELOPMENT OF AN APPROPRIATE TARGETING VECTOR

10.3.1 General Considerations

As shown, the scope of potential molecular targets is extensive, and this diversity entails a wide spectrum of respective target-specific requirements for an "optimal" targeted radiopharmaceutical. Nevertheless, some aspects and requirements for the selection of an appropriate targeting vector as a scaffold for radiopharmaceutical development are quite universal, including

- maximum specificity and selectivity for the molecular target,
- a sufficiently efficient accumulation mechanism in target-expressing cells,
- adequate accessibility of the molecular target, and
- high affinity to the target (dissociation constant (K_D) or IC_{50} in the low nM range or below).

How well the first two criteria may be fulfilled already at the stage of tracer design very much depends on how well characterized a given molecular target is (see previous section). For example, if an endogenous ligand (e.g. for receptors) or substrate (e.g. for enzymes and transporters) is known, and its binding or uptake mechanism to/into the target cell is established, tracer development can build upon these natural lead structures, and major focus can be directed towards adaptation of this molecular scaffold for radiolabelling and optimal tracer performance. This is the case for example, for radiolabelled GLP receptor ligands (Exendin-40 [3]) or GRP receptor ligands (bombesin analogues [4]). Ideally, when the molecular target of interest has already been identified as a relevant therapeutic target, tracer design can build upon the results from pharmaceutical research; in those cases, highly optimized, oftentimes orally available, and thus highly stable and potent receptor ligands or inhibitors with high target affinity constitute the molecular basis for tracer development. Paradigmatic examples for such an approach are peptidic octreotide- and octreotate-based somatostatin receptor ligands [5, 6] and tracers targeting the chemokine receptor 4 (CXCR4) [7] as well as small-molecule inhibitors for prostate specific membrane antigen (PSMA) [8] or fibroblast associated protein (FAP) [9].

In other cases, where there is no natural ligand or substrate known for a given disease-specific molecular target, high throughput screening approaches (e.g. phage display or combinatorial chemistry) or the development of specific monoclonal antibodies (mAb) against the molecular target oftentimes represents the only accessible route towards targeting vectors for tracer development. By nature, these approaches require a substantial number of optimization and refinement steps until a targeting vector with suitable features, in particular with respect to affinity, selectivity and in vivo stability, is obtained. And despite the undisputed value of full-sized radiolabelled mAbs for preclinical (and sometimes clinical) target evaluation, their molecular size (app. 150 kDa) conveys a number of disadvantages – for example, very slow clearance from the circulation and high non-specific uptake in the excretion organs, which challenge their suitability as targeting vectors.

Thus, the selection of an appropriate targeting vector is to a great extent determined by the current "state of the art" with respect to target characterization and target-related pharmaceutical development, providing a very restricted (mAb only) to highly diverse (high-end therapeutics) selection of targeted molecules as scaffolds for tracer developments. In the latter case, radiopharmaceutical developers have the "agony of choice" between alternative targeting vectors, and further selection criteria need to be taken into account to identify the most promising candidates for tracer development.

One of these criteria, as has been mentioned above, is an adequate accessibility of the molecular target by the targeting vector. Depending on the localization of the molecular target (vascular, stromal, membrane associated, cytosolic, nuclear), requirements for the targeting vector differ substantially. While full-sized mAbs may be an appropriate choice for targeting extracellular structures, small molecules (in some cases lipophilic and membrane-penetrating: cerebral tracers) with much higher diffusion rates are needed for reaching their target in poorly diffused tissue or to reach intracellular targets.

Another key feature, which also needs to be taken into account when selecting the targeting vector, is the need for a very high affinity (low nM range) towards the molecular target. The "tracer" principle relies on visualizing molecular processes without disturbing the respective biochemical or kinetic equilibrium of the observed process. To achieve this, the amount of targeting vector present in the radiopharmaceutical preparation must lie far beyond pharmacologically active levels. Typically, administration of tracer (no carried added) doses will optimally lead to ligand concentrations in the pico-molar to femto-molar range at the target site. Consequently, according to the law of mass action, binding of a compound in such low concentrations to a target structure (e.g. a receptor – a preparation of a radioactive isotope which is essentially free from stable isotopes of the element in question) requires a pico-molar to nano-molar binding constant (K_D) [10]. Especially when a relevant molecular target with comparably low expression levels is addressed, tracer affinity is the key determinant for the detection sensitivity using nuclear imaging techniques.

10.3.2 Factors Determining Translatability

Further criteria that reach beyond the selection process of a suitable targeting vector, but also ultimately determine tracer performance and need careful consideration during tracer design and development, are summarized in Figure 10.1.

As shown, the ultimate potential of a radiotracer for clinical translation depends on "tracer performance", that is, an appropriate pharmacokinetic profile as well as high and specific accumulation in the target tissue. This, however, is not only dependent on the selection of the targeting vector – that is, the molecular weight of the tracer (mAb down to small-molecule inhibitor or substrate) – but on many other physicochemical aspects. These, in turn, are primarily determined by the chemical modifications required for implementing one or the other (radio)labelling methodology, which of course is dependent on the choice of the radionuclide for a given application. Furthermore, certain characteristics of the radiolabelling precursor used for the preparation of a given tracer will also have impact on the ease and, ultimately, the cost of translation of a novel radiopharmaceutical into clinical application. Thus, besides including purely biological considerations, the choice of an appropriate targeting vector also has to include these diverse aspects.

10.4 DIAGNOSTIC IMAGING AGENTS

For the development of diagnostic nuclear tracers, the requirements for a "good" tracer deserving clinical translation are clearly defined. High-contrast imaging using PET or SPECT requires

- relatively fast clearance from the circulation, resulting in
- low background activity levels,
- renal excretion (except cerebral tracers), and
- fast and high accumulation in target tissue, leading to
- high target/non-target ratios at early time points.

If fulfilled to an optimal achievable degree, these characteristics result in not only excellent imaging results and clinically suitable imaging protocols (imaging at early time points post injection), but also result in a low radiation exposure for the patient.

Of course, reaching this balanced tracer profile mostly requires extensive tracer optimization. A first, and decisive, aspect in the design of novel tracers for diagnostic imaging is choosing a radionuclide with a half-life that correlates well with the pharmacokinetics of the targeting vector, that is, using ^{18}F ($t_{1/2}$ = 110 min) or ^{68}Ga ($t_{1/2}$ = 68 min) for small proteins, peptides, and small-molecule PET tracers, and longer-lived isotopes, such as ^{64}Cu ($t_{1/2}$ = 12.7 h) or ^{89}Zr ($t_{1/2}$ = 78.4 h) for mAb-based and other macromolecular radiopharmaceuticals. With respect to half-life, the use of the SPECT radionuclides ^{99m}Tc ($t_{1/2}$ = 6 h) or ^{123}I ($t_{1/2}$ = 13.2 h) is compatible with a slightly broader range of targeted biomolecules. However, another aspect that needs to be taken into account when selecting a given radioisotope is the sometimes-limited compatibility of the radiolabelling chemistry with the targeting vector. Many radiolabelling procedures, in particular direct ^{18}F- as well as ^{99m}Tc-labelling, require elevated temperatures and/or harsh chemical conditions (anhydrous organic solvents, basic conditions), which are incompatible with sensitive biomolecules such as proteins. Thus, certain combinations of targeting vector and radionuclide are practically "mutually exclusive."

Furthermore, as demonstrated in Figure 10.1, the physicochemical characteristics as well as the binding affinity of a tracer are strongly influenced by the radionuclide and, consequently, the radiolabelling method used. There are innumerable examples where even minor structural changes in a targeted tracer – for example, exchange of ^{68}Ga for ^{177}Lu or ^{90}Y in a DOTA complex – have been shown to induce major changes in affinity and also in vivo behaviour [11, 12]. Thus, it is not surprising that especially in ^{99m}Tc- and ^{18}F-labelling chemistry, where fundamentally different structural principles may be alternatively applied for radionuclide introduction into the target vector, the choice of the radiolabelling method may have considerable impact on tracer performance in vitro and in vivo. For example, ^{99m}Tc-labelling of the same peptide ligands using the HYNIC/EDDA co-ligand system, the ^{99m}Tc-tricarbonyl method, MAG_3- and MAS_3-chelators or the N_4-chelator has been shown to yield tracers with fundamentally different pharmacokinetic profiles [13–15].

Similar effects are also observed in the case of alternative radiofluorination methods. Modern direct labelling methods such as ^{18}F-labelling via [^{18}F]AlF complexation [16] or isotopic exchange reactions using BF_3- [17, 18] or silicon-fluoride acceptor-conjugated precursors (SiFA) [19] have the advantage of proceeding in one step under very mild conditions, leading to ^{18}F-labelled tracers with high molar activities and in vivo stabilities. However, especially in the case of the SiFA-technology, the unprecedented ease of direct ^{18}F-labelling at room temperature is accompanied by a substantial, pharmacokinetically highly unfavourable increase in tracer lipophilicity.

Thus, when designing tracers for imaging purposes, all the different factors influencing tracer performance and their relevance for later clinical translation must be carefully weighed against each other and counterbalanced. Fortunately, if a given radiolabelling technique negatively influences tracer performance, either by altering tracer targeting or by inducing unwanted pharmacokinetic effects, individual optimization is feasible, using dedicated chemical modifications of the tracer. This is particularly true for peptide-based tracers, which offer exceptional flexibility with respect to diverse chemical modifications. These include, amongst others, conjugation with polar groups such stretches of charged amino acids [20], carbohydrate [21], or PEG moieties [22] as well as chelators/chelates as pharmacokinetic modifiers [23, 24]. The effects of such modifications are manifold, including modulated circulation times, altered excretion patterns, reduced non-specific tracer uptake and tissue retention, to name only a few. Usually, great care is taken to introduce structural modifications in such a way as to minimize interference with tracer binding to the target; in some cases, however, these entail beneficiary effects on the targeting efficiency of the tracer.

10.5 THERANOSTIC AND THERAPEUTIC AGENTS

10.5.1 Therapeutic Tracers

By nature, the development of targeted tracers for the delivery of therapeutic radioisotopes (β^-, α- or Auger-emitters) faces a slightly altered spectrum of requirements compared to tracer development for diagnostic imaging. Here, achieving

- an optimal "area under the curve" and high absorbed dose to the tumour,
- efficient activity retention in the target tissue and
- minimum radiation burden to other tissues

are the primary objectives. Since the currently used therapeutic radionuclides in targeted radioligand therapy (RLT) are primarily M^{3+} radiometals (e.g. $^{177}Lu^{3+}$, $^{90}Y^{3+}$, $^{225}Ac^{3+}$) – stably complexed by DOTA or DOTAGA – therapeutic tracers are predominantly based on a well-established labelling chemistry, and thus no noteworthy influence of the radiolabelling procedure on tracer performance is to be expected. Attempts towards improving or modulating the above characteristics of therapeutic tracers are thus concentrated on introducing additional pharmacokinetic modifiers into the targeted biomolecule. One approach consists in enhancing circulation time and thus absolute tumour uptake of small, targeted tracers by integrating either PEG moieties (to increase the hydrodynamic radius of the molecule and slow excretion) or specific plasma protein binders (to enhance binding to serum albumin and slow excretion) [25]. Common strategies to decrease non-specific uptake and residualization of therapeutic tracers in non-target (excretion) organs consist in modulation of net charge [20, 26] or introduction of cleavable linkers [27].

10.5.2 Theranostics

Paradigmatically exemplified by the clinical success of using the somatostatin receptor (sst) ligands DOTATOC and DOTATATE, both for imaging (labelled with ^{68}Ga) and RLT (labelled with ^{177}Lu or ^{90}Y) of neuroendocrine tumours, the term "theranostics" is oftentimes reduced to the very narrow definition of "using the same radiolabelling precursor for different clinical applications, that is, diagnosis and subsequent RLT."

However, the theranostic concept is a more general concept of personalized medicine, generally based on "using targeted diagnostic imaging to identify appropriate disease-specific molecular targets, to quantify expression levels (diagnostic tool), to subsequently allow personalized management of the disease (therapeutic tool) and to monitor treatment response (diagnostic tool)" [28]. Thus, this concept also comprises approaches where specifically optimized (see requirements for diagnostic and therapeutic tracers above), but potentially structurally quite different tracer molecules may be used as the respective "companion diagnostic" and "companion therapeutic" within the setting of a theranostic approach. One recent example for such a strategy is the complementary use of [^{68}Ga]PSMA-11 or [^{18}F]DCFPyl as diagnostic agents for PET imaging and of the therapeutic tracers [^{177}Lu]PSMA-617 or [^{177}Lu]PSMA-I&T for subsequent RLT of advanced metastatic prostate cancer [29]. Another example are CXCR4-targeted theranostics using the structurally related, but specifically optimized ligands [^{68}Ga]PentixaFor and [^{177}Lu]PentixaTher for PET and RLT, respectively [30].

Of course, there are also other examples besides the aforementioned sst-targeted tracers, where the same radiolabelling precursor shows all the necessary characteristics for an excellent "one-for-all" theranostic agent, for example, the recently introduced radiohybrid PSMA-ligands. They allow alternative radiolabelling with ^{18}F and ^{68}Ga for PET, depending on the radionuclide availability, as well as ^{177}Lu-labelling for therapeutic application [24]. Besides nearly optimal in vivo performance of both the diagnostic and the therapeutic counterpart for the intended application, such an approach certainly also has practical and regulatory advantages concerning clinical translation, given that only one precursor molecule needs to be fully characterized and evaluated to allow application in clinical studies.

However, also in this context, careful tracer selection and optimization with the aim of providing optimal personalized care to the patients should be well equilibrated with the practical aspects related to tracer translation.

10.6 EVALUATION STEPS NEEDED FOR CLINICAL TRANSLATION

After target selection, design and development of an appropriate targeting vector, and optimizing radiolabelling protocols, the newly established radiolabelled compounds proceed into a series of in vitro and in vivo evaluation studies. In vitro studies comprise assessments regarding molecular targeting, lipophilicity and plasma protein binding as well as metabolic stability, while in vivo studies evaluate tracer biodistribution and pharmacokinetics, also allowing for dosimetry calculations, as well as metabolic stability in vivo. Furthermore, they include animal toxicity studies and microdosing experiments. On top of leading to full proof-of-concept validation of the radiolabelled compound, the results of these steps can be used to build the Investigational Medicinal Product Dossier (IMPD) that is required for approval of clinical trials by the competent authorities in the EU.

10.6.1 In Vitro Evaluation

10.6.1.1 Molecular Targeting

First and foremost, the affinity of a tracer for its target needs to be determined in vitro. Commonly this is done by determining the IC_{50} in competition assays or K_d values. In the first category of assay (determination of IC_{50}), the new tracers are compared to a known binder for the same target (with known affinity), so that a relative increase or decrease in affinity can be defined, whereas K_d studies are used to characterize a novel radioligand in conjunction with target expression, that is, a given cancer cell line. In both assays, it also becomes possible to obtain insight in the target-specificity of the tracer.

Although these assays are usually very accurate and reproducible, they may be highly influenced by the cell lines used and the conditions wherein they are used, and thus, for example (over-)expression levels may fluctuate. This means that it can be difficult to compare affinities between studies performed by different research groups. When tracers contain a fluorescent component, technologies such as confocal fluorescence microscopy can be used to define the cellular localization of the tracer, something that should of course relate to the cellular regions where the target receptor is expressed.

Furthermore, the efficiency of radioligand uptake is determined by in vitro uptake studies and internalization studies and complemented by externalization/retention assays quantifying the amount of activity retention in the target cells over time [21].

10.6.1.2 Lipophilicity and Plasma Protein Binding

Before assessing the biodistribution of a compound in vivo, lipophilicity measurements and plasma protein binding experiments should be performed. Both features are related, at least to a certain extent, and the degree of binding to plasma protein influences key features such as blood activity concentration, biological half-lives and hepatic uptake/clearance. All these values increase when plasma binding increases. Hydrophilicity, that is, via a high number of polar or charged groups in a molecule, increasing its water solubility, also influences renal reabsorption. Thus, a balanced physicochemical profile has to be found to achieve optimal renal clearance. Very lipophilic compounds may be prone to none-specific uptake in lipids (e.g. cell membranes and fat), but can also facilitate diffuse across the blood brain barrier. Because the biodistribution of compounds is influenced by many features it is often impossible to predict, which lipophilicity or degree of protein binding is optimal for new tracers. Therefore, documenting these values helps increase our understanding of these features for future applications.

10.6.1.3 Metabolic Stability

When assessing the biodistribution of a tracer based on its imaging label – for example radioactive or fluorescence – it is essential that this label represents the distribution of the intact administered compounds and not a label-containing metabolite ("fate of the label"). Hence, tracer developers have to ensure that the compounds tested are stable under in vivo conditions, for example, in serum. Key stability elements that should be determined are the dissociation of isotopes from chelates, reactivity of tracers with endogenous proteins, and enzymatic cleavability of tracers. It should be noted that well-defined metabolic instabilities can be exploited in so-called "activatable" tracers.

10.6.2 In Vivo Evaluation

After determination of in vitro targeting efficiency, a radiolabelled compound that also fulfils the previously mentioned criteria (suitable plasma protein binding, suitable lipophilicity for the intended purpose, appropriate metabolic stability to guarantee sufficient bioavailability for targeting) can proceed into in vivo studies, where biodistribution, metabolism and toxicity are further evaluated in the complexity of the living organism.

Suitable animal models that reflect a specific diseased state are used to evaluate the properties of the radiolabelled compounds in vivo. To assess the properties of a radiopharmaceutical product for oncological application, rodents can be inoculated with syngeneic or xenograft tumours via subcutaneous or orthotopic (e.g. breast, prostate, brain) injection of cancer cells. Animal models of some neurological disorders can be generated by a unilateral injection of cytotoxic agents in a brain region, using the contralateral site as control. Similarly, cardiovascular disorders can be studied in animals where stroke or hypertension is induced via surgical procedures or cytotoxic agents. Finally, also transgenic animal models that lack expression of the target (knock-out) or overexpress a disease-related protein can be used to evaluate the radiolabelled compound [31].

10.6.2.1 Biodistribution Studies

Absorption, distribution, metabolism, and elimination define the fate of the tested radiolabelled compound in the organism. Studying these parameters provides crucial information on the rate at which the radiopharmaceutical is distributed, accumulated, and eliminated (pharmacokinetics). These parameters will also help in determining whether the tracer should preferably be administered locoregionally or intravenously. Depending on these conclusions, it is the galenist's responsibility to find a suitable formulation [32].

Non-invasive imaging with small-animal PET or SPECT scanners allows visualization of the distribution of the radiolabelled compound. Further, dynamic imaging can be combined with blood sampling and quantification of radiometabolites for exact quantification of (pharmaco)kinetic parameters such as plasma half-life or association/dissociation constants. Pharmacokinetics of compounds that are labelled with therapeutic radionuclides, that do not allow SPECT or PET imaging due to a lack of gamma (co)emission, can be determined by ex vivo biodistribution studies in which radioactivity is quantified using scintillation counting [31]. To assess the in vivo target specificity of tracer uptake, in vivo blocking studies are conducted, where small animals are pre-treated or co-injected with a molar excess of a standard ligand for the respective target, preventing target-specific tracer accumulation A displacement experiment where a blocking agent is administered after injection of the radiopharmaceutical, can be used to determine whether the radiopharmaceutical binds reversibly or irreversibly to its target [31].

10.6.2.2 Animal Toxicity Studies

Biodistribution studies performed by imaging or by gamma counting of harvested organs enable the determination of tracer accumulation in target and non-target tissues (specific/unspecific), permitting the prediction of a possible toxicity in tissues and organs in which the radiolabelled compound shows significant uptake.

For example, a strong hepatic and biliary uptake of a radiotracer indicates a hepatobiliary excretion of radioactivity, but also some risk of hepatic radiotoxicity. Thus, in addition to the evaluation of metabolism and excretion rates, the excretory tissues (liver, kidney, bladder) and all high uptake tissues are oftentimes subjected to macroscopic and histological analysis, to confirm the absence of abnormalities or to predict the radiotoxicity. Indeed, biodistribution studies can also be used to calculate the effective dose and thus the dosimetry in each of the target tissues.

However, these preliminary studies are not sufficient to translate radiolabelled compounds to the clinic. Extensive safety testing in preclinical studies [33] must be undertaken to meet the investigational new drug requirements prior to a phase I clinical trial in humans.

Radiopharmaceuticals are a special class of drugs, composed in two parts, one "cold" or nonradioactive (the respective targeting vector) and the radionuclide used for labelling (e.g. ^{18}F, ^{68}Ga, ^{64}Cu, ^{89}Zr, ^{99m}Tc, ^{177}Lu, ^{90}Y, …). Thus, toxicity of radiopharmaceutical can be caused by the nonradioactive as well as the radioactive components. Considering the first version of the new guideline on non-clinical requirements for radiopharmaceutical [33], three schemes are possible:

- If the nonradioactive part of the radiopharmaceutical is a known compound, and if preclinical studies are available, then no need for additional toxicity studies if there is available information or data demonstrating that the radioactive atom did not change the pharmacology of the compound.
- If minimal modification has been performed in the structure of the nonradioactive compound (sometime necessary for radiochemistry), then the possible risk related to that modification needs to be considered.
- If the nonradioactive compound is unknown and no preclinical toxicity data are available, then full toxicity studies have to be performed and conducted under Good Laboratory Practice (GLP) regulations (see section 10.8).

In this case, toxicity tests are carried out in two different mammalian species (one rodent and one non-rodent) and aim at evaluating the risk of radiopharmaceutical overdose. This approach can be achieved by injecting an acute dose of radiopharmaceutical and monitoring animals for abnormal clinical signs, changes in body weight and food intake, followed by post-mortem examination. This approach is today mainly replaced by so-called extended single dose toxicity studies where toxicology studies consist in assessment in only one mammalian species; usually rodent, with evaluation at day 1 and 14-days post dose administration to assess acute and delayed toxicity and/or recovery.

Recently, a new approach (summarized in Figure 10.2) has been proposed, based upon the definition of three distinct toxicological limits [34]:

- Toxicological Limit 1 <1.5 μg
- Toxicological Limit 2 <100 μg
- Toxicological Limit 3 >100 μg

Concerning therapeutic tracers in clinical trials, they are regarded as any other medicinal product and therefore have to follow regulations regarding investigational medicinal products (IMPs). Mutagenic and carcinogenic potential effect of the non-radioactive component can be evaluated, and dosimetry studies should be performed to predict the radioactive exposure in humans and to mitigate radiation induced toxicity. Also, considering that radiation is a contributor to cancer induction, dosimetry can, to some extent, predict the carcinogenic potential effect of the radiotherapeutic compound. Note that radiation induced clinical toxicity is covered by Directive 2013/59/Euratom [35].

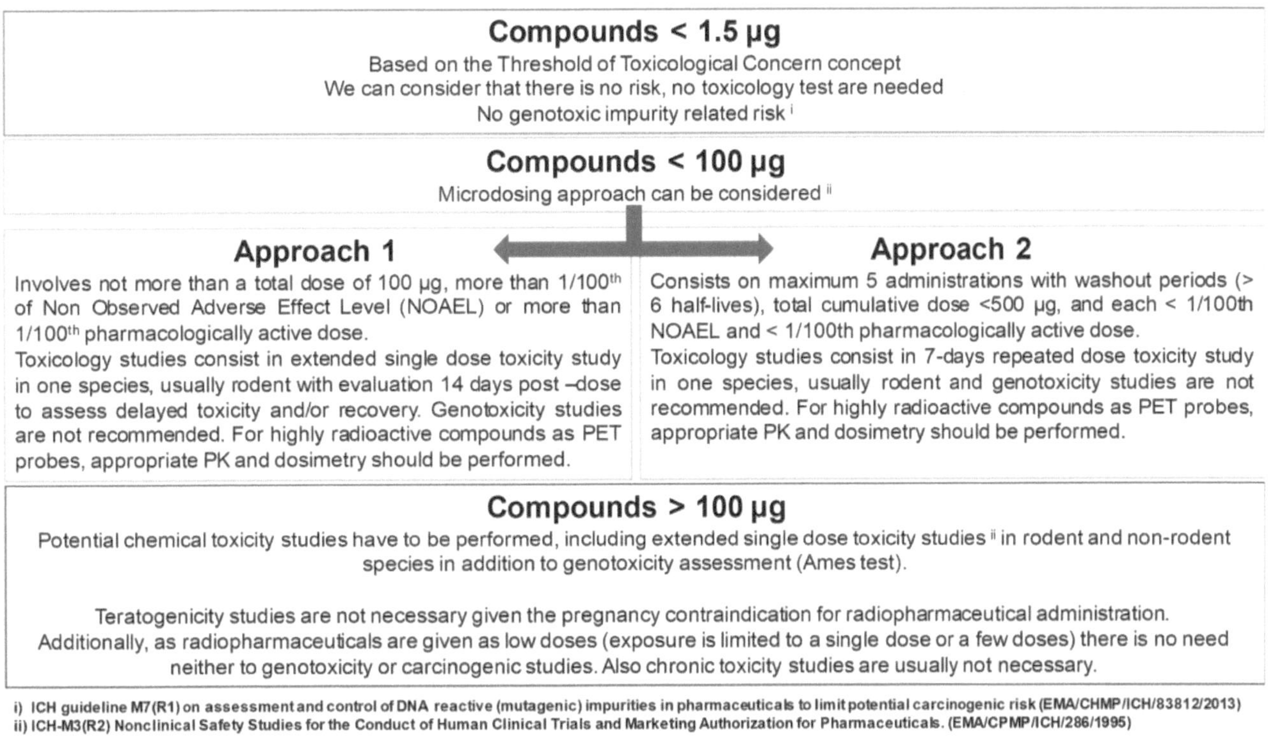

FIGURE 10.2 Possible approaches for toxicity evaluation depending on the compound mass.

10.7 GOOD MANUFACTURING PRACTICE (GMP) FOR RADIOPHARMACEUTICALS

Given that ultimately labelled agents are aimed at application in humans, the drug ready for administration must comply with the principles and guidelines of good manufacturing practice (GMP), so as to ensure that the manufactured agent is of high quality and poses no risk to patients and public.

GMP is a series of requirements and provisions ensuring that products are consistently produced and controlled according to the quality standards appropriate to their intended use and as required by the marketing authorization, clinical trial authorization, or product specification [36]. GMP applies to the life-cycle stages from the manufacture of investigational medicinal products, technology transfer, and commercial manufacturing, through to product discontinuation [36].

Given the unique characteristics of radiopharmaceuticals, there are additional requirements to be considered when bringing candidates into the clinic. In general, GMP will refer to quality management, equipment, personnel, documentation, production quality control, self-inspection, and outsourced activities [37, 38].

In terms of systematic organization, GMP dictates the need for a quality management framework, where all the operations of production and manufacturing and quality control must be accurately and transparently documented. This is achievable by setting up standard operating procedures (SOP) documentation, ensuring that practices are harmonized, traceable, and in compliance with national law and current standards.

All preparation and quality control of radiopharmaceuticals should be conducted only by properly trained and qualified personnel. The regime of involved personnel will adapt to the clinical application of the radiopharmaceutical. For instance, PET facilities will allow single-person oversight of production and quality assurance, while for therapeutic radiopharmaceuticals dual verification must be applied in critical production steps, such as production and quality control that must be performed by separate operators [39].

Another aspect that is inherent to radiopharmaceuticals is the fact that the physical radioactive vector (targeting vector) will continuously decrease with time, as a consequence of radioactive decay. Therefore, leading to the use of semi-manufactured products such as radionuclide generators, radioactive precursors and kits. It is fundamental to develop straightforward GMP-compliant synthesis modalities (e.g. fully automated synthesis modules and suitable procedures for quality control) that will enable an optimal access of radiolabelled drugs to the patient.

GMP compliance is one major aspect to take into account in the translational process of radiopharmaceuticals, as it will, in some cases, introduce considerable obstacles to the availability of radiotracers for human use. Nevertheless, GMP is the most widely used method to ensure adequate quality of a medicinal product and, with this, conserve the integrity of investigational medicinal products (and their development) used in clinical trials.

10.8 FIRST-IN-HUMAN IMAGING STUDIES

First-in-human studies constitute the first opportunity to evaluate a drug candidate in a human, and can be considered as the gateway between fundamental research and clinic (the so-called move from bench to bedside) with the primary objective of providing information on feasibility, safety biodistribution, and pharmacokinetics of the compound in humans.

10.8.1 Human Microdosing Studies

As described above, in vitro studies and in vivo evaluation in animal models aim to predict behaviour and fate of the radiolabelled compound in humans. However, drug metabolism pathways and pharmacokinetics can differ substantially from those predicted from preclinical studies. Hence, based on the concept that the "human is the best model for human" the "phase 0" or "exploratory studies" emerged, using the approach of microdosing (low dose of the compound).

In Europe, this approach is known as "human microdosing studies," according to the European Medicines Agency (EMA) or "exploratory clinical trials" when it concerns the Food and Drug Administration (FDA). Both EMA in 2004 [40], and the US FDA in 2006 [41] provided recognition to the concept and legitimacy to the conduct of such studies.

Human microdosing studies can be considered for radiolabelled compounds. In this case, very low single doses of the tested compound are administered to very few human subjects (healthy volunteers or patients) to investigate target uptake (e.g. target receptor binding) or tissue distribution and/or to obtain pharmacokinetics parameters (such as volume of distribution, and clearance).

According to the EMA, a microdose is defined as less than 1/100th of the dose calculated to yield a pharmacological effect of the test substance. This calculation is based on primary pharmacodynamic data obtained by in vitro and in vivo studies. Typically, the maximum dose must be less than 100 μg or 30 nMol [41]. Using such a low amount of the compound, human body exposure is expected to be very limited.

It is important to note that human microdosing studies allow providing preliminary pharmacokinetics without exposing excessive numbers of subjects to the radiolabelled compound, but not meant to replace the traditional phase 1 clinical trial.

10.8.2 Phase 1 Clinical Trial

When designing a phase 1 clinical trial, the overall study design has to consider many factors, including study size, study population, patient selection, inclusion criteria, dose escalation scheme (in case of radiotherapeutic compounds), specification of dose limiting toxicities, and secondary study objectives.

Before its use in phase 1 trials, the radiopharmaceutical has to be classified as "Investigational Medicinal Product (IMP)", and several mandatory documents, such as an investigators brochure (IB) and an Investigational Medicinal Product Dossier (IMPD) have to be prepared and submitted to the competent authorities and the ethics committee to obtain written approval. The IMPD should include all the useful information related to chemical and radiopharmaceutical quality of the compound (molecular weight, affinity, physico-chemical properties ...), as well as non-clinical data related to pharmacology, pharmacokinetics, toxicology, and dosimetry [42].

A guidance on the preparation of IMPD for radiopharmaceuticals has been published by the Radiopharmacy Committee of the EANM in 2014 [43]. Furthermore, a detailed study protocol has to be prepared, where every step of the protocol is well documented. Furthermore, an informed consent of the healthy volunteers or patients is also mandatory.

Phase 1 Study Design

As radiopharmaceutical imaging agents are usually administered at an extremely small mass dose with no pharmacologic effects and a short half-life, a very low incidence of adverse events is generally observed. Phase 1 clinical studies aim to provide information on the general feasibility of imaging/therapy with a given tracer, its target specificity, stability, safety, biodistribution, pharmacokinetics, and metabolism. Dosimetry studies of the radiolabelled compound in humans is also performed to be able to anticipate/exclude radiation-related side effects.

The following are key aspects for the design of a FIH study:

- Study population: Healthy volunteers and/or patients can be enrolled in one or multiple cohorts.
  - Demographic information from each subject (weight, body surface, age, gender ...).
- Test radiolabelled compound and reference radiotracers (e.g. test compared to [^{18}F]FDG) or a standard of reference (histology, or radiological imaging).
- Administered dose: usually, a single dose is used by intravenous route.
- Choice of imaging parameters (scan duration, acquisition mode, number of scans per subject) for biodistribution and dosimetry.
- Blood/urine sampling intervals for pharmacokinetic study.
- Safety profile.

The phase 1 study design for diagnostic radiotracers is quite straightforward compared to, for example, an interventional drug FIH clinical trial, where for instance, healthy volunteers or patients receive a single dose of the investigational drug or a placebo, starting with a very low dose for the first cohort. Thereafter, the dose is escalated in the next cohorts (or stopped depending on the tolerability and the safety). Single ascending studies (SAD) are usually followed by multiple ascending dose (MAD) studies in a very similar design, whereby the subjects receive multiple doses of the drug (or a placebo). Thus, the EMA advises that it is usually appropriate to design the administration of the first dose in a way that a single subject receives a single dose of the active IMP, with justification of the period of observation before the next subject receives a dose. This procedure allows for mitigating the risks associated with exposing all subjects in a same cohort simultaneously [44]. Note that there are several possible designs for interventional drugs in phase 1 study trials.

Contrary to imaging studies, interventional drug studies take into consideration the pharmacological effect of the drug, so the starting dose, maximum dose and exposure, and maximal duration of treatment are carefully considered. Also, besides the route and frequency of administrations, half-time and washout times of the IMP (if the same subjects are participating in multiple cohorts or accumulation for multiple dosing parts) are determined as well as sequence and interval between dosing of subjects within the same cohort. In the case of dose escalation increments, transition to the next dose increment cohort or next study part (if the FIH study include several parts) has to be decided depending on the tolerability and safety, and stopping rules have to be clearly established as well as the safety parameters to monitor.

Note that even if phase 1 clinical trials are primarily designed to assess the safety and tolerability of an interventional drug, the pharmacokinetics and, when appropriate and feasible, a pharmacodynamic measure is often included in order to facilitate the link with the non-clinical data and to support dose escalation decisions.

For radiotracers as well as for interventional drugs, the study design should take into consideration all preclinical findings, integrating all available toxicology and pharmacology information on the compound candidate to ensure the safety of the subjects.

10.9 ROLE OF PHYSICISTS IN RADIOPHARMACEUTICAL TRANSLATIONAL RESEARCH

It is common that during translational research a variety of health-related scientists with different backgrounds will gather for specific projects. The strong multidisciplinary component in this field requires additional training for each discipline and presents a unique opportunity to gain insight outside each individual's field of expertise.

For physicists or, more commonly working in health research, medical physics experts (MPE) involved in the different steps of drug discovery, there are typical areas in which to play a key role. In fact, core disciplines of MPE traditional curriculum will be directly applicable to translational research, including radiation protection, instrumentation calibration, image quantification, mathematical modelling, dosimetry, and application of novel radiomics models to preclinical data.

10.9.1 Radiation Protection

As almost all compounds to be tested are treated as unsealed radioactive sources, it is the MPE duty to establish a suitable radioprotection model, taking into account the amount, quality, and frequency of procedures. Use and disposal of radionuclides must be in conformity with the local radiation-protection law, and the same applies to dosimetric surveillance of occupationally exposed personnel.

10.9.2 Instrument Calibration

Most in-vivo and in-vitro sampling for radioactivity quantification (be it in form of autoradiography, gamma counter, SPECT, PET imaging) should be cross-calibrated to a reference dose calibrator. This will allow a more accurate assessment of interest metrics relevant for the radiopharmaceutical performance. In general terms, cross-calibration factors are nuclide-specific and will be valid for a specific range, provided by the manufacturer of the respective dose calibrator, in which signal response is linear. Animal imaging infrastructures are not bound by the same quality control requirements as human scanners, as such, there are many features that must be user-customized, requiring first line support that can be delivered by MPEs.

10.9.3 In Vivo Imaging

Nuclear medicine preclinical in-vivo imaging, image analysis and acquisition protocol modality are additional subjects with close cooperation with MPEs [45]. It is advisable to define a standardized strategy for quantification based on acceptance testing of dedicated phantoms (such as the NEMA methodology), with particular importance if multi-centre preclinical trials are the ultimate objective [46]. This includes efforts in determining the ideal tracer activity administered based on the scanner's sensitivity, evaluating the need of CT-based attenuation correction for different animal anatomical regions, defining volumetric approaches of image segmentation that minimize partial volume effects, or even developing dedicated DICOM analysis platforms that will facilitate the access to preclinical quantitative data [47].

10.9.4 Radiotracer Kinetics

Radiotracer kinetics is also intimately related with image quantification, as it will allow for determining the rate of metabolism of a radiopharmaceutical. The development and application of an adequate mathematical model that describes the expected in vivo biochemical behaviour is an important contribution of physics in translational research. For a thorough description of the available methods and techniques, refer to Volume 2, Chapter 5: Tracer Kinetic Modelling and Its Use in PET Quantification.

10.9.5 Radiation Dosimetry

Radiopharmaceutical accurate quantification and kinetic description constitute a fundamental step for radiation dosimetry, as these allow the estimation of the biological parameters considered in the Medical Internal Radiation Dose (MIRD) system. In other words, PET or SPECT animal biodistribution studies are acquired during an interval compatible with the expected tracer kinetics. The regions defined on the reconstruction data correspond to compartments that are mathematically described as a system of linear equations, defined in the MIRD scheme. Taking the physical half-life into consideration, the time-integrated activity (or cumulated activity) can be computed for each of the compartments. Finally, a model for radiation transport (S value in MIRD formalism) will describe the amount of absorbed dose per unit of time-integrated activity for specified radionuclide and reference anatomical model. With these tools, widely described in clinical dosimetry, an MPE is ready to contribute in radiopharmaceutical dosimetry investigations applied to small-animal data and extrapolation to humans [48].

10.9.6 Radiomics

Finally, in the context of precision medicine, radiomics is expected to provide a significant contribution to both clinical and translational research [49]. The construction of predictive models is based on image acquisition, computation of radiomics features, and statistical analysis, all of which are generally accepted MPE domains. All translational steps could be subject to data mining and computer aided decision-making, potentially increasing the speed and possibly the efficacy of the global process [49]. One of the key aspects for the success of radiomics workflows is the standardization of computation methods [50], emphasizing the fact that high-quality, properly designed, and transparent radiomics research should be perused by responsible researchers for whom new guidelines start to become available [51].

REFERENCES

[1] E.B. Podgorsak, "Rutherford-Bohr Model of the Atom. In: Radiation Physics for Medical Physicists, 3rd edn." in: *Graduate Texts in Physics*. Switzerland: Springer, Cham, 2016, 180–181.

[2] W. Cai, *Engineering in Translational Medicine*. Springer, 2014.

[3] M. Gotthardt *et al.*, "A new technique for in vivo imaging of specific GLP-1 binding sites: First results in small rodents," *Regul Pept,* vol. 137, no. 3, pp. 162–67, 2006, doi: 10.1016/j.regpep.2006.07.005.

[4] W. A. Breeman *et al.*, "Evaluation of radiolabelled bombesin analogues for receptor-targeted scintigraphy and radiotherapy," *Int J Cancer,* vol. 81, no. 4, pp. 658–65, 1999, doi: 10.1002/(sici)1097-0215(19990517)81:4<658::aid-ijc24>3.0.co;2-p.

[5] E. P. Krenning *et al.*, "Somatostatin receptor scintigraphy with indium-111-DTPA-D-Phe-1-octreotide in man: Metabolism, dosimetry and comparison with iodine-123-Tyr-3-octreotide," *J Nucl Med,* vol. 33, no. 5, pp. 652–8, 1992.

[6] M. de Jong *et al.*, "Comparison of (111)In-labeled somatostatin analogues for tumor scintigraphy and radionuclide therapy," *Cancer Res,* vol. 58, no. 3, pp. 437–41, 1998.

[7] E. Gourni *et al.*, "PET of CXCR4 Expression by a Ga-68-Labeled Highly Specific Targeted Contrast Agent," *Journal of Nuclear Medicine,* vol. 52, no. 11, pp. 1803–10, 2011, doi: 10.2967/jnumed.111.098798.

[8] M. Eder *et al.*, "68Ga-complex lipophilicity and the targeting property of a urea-based PSMA inhibitor for PET imaging," *Bioconjug Chem,* vol. 23, no. 4, pp. 688–97, 2012, doi: 10.1021/bc200279b.

[9] T. Lindner *et al.*, "Development of Quinoline-based Theranostic ligands for the targeting of Fibroblast Activation Protein," *J Nucl Med,* vol. 59, no. 9, pp. 1415–22, 2018, doi: 10.2967/jnumed.118.210443.

[10] H. J. Wester, "Nuclear imaging probes: from bench to bedside," *Clin Cancer Res,* vol. 13, no. 12, pp. 3470–81, 2007, doi: 10.1158/1078-0432.CCR-07-0264.

[11] M. Fani *et al.*, "Unexpected sensitivity of sst2 antagonists to N-terminal radiometal modifications," *J Nucl Med,* vol. 53, no. 9, pp. 1481–89, 2012, doi: 10.2967/jnumed.112.102764.

[12] J. C. Reubi *et al.*, "Affinity profiles for human somatostatin receptor subtypes SST1-SST5 of somatostatin radiotracers selected for scintigraphic and radiotherapeutic use," *Eur J Nucl Med,* vol. 27, no. 3, pp. 273–82, 2000, doi: 10.1007/s002590050034.

[13] S. Ray Banerjee *et al.*, "Effect of chelators on the pharmacokinetics of (99m)Tc-labeled imaging agents for the prostate-specific membrane antigen (PSMA)," *J Med Chem,* vol. 56, no. 15, pp. 6108–21, 2013, doi: 10.1021/jm400823w.

[14] M. Gabriel *et al.*, "99mTc-N4-[Tyr3]Octreotate Versus 99mTc-EDDA/HYNIC-[Tyr3]Octreotide: An intrapatient comparison of two novel Technetium-99m labeled tracers for somatostatin receptor scintigraphy," *Cancer Biother Radiopharm,* vol. 19, no. 1, pp. 73–9, 2004, doi: 10.1089/108497804773391702.

[15] C. Decristoforo and S. J. Mather, "The influence of chelator on the pharmacokinetics of 99mTc-labelled peptides," *Q J Nucl Med,* vol. 46, no. 3, pp. 195–205, 2002.

[16] P. Laverman *et al.*, "A novel facile method of labeling octreotide with (18)F-fluorine," *J Nucl Med,* vol. 51, no. 3, pp. 454–61, 2010, doi: 10.2967/jnumed.109.066902.

[17] Z. Liu, M. Pourghiasian, F. Benard, J. Pan, K. S. Lin, and D. M. Perrin, "Preclinical evaluation of a high-affinity 18F-trifluoroborate octreotate derivative for somatostatin receptor imaging," *J Nucl Med,* vol. 55, no. 9, pp. 1499–505, 2014, doi: 10.2967/jnumed.114.137836.

[18] R. Ting, M. J. Adam, T. J. Ruth, and D. M. Perrin, "Arylfluoroborates and alkylfluorosilicates as potential PET imaging agents: High-yielding aqueous biomolecular 18F-labeling," *J Am Chem Soc,* vol. 127, no. 38, pp. 13094–95, 2005, doi: 10.1021/ja053293a.

[19] E. Schirrmacher *et al.*, "Synthesis of p-(di-tert-butyl[(18)F]fluorosilyl)benzaldehyde ([(18)F]SiFA-A) with high specific activity by isotopic exchange: A convenient labeling synthon for the (18)F-labeling of N-amino-oxy derivatized peptides," *Bioconjug Chem,* vol. 18, no. 6, pp. 2085–89, 2007, doi: 10.1021/bc700195y.

[20] M. Behe, G. Kluge, W. Becker, M. Gotthardt, and T. M. Behr, "Use of polyglutamic acids to reduce uptake of radiometal-labeled minigastrin in the kidneys," *J Nucl Med,* vol. 46, no. 6, pp. 1012–15, 2005.

[21] M. Schottelius, F. Rau, J. C. Reubi, M. Schwaiger, and H. J. Wester, "Modulation of pharmacokinetics of radioiodinated sugar-conjugated somatostatin analogues by variation of peptide net charge and carbohydration chemistry," *Bioconjug Chem,* vol. 16, no. 2, pp. 429–37, 2005, doi: 10.1021/bc0499228.

[22] M. Jamous *et al.*, "PEG spacers of different length influence the biological profile of bombesin-based radiolabeled antagonists," *Nucl Med Biol,* vol. 41, no. 6, pp. 464–70, 2014, doi: 10.1016/j.nucmedbio.2014.03.014.

[23] A. Roxin *et al.*, "A metal-free DOTA-conjugated (18)F-labeled radiotracer: [(18)F]DOTA-AMBF3-LLP2A for imaging VLA-4 over-expression in murine melanoma with improved tumor uptake and greatly enhanced renal clearance," *Bioconjug Chem,* vol. 30, no. 4, pp. 1210–19, 2019, doi: 10.1021/acs.bioconjchem.9b00146.

[24] A. Wurzer *et al.*, "Radiohybrid ligands: A novel tracer concept exemplified by (18)F- or (68)Ga-labeled rhPSMA-inhibitors," *J Nucl Med,* 2019, doi: 10.2967/jnumed.119.234922.

[25] J. Lau, O. Jacobson, G. Niu, K. S. Lin, F. Benard, and X. Chen, "Bench to bedside: albumin binders for improved cancer radioligand therapies," *Bioconjug Chem,* vol. 30, no. 3, pp. 487–502, 2019, doi: 10.1021/acs.bioconjchem.8b00919.

[26] A. M. Farahani, F. Maleki, and N. Sadeghzadeh, "The influence of different spacers on biological profile of peptide radiopharmaceuticals for diagnosis and therapy of human cancers," *Anticancer Agents Med Chem,* 2019, doi: 10.2174/1871520620666191231161227.

[27] G. Vaidyanathan *et al.*, "Brush border enzyme-cleavable linkers: Evaluation for reducing renal uptake of radiolabeled prostate-specific membrane antigen inhibitors," *Nucl Med Biol,* vol. 62–63, pp. 18–30, 2018, doi: 10.1016/j.nucmedbio.2018.05.002.

[28] J. Zhang *et al.*, "From bench to bedside-The bad Berka experience With first-in-human studies," *Semin Nucl Med,* vol. 49, no. 5, pp. 422–37, 2019, doi: 10.1053/j.semnuclmed.2019.06.002.

[29] A. Farolfi *et al.*, "Theranostics for advanced prostate cancer: Current indications and future developments," *Eur Urol Oncol,* vol. 2, no. 2, pp. 152–62, 2019, doi: 10.1016/j.euo.2019.01.001.

[30] M. Kircher *et al.*, "CXCR4-directed theranostics in oncology and inflammation," *Ann Nucl Med,* vol. 32, no. 8, pp. 503–11, 2018, doi: 10.1007/s12149-018-1290-8.

[31] K. Vermeulen, M. Vandamme, G. Bormans, and F. Cleeren, "Design and Challenges of Radiopharmaceuticals," in *Seminars in Nuclear Medicine*, 2019: Elsevier.

[32] R. Zimmermann, *Nuclear Medicine: Radioactivity for Diagnosis and Therapy*. EDP Sciences, 2019.

[33] "Guideline on the non-clinical requirements for radiopharmaceuticals," (2018), in "EMA/CHMP/SWP/686140/2018,"

[34] J. Koziorowski *et al.*, "Position paper on requirements for toxicological studies in the specific case of radiopharmaceuticals," *EJNMMI Radiopharmacy and Chemistry,* vol. 1, no. 1, p. 1, 2017.

[35] D. M. Ramsey and S. R. McAlpine, "Halting metastasis through CXCR4 inhibition," *Bioorg Med Chem Lett,* vol. 23, no. 1, pp. 20–25, 2013, doi: 10.1016/j.bmcl.2012.10.138.

[36] "Good manufacturing practices for pharmaceutical products," in *WHO Expert Committee on Specifications for Pharmaceutical Preparations. Thirty-second Report.* Geneva, World Health Organization, 2014, Annex 2 (WHO Technical Report Series, No. 986).

[37] M. Schottelius, M. Konrad, T. Osl, A. Poschenrieder, and H.-J. Wester, "An optimized strategy for the mild and efficient solution phase iodination of tyrosine residues in bioactive peptides," *Tetrahedron Letters,* vol. 56, no. 47, pp. 6602–05, 2015.

[38] P. Elsinga *et al.*, "Guidance on current good radiopharmacy practice (cGRPP) for the small-scale preparation of radiopharmaceuticals," *European Journal of Nuclear Medicine and Molecular Imaging,* vol. 37, no. 5, pp. 1049–62, 2010.

[39] J. S. Lewis, A. D. Windhorst, and B. M. Zeglis, *Radiopharmaceutical Chemistry*. Springer, 2019.

[40] E. Tavernier-Tardy *et al.*, "Prognostic value of CXCR4 and FAK expression in acute myelogenous leukemia," *Leukemia Research,* vol. 33, no. 6, pp. 764–68, 2009.

[41] "Center for Drug Evaluation and Research. Guidance for Industry, Investigators, and Reviewers – Exploratory IND Studies," Center for Drug Evaluation and Research (CDER), Food and Drug Administration, US Department of Health and Human Services, 2006,

[42] G. L. Uy *et al.*, "A phase 1/2 study of chemosensitization with the CXCR4 antagonist plerixafor in relapsed or refractory acute myeloid leukemia," *Blood,* vol. 119, no. 17, pp. 3917–24, 2012.

[43] S. Todde *et al.*, "EANM guideline for the preparation of an Investigational Medicinal Product Dossier (IMPD)," *European Journal of Nuclear Medicine and Molecular Imaging,* vol. 41, no. 11, pp. 2175–85, 2014.

[44] K. N. Weilbaecher, T. A. Guise, and L. K. McCauley, "Cancer to bone: A fatal attraction," *Nature Reviews Cancer,* vol. 11, no. 6, pp. 411–25, 2011.

[45] C. Vanhove, J. P. Bankstahl, S. D. Kramer, E. Visser, N. Belcari, and S. Vandenberghe, "Accurate molecular imaging of small animals taking into account animal models, handling, anaesthesia, quality control and imaging system performance," *EJNMMI Phys,* vol. 2, no. 1, p. 31, 2015, doi: 10.1186/s40658-015-0135-y.

[46] W. McDougald *et al.*, "Standardization of preclinical PET/CT imaging to improve quantitative accuracy, precision, and reproducibility: A multicenter study," *Journal of Nuclear Medicine,* vol. 61, no. 3, pp. 461–68, 2020, doi: 10.2967/jnumed.119.231308.

[47] A. K. Tahari *et al.*, "Absolute myocardial flow quantification with 82 Rb PET/CT: Comparison of different software packages and methods," *European Journal of Nuclear Medicine and Molecular Imaging,* vol. 41, no. 1, pp. 126–35, 2014.

[48] T. Funk, M. Sun, and B. H. Hasegawa, "Radiation dose estimate in small animal SPECT and PET," *Medical Physics,* vol. 31, no. 9, pp. 2680–86, 2004.

[49] J. Vamathevan *et al.*, "Applications of machine learning in drug discovery and development," *Nature Reviews Drug Discovery,* vol. 18, no. 6, pp. 463–77, 2019.

[50] S. Chatterjee, B. B. Azad, and S. Nimmagadda, "The intricate role of CXCR4 in cancer," *Advances in Cancer Research,* vol. 124, p. 31, 2014.

[51] M. Vallières, A. Zwanenburg, B. Badic, C. Cheze Le Rest, D. Visvikis, and M. Hatt, "Responsible radiomics research for faster clinical translation," *Journal of Nuclear Medicine,* vol. 59, no. 2, pp. 189–93, 2018, doi: 10.2967/jnumed.117.200501.

11 Radionuclide Bone Scintigraphy

Kanhaiyalal Agrawal and Gopinath Gnanasegaran

CONTENTS

Radionuclide bone scintigraphy is one of the most common Nuclear Medicine investigations. There have been significant advances in both the instrumentation and in radiopharmaceuticals. However, conventional bone scan remains to be an extremely valuable diagnostic test in several benign and malignant bone diseases.

DOI: 10.1201/9780429489501-11

11.1 RADIOPHARMACEUTICALS USED IN BONE SCAN

Single Photon Emission Computed Tomography (SPECT) tracers used in bone scintigraphy are Technetium-99m (^{99m}Tc) based. The tracers are ^{99m}Tc methylene diphosphonate (^{99m}Tc MDP), ^{99m}Tc hydroxymethylene diphosphonate (^{99m}Tc HDP or HMDP), ^{99m}Tc hydroxyethylidene diphosphonate (^{99m}Tc HEDP), and ^{99m}Tc pyrophosphate. The positron emission tracer is sodium fluoride-18 (^{18}F -NaF).

The bone imaging agents are either phosphates or diphosphonates. The P-O-P bond in phosphate is easily broken down by phosphatase enzyme and are less stable in vivo compared to stable P-C-P bond in diphosphonate. Therefore, diphosphonate is a commonly used radiopharmaceutical in bone imaging [1]. The diphosphonate agents are usually available in kit form. Labelling occurs by adding pertechnetate to the cold kit and mixing. Generally, the oxidation state of ^{99m}Tc in bone kits is 3+. The uptake in bone is due to the ion exchange phenomena. ^{99m}Tc diphosphonate binding occurs by chemisorption in the hydroxyapatite mineral component of the bone matrix. Similarly, ^{18}F fluoride after diffusion through capillaries into bone extracellular fluid, exchanges with hydroxyl groups in hydroxyapatite crystal to form fluorapatite, which is deposited mainly at the bone surface. Both tracers are primarily excreted through the kidneys. ^{18}F fluoride is a better imaging agent, being a positron emitter agent, with faster blood clearance due to no protein binding and two-fold higher uptake (nearly 100% first pass extraction) in bone compared to ^{99m}Tc MDP [2].

11.2 RADIOPHARMACEUTICAL DOSE

^{99m}Tc agents: The recommended activity for an adult is intravenous injection of 740–1,110 MBq (20–30 mCi) of ^{99m}Tc labelled agents. For obese patients, the administered tracer activity may be increased to 11–13 MBq/kg (300–350 μCi/kg). In children, the recommended dose is 9–11 MBq/kg (250–300 μCi/kg), with a minimum of 20–40 MBq (0.5–1.0 mCi) [3].

^{18}F fluoride: The typical adult dose is 185–370 MBq (5–10 mCi). Higher dose should be considered in obese patients. Paediatric patients' dose should be weight-based (2.22 MBq/kg, 0.06 mCi/kg) with a minimum activity of 18.5 MBq (0.5 mCi) [4].

Dosimetry of ^{99m}Tc diphosphonates is mentioned in Table 11.1 [5].

11.3 IMAGE ACQUISITION IN ^{99M}Tc LABELLED BONE SCAN

Image acquisition in a bone scan depends on the different methods of bone scan as described below [6]:

11.3.1 Types of Bone Scan

Whole-body scan: Planar imaging of anterior and posterior projections

Focal static planar images: imaging specific part of the skeleton.

Three-phase bone scan: Multiphase bone scan produces planar images of the vascular phase (phase 1), the soft tissue/blood pool phase (phase 2), and delayed phase (phase 3) images over a specific region of interest.

TABLE 11.1
Dosimetry of ^{99m}Tc Phosphates and Phosphonates [5]

	Effective Dose mSv per MBq	Organ Receiving the Largest Radiation Dose mGy per MBq
Adult	0.0057 mSv/MBq	Bone Surfaces 0.063 mGy/MBq
Child (5 years)	0.014 mSv/MBq	Bone surfaces 0.22 mGy/MBq
Child (1 year)	0.027 mSv/MBq	Bone surfaces

Single photon emission computed tomography (SPECT): Three-dimensional distribution of the radiopharmaceutical in the skeleton.

SPECT/CT: SPECT acquisition combined with CT

11.3.2 Whole-body Bone Scan

11.3.2.1 Patient Preparation

No specific patient preparation is needed for a bone scan except that the patient must be well hydrated on the day of the study. Patient should void frequently after administration of tracer and just before the study. All metallic objects must be removed from the patient during imaging if possible.

11.3.2.2 Imaging Procedure

Anterior and posterior view images are acquired. A whole-body scan is acquired such that more than 1.5 million total counts may be obtained. The table rate should be adjusted depending on the equipment manufacturer's specification. Energy window should be centred at 140 keV. Window width is kept at 15–20 per cent. Matrix size is 256 x 1024 or higher. If needed, a spot view may be acquired depending upon the clinical indications and should be 500K–1000K per image (less for skull and extremities). Additional lateral, oblique or special (frog-leg views of the hips) views may be acquired if necessary.

11.3.3 Three-phase Bone Scan

The patient preparation is similar to the whole-body scan. The gamma camera detector should be positioned over the area of interest before radiotracer administration. Flow phase is the first phase of the triphasic bone scan, which is a dynamic image acquisition. It is acquired for 60 seconds with a frame rate of 1–3 sec/frame in a matrix size 64 x 64 or higher. The second phase is the blood pool phase, which is a static image following flow phase within 10 minutes of tracer administration. Time-based acquisition for 5 minutes or counts-based 300K counts/image is acquired. Matrix size is 256 x 256. Finally, the last phase, that is, the delayed phase, is acquired as a static image 2–4 hours after tracer administration, for approximately 1000 K counts.

11.3.4 SPECT-CT

SPECT acquisition has the following parameters: Contoured orbit, 64 x 64 or greater matrix, 3–6° intervals, 10–40 sec/stop. SPECT image reconstruction is done with 3D iterative OSEM (ordered-subsets expectation maximization) with ideally 3–5 iterations and 8–10 subsets. CT acquisition parameters are 512 x 512 matrix, 80–130 kV; the intensity–time product depends upon the body part being imaged and ranges from 10–300 mAs.

11.4 NORMAL VARIANTS OF RADIOTRACER DISTRIBUTION IN A BONE SCAN

The normal tracer distribution is homogeneous throughout the axial and appendicular skeleton with tracer excretion through kidneys into the urinary bladder (Figure 11.1). However, there are many normal variants which should be considered while reporting a bone scan.

11.4.1 Skull

Increased uptake at the confluence of sutures in the skull, such as at the pterion, is usually normal physiological uptake (Figure 11.2) [7]. Focal uptake at the superior lateral margin of the orbits is normal uptake at fronto-zygomatic suture (Figure 11.3A) [8]. Hot skull sign or diffuse increased uptake in the skull is a normal variant mainly seen in post-menopausal women (Figure 11.4) [9].

11.4.2 Thorax

Two or three linear bands of high-grade tracer uptake in the sternum corresponding to synchondroses is a normal variation [10]. The uptake in paired ossification centre in the sternum in children appears as a double sternum is physiological [10].

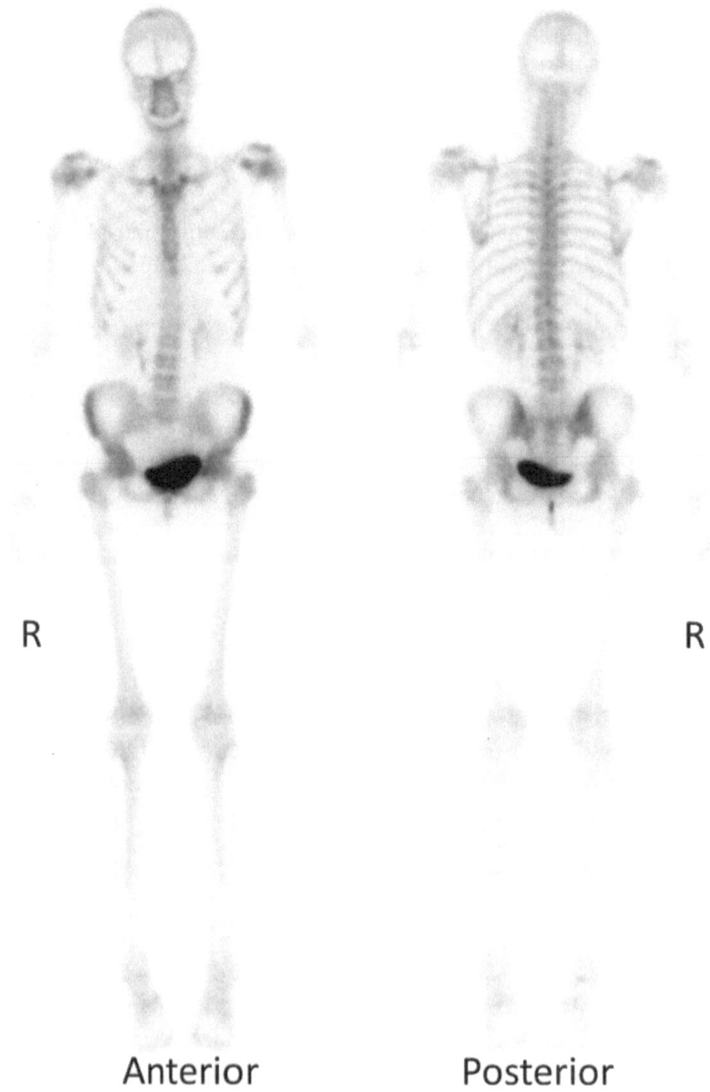

FIGURE 11.1 Normal bone scan showing homogeneous tracer distribution in the axial and appendicular skeleton and tracer excretion through kidneys and ureters into urinary bladder.

The cold area above the xiphoid is due to sternal foramina and is a normal variation seen in approximately 9 per cent of patients on a planar bone scan (Figure 11.5) [11]. Low-grade focal uptake at a posterior angle of three or more consecutive ribs is known as *Stippled ribs* and is due to the insertion of the iliocostalis thoracis portion of the erector spinae muscles to the ribs [12].

11.4.3 Abdomen and Pelvis

Focal tracer uptake at the ossifying ischiopubic synchondrosis in children between 4 and 12 years of age is normal [10]. In postpartum women, increased uptake in the pubic symphysis could be due to increased stress reaction or pelvic diastases [7]. Transient tracer uptake in the uterus on early phases of the three-phase bone scan in women of the reproductive age group is physiological. Sometimes altered tracer distribution in bowel is confused with bone uptake.

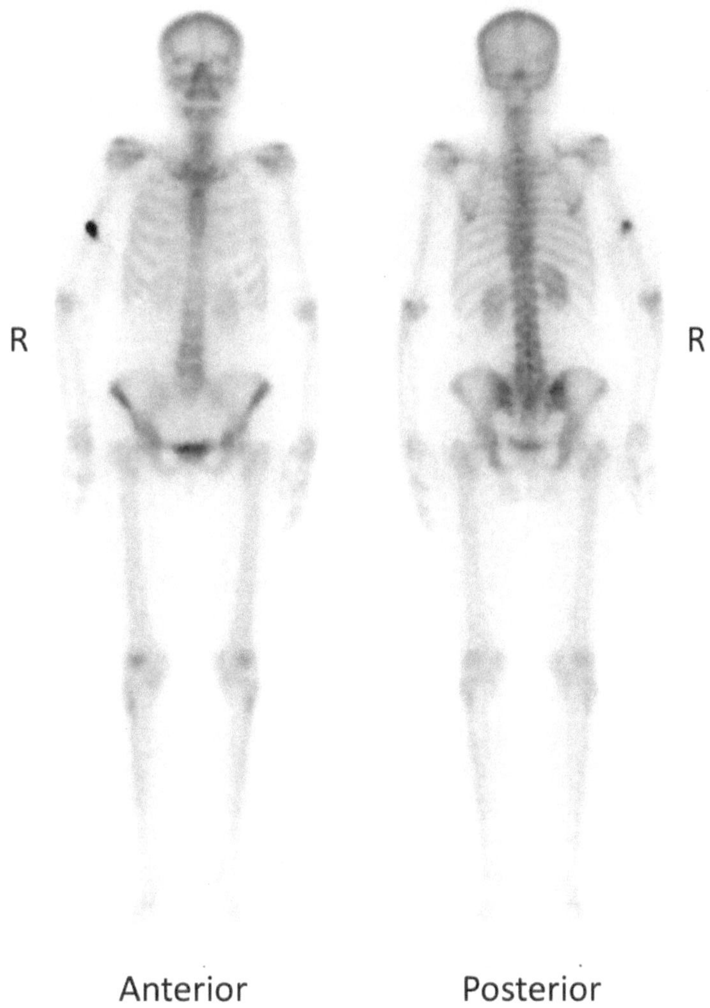

FIGURE 11.2 Whole body bone scan showing prominent tracer uptake in the skull posteriorly at the confluence of sutures. Injection site tracer extravasation in the right arm.

11.4.4 Benign Conditions Mimicking Sinister Pathology on Bone Scan

11.4.4.1 Skull

Solitary focal uptake is not always malignant or metastases and may be due to sutural foramina, small cartilaginous rests, or enlarged Pacchionian granulation [9]. Diffuse increased skull uptake is seen in metabolic bone disorder [9]. Focal uptake in mandible is commonly due to benign local dental pathology (Figure 11.6). Increased uptake in sinuses is usually due to infection/inflammation.

11.4.4.2 Thorax

Sternoclavicular joints uptake is mostly due to degenerative changes (Figure 11.6). Midline vertical linear area of increased uptake is seen in the sternum in patients who have undergone sternotomy in cardiothoracic surgery (sternal split sign). Focal uptake in a few consecutive ribs is seen in linear fashion due to traumatic fractures (Figure 11.7).

11.4.4.3 Abdomen and Pelvis

Transplant and ectopic pelvic kidney may resemble abnormal uptake in the pelvic bones (Figure 11.5). Non-osseous tracer uptake is seen in splenic infarct, liver metastasis (Figure 11.8), lung metastasis, ascites, urinary bladder stones,

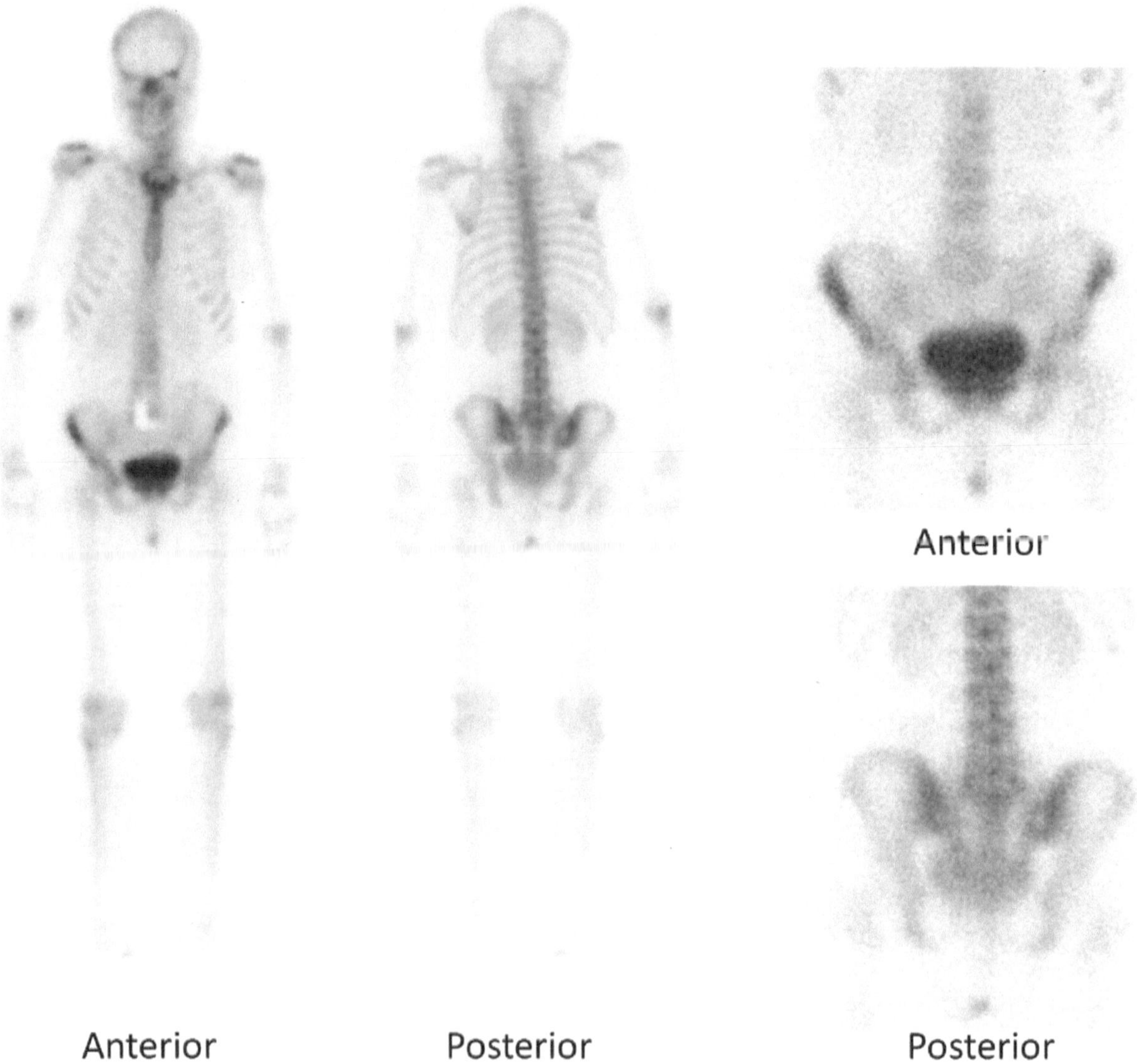

FIGURE 11.3 Whole body bone images show photopenic area in the lower lumbar region anteriorly, which is due to metallic belt buckle and normalises in the image taken after removing the belt (right column in the figure). The focal uptake at the left frontozygomatic suture is a normal variant.

neonatal necrotizing enterocolitis, sanitary tampons in the vagina [9]. Sometimes the primary tumour shows tracer uptake and should be differentiated from metastases, for example, neuroblastoma primary, lung cancer (Figure 11.9–10).

11.5 ARTEFACTS ON BONE SCAN

The artefacts can be categorized into instrumental, technical, radiopharmaceutical-related, treatment-related, and patient-related. Understanding these artefacts help in avoiding misinterpretation of a study.

11.5.1 Instruments-related Artefacts

Improper photopeak window setting leads to loss of photopeak counts and more of scatter, leading to poor quality of image. Therefore, reset of photopeak for ^{99m}Tc must be done. Photomultiplier tube defect gives a cold area in the image. Although it is rare and generally obvious on the image as an artefact, sometimes it may be mistaken as true abnormal uptake.

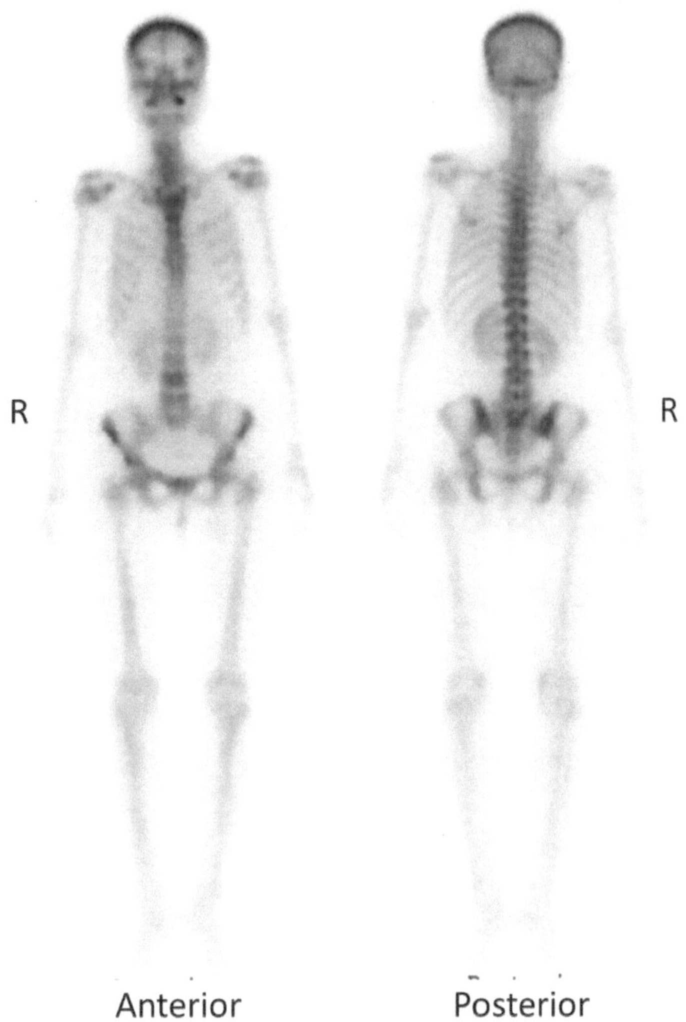

FIGURE 11.4 Bone scan of a 61-year-old lady with lung cancer shows diffuse increased uptake in the skull which is a normal variant (hot skull).

11.5.2 Technical Artefacts

Injection site tracer extravasation may give a poor image due to lesser injection dose to bones. Injection site tracer extravasation may also lead to lymph node visualization. Arterial injection leads to diffuse intense uptake at the site of injection. Therefore, careful tracer administration is very important in all Nuclear Medicine studies.

11.5.3 Patient Related Artefacts

Patient motion artefact may lead to blurring of image with loss of resolution. Also, if patient moves hand from one position to other, there may a dual image mimicking abnormal uptake. This may be avoided by proper patient positioning and instruction [9]. Metals (e.g. jewellery, pacemaker, belt buckle) lead to a cold defect on a planar bone scan and may be misinterpreted as a lytic lesion. This can be avoided by removal of the metal before the scanning if possible or documenting metal that cannot be removed, like pacemaker (Figure 11.3). Skin tracer contamination should be considered, particularly in the pelvis and lower extremity region. This may be mistaken as metastatic lesion or abnormal pathology. Usually, the uptake due to contamination is noted in only one image, either anterior or posterior. If suspected, repeat imaging after cleaning the particular area or changing the cloth should be performed (Figure 11.11)

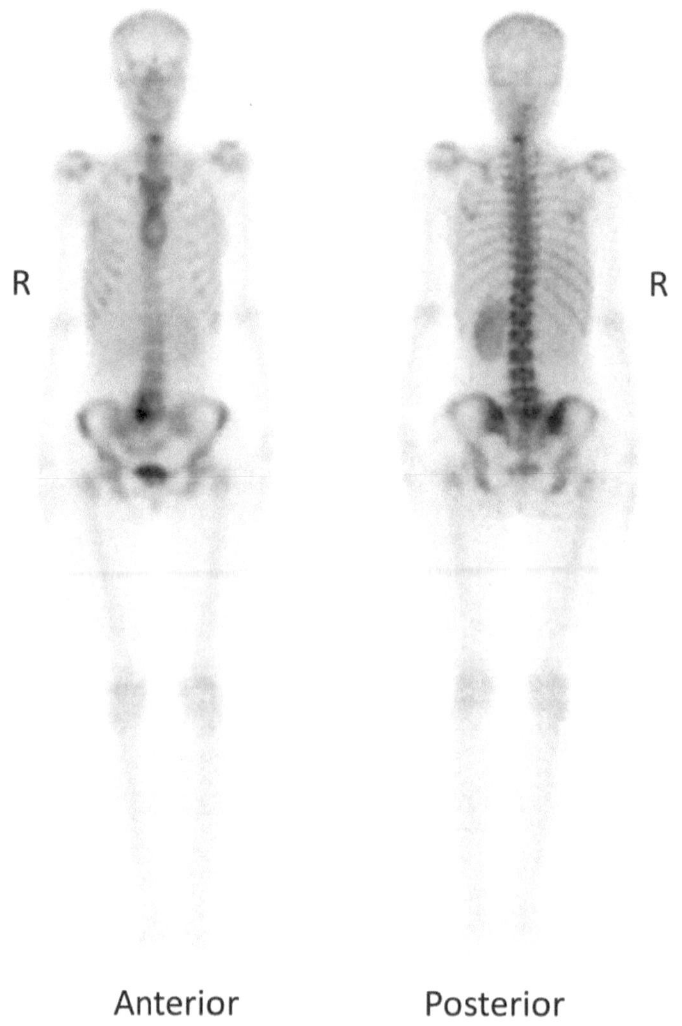

FIGURE 11.5 Bone scan of 34-year-old lady with breast carcinoma shows sternal foramina which is noted at the inferior part of sternum and is within normal limits. Right side ectopic pelvic kidney should not be confused with increased uptake in the lumbar vertebrae.

11.5.4 Treatment-related Artefacts

Abnormal bone uptake persists for a long time after surgical manipulation of particular bone. Therefore, careful history is essential to avoid a wrong diagnosis. Radiation therapy to a particular area leads to reduced uptake in bones within the radiation field and proper history helps in making a correct interpretation of the study. In patients with a recent history of chemotherapy, there may be increased tracer uptake (flare) in the existing lesions due to good response to therapy and should not be interpreted as disease progression. The tracer activity reduces after 4–6 months of flare.

11.5.5 Radiopharmaceutical-related Artefacts

Excess stannous in the cold kit remains available for labelling of red blood cells, resulting in reticuloendothelial system and soft tissue tracer localization. However, this is not a problem nowadays with good quality kit preparation.

11.6 INDICATIONS OF BONE SCAN

Bone scan is an extremely sensitive modality to detect osteoblastic reaction in the bone. The common clinical indications are as follows:

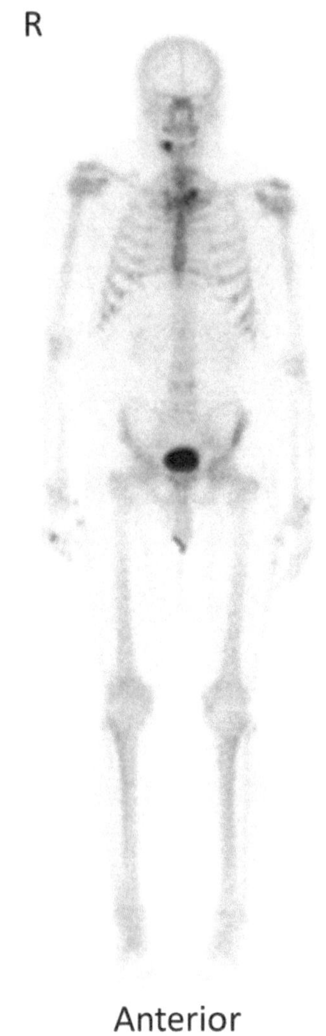

FIGURE 11.6 Focal tracer uptake in the right hemi-mandible is likely due to local dental pathology. However, rarely this could be metastatic and should be interpreted cautiously. The uptake seen at the left sternoclavicular joint is due to degenerative changes.

11.6.1 Malignant Conditions

- To detect osteoblastic bone metastasis
- Primary bone tumour

11.6.2 Benign Bone Diseases

- Bone dysplasia, for example, fibrous dysplasia
- Benign bone tumour like osteoid osteoma
- Osteomyelitis
- Stress fracture/shin splint
- Metabolic bone disease and Paget's disease
- Avascular necrosis
- Complex regional pain syndrome
- Painful knee and hip prosthesis
- Unexplained bone pain, for example, low back pain, hip pain, foot pain
- Bone-graft viability
- Non-union fracture

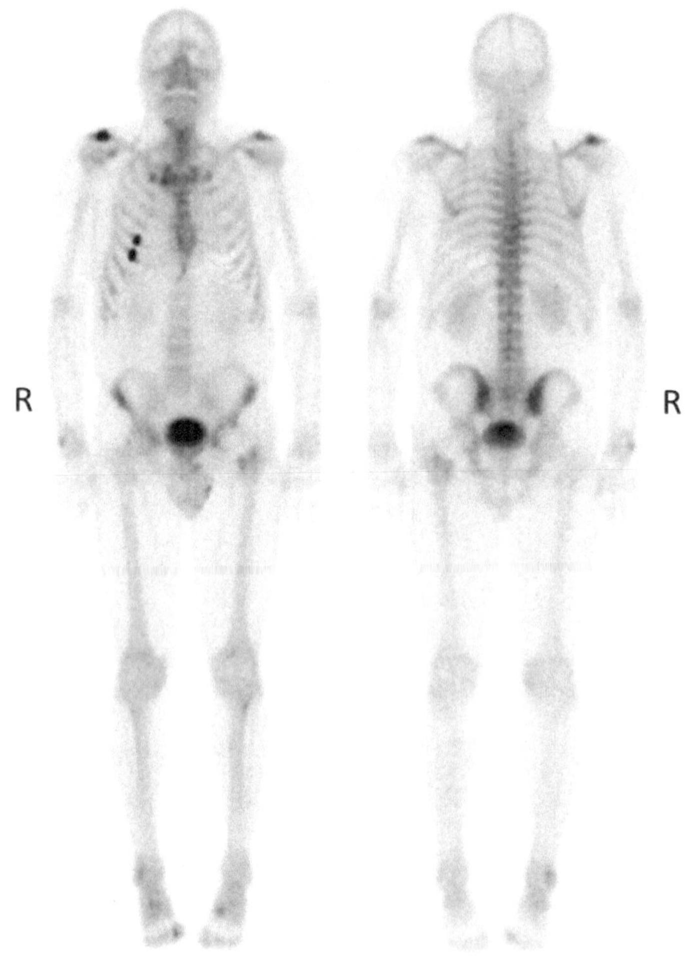

FIGURE 11.7 Patient with left revision total hip replacement underwent bone scan to rule out infection and loosening. The whole-body images show mild tracer uptake in the femoral component of hip left prosthesis. Additionally, there is focal increased uptake in the right acromioclavicular joint, right fifth and sixth ribs compatible with fractures in context of history of fall.

11.7 BONE SCAN IN METASTASES

Bone is the third most common site of metastasis after lung and liver [13]. The majority of metastases in the bone are hematogenous in nature. Mostly, the spread to bone occurs through the normal venous system or through Batson's plexus. Through this, tumour cells enter into the red marrow of the medullary bone. These cells attach to endothelial surfaces, multiply, and invade the bony structures to include the cortical bone [14]. These cells then induce osteolytic activity by secreting various factors. The parathyroid hormone-related peptide (PTHrP) has a dominant role in the development of lytic metastasis. The receptor activator of NF-kappaB (RANK) ligand has a major role in the formation of osteoclasts by stimulating precursor cells when it binds to the receptor activator of RANK on the cell membrane of osteoclast precursors [15]. The ongoing bone resorption induces a reparative response in the adjacent normal bone, that is, an osteoblastic reaction. However, some of the metastatic cells directly secrete local substances, like transforming growth factor, bone morphogenic proteins (BMP) and endothelin-1: Those are associated with osteoblast generation. In general, malignancies that are rapidly growing produce osteolytic lesions. Hence, the metastatic cells can induce either osteolytic, osteoblastic, or mixed response in bone. Metastasis from different types of malignancies can produce predominantly blastic, lytic or mixed bone metastases [Table 11.2]. Prostate and breast primarily cause the majority of the bone metastases, that is, up to 70 per cent [16].

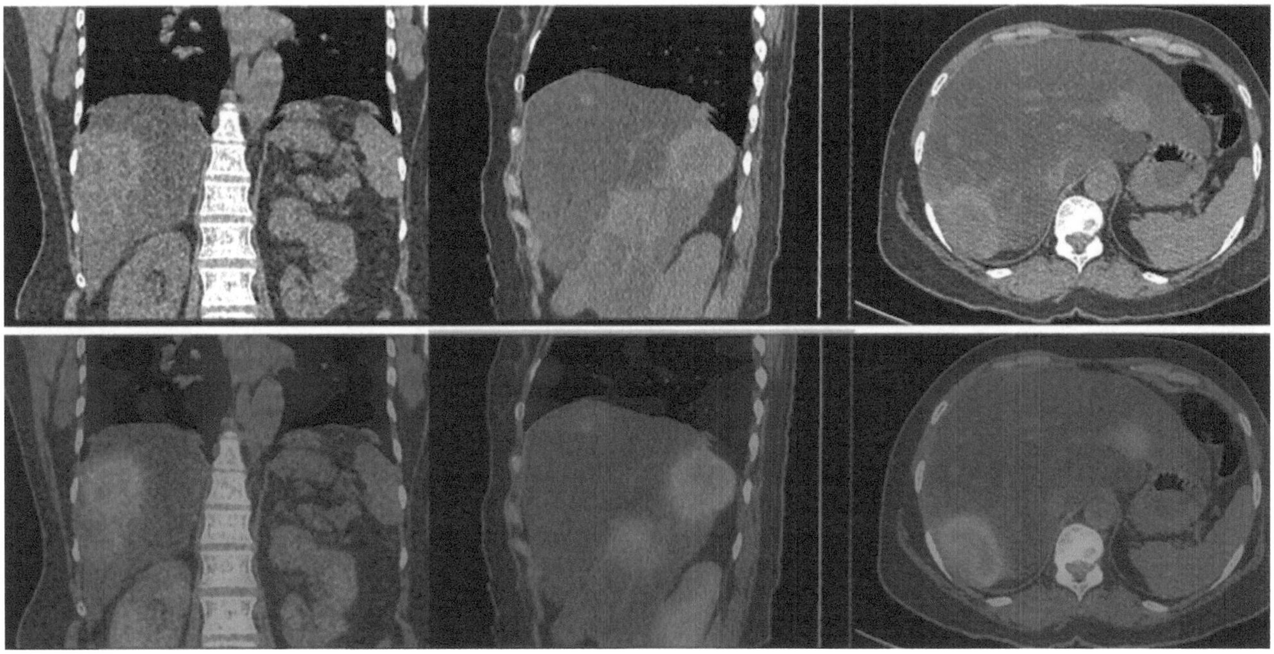

FIGURE 11.8 46-year-old lady with breast cancer showed prominent tracer uptake in the right thoracic base region (not shown), which on SPECT-CT of the lower thorax localizes to the hyperattenuating lesions in the liver suggestive of metastasis. This image highlights that bone specific tracers can localize to the non-osseous sites. Colour image available at www.routledge.com/9781138593312.

The radiopharmaceuticals used in bone scans are usually incorporated into the hydroxyapatite crystal matrix in the newly formed bone. Hence, a bone scan is ideal for a metastasis that induces osteoblastic reaction. The lytic lesions may be detected when there is an associated osteoblastic reaction or when lytic lesion itself is large enough to be detected as a photopenic defect. However, the lytic lesions may be missed when there is no associated osteoblastic reaction. The different patterns of metastatic disease on bone scan are mentioned in Table 11.3.

11.8 BONE SCAN IN BENIGN BONE DISEASES

Bone scans, particularly bone SPECT-CT, are very helpful in the diagnosis of benign bone diseases. The planar bone scan is highly sensitive in detecting uptake in the benign bone disease; however, the findings can be non-specific and non-contributary in most of such cases. SPECT-CT helps in localization of the uptake seen on the planar bone scan and also characterization of pathology at the uptake site. In most patients with shin splints, stress fractures, Paget's disease, and Fibrous dysplasia, the diagnosis can be easily made on a planar bone scan based on the tracer uptake pattern. The role of SPECT-CT is complementary and, hence, can be additionally used to increase diagnostic confidence. However, in unexplained pain in the wrist, back, hip, knee, ankle, and foot, the uptake pattern is non-specific on the planar scans. SPECT-CT helps immensely in these conditions to reach an accurate diagnosis.

11.8.1 Fibrous Dysplasia

Fibrous dysplasia (FD) can be monostotic or polyostotic. A whole-body bone scan is performed to differentiate monostotic FD from the polyostotic one. The tracer uptake in FD is very high. It can be differentiated from Paget's disease by the patient's age and the tracer uptake pattern on the bone scan. Paget's disease typically involves the end of the affected bone, whereas FD does not involve the bone's ends.

11.8.2 Osteoid Osteoma

Osteoid osteoma is a benign bone tumour usually present in the adolescent age group, typically as severe night pain that is relieved using salicylate analgesics. It is a cortex-based tumour. In general, a three-phase bone scan is positive

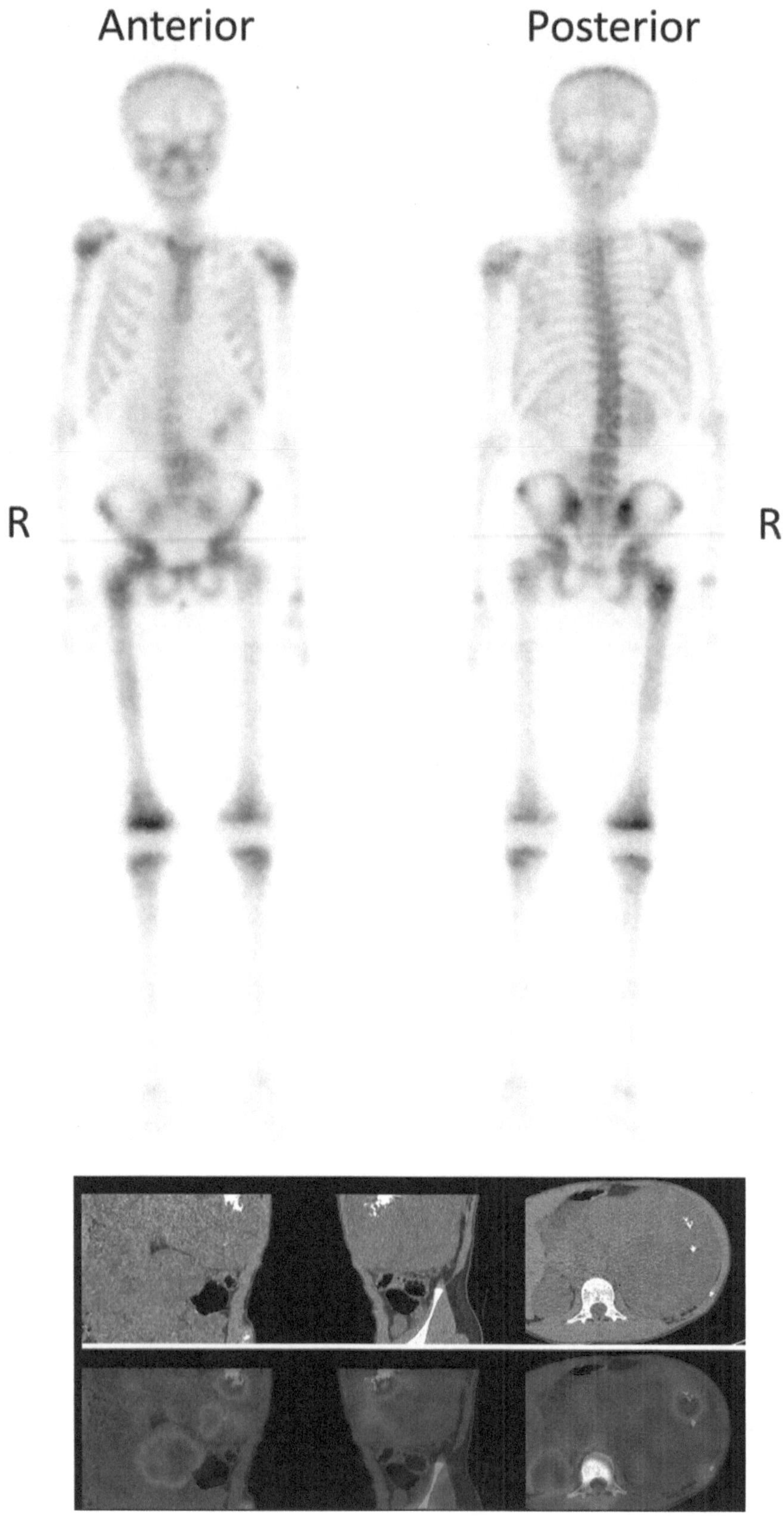

FIGURE 11.9 12-year-old boy with diagnosed neuroblastoma shows patchy uptake in the left side of abdomen and pelvis, which is due to some degree of uptake in the primary soft tissue mass in abdomen (lower images) pushing the left kidney inferomedially. There is also asymmetrical uptake in the femora suggestive of bone metastasis. Colour image available at www.routledge.com/9781138593312.

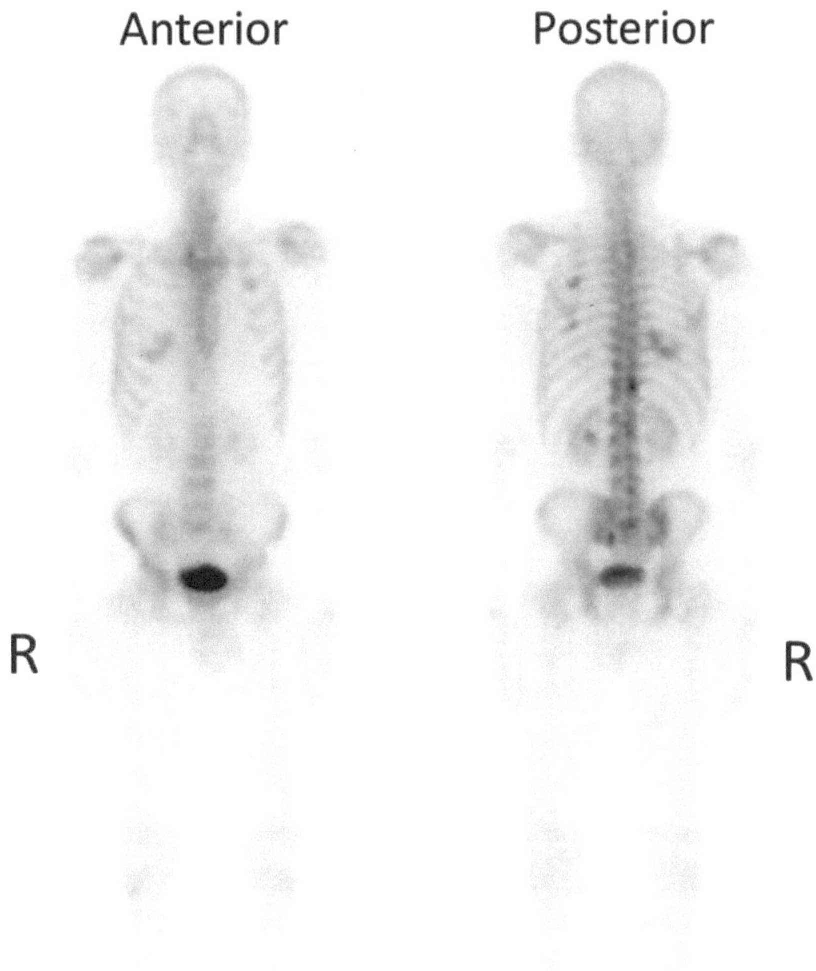

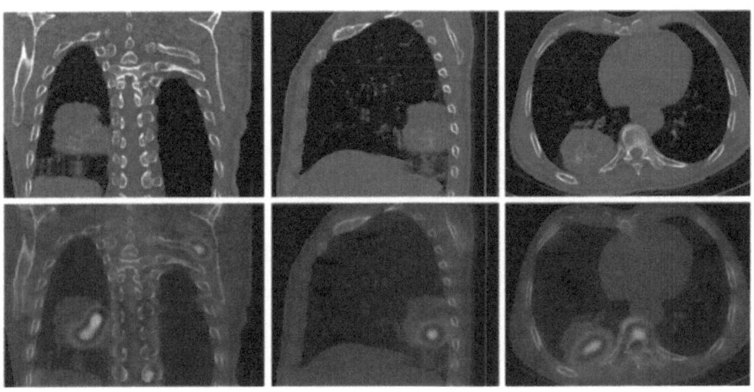

FIGURE 11.10 Patient with lung cancer shows a few foci of tracer uptake in the bones suggestive of metastases, along with tracer uptake in the right hemithorax. Lower images: The tracer uptake in the right hemithorax localizes to calcification within the primary lung mass. Colour image available at www.routledge.com/9781138593312.

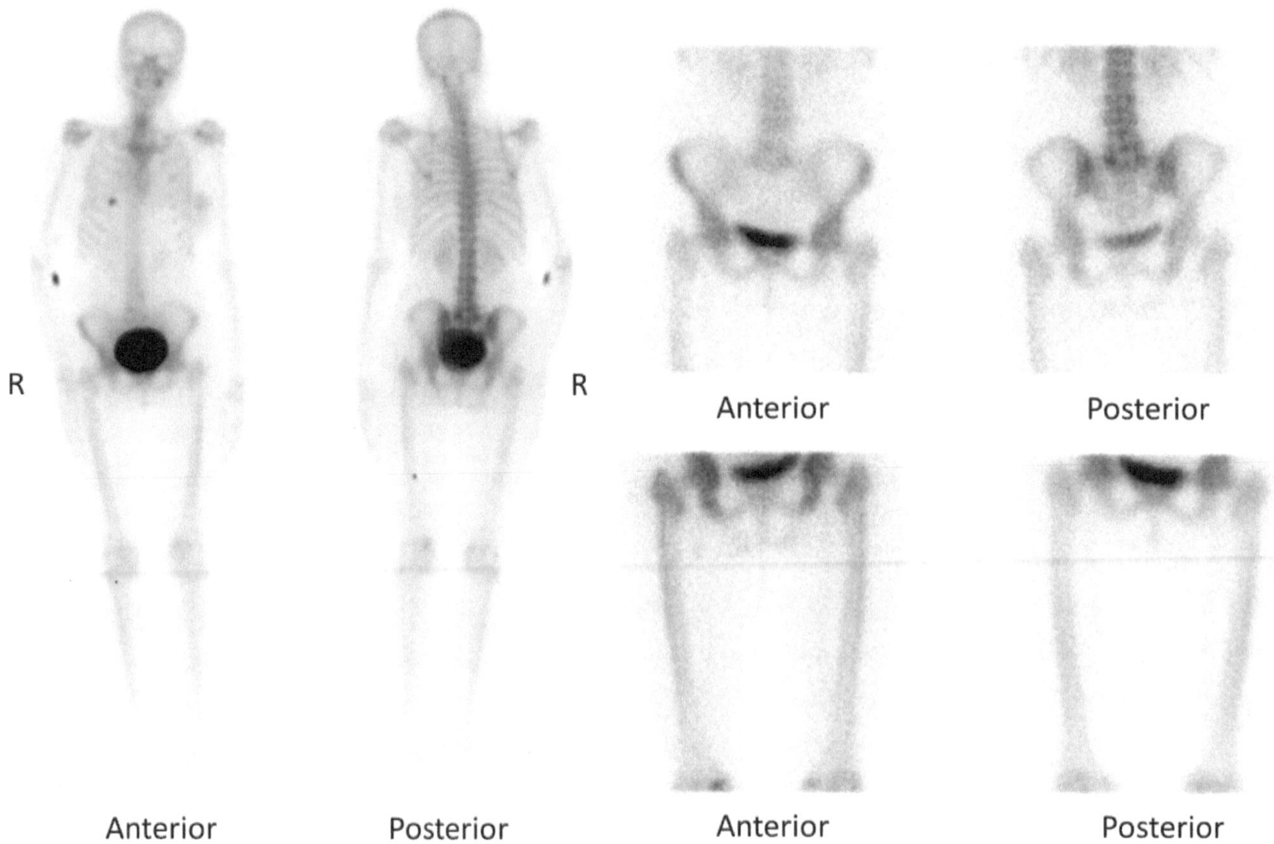

FIGURE 11.11 53-year-old lady with breast cancer shows focal uptake in the right 5th rib, which on SPECT-CT localizes to the callus formation suggestive of healing fracture (not shown). Further on whole body images, patient had urinary bladder filled with radioactive urine which is masking the sacrum and also a tiny focal uptake in the left mid-femur only on posterior image. The focal uptake in left femur is not seen after cleaning the area and re-imaging, suggestive of local surface contamination.

TABLE 11.2
Bone Metastasis in Malignancy Producing Different Responses

Predominantly osteoblastic	Predominantly osteolytic	Mixed reaction
Prostate cancer	Renal cell carcinoma	Breast cancer
Carcinoid	Thyroid cancer	Stomach cancer
Small cell lung cancer	Multiple myeloma	Colon cancer
Hodgkin lymphoma	Melanoma	Urinary bladder cancer
Medulloblastoma	Non-small cell lung cancer	Squamous cancers
	Langerhans-cell histiocytosis	

with focal uptake and double density sign, that is, intense tracer accumulation in the central nidus and moderate uptake of tracer in surrounding reactive sclerosis. SPECT-CT is highly sensitive and specific in the identification of osteoid osteoma, particularly in the spine.

11.8.3 Osteomyelitis and Septic Arthritis

A three-phase bone scan's typical pattern is increased tracer accumulation at the infection site in all three phases. Although the bone scan is highly sensitive with more than 95 per cent sensitivity, the specificity is low. Other conditions

TABLE 11.3
Metastatic Patterns on Bone Scan

Metastasis	Benign uptake mimicking metastasis
Randomly scattered multifocal uptake (Figure 11.12)	Multifocal uptake in contiguous ribs could be post-traumatic (Figure 11.7)
Diffuse heterogeneous uptake in skeleton (Figure 11.13)	Multifocal uptake may be seen in polyostotic bone dysplasia
A cold defect with peripheral rim uptake (lytic lesion)	Multifocal uptake in the facet joints/end plate region of spine could be degenerative changes
Photopenic defect (lytic lesion)	Diffuse uptake in the long bones due to hypertrophic pulmonary osteoarthropathy (Figure 11.14)
Asymmetric uptake in growth plates of long bones in children in neuroblastoma (Figure 11.9)	

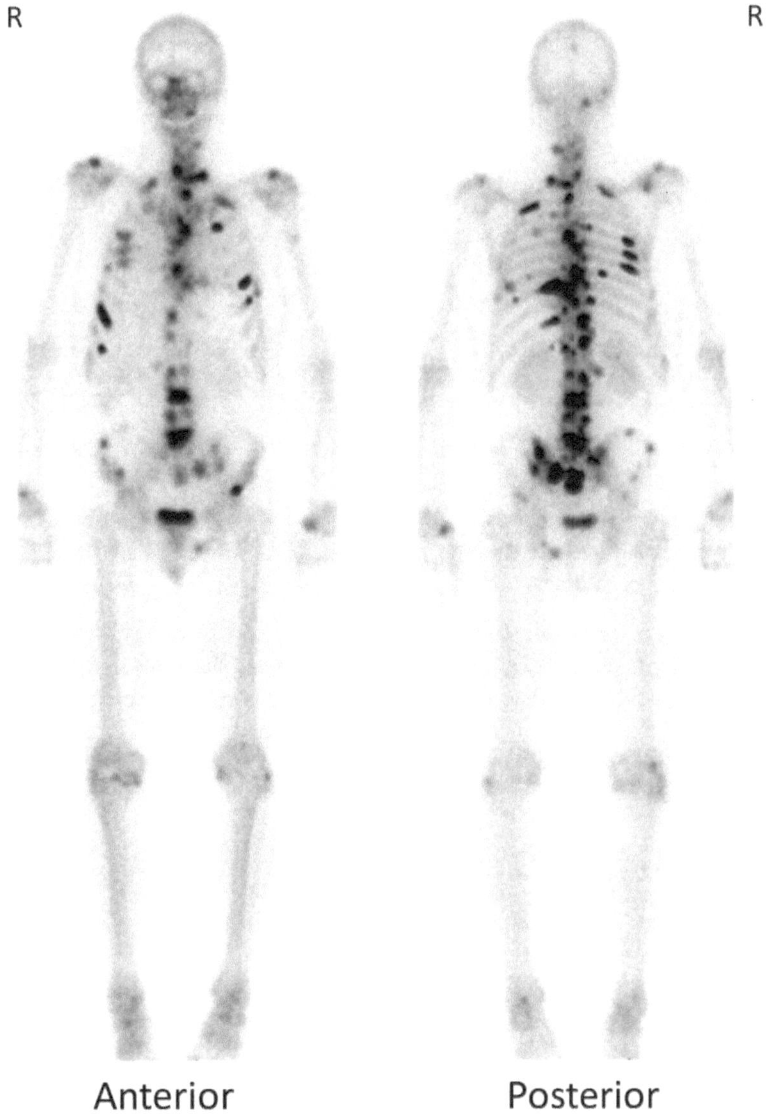

FIGURE 11.12 Patient with prostate cancer shows multiple foci of abnormal increased tracer uptake in multiple bones suggestive of widespread metastatic disease.

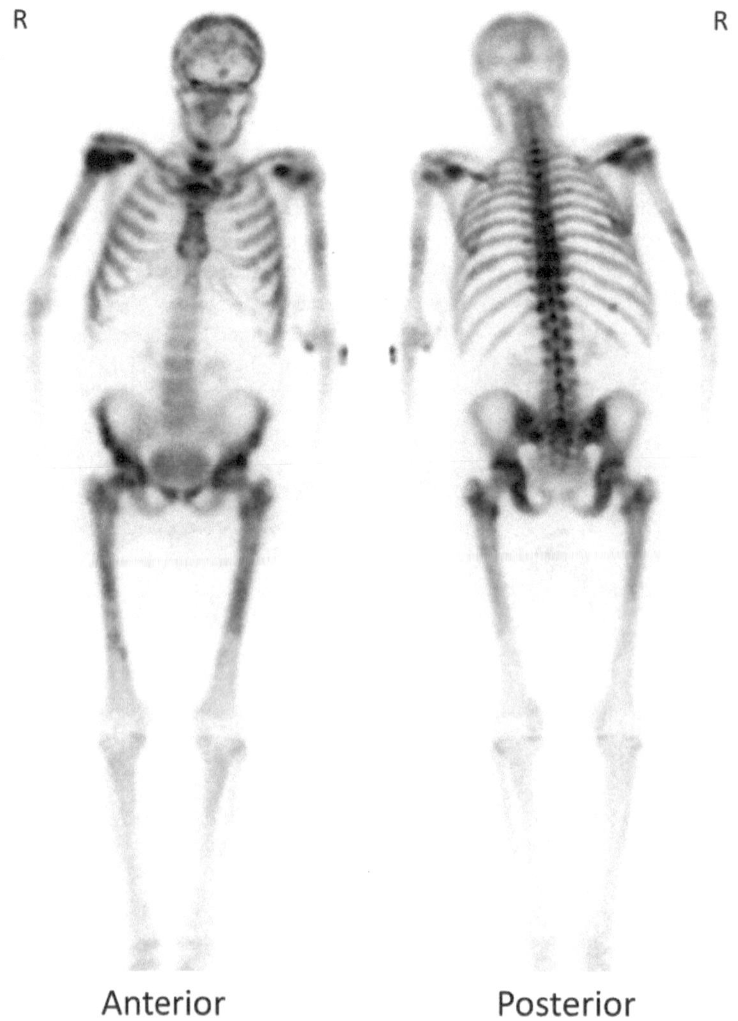

FIGURE 11.13 Diffuse heterogeneously increased tracer uptake is seen throughout the axial and appendicular skeleton with faintly visualized kidney, typical of superscan of malignancy. This pattern is usually suggestive of widespread skeletal metastases.

also show positive three-phase uptake, for example, malignancy, stress fracture, and osteoid osteoma. However, in non-violated bone with clinical suspicion of osteomyelitis, three phase bone scan detects bone infection with good specificity. False-negative findings may be seen in chronic osteomyelitis and vertebral osteomyelitis. Indium-111 labelled WBC, along with marrow imaging, is the radionuclide imaging of choice in violated bone infection. Septic arthritis, like osteomyelitis, shows increased tracer accumulation in the involved joints in all three phases.

11.8.4 Stress and Insufficiency Fracture

Stress fractures are fractures that occur in normal bone due to repetitive activity or overuse. In contrast, an insufficiency fracture is due to normal stress in a pathological or weak bone. Early diagnosis helps in early recovery and prevention of frank fractures. The bone scan is positive from the very early, or initial, phase of stress reaction and is extremely sensitive. Further, as stress reactions may involve multiple sites, the whole-body assessment helps detect additional areas. The uptake is usually focal intense with a shape ranging from oval to fusiform, depending upon the severity (Figure 11.15). The first two phases on the bone scan typically show increased hyperaemia in a stress fracture. In shin splint or tibial periostitis, there is micro-avulsion of fibres connecting periosteum to bones. Hence, periostitis is seen as linear tracer uptake, typically in the posterior medial aspect of the tibia involving more than a third of the tibia. The first two phases of the scan are usually normal.

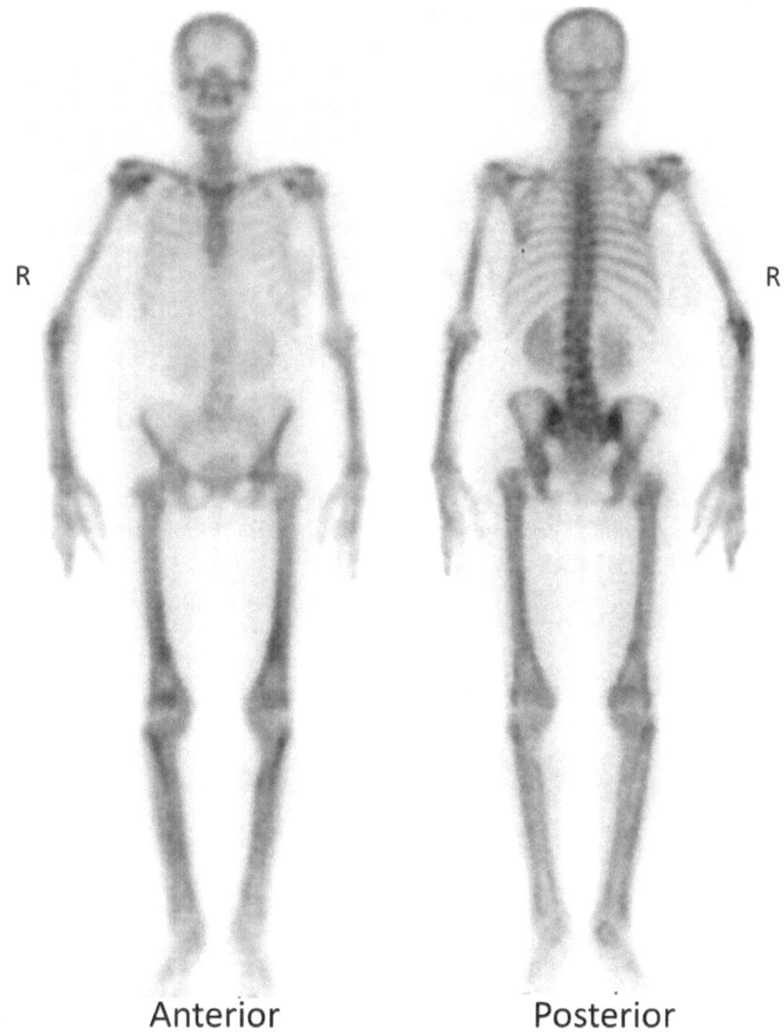

FIGURE 11.14 59-year-old male with lung cancer shows diffuse increased tracer uptake in the long bones with no other lesions suggestive of hypertrophic pulmonary osteoarthropathy (HPOA), which is due to periosteal reaction and considered as a paraneoplastic syndrome.

11.8.5 Metabolic Bone Disease

Metabolic bone disease due to primary hyperparathyroidism and renal failure produces a metabolic super scan pattern that shows homogenously increased tracer uptake in the axial and appendicular skeleton with faintly visualized kidneys. Sometimes, focal increased tracer uptake is seen in the brown tumours in hyperparathyroidism. Due to hypercalcemia, sometimes there is soft tissue uptake in the lungs, stomach, and kidneys. In rickets, there are changes in endplates of the long bones and beading at costochondral junctions. There may be super scan pattern, increased uptake in the sternum (Tie pattern), pseudofractures, or true fractures.

In Paget's disease, abnormal bone remodelling is seen. A bone scan is useful for confirming the diagnosis, and assessing the disease extent. However, most often, it is seen incidentally on a bone scan performed due to other causes. The typical scintigraphic pattern is intense florid tracer uptake in the involved bones with a blooming uptake appearance (Figure 11.16).

11.8.6 Osteonecrosis or Avascular Necrosis

The causes of osteonecrosis include trauma, long-term steroid intake, sickle cell disease, other vascular diseases, and so forth. During the initial phase, there is a photopenic area in the involved region due to the cut-off of blood supply.

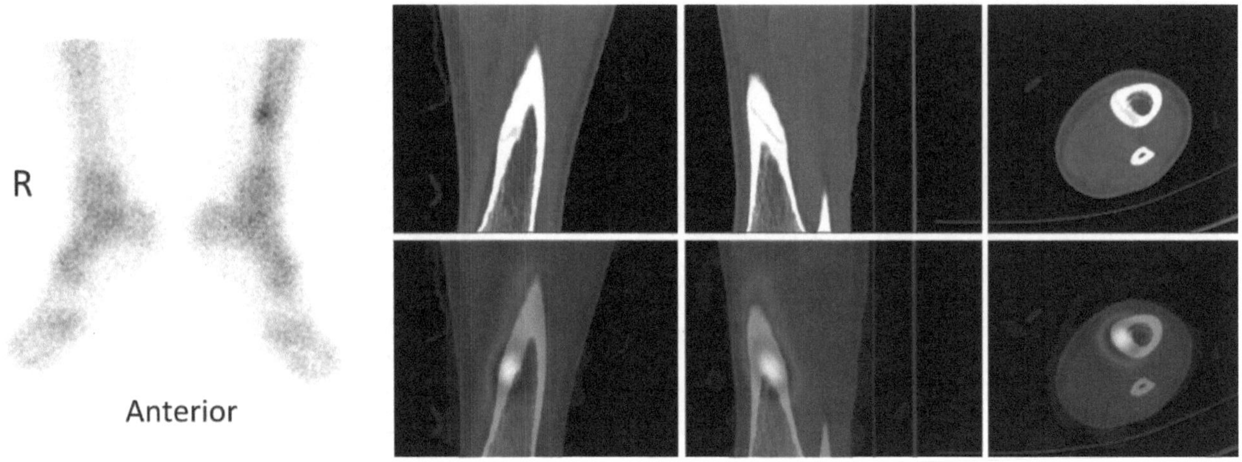

FIGURE 11.15 Static planar image of a marathon runner with history of pain left leg shows focal uptake in the left mid leg. This on SPECT-CT localizes to the cortex showing a fracture line suggestive of stress fracture in this particular context. Colour image available at www.routledge.com/9781138593312.

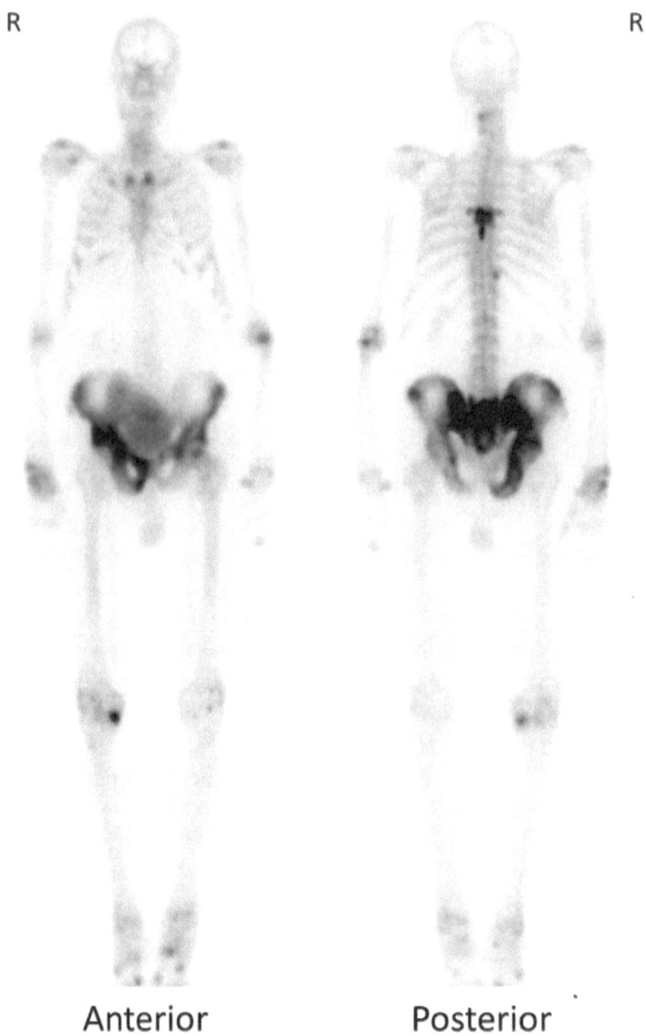

FIGURE 11.16 Intense tracer uptake is seen throughout much of the pelvis. Intense uptake is seen in D7 vertebra involving entire vertebra and spinous process (Mickey mouse or mouse face sign). This pattern is consistent with metabolically active Paget's disease. Colour image available at www.routledge.com/9781138593312.

Later, there is increased uptake due to reactive osteoblastic activity. MRI is usually the gold standard imaging modality in AVN. However, the bone scan shows changes quite early during the disease process.

11.8.7 Complex Regional Pain Syndrome (CRPS)

CRPS, or reflex sympathetic dystrophy (RSD), is a response to trauma and immobilization. There is variable presentation; however, typically, there is pain, edema, and muscle wasting. A three-phase bone scan is helpful, and the most specific finding is the periarticular uptake of tracer. Classical increased tracer accumulation in all three phases with periarticular uptake in the delayed phase is seen only in early disease. During the later course of the disease, the blood flow phase may be normal.

11.8.8 Miscellaneous Indications

A bone scan is useful in evaluating sacroiliitis (Figure 11.17), bone non-union (Figure 11.18), bone graft viability and, usually, SPECT-CT is more specific in such scenarios. A Bone scan is also extremely sensitive in identification of a pain generator in unexplained bone pain in the back, hip, knee, and foot (Figure 11.19) [17–21]. Often the fractures are seen early in a bone scan in comparison to conventional radiographs. Aseptic loosening and hip and knee prostheses infection can be differentiated with high sensitivity on a three-phase bone scan.

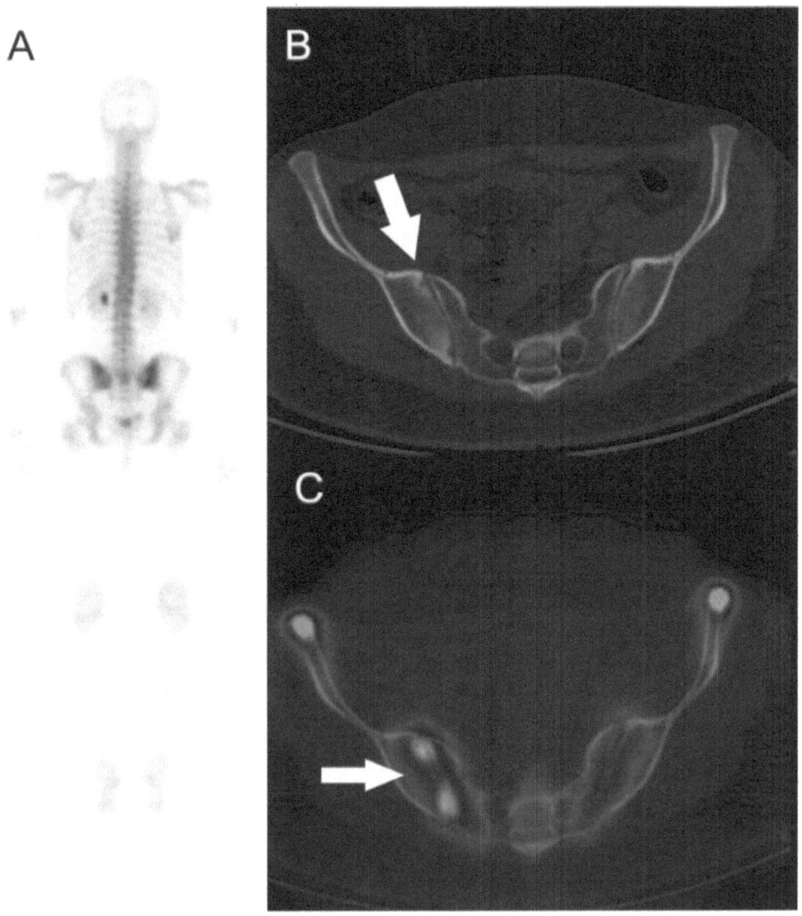

FIGURE 11.17 Patient with chronic intermittent low back pain. Posterior view of whole-body bone scan (A) shows relatively increased tracer activity in right sacroiliac joint compared to the left side, suspicious of right side sacroilitis. On SPECT-CT of hip joint (B, C), the tracer uptake localizes to sclerosis and subarticular cyst formation in the right iliac bone adjacent to sacroiliac joint. The SPECT-CT findings are indicative of right side sacroilitis, which was later confirmed on MR imaging. Colour image available at www.routledge.com/9781138593312.

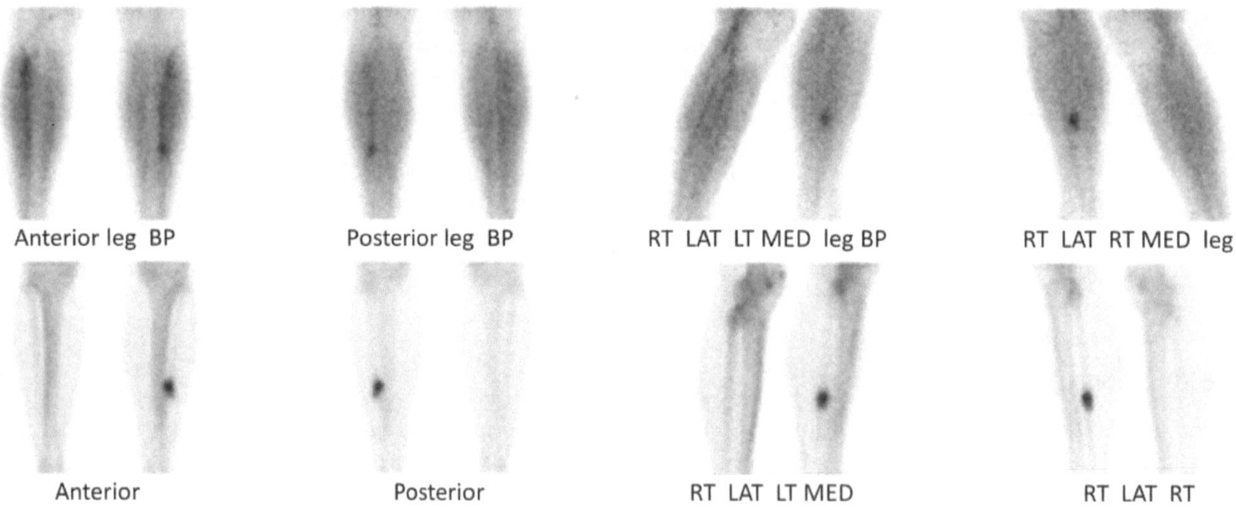

FIGURE 11.18 History of previous left fibula fracture with persistent pain. Three phase bone scan shows mildly increased vascularity in the left lower leg with corresponding increased uptake in the left distal fibula on delayed phase, suggestive of non-union fracture.

11.9 BONE SCAN INDEX

The bone scan index (BSI) is an imaging biomarker of the total extent of the whole-body bone metastasis derived from the bone scintigraphy. It has been developed with the help of artificial intelligence and is more reproducible than visual subjective quantification of disease burden. The commercially available softwares are EXINI Bone, developed by EXINI Diagnostics, Lund, Sweden, and BONENAVI, developed by Fujifilm RI Pharma, Tokyo [22]. BSI is automatically processed in seven steps, namely: (1) Segmentation of the whole-body image into 12 regions. (2) Detection of the hot spots based on a specific algorithm. (3) Standardization of images with different timing and shown in same count density format. (4) Hot spot quantification. (5) Hot spot classification into high and low probabilities of metastasis. (6) Calculation of BSI, and (7) Display of results into three major indices probability of abnormalities, total amount of the bone metastasis, and number of hot spots [22]. The false positives uptake on a bone scan may be classified by software as high probability and can be reclassified manually. BSI is helpful in diagnosis, determining the extent of the bone metastasis, treatment response assessment, and prognosis. Study has shown that in patients with castration-resistant prostate cancer, a doubled BSI leads to 1.9-fold increase in risk of death [23]. Further, the Gleason score and BSI have been shown as associated with survival of patients with high-risk prostate carcinoma on hormonal treatment [22]. Some researchers have shown that BSI is dependent on counts in the anterior and posterior images, and it is recommended that total count should be more than or equal to 1.5 million per image [22].

11.10 CONCLUSION

Radionuclide bone imaging is a useful technique in the assessment of benign and malignant bone disease. The addition of Bone SPECT-CT improves sensitivity, specificity, accuracy, localization, and diagnostic confidence.

REFERENCES

[1] G. B. Saha, *Fundamentals of Nuclear Pharmacy*, 5th ed. New York: Springer, 2004, pp. xvii, 383.

[2] E. Even-Sapir, "Imaging of malignant bone involvement by morphologic, scintigraphic, and hybrid modalities," *J Nucl Med,* vol. 46, no. 8, pp. 1356–67, 2005.

[3] K. Donohoe, M. Brown, B. Collier, R. Carretta, R. Henkin, and H. Royal, *Society of Nuclear Medicine Procedure Guideline for Bone Scintigraphy*, 2003.

[4] G. Segall *et al.*, "SNM practice guideline for sodium 18F-fluoride PET/CT bone scans 1.0," *J Nucl Med,* vol. 51, no. 11, pp. 1813–20, 2010, doi: 10.2967/jnumed.110.082263.

[5] ICRP, 1998. *Radiation Dose to Patients from Radiopharmaceuticals Addendum 2 to 53,* ICRP Publication 80, Ann. ICRP 28 (3).

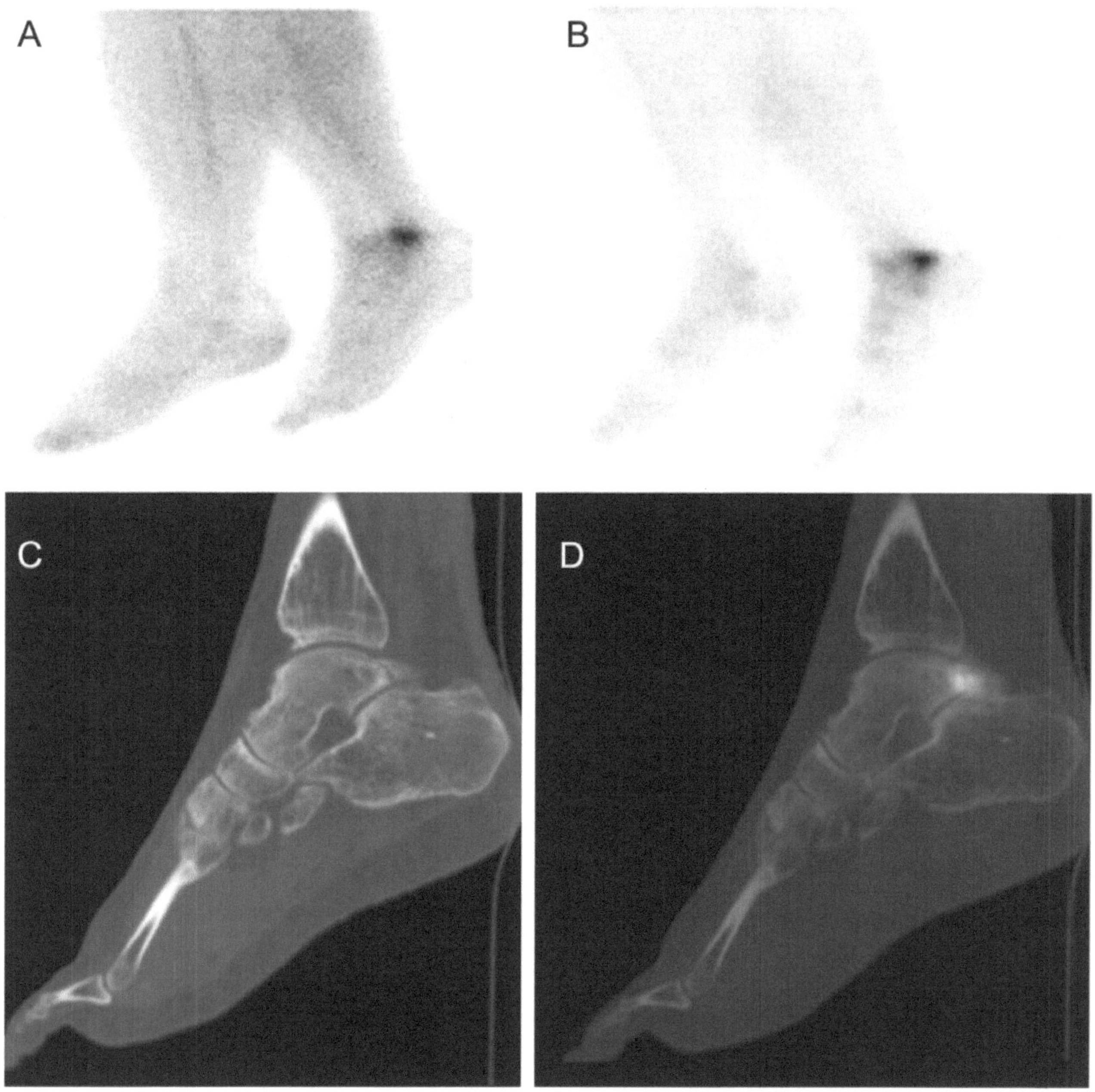

FIGURE 11.19 Blood pool and delayed phase bone scan (A, B) of a patient with painful right ankle following trauma shows increased tracer localization in the right ankle joint in both phases of study. On SPECT-CT (C, D), the epicenter of tracer uptake localizes along the syndesmosis of an os trigonum. This reflects associated posterior impingement of the ankle associated with biomechanical change post trauma. Colour image available at www.routledge.com/9781138593312.

[6] T. B. Bartel *et al.*, "SNMMI procedure standard for bone scintigraphy 4.0," *J Nucl Med Technol,* vol. 46, no. 4, pp. 398–404, 2018.

[7] G. Gnanasegaran, G. Cook, K. Adamson, and I. Fogelman, "Patterns, variants, artifacts, and pitfalls in conventional radionuclide bone imaging and SPECT/CT," *Semin Nucl Med,* vol. 39, no. 6, pp. 380–95, 2009, doi: 10.1053/j.semnuclmed.2009.07.003.

[8] S. P. Thang, A. E. H. Tan, and A. S. W. Goh, "Bone scan 'Hot Spot' at the superior lateral orbital margin fronto-zygomatic suture uptake characterized with Tc-99m MDP SPECT/CT," *World J Nucl Med,* vol. 10, no. 2, pp. 139–40, 2011, doi: 10.4103/1450-1147.89782.

[9] K. Agrawal, F. Marafi, G. Gnanasegaran, H. Van der Wall, and I. Fogelman, "Pitfalls and limitations of radionuclide planar and hybrid bone imaging," *Semin Nucl Med,* vol. 45, no. 5, pp. 347–72, 2015, doi: 10.1053/j.semnuclmed.2015.02.002.
[10] L. P. Connolly, L. A. Drubach, S. A. Connolly, and S. T. Treves, "Bone," in *Pediatric Nuclear Medicine/PET*, S. T. Treves, Ed. New York: Springer, 2007, pp. 312–403.
[11] G. M. Syed, H. W. Fielding, and B. D. Collier, "Sternal uptake on bone scintigraphy: age-related variants," *Nucl Med Commun,* vol. 26, no. 3, pp. 253–57, 2005, doi: 10.1097/00006231-200503000-00010.
[12] D. Fink-Bennett and J. Johnson, "Stippled ribs: a potential pitfall in bone scan interpretation," *J Nucl Med,* vol. 27, no. 2, pp. 216–18, 1986.
[13] R. E. Coleman, "Metastatic bone disease: clinical features, pathophysiology and treatment strategies," *Cancer Treat Rev,* vol. 27, no. 3, pp. 165–76, 2001, doi: 10.1053/ctrv.2000.0210.
[14] A. Z. Krasnow, R. S. Hellman, M. E. Timins, B. D. Collier, T. Anderson, and A. T. Isitman, "Diagnostic bone scanning in oncology," *Semin Nucl Med,* vol. 27, no. 2, pp. 107–41, 1997, doi: 10.1016/s0001-2998(97)80043-8.
[15] F. Macedo *et al.*, "Bone metastases: An overview," *Oncol Rev,* vol. 11, no. 1, p. 321, 2017, doi: 10.4081/oncol.2017.321.
[16] M. Cecchini, A. Wetterwald, G. Pluijm, and G. Thalmann, "Molecular and biological mechanisms of bone metastasis," *EAU Update Series,* vol. 3, pp. 214–26, 2005.
[17] F. S. Schleich *et al.*, "Diagnostic and therapeutic impact of SPECT/CT in patients with unspecific pain of the hand and wrist," *EJNMMI Research,* vol. 2, no. 1, p. 53, 2012, doi: 10.1186/2191-219X-2-53.
[18] G. Gnanasegaran, T. Barwick, K. Adamson, H. Mohan, D. Sharp, and I. Fogelman, "Multislice SPECT/CT in benign and malignant bone disease: when the ordinary turns into the extraordinary," *Semin Nucl Med,* vol. 39, no. 6, pp. 431–42, 2009, doi: 10.1053/j.semnuclmed.2009.07.005.
[19] S. J. Lu, F. Ul Hassan, S. Vijayanathan, I. Fogelman, and G. Gnanasegaran, "Value of SPECT/CT in the evaluation of knee pain," *Clin Nucl Med,* vol. 38, no. 6, pp. e258–60, 2013, doi: 10.1097/RLU.0b013e31826390b2.
[20] H. K. Mohan, G. Gnanasegaran, S. Vijayanathan, and I. Fogelman, "SPECT/CT in imaging foot and ankle pathology-the demise of other coregistration techniques," *Semin Nucl Med,* vol. 40, no. 1, pp. 41–51, 2010, doi: 10.1053/j.semnuclmed.2009.08.004.
[21] K. Agrawal and G. Gnanasegaran, "Painful knee prosthesis: Is there a role for bone SPECT/CT?," *Nuclear Medicine Communications,* vol. 35, no. 7, 2014.
[22] K. Nakajima, L. Edenbrandt, and A. Mizokami, "Bone scan index: A new biomarker of bone metastasis in patients with prostate cancer," *Int J Urol,* vol. 24, no. 9, pp. 668–73, 2017, doi: 10.1111/iju.13386.
[23] E. R. Dennis *et al.*, "Bone scan index: A quantitative treatment response biomarker for castration-resistant metastatic prostate cancer," *Journal of Clinical Oncology: Official journal of the American Society of Clinical Oncology,* vol. 30, no. 5, pp. 519–24, 2012, doi: 10.1200/jco.2011.36.5791.

12 Radionuclide Examination of the Kidneys

Martin Šámal and Jiří Trnka

CONTENTS

12.1 INTRODUCTION

Radionuclide examinations of the kidneys include in vitro measurement of plasma clearance of radiopharmaceuticals excreted by the kidneys and imaging examinations of the kidneys and urinary tract: static and dynamic renal scintigraphy and radionuclide cystography. Measurement of plasma clearance provides accurate values of glomerular filtration rate, a basic index of kidney function. Static renal scintigraphy visualizes functional renal parenchyma, provides

DOI: 10.1201/9780429489501-12

information on relative renal function (percentage of total renal function covered by the left and right kidneys), and on the size, shape and location of the kidneys. Dynamic renal scintigraphy provides quantitative data on relative and absolute renal function (depending on the radiopharmaceutical, either glomerular filtration or tubular secretion), renal perfusion, tracer transit through the kidneys, and urine outflow. Radionuclide cystography is used to detect vesico-ureteral reflux. Respective examinations are described in national and international procedure guidelines (www.eanm.org, www.snmmi.org), and in continuously updated handbooks published on-line by the European Association of Nuclear Medicine – "European Nuclear Medicine Guide" and "Nuclear Medicine Clinical Decision Support" (www.eanm.org) – and in the literature reviews [1-4].

The main advantage of radionuclide examinations of the kidneys is accuracy of measurement of kidney function. Examination of any physiological function requires that the function is not affected by its measurement. This requirement is fully satisfied by extremely low amount and concentration of renal tracers administered to the patients and by the high sensitivity of their external detection. A related benefit is the safety of examination. Renal radiopharmaceuticals have neither pharmacodynamic nor osmotic effects and are neither toxic nor allergenic (consequently, they have no absolute contraindications). Even the patient with iodine allergy can be examined with the tracers labelled by radioactive iodine. The injected volume of radiopharmaceuticals is substantially smaller than the volume of contrast agents used in radiology examinations, including computed tomography (CT) and magnetic resonance (MR). In the patients with normal renal function, most of the radiopharmaceutical is excreted out of the body within several hours post examination. An effective dose from radionuclide examinations is comparable with x-ray examinations or lower. In plasma clearance measurement, the dose is below 0.1–0.2 mSv; in static renal scintigraphy, below 1 mSv; in dynamic renal scintigraphy, below 1–2 mSv (depending on the tracer, weight, gender, and age of the patient). In direct radionuclide cystography, the dose is as low as 0.05–0.1 mSv. The dose can be significantly reduced by hydration of the patient and timely voiding of the urinary bladder after examination.

A disadvantage of renal scintigraphy is its relatively long data-acquisition time and poor spatial resolution. Anatomy of pelvicalyceal system and of the ureters cannot be examined in detail. Due to prolonged acquisition, there is a risk of motion artifacts (especially in children) that are difficult to correct. They can be avoided or reduced by gentle immobilization of the child (e.g. using a vacuum cushion), contact with the parents, and a favourite toy during examination, a quiet environment and friendly behaviour of the staff, and by superficial local anesthesia of injection site before administration of the tracer. Introduction of intravenous cannula may help, especially with the methods that require collection of blood samples. Sedatives and general anesthesia are only required in exceptional circumstances and may affect kidney function.

Radionuclide examinations of the kidneys belong to the oldest nuclear medicine examinations. Radiopharmaceuticals that are still in common use were introduced decades ago: ^{51}Cr-EDTA in 1966, ^{99m}Tc-DTPA in 1970, ^{99m}Tc-DMSA in 1974, and ^{99m}Tc-MAG3 in 1986. Despite their longevity, radionuclide examinations of the kidneys belong to the least-standardized examinations in nuclear medicine. Exploitation of their full potential requires unification of data-acquisition schemes, image-processing techniques, interpretation, and reporting.

Medical physicists may significantly help to consolidate vitally important standardization in the field. They should understand relative benefits and problems of individual methods and support physicians in their choice. A key role of a medical physicist in radionuclide examinations of the kidneys is to design a proper data-acquisition scheme, introduce, validate, and supervise the methods of quantitative data processing and to apply appropriate measures of dosimetry and quality control.

12.2 REMARKS ON RENAL ANATOMY AND PHYSIOLOGY

At first glance, the kidneys filter blood plasma, reabsorb useful substances, and excrete waste matters out from the body. In fact, the kidney function is much more complex. Kidneys participate in the maintenance of overall fluid, mineral, and acid–base balance, regulation of blood-pressure and secretion of hormones. The details of renal physiology, including physiological models of kidney function, are described in dedicated monographs [5–9] and review papers [10, 11].

The kidneys are located retroperitoneally (outside peritoneal cavity) at the lower back of the body, one on each side of the spine. The upper poles of the kidneys are approximately at the level of the 12th thoracic vertebra and the lower poles at the level of the 3rd lumbar vertebra. Due to the liver position, the right kidney is usually a bit lower than the left one. An adult kidney is about 10–12 cm long. Basic anatomical information is important for proper positioning of the patient under the camera (in dynamic renal scintigraphy, the position of the camera detector is fixed prior to tracer injection before the kidneys become visible). The kidneys should be displayed in the central part of the image. Ideally, the image should contain also the heart (to derive input function for subsequent calculations) and urinary bladder (to assess its filling and emptying during potentially recorded descending or indirect radionuclide cystography, especially in children). This can be easily achieved in babies and small children but becomes difficult or impossible in older children and adults. Then the physician should decide what is clinically more important to see in the specific examination (the heart or the bladder). In calculation of relative renal function using the Patlak-Rutland plot [12–14], the input function from the heart region of interest (ROI) can be substituted by that from the liver or spleen ROIs with identical results [15–17]. The requirement of relatively large field of view in dynamic renal scintigraphy also limits the utility of camera zoom to very small children.

12.3 MEASUREMENT OF RENAL FUNCTION IN CLINICAL PRACTICE

A general index of kidney function is its plasma filtration ability represented by glomerular filtration rate (glomeruli or renal corpuscles are the sites where the filtration occurs). Reduced glomerular filtration rate (GFR) is a common symptom of many diseases of the kidneys.

12.3.1 Clearance

Glomerular filtration rate is measured as clearance of substances that are excreted solely by the kidneys where they are filtered in the glomeruli and neither metabolized, reabsorbed, nor secreted in subsequent segments of the urine excretion pathway [5, 18]. The clearance Z [ml/min] is the imaginary volume of blood plasma that is cleared of the substance per unit time. It depends on renal plasma flow Q [ml/min] through the kidneys and extraction fraction E of the substance (the fraction of incoming substance, which is removed by the kidneys).

$$Z = EQ \tag{12.1}$$

For example, if renal plasma flow through both kidneys is 600 ml/min and extraction fraction of the substance is 0.20 (20%) then the clearance of the substance is 120 ml/min. Millilitres per minute are the units of clearance most frequently used in clinical practice and in the literature, though millilitres per second (120 ml/min = 2 ml/s) can be also found.

Clearance can be measured as plasma clearance, which is the volume of plasma cleared of the substance per unit time (not necessarily just by the kidneys), or renal clearance, which is the volume of plasma cleared per unit time by the kidneys (or cleared of the substance taken up by the kidneys). Urinary clearance is the volume of plasma cleared of the substance that is excreted into the urine. With the substances used to measure GFR, plasma, renal, and urinary clearances are equal because they are excreted exclusively by glomerular filtration in the kidneys and ultimately appear in the urine. The term plasma (instead of blood) clearance is used because the concentration of the test substance is usually measured in plasma. Whole blood (plasma plus blood cells) clearance is used less frequently. It can be the same as plasma clearance (with the substances entering blood cells) or greater than that, with the substances confined to plasma [19].

Equation 12.1 illustrates the concept of clearance. In practice, urinary clearance Z [ml/min] of the substance is obtained under steady-state conditions (when plasma concentration of the substance, renal blood flow, and clearance do not change with time) as urine flow scaled to plasma flow by the ratio of the substance concentrations in the urine and in plasma,

$$Z = V_u u / c \tag{12.2}$$

where V_u [ml/min] is the volume of urine excreted per unit time (urine flow), u [mg/ml] is a concentration of the substance in the collected urine, and c [mg/ml] is concentration of the substance in plasma at the same time.

The substance used for clearance measurement can be exogenous (injected into the patient's blood in continuous infusion or as bolus injection) or endogenous (normally present in plasma in relatively stable concentration and excreted into the urine by glomerular filtration). In clinical practice, the most frequently used indicator of GFR is endogenous creatinine, a product of muscle metabolism. GFR is thus measured as creatinine clearance. Eqn. 12.2 can be used also for exogenous substances administered as bolus injection: decreasing concentration of the substance in plasma is then measured in blood samples obtained in the middle of several 30–60 minute intervals of urine collection, to which the Eqn. 12.2 is applied separately. Complete review of clearance methods can be found elsewhere [7, 18, 20, 21].

Accurate measurement of creatinine clearance requires urine collection over an extended time interval, often 24 hours, which is not always possible with all patients and in all clinical circumstances. Kidney function is thus frequently only estimated by simple methods (creatinine concentration in plasma and prediction equations) developed for the purpose. In comparison with creatinine clearance, they are less accurate, and their interpretation should be based on adequate knowledge of their limits [22].

12.3.2 Simplified Methods for GFR Estimation

The simplest way to estimate kidney function is to measure plasma concentration of creatinine in a single blood sample. If the kidneys work well, the creatinine produced in the body is continuously excreted, and its plasma level is maintained low and constant. The problem with creatinine is that the relationship between its plasma concentration and clearance (equivalent to GFR) is not linear but hyperbolic. In consequence, its plasma concentration increases only when the kidney function is already significantly reduced. This problem can be partly overcome by using an inverse value of creatinine concentration [23].

Efforts to find a simple, cheap, and generally available index of kidney function (better than plasma level of creatinine) resulted in introduction of so-called prediction equations. Prediction equations are regression equations in which GFR is estimated as a dependent variable (often marked as eGFR) from independent variables – plasma creatinine, anthropometric indices (the patient's weight, height, gender, age, ethnicity, lean body mass, etc.) and laboratory data (as plasma level of cystatin C, etc.). Individual prediction equations differ in the spectrum of independent variables and respective regression coefficients. Due to their low cost, availability and utility, prediction equations have become popular and widely applied in clinical practice. However, their users should be aware of their extremely low accuracy (somewhat better than that of inverse plasma creatinine concentration but lower than the accuracy of creatinine clearance and much lower than the accuracy of radionuclide examinations). Recent studies and procedure guidelines recommend avoiding prediction equations for estimation of numerical values of GFR in individual patients because of their large prediction errors, and apply them instead (with good diagnostic accuracy) to binary decisions as to whether the patient's GFR falls below or above a specified value [22, 24, 25].

12.3.3 GFR Adjustment to Standard Body Size

An important aspect of clearance and GFR measurement is scaling (normalization, indexation) its results to body size. Kidneys clear the wastes of the body so it is intuitively obvious that interindividual variation attributable to body size should be removed and GFR converted to interindividually comparable values. Despite the fact that their kidneys perform equally well, large individuals have greater GFR than small individuals. A frequently used example demonstrates that an elephant has a much higher GFR than a mouse, but both numbers become equal when they are scaled to the same body size. Traditional GFR scaling converts the values measured or estimated in the patient to "standard patient" body surface area (BSA) of 1.73 m^2. In practice, BSA is estimated by regression from the patient's weight and height, using (for example) the Haycock formula [26],

$$BSA = 0.024265\, W^{0.5378}\, H^{0.3964} \tag{12.3}$$

where BSA is body surface area in m^2, W the patient's weight in kg, and H the patient's height in cm. Individual GFR is then normalized to GFR_n of a person with standard BSA of 1.73 m^2 as

$$GFR_n = 1.73\, GFR / BSA \tag{12.4}$$

In some clinical applications (as the measurement of kidney function before nephrotoxic therapy to ensure safe treatment), it is necessary to work with the raw (true, individual, i.e. non-scaled) performance of the patient's kidneys. Then the measured GFR is applied directly without scaling, or the normalized value of GFR is denormalized to reflect the kidney function in the patient with specific body size. The best practice is to report both the raw and normalized GFR values.

For calculation of BSA, many alternative regression equations have been developed. The most frequently used equations produce similar results, while others should be used with caution [27, 28]. In several respects, BSA as index of body size is not ideal. Its critics argue that body surface area calculated from the patient's weight and height is not essentially different from the measurement of weight alone and that renal function is linked to metabolic rate rather than body weight or BSA, although both parameters have loose correlation with metabolic rate. Indexing to BSA has little consequence in a population with normal body size, but potentially significant effects in children and obese patients. In the past, alternative indices other than BSA have been proposed (the patient's age, weight, height, body mass index, lean body mass, total body water, plasma volume, extracellular fluid volume, cardiac output, fractal volume of distribution, liver size, metabolic rate, or tabulation of normal GFR values for specific groups of patients without scaling). Up to now, the clinical community has not accepted any of them. Insistence on BSA seems to be reasonable because the merits of its abandoning in favour of more logical indices would not justify the upheaval it would involve [29]. It should be realized that GFR indexed to BSA may underestimate true renal function in obese patients and underestimate or overestimate true renal function in children. In these patients, the normalized values should be interpreted with caution.

12.3.4 Fractional Clearance

An interesting way to avoid estimation of BSA is to present clearance as fraction of the body fluid volume Z / V (most often extracellular fluid volume or plasma volume) that is cleared of the substance per unit time [30, 31]. In fact, such a number (called fractional clearance) is the rate constant with units of reciprocal time [1/min] rather than clearance. It has intuitive physiological meaning, indicating the rate or efficiency of cleaning body fluids by the kidneys. Its measurement is simple and potentially more accurate than the measurement of clearance because it is not necessary to quantify any volume. The reciprocal value of fractional clearance is the mean residence (also transit or waiting) time of the substance in the respective volume before it is cleared. Fractional clearance can be converted into a usual expression of clearance (in ml/min normalized to standard BSA of 1.73 m^2) by multiplication of Z / V by normal volume V_n (as $Z\, V_n / V$) using relationship $V_n = k\, BSA_n$ between BSA and the respective body fluid volume where k is proportionality constant [32, 33]. Despite its potential advantages, fractional clearance has not been adopted in clinical practice, with the argument that it may reflect the changes of body fluid volumes rather than renal function. However, fractional clearance can be still used and reported as an interesting additional, potentially useful information and quality control index [7, 19, 31, 34, 35]. The term fractional clearance described above should not be confused with fractional clearance of urea used in the kidney dialysis control, and with fractional clearance (also fractional excretion) specifying the amount of the substance that appears in the urine as the percentage of its amount filtered by the kidneys.

The historical gold standard of the GFR measurement was the clearance of inulin, a polysaccharide that is not a normal constituent of the body. The method was accurate but labour intensive and time consuming. In clinical practice, it has been replaced by the tests based on plasma creatinine, and as the new gold standard (with respect to their accuracy) the radionuclide methods have been established.

12.4 MEASUREMENT OF RENAL FUNCTION WITH RADIONUCLIDES

Radiopharmaceuticals excreted by glomerular filtration in the kidneys and used for GFR measurement are technetium labelled diethylenetriaminepentaacetic acid (^{99m}Tc-DTPA) and chromium labelled ethylenediaminetetraacetic acid (^{51}Cr-EDTA). The latter (favoured for its low injected activity and possibility of extended blood sampling due to chromium-51 half-life of nearly 28 days) disappears from the market and its future remains uncertain [36]. Alternatively to ^{99m}Tc-DTPA and ^{51}Cr-EDTA, filtered iodinated x-ray contrast agents (like iohexol or iothalamate) that are also excreted by GFR, can be used either with radioactive iodine (then they are treated and detected as radiopharmaceuticals) or with non-radioactive iodine (then they are treated as chemicals, and their concentration in plasma is determined by laboratory methods). In contrast to non-radioactive substances, radiopharmaceuticals are applied in smaller amounts, and smaller blood samples are required for activity measurements. Due to very low injected activity, the examination (injection of the radiopharmaceutical and subsequent blood sampling) can be performed outside the nuclear medicine department after a simplified certification procedure. Several studies have demonstrated that GFR measured by DTPA, EDTA, iothalamate, or iohexol is close to GFR measured by inulin [36–39].

12.4.1 Procedure

Clearance examinations with radionuclides consist in intravenous administration of a radiopharmaceutical (either as single bolus or continuous infusion) and subsequent blood sampling. In clinical practice, a single bolus injection followed by collection of 2–4 blood samples have been used most often. Recently, single-sample techniques became popular for their simplicity and repeatedly confirmed accuracy [40–45]. Review of a wide spectrum of clearance methods can be found in respective procedure guidelines and in the literature [46–50].

Calculation of GFR (as clearance Z of the substances excreted by glomerular filtration) can be derived from the Eqn. 12.2 after its integration over time [20]

$$Z\int_0^\infty c(t)dt = \int_0^\infty u(t)V_u(t)dt$$
$$Z = \int_0^\infty u(t)V_u(t)dt / \int_0^\infty c(t)dt = D / \int_0^\infty c(t)dt \tag{12.5}$$

where

$$D = \int_0^\infty u(t)V_u(t)dt$$

is the injected amount of a tracer [Bq] that is eventually completely excreted into urine. Alternatively, formula equivalent to Eqn. 12. 5 can be derived from the general equation of the first order kinetics [7] or Stewart–Hamilton method for the measurement of flow [51] according to which an uptake rate of the tracer by the kidneys *dR(t)/dt* [Bq/min] is proportional to its concentration in plasma *c(t)* [Bq/ml] with proportionality constant equal to clearance Z [ml/min]

$$dR(t)/dt = Zc(t) \tag{12.6}$$

$$R(t) = Z\int_0^t c(t)dt \tag{12.7}$$

$$Z = R(t)/\int_0^t c(t)dt = R(\infty)/\int_0^\infty c(t)dt = D/\int_0^\infty c(t)dt \tag{12.8}$$

where again, at the end, all the injected amount of *D* is excreted by the kidneys, $D = R(\infty)$. As the injected amount can be properly determined (after correction for residual activity in an empty syringe), the measurement of plasma clearance relies on accurate measurement of the area under the plasma concentration curve of a radiopharmaceutical which, in turn, depends on the number and timing of blood samples as well as on the model used for extrapolation and calculation of total area under the curve. Ideally, one should collect enough blood samples to numerically integrate the observed part of the curve, which is then extrapolated to infinity using mathematical models. That is well possible in research studies but not necessary in routine clinical practice. However, reduction of the number of blood samples requires additional information [20].

12.4.2 Simplified Procedures

After intravenous injection, radiopharmaceutical is mixed with the blood, excreted by the kidneys, and simultaneously transported from blood vessels (capillaries) into extravascular space in the body by diffusion. Diffusion lasts until the tracer concentrations in extravascular and intravascular spaces reach transient equilibrium. With GFR tracers and normal kidney function it takes about two hours. The lower the kidney function and the bigger the extravascular space, the longer the time to transient equilibrium.

In the patients without fluid retention (in extra spaces collecting fluids as edema or ascites also called as the "third" spaces – third or additional to physiological intravascular and extravascular spaces), the process of plasma clearance can be represented by the two-compartment model (the two compartments corresponding to intravascular and extravascular spaces) with plasma clearance curve approximated by the sum of two exponentials [7]. The first segment of the curve corresponds to the fast removal of the tracer from plasma by both diffusion and renal excretion, the second one represents removal of the tracer by the kidneys. The latter one, which reflects kidney function, is sometimes called "terminal exponential" (Figure 12.1).

Using a two-compartment model, the number of blood samples required for the curve fitting can be reduced from many to four (two for each exponential). Further simplification is based on observation that the area under the first (or "fast") exponential can be accurately estimated by regression from the GFR value given by the terminal exponential alone. Several formulas have been developed to apply such a correction to the area under the curve calculated only from the terminal exponential [31, 52]. Using the correction for the fast exponential, the minimum number of required blood

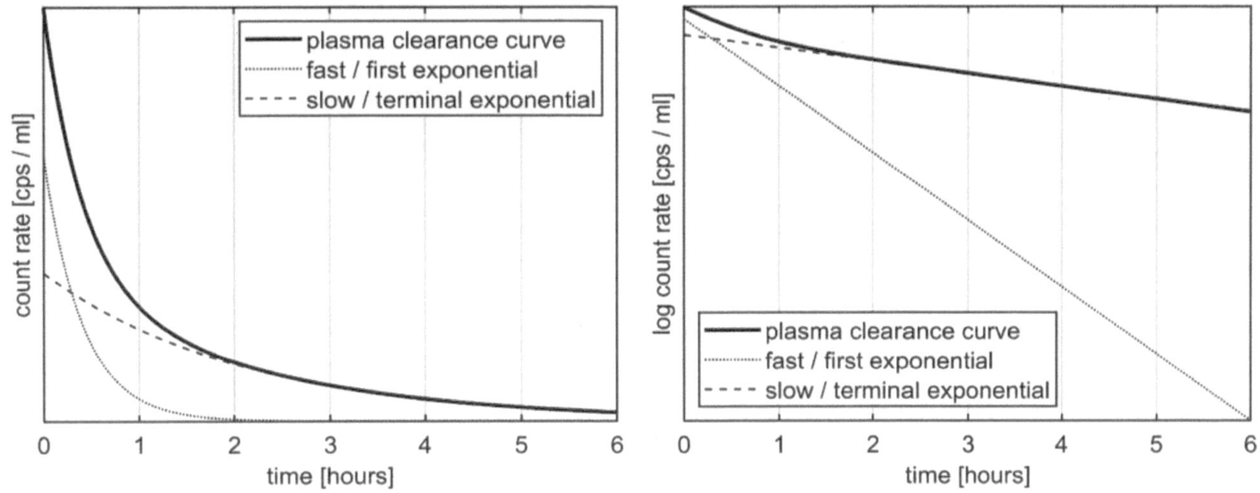

FIGURE 12.1 The most frequently used biexponential model of plasma clearance takes into account diffusion of the tracer from plasma to extravascular space (fast or the first exponential) and excretion of the tracer by the kidneys (slow or terminal exponential). Terminal exponential is usually estimated by fitting the count rate in 2 blood samples by a straight line (in logarithmic scale on the right) while the area under the first exponential is subsequently estimated by regression.

samples can be reduced to two. With GFR tracers, the two blood samples are usually collected between 2–3 and 4–6 hours post injection with at least 2 hours interval between the samples.

In the patients without fluid retention in the third spaces, determination of GFR with two blood samples represents a good compromise between required accuracy of the measurement, quality control and difficulty of examination. Recently, it has been shown that in these patients, the number of blood samples can be further reduced to one, using regression equations estimating the area under the plasma concentration curve from a single blood sample [53, 54]. From early single-sample methods, international consensus recommended the method by Christensen and Groth, later modified by Watson [45, 46, 55]. Newer single-sample methods are reported also to measure accurately very low GFR providing blood samples are collected late enough [41, 42, 47, 53, 56]. However, they should not be applied to the patients with fluid retention. Some of the single-sample methods require collection of the blood sample exactly at a specific time after the tracer injection, with others the collection time is flexible and represents one of the regression variables.

If the kidney function is poor (with estimated GFR below 30 ml/min) and the tracer removal from plasma is slow, or if there is retention of fluid in additional compartments, equilibration of tracer concentration between intravascular and extravascular spaces lasts longer than two hours (Figure 12.2). Early blood samples then reflect a quickly decreasing concentration of the tracer in plasma by both GFR and continuing diffusion. Using these samples, the area under the plasma curve is underestimated and clearance overestimated. Terminal exponential, if any, is reached only after many hours post injection. In these patients, it is necessary to extend blood sampling to 8, 12, or even 24–48 hours or use the clearance method with the urine collection instead [21, 37].

12.4.3 Alternative Models

Approximation of the plasma clearance curve by the sum of two exponentials is sometimes considered as oversimplified model even in the patients with relatively good renal function and alternative models are proposed. One such model is based on regularized gamma variate fitting [57]. The first clinical experience with the model is interesting but limited [58]. The cost of better fitting is the higher number of blood samples (4 or more between 10–240 minutes post injection).

Regardless of the model, a crucial requirement in radionuclide measurement of plasma clearance is accuracy of activity measurement in the syringe (both before and after injection, including background subtraction, decay correction, etc.) and in blood samples.

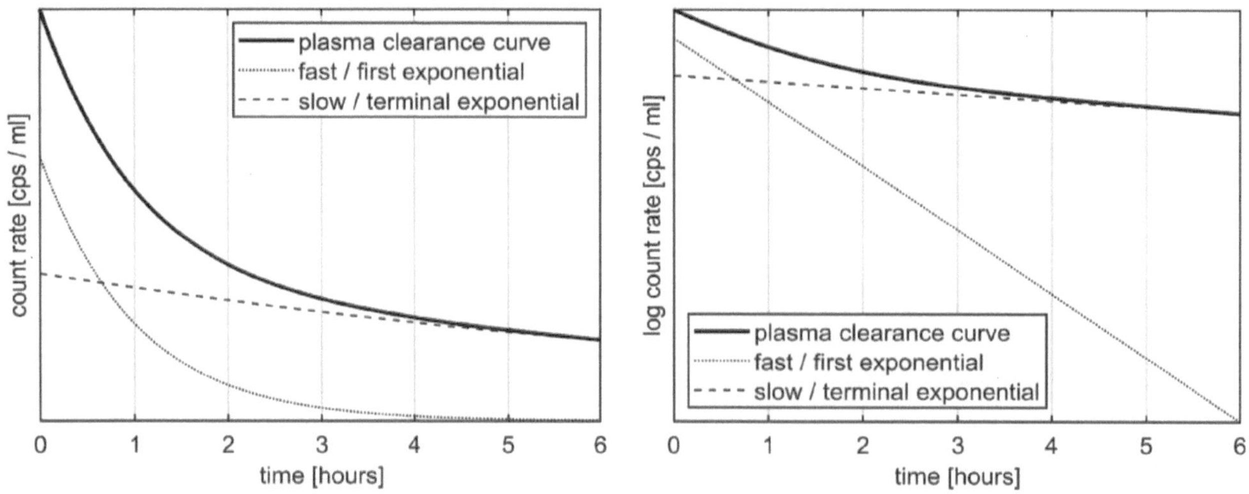

FIGURE 12.2 In the patients with poor kidney function, approximate equilibrium of tracer concentration between plasma and interstitium appears later (here at about 5 hours post injection). Early sampling would overestimate the slope of terminal exponential, underestimate the area under the curve, and thus overestimate clearance.

12.4.4 Radiation Dose

The effective radiation dose related to in vitro examination of GFR with ^{99m}Tc-DTPA is less than 0.1 mSv in adults and less than 0.2 mSv in children (which is less than 4% of the dose from natural radiation background of 5 mSv/yr.). With ^{51}Cr-EDTA the dose is yet about ten times lower [47]. Despite low doses, radiation is often the reason for replacement of radionuclide methods by those based on non-radioactive x-ray contrast agents (iothalamte, iohexol), which are more expensive and require greater injected amounts and greater volumes of blood for chemical analysis.

12.4.5 Renal Perfusion and Tubular Function

In parallel with the measurement of GFR, renal plasma or blood flow can be measured as plasma clearance of the substances with their extraction fraction close to one (see Eqn. 12.1). A classic example of such substance is paraaminohippuric acid (PAH) and the related radiopharmaceutical orthoiodohippurate (^{123}I- or ^{131}I-OIH), with a bit lower extraction fraction indicating lower renal plasma flow, referred to as effective renal plasma flow (ERPF). Recently, the use of OIH has been rare, and the new radiopharmaceuticals labelled by ^{99m}Tc for the measurement of renal plasma flow are investigated, such as ^{99m}Tc-(CO)$_3$-tricarbonylnitriloacetic acid with clearance equivalent to that of OIH [59]. In contrast to them, common radiopharmaceuticals used for kidney imaging (^{99m}Tc-mercaptoacetyltriglycin, ^{99m}Tc-MAG3, and ^{99m}Tc-ethylenedicysteine, ^{99m}Tc-EC) excreted by tubular secretion have a much lower extraction fraction of about 50–60 per cent, although their renal uptake can be still considered as proportional to renal plasma flow [1]. Plasma clearance of ^{99m}Tc-MAG3 is also known as tubular extraction rate or TER [60, 61].

12.5 IMAGING METHODS OF THE KIDNEYS

Routine imaging methods of the kidneys in clinical practice are ultrasound examination and x-ray computed tomography (CT). Radionuclide imaging methods are used in specific indications and include (static) renal cortical scintigraphy, dynamic renal scintigraphy, and radionuclide cystography. Recently, positron emission tomography (PET/CT) and magnetic resonance (MR) tomography are introduced into the kidney imaging in both research and clinical settings. Potential acceptance of non-oncological clinical PET/CT imaging in nephrology and urology will depend on clinical information provided by the new radiopharmaceuticals, their cost and relative radiation burden, especially in children. In contrast to radionuclide methods that accurately assess kidney function, radiological methods predominantly provide anatomical information with high spatial resolution, although they also may (with or without contrast agents excreted by the kidneys) reflect some kidney functions such as perfusion, urine excretion and outflow, diffusion, oxygenation, and so forth. In the choice of the appropriate imaging method, both its radiation and non-radiation risks (related to contrast agents) should be considered besides its availability and diagnostic accuracy for a given problem.

12.6 STATIC RENAL SCINTIGRAPHY (RENAL CORTICAL SCINTIGRAPHY)

Static renal scintigraphy (or renal cortical scintigraphy) provides static images of the distribution of ^{99m}Tc-dimercaptosuccinic acid (^{99m}Tc-DMSA) that binds itself in the cells of proximal renal tubules [62]. As the deposit of DMSA requires intact renal function (renal perfusion, glomerular filtration, and tubular reabsorption), the resulting images demonstrate the extent of renal parenchyma with normal function. Most frequently examined pathological conditions are inborn malformations (as horse-shoe kidney), ectopic kidneys, pyelonephritis, and assessment of its late outcome (renal scarring), renal trauma, and measurement of relative renal function (contribution of a single kidney to global renal function in percent). Detection of cortical lesions by renal cortical scintigraphy is highly sensitive (sensitivity is higher than that of ultrasound and comparable with CT and MR), but not specific (usually it is not possible to differentiate between acute pyelonephritic lesions and older scars, both manifested as the absence of normal tissue).

12.6.1 Procedure

Imaging starts 2–4 hours after injection of ^{99m}Tc-DMSA, when about 40–50 per cent of the tracer is retained in renal tubules. The rest is excreted into urine in about 24 hours. To reduce radiation burden and avoid retention of radioactive urine in the collecting system of the kidneys (especially in the dilated calyces and pelves), which may complicate image interpretation and invalidate the measurement of relative renal function, good hydration and frequent emptying of the patient's urinary bladder is required before image acquisition, potentially supported by diuretics. In hydronephrotic kidneys (with urine retention in the dilated outflow tract) it may be necessary to repeat image acquisition later when there is no active residual urine in the collecting system.

Examination is performed in supine position, preferably using a double-head gamma camera with LEHR collimators. Static images are recorded in posterior projection and in both left and right oblique posterior projections into 256 x 256 image matrix (Figure 12.3). Anterior projection is completed in the patients with suspected malformations, ectopic kidneys, and with the left and right kidneys apparently placed at a different kidney "depth" (different position of the kidneys in anterio-posterior direction). In very small children, zoom helps to recognize details. In some departments, data acquisition is performed with LEUHR or pinhole collimators and with SPECT or with their combinations (Figure 12.4).

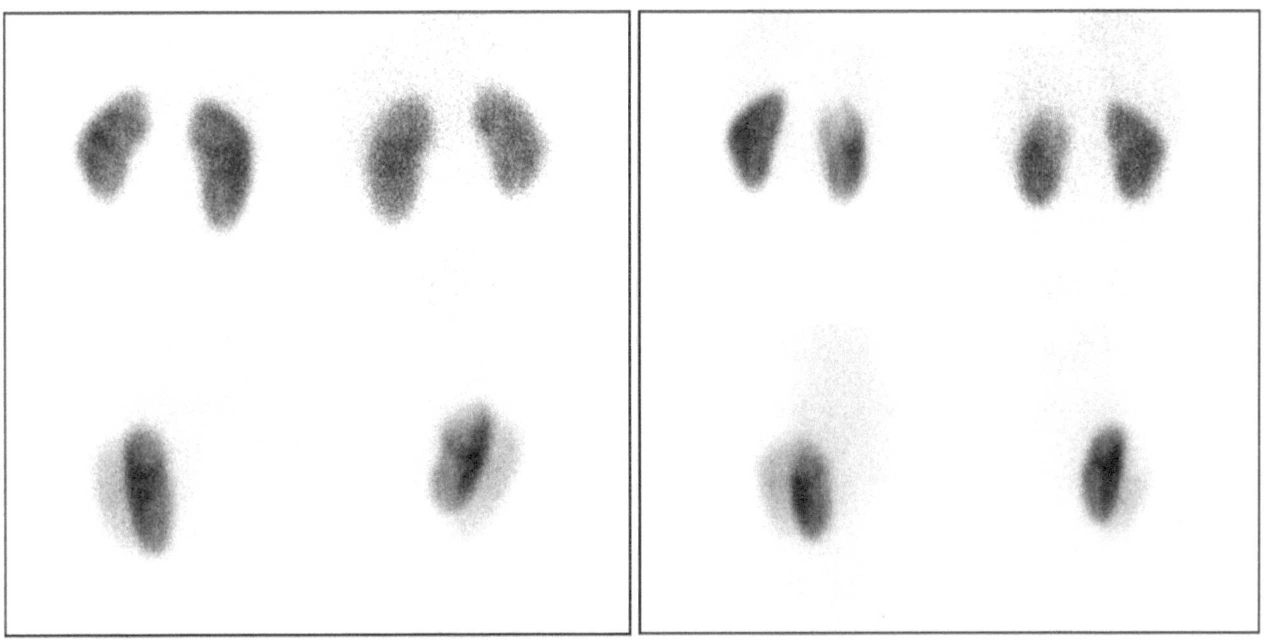

FIGURE 12.3 Static renal scintigraphy in the patient with normal kidneys (left) and in the patient with parenchymal defect in the upper pole of the right kidney (right). Posterior, anterior, and the two oblique posterior projections.

Source: Images courtesy of Daniela Chroustová, M.D., Ph.D., Institute of Nuclear Medicine, General University Hospital in Prague, Czech Republic.

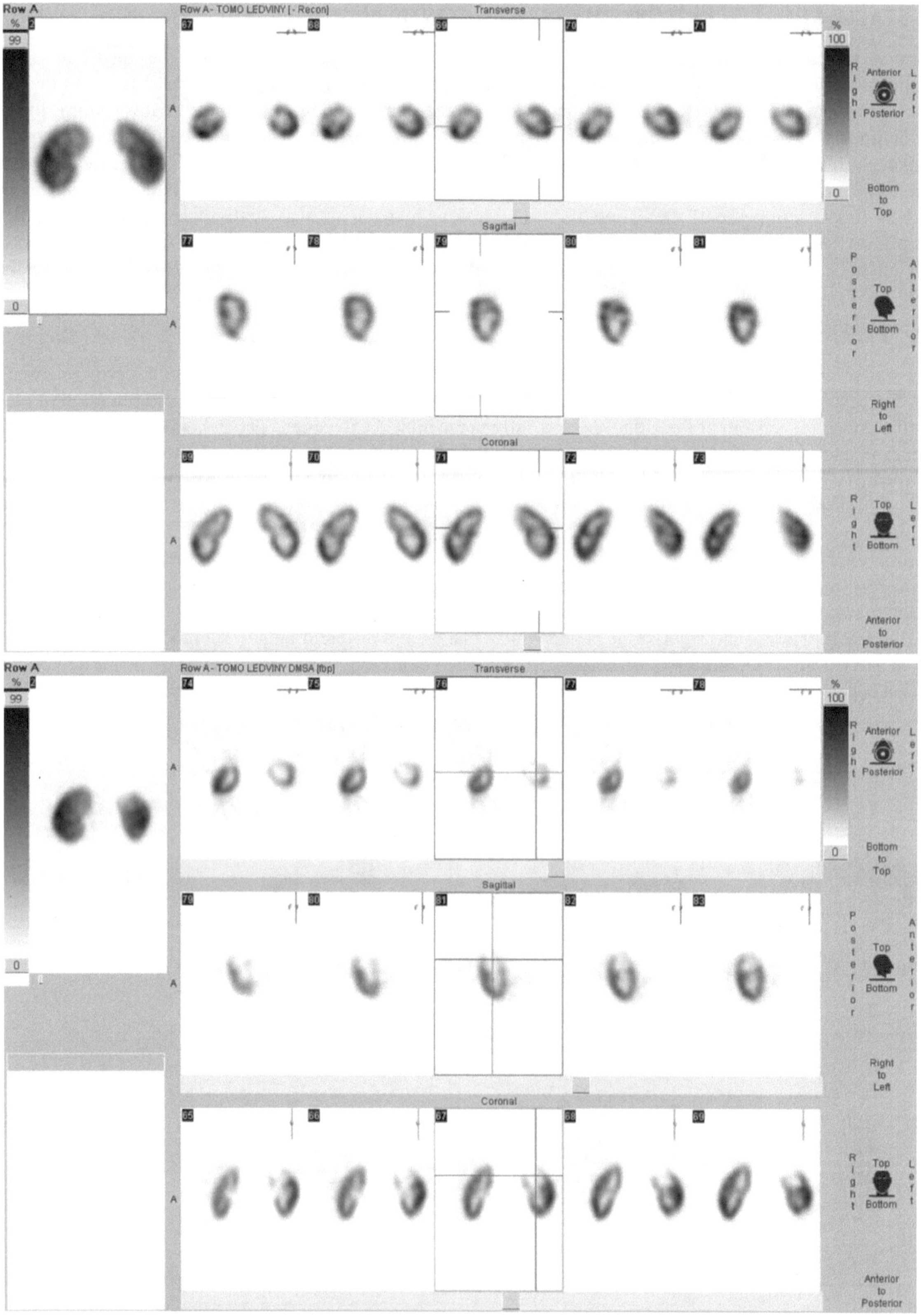

FIGURE 12.4 ^{99m}Tc-DMSA SPECT in the patient with normal kidneys (top) and in the patient with parenchymal defect in the upper pole of the left kidney (bottom).

Source: Images courtesy of Kateřina Táborská, M.D., Department of Nuclear Medicine and Endocrinology, Motol University Hospital, Prague, Czech Republic.

However, these techniques are not generally accepted, mainly because they prolong the time of data acquisition and increase the risk of motion artifacts. In restless children, pharmacological sedation can be sometimes avoided using dynamic image acquisition and subsequent summation of selected, motion-corrected and registered images.

12.6.2 Data Processing

Quantitative parameters derived from data of renal cortical scintigraphy are dimensions of the kidneys (length, width) and relative renal function (sometimes referred to as split or differential renal function). Relative renal function is measured in posterior projection. Regions of interest (ROIs) are drawn around the left and right kidneys and around extrarenal tissue next to each kidney for extrarenal background subtraction. From the ROI count of each kidney, the sum of counts in the background ROI (after its scaling to the size of the kidney ROI), is subtracted and relative function of the left (*LK* %) and (or) right kidneys calculated

$$LK\ \% = 100 N_{LK} / \left(N_{LK} + N_{RK}\right) \tag{12.9}$$

where N_{LK} and N_{RK} are background-corrected counts in the left and right kidney ROIs. This simple and commonly used method does not take into account the fact, that the background counts should first be reduced for the "kidney thickness" – the space in the kidney ROI occupied by the kidney itself, in adults normally 0.2–0.3 of antero-posterior body thickness [63]. Correction for the kidney thickness is often neglected because its accurate value in individual patients is unknown.

12.6.3 Conjugate View

In most patients, the position of the left and right kidneys in antero-posterior direction differs by less than 1 cm, which does not introduce clinically significant error in relative renal function greater than 5 per cent. However, in about 5 per cent of children and 25 per cent of adults, the difference in the kidney depth exceeds 1 cm and should be corrected. Kidney depth can be measured in additional lateral scintigraphic views or in CT and MR transversal slices through the kidneys, determined by ultrasound or estimated by regression. Regression equations do not account for individual differences in the left and right kidney depths and are more useful in dynamic renal scintigraphy to approximate attenuation correction in the assessment of kidney function. Lateral images are useful but represent additional time and workload, as does ultrasound measurement. Tomographic images are rarely available.

The best, fastest, and easy way to measure relative renal function accurately is to use conjugate posterior and anterior views or geometric mean images [8, 63–65]. Data acquisition with a double-head camera is performed simultaneously in posterior and anterior projections

$$\begin{aligned} N_p &= N\,exp\left(-\mu d_p\right) \\ N_a &= N\exp\left(-\mu d_a\right) \end{aligned} \tag{12.10}$$

where N_p and N_a [cps] is count rate in the kidney ROI recorded in posterior and anterior projections, μ is linear attenuation coefficient of the attenuating tissue for a given radionuclide [cm^{-1}], and d_p and d_a are the distances of the kidney centre from posterior and anterior surfaces of the body in cm. True kidney counts *N* are then estimated as the geometric mean of posterior and anterior counts

$$\begin{aligned} N &= \sqrt{N_p N_a}\exp\left(\mu D / 2\right) \\ D &= d_p + d_a \end{aligned} \tag{12.11}$$

where *D* is body thickness. As *D* is approximately equal on the left and right sides of the body, calculation of relative renal function using true counts is independent of kidney depth (in the Eqn. 12. 9, the attenuation factor cancels out). Before introduction of geometric mean calculation, gamma camera should be tested for accurate registration of its posterior and anterior views. Geometric mean images are strongly attenuated and should be used only for the measurement of relative renal function, not for visual evaluation of image details. It is also not recommended to record opposite projections in oblique projections due to unequal attenuation thickness and greater distances of the detectors from the source.

Renal cortical scintigraphy is a reliable tool to measure relative renal function [66], but it should not be used to estimate global and individual kidney function because renal uptake of ^{99m}Tc-DMSA depends not only on the kidney function but also on plasma input function and DMSA deposits in extrarenal tissues [67, 68]. The error may reach tens of percent. Despite the fact that some studies reported on good correlation between DMSA uptake and creatinine clearance [69], good correlation itself does not establish equivalence of the two methods [70]. If the kidney function has to be known in both relative and absolute terms, then in vitro plasma clearance examination with ^{99m}Tc-DTPA or ^{51}Cr-EDTA should be measured to assess global renal function and ^{99m}Tc-DMSA scintigraphy used to measure relative renal function.

An alternative radiopharmaceutical for static renal scintigraphy is ^{99m}Tc-glucoheptonate (^{99m}Tc-GH) cleared primarily by glomerular filtration with part of the injected activity retained in the renal tubules. It allows delayed, high-resolution images of renal parenchyma. Glucoheptonate is used less frequently, mostly to substitute ^{99m}Tc-DMSA if it is unavailable [1].

12.7 DYNAMIC RENAL SCINTIGRAPHY

Dynamic renal scintigraphy is performed with radiopharmaceuticals excreted by glomerular filtration (^{99m}Tc-DTPA) or tubular secretion (^{99m}Tc-MAG3). While DTPA reflects more desirable renal function (GFR), MAG3 has better imaging properties for which it is currently used more frequently. A substantial part of MAG3 is bound to plasma proteins that slows down its diffusion into extravascular space, reduces intensity of extrarenal background, but also reduces its extraction fraction. With DTPA, weak kidneys are often difficult to separate from high extrarenal background (due to fast DTPA diffusion into extravascular space and lower extraction by the kidneys). With MAG3, even weak kidneys are visualized clearly, and their split function thus can be measured more accurately. On the other hand, MAG3 does not reflect the accepted common index of kidney function (that is GFR) but tubular secretion that may vary in wider limits and involve competition of various substances for secretion mechanisms in the tubules. However, with this limitation in mind, in most cases the tubular and glomerular functions are parallel (also due to physiological regulation and tubuloglomerular feedback) so the split renal function can be assessed with MAG3 as well as with DTPA. Exceptions are acute obstruction and renovascular hypertension. In either case, MAG3 secretion can be preserved even when glomerular filtration is severely decreased [1, 2]. An alternative radiopharmaceutical for examination of tubular function is ^{99m}Tc-ethylenedicysteine (^{99m}Tc-EC). It has a higher extraction fraction than ^{99m}Tc-MAG3, though not as high as ortho-iodohippurate (^{123}I / ^{131}I-OIH). In comparison with ^{99m}Tc-MAG3, ^{99m}Tc-EC is less commercially available [1, 71]. A technetium-labelled alternative to ortho-iodohippurate is ^{99m}Tc-$(CO)_3$-tricarbonylnitriloacetic acid, which is still subject to experimental and clinical study [59].

In nephrological and urological diseases and in transplanted kidneys, dynamic renal scintigraphy is used to examine blood flow through the kidneys (renal perfusion), uptake of radiopharmaceutical in renal parenchyma (relative and absolute kidney function), transport of radiopharmaceutical through the kidney (renal transit), and transport of urine through outflow structures of the kidney, renal calyces, and pelves into the ureters and urinary bladder (renal drainage or outflow). Standard examination can be modified by pharmacological intervention to differentiate between renal obstruction and an unobstructed, dilated renal pelvis that mimics obstruction by application of diuretics, and between renovascular hypertension and coincidental hypertension with renal artery stenosis by application of an inhibitor of angiotensin-converting enzyme [2].

12.7.1 Procedure

Dynamic renal scintigraphy is performed in supine position of the patient with gamma camera detector under the table in posterior projection. Despite numerous procedure guidelines and consensus reports (clinically validated, reflecting long-term experience and good clinical practice), data acquisition is not generally standardized, and many individual departments follow their own traditional acquisition protocols.

Data acquisition should start before or (at the latest) simultaneously with the injection of a radiopharmaceutical. It is the only way to ensure that the peak of the heart ROI curve is not missed. That is important because the peak of the heart ROI curve is a natural mark of "time zero" (beginning of the study) from which diagnostic time intervals are then counted (time to peak, transit times, etc.). An alternative zero point, the beginning of renal uptake, is much less easy to recognize. If necessary, initial empty images and the images before the heart peak can be deleted before further processing. Activity injected intravenously as a bolus should be sufficient to get good renal count rate, that is, kidney ROI peak count rate at least 200–250 cps after background subtraction for visual assessment of basic renogram and

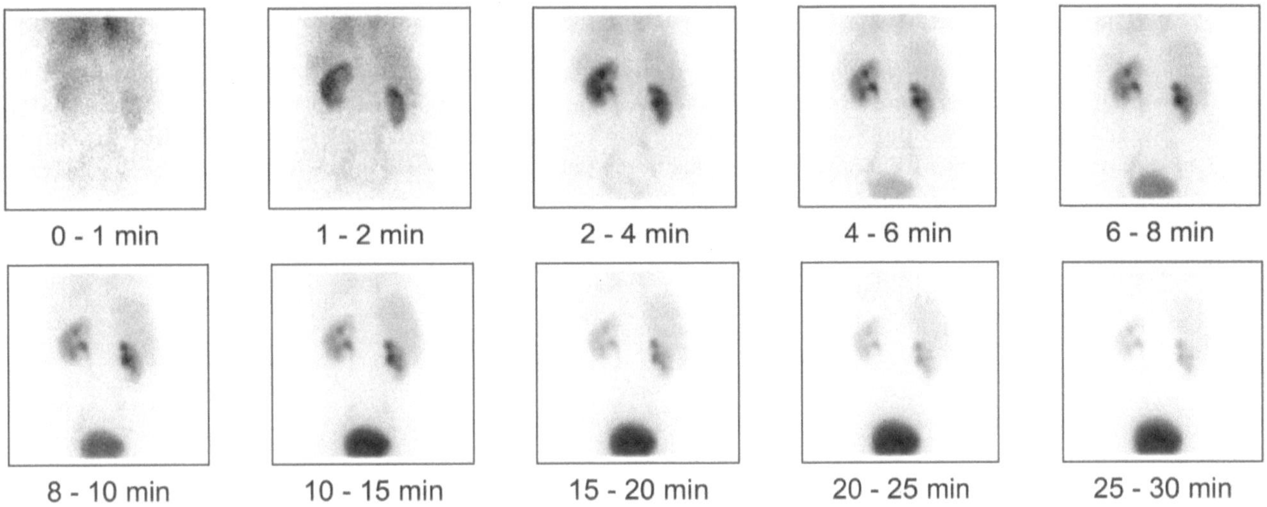

FIGURE 12.5 ^{99m}Tc-MAG3 dynamic renal scintigraphy demonstrating blood flow (0 – 1 min), renal uptake (1 – 2 min), and renal outflow (2 – 30 min). In this patient and LFOV gamma camera, both the heart and the urinary bladder are included in a single frame. Image data from the study drsprg_001 (www.dynamicrenalstudy.org).

calculation of relative renal function [72–74]. Quantitative analysis such as deconvolution occasionally requires somewhat higher counts. Images should be recorded with good resolution (128 x 128). The field of view ideally includes the kidneys (at the centre), the heart and the urinary bladder (Figure 12.5). That is not always possible with larger patients. Then the physician should decide what should be included preferably (the heart or the bladder). Unlike with renal cortical scintigraphy, zoom can be used only if it does not significantly reduce the field of view and exclude the heart or the urinary bladder or both.

Images should be recorded in a single group with 10 s frames for 20–30 minutes. Historical acquisition schemes (2–3 groups, each with a different frame rate, 10–20 s frames during uptake and 20–30 s frames during outflow phase, initial group with fast frame rate of 30–60 frames per minute) that are still used and even recommended were justified by technical limitations of historical instrumentation. There is no reason (medical, physical or technical) to continue in the practice that increases variability between departments. An initial fast acquisition group (1–2 s frames for 1–2 minutes) is required only in the examination of transplanted kidneys [1]. If it should meet its purpose and provide good quality perfusion data, it requires injection of substantially higher activity to ensure a good count rate in very short frames (in adults 370–500 MBq in comparison to the usual 90–200 MBq without recording the perfusion phase). Due to that and because in other clinical indications the perfusion phase is not interpreted, the initial fast group can be abandoned and substituted by standard 10 s frames. Recently, transplantation medicine benefits also from nuclear medicine techniques (SPECT, PET) other than dynamic renal study [166].

A fast initial frame rate has been also used in deconvolution analysis to check the quality of injected bolus [75]. However, robust deconvolution methods work well with a standard 10 s frame rate, too, providing the input curve is monotonically decreasing [76, 77].

The minimum length of the study is 20 minutes. If a diuretic is applied during examination to increase urine flow that may help to differentiate between renal obstruction and unobstructed dilated renal pelvis with delayed outflow, acquisition is prolonged to 30–40 minutes. Overall length of the study depends on the timing of diuretic injection, after which the acquisition should continue for additional 10–15 minutes. If the kidneys are not "empty" at the end of the study (still containing significant activity), the dynamic study is followed by several additional static images of the kidneys recorded after the patient assumes an upright position, makes a few steps, and empties the urinary bladder – the factors potentially facilitating "gravity assisted" drainage and urine outflow (in the literature, these images are sometimes called post-void, post-micturition, or post-erect images).

Besides scintigraphic data, it is important to record injected activity (measured as difference between full and empty syringe after injection) and the weight and height of the patient for potential calculation of BSA, estimation of the kidney depth, and so forth. It is important to record weights and heights that are really measured in the department because the values reported by the patients may not be reliable.

12.7.2 Preprocessing and Quality Control

Before processing, image data of dynamic renal scintigraphy should be first checked for a sufficient number of counts (signal-to-noise ratio), extravasation, appearance of activity in the heart, position of the patient and the kidneys in the field of view, and for motion. A simple means of quality control is to run the study in a cine mode. Patient movement, renal uptake of the tracer, transit from renal parenchyma to the pelvis, as well as drainage of the collecting system are then easily recognized [72, 78, 79].

It is assumed that in a normal kidney, peak renal count rate of approximately 200-250 cps (after background subtraction) will result in a renogram that requires no or little smoothing prior to visual interpretation and estimation of relative function. For time-activity curves from the kidney and background ROIs, a formula for the number n of passes of a (1-2-1) filter, subject to a minimum of two, has been derived by Fleming [63]

$$n = 12 - \sqrt{c}/15 \quad (12.12)$$

where c is the count per frame in the kidney ROI at 2 min.

Extravasation at the site of injection may give rise to difficulties in quantitative data processing and lead to incorrect interpretation of the study. Assessment of total renal function requires measurement of count rate in the kidneys, which is often related to injected activity and expressed as its fraction. If part of administered activity is injected paravenously, or if it is delayed at the site of injection, the measurement is inaccurate. Some authors therefore recommend scanning the injection site after the study. If the count rate at the injection site exceeds 1–2 per cent of injected counts, calculation of total renal function should be avoided.

Motion can be detected either visually (checking that the kidneys remain within the renal ROIs) or using dedicated software. Small motion usually can be well corrected by motion-correction software or simply compensated for by drawing kidney ROIs large enough to encompass motion. Large and complex motion of the patient (more likely in small children), movements of the kidneys due to deep breathing and other physiological functions (often of different size and direction on the left and right sides), and especially an intra-frame motion is difficult or impossible to correct properly with the tools routinely available. Therefore, considerable effort should be made to avoid motion during data acquisition. Clear explanation of the procedure to adult patients and vacuum cushions or similar gentle immobilization aids for small children are often sufficient to avoid motion so that neither sedation nor anesthesia is usually required.

Basic quantitative diagnostic indices extracted from dynamic renal study are the value of relative renal function and an index quantifying urine outflow. They are derived from time-activity curves (Figure 12.6) generated by the regions of interest (ROIs) drawn by a user around examined organs.

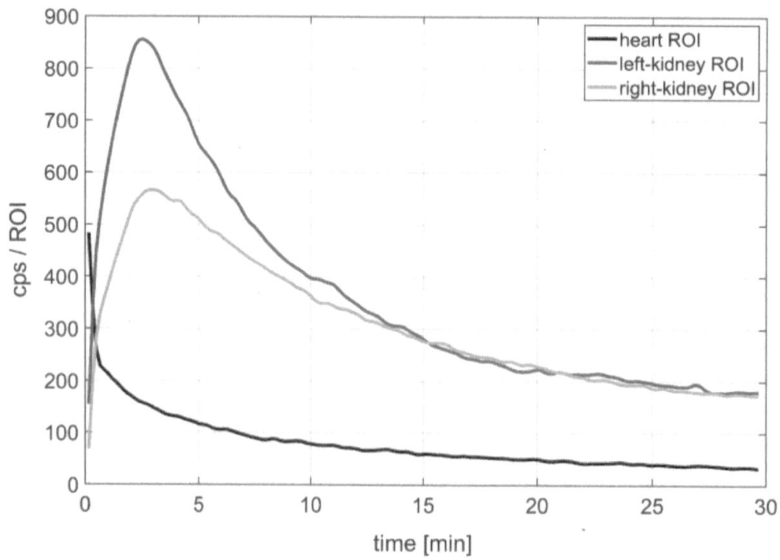

FIGURE 12.6 Time-activity curves extracted from the regions of interest with the heart and the left and right kidneys applied to dynamic renal study in the Fig. 12.5.

Software for automatic definition of ROIs exists, but its performance is not perfect. Comparison of available methods has shown that in over 10 per cent of the patients the automatic methods fail and the ROIs need to be drawn manually by a user [80, 81]. Commercial software often includes semi-automatic procedures facilitating definition of ROIs that, however, require user's interaction. Manual definition of ROIs can be facilitated by parametric images demonstrating individual pixel values of the selected functional parameters (kidney uptake, time to the curve maximum, mean transit time, etc.). Another alternative is factor analysis that simultaneously extracts both the images and time-activity curves of so called "dynamic" or "physiological" structures – the image segments with identical dynamics (vascular structures, renal parenchyma, renal pelvis). Factor images can be interpreted as "fuzzy" ROIs with weighted pixel contribution to several ROIs that can overlap each other. Counts measured in each pixel are thus proportionally distributed among several different ROIs [82–85]. Presenting time-activity curves from different patients on the graphs with identical count-rate scale (identical range of vertical axis) may help to assess the kidney function easily immediately on the first inspection.

Accuracy and reproducibility of the measurement of split renal function depend on kidney size and kidney function. The smaller the kidneys and the lower their function, the lower the accuracy of the measurement of split renal function. Other factors affecting accuracy are background subtraction and attenuation correction. Main sources of error in the measurement of renal function are indeed background activity and attenuation.

12.7.3 Background Subtraction

For several reasons, proper subtraction of background counts from the kidney ROIs in dynamic renal scintigraphy is more important (and more difficult) than in static renal cortical scintigraphy. Background is more intensive and its variability with time affects relative renal function in a more complex way [86, 87]. Counts in the kidney ROI originating outside the renal tissue can be divided into three components: intrarenal vascular background (from the blood inside the kidney), extrarenal vascular background (from the blood outside the kidney), and extrarenal extravascular background (from interstitial fluid and the tissues outside the kidney). All background counts increase the count rate in the kidney ROI. However, intensity of intravascular background (both intra- and extrarenal) decreases with time while extravascular background first increases and then decreases with time (its time-activity curve is similar to that of the kidney with low function). In the kidney ROI, even accurate subtraction of the background estimated in extrarenal ROI next to the kidney (normalized to the size of the kidney ROI and corrected for the kidney thickness) still leaves significant intrarenal intravascular activity that may affect the measurement of kidney function.

Accurate subtraction of background counts from the kidney ROI is vital for subsequent calculations of both relative and absolute kidney function. During the time interval of renal uptake, the vascular background dominates over the extravascular one and it is more important to subtract it [88, 89]. Extravascular tissue background affects accuracy of the measurement mainly in kidneys with very weak function – without subtraction, their function is overestimated [90].

Vascular background (both intrarenal and extrarenal) can be subtracted by weighted subtraction of the heart ROI curve from the kidney ROI curve (renogram) by gradually increasing subtraction weight and checking the first (perfusion) segment of the renogram. Disappearance of the perfusion phase and overall straightening of the upward segment of the renogram mark proper subtraction of vascular background. To avoid over-subtraction (indicated by negative values of the resulting kidney ROI curve), vascular background should be subtracted after previous subtraction of extrarenal background (containing both intravascular and extravascular components). Providing the first vascular phase of the renogram is well expressed (which is not always the case) this method is easy though subjective.

A more accurate and objective method to subtract vascular background is deconvolution. Time-activity curve from the kidney ROI can be considered as convolution of the input function (the curve from the heart ROI) and renal retention function, representing time-activity changes in the kidney ROI after bolus injection of the tracer into the renal artery [91, 92]. The plot of retention function starts with the peak activity in renal vessels (vascular peak), followed by a plateau (Figure 12.7). During transit of the tracer through the kidney, the plateau is horizontal with the height reflecting kidney function and the length indicating minimum transit time. When the tracer starts to leave the kidney, the retention function gradually falls to zero. After maximum transit time, the kidney is empty. Renal retention function is a useful tool for quantitative description of kidney function. Its use for subtraction of vascular background consists in cutting off its vascular peak (by backward extrapolation of the plateau to zero time) and reconvolution of the modified renal retention function with original input function. Reconvolved renogram then does not involve vascular background.

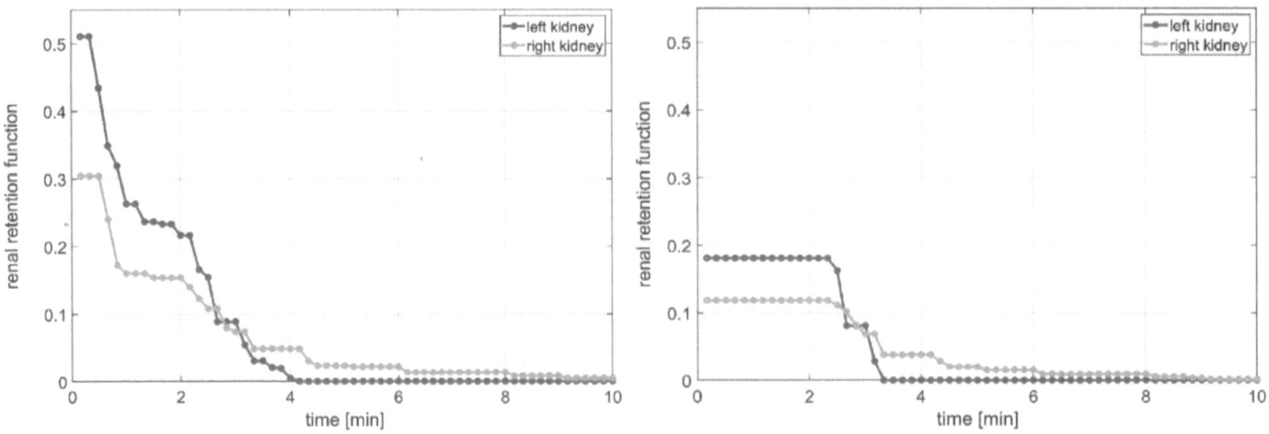

FIGURE 12.7 Renal retention function of the left and right kidneys demonstrated in Fig. 12.5: Deconvolution of the raw kidney curves (without background subtraction) results in renal retention function with initial vascular peak and higher plateau due to the contribution of tissue background (left). After both vascular and tissue background subtraction, plateau can be seen more clearly and used for accurate measurement of renal function and transit times

An excellent tool for objective subtraction of vascular background from the renogram is the Patlak-Rutland plot [12–14, 93]. The name of this technique has been adopted in radionuclide nephrology and urology although the method itself has a more complicated parentage [94–97]. The model can be derived from Eqn. 12.7 above

$$R(t) = Z\int_0^t c(t)\,dt$$

where *R(t)* [cps] are "pure" renal counts in the kidney ROI (without background) at time *t* during uptake interval before activity starts to leave the kidney, Z [ml/min] is renal clearance and *c(t)* [cps/ml] tracer concentration in plasma. Substituting plasma concentration *c(t)* by the heart ROI curve *P(t)* [cps]

$$P(t) = c(t)V_h \tag{12.13}$$

where V_h [ml] is the plasma volume in the heart ROI, the equation becomes

$$R(t) = Z/V_h\int_0^t P(t)\,dt \tag{12.14}$$

Observed renal counts (renogram) $R_o(t)$ are contaminated by intrarenal vascular background *hP(t)* where *h* is the ratio of plasma volumes in the kidney and the heart V_k/V_h, and by extrarenal background *T(t)* in the kidney ROI (containing both intravascular and extravascular components)

$$R_o(t) = Z/V_h\int_0^t P(t)\,dt + hP(t) + T(t) \tag{12.15}$$

Subtracting extrarenal background *T(t)* to obtain corrected renal counts $R_c(t)$ and dividing the equation by *P(t)* results in the equation of a straight line

$$R_c(t)/P(t) = k\int_0^t P(t)\,dt/P(t) + h \tag{12.16}$$

with the slope $k = Z/V_h$ proportional to renal clearance Z, and with the intercept *h* quantifying the ratio of plasma volumes in the kidney and the heart. The slope *k* and the intercept *h* are derived from plotting $R_c(t)/P(t)$ against $\int P(t)dt/P(t)$ and fitting the plot by a straight line (Figure 12.10). Vascular background then can be subtracted from the corrected renogram as

$$R(t) = R_c(t) - hP(t) \quad (12.17)$$

However, with the Patlak-Rutland plot, relative renal function unaffected by vascular background can be calculated directly from the slopes of the left and right kidneys k_L and k_R

$$LK\,\% = 100 k_L / (k_L + k_R) \quad (12.18)$$

without a need to explicitly subtract vascular background from the renogram.

In most departments, extrarenal background is usually subtracted routinely by simple definition of the background ROI next to the kidney and subtracting the background time-activity curve (scaled to the size of the kidney ROI) from the renogram. With respect to the kidney ROI, the most frequently used background ROIs are lateral, circumferential, subrenal, and central (a single background ROI between the two kidneys). Effective but infrequently used approach is interpolated background subtraction with the background inside the kidney ROI estimated (interpolated) from pixel values around the kidney [98]. A multiple regression method introduced to reduce the dependence of split renal function on the choice of specific background ROI developed by Middleton and colleagues [99] has not met expectations due to multicollinearity of involved variables [90].

It is usually understood that the background ROI should not include kidney hiluses and ureters to avoid counts from active urine. However, it does not affect measurement of renal function as it is carried out strictly in the interval of renal uptake before the tracer starts to leave the kidney. The problem with extrarenal background is that its respective locations include different ratios of vascular and extravascular counts. Despite several studies clearly demonstrating the deleterious effect of using various background ROIs on accuracy and reproducibility of measured renal function, clinical practice is not unified [90, 100–104].

If background subtraction procedure includes correction for both extrarenal and intrarenal vascular background, extrarenal background should be subtracted first to avoid oversubtraction of vascular background. A careful approach should also include approximate correction for the kidney "thickness" and subtraction of extravascular background from the heart ROI in case the heart ROI curve is involved in the method [63, 105]. The latter, however, is not critical in the measurement of relative renal function where the count rate from plasma cancels out. After appropriate background subtraction, renal counts should remain non-negative.

12.7.4 Attenuation Correction

In the measurement of relative renal function, it is recommended to exclude an effect of different kidney depth (skin to kidney distance) of the left and right kidneys, especially in adults. As with static renal scintigraphy, the simplest way to correct for different kidney depth is to record both posterior and anterior views and measure split function in the geometric mean image. In the measurement of individual kidney or global renal function (renal clearance), it is necessary to apply the proper attenuation correction to each kidney count. Unlike in static examination, in dynamic renal scintigraphy direct measurement of kidney depth in the lateral images may not be easy if the kidney counts at the end of the study are too low. Kidney depth is thus either estimated by regression, measured by ultrasound or CT (low dose or in existing diagnostic images, if available), or the attenuation is corrected using transmission maps.

Simple, frequently used, but not very accurate methods estimate kidney depth by regression equations originally derived from the measurement in lateral scintigraphic projections and CT transversal slices through the kidneys. In children, an often used equation was derived by Tonnesen and colleagues [106–108]

$$\begin{aligned} d_L &= 13.2\frac{w}{h} + 0.7 \\ d_R &= 13.3\frac{w}{h} + 0.7 \end{aligned} \quad (12.19)$$

where d_L, d_R [cm] are the depths of the left and right kidneys, and w [kg] and h [cm] are the weight and height of the patient. The Tonnesen equation estimates kidney depth with the mean absolute prediction error of 4–5 mm and with about 90 per cent of individual errors lower than 1 cm [109]. In adults, a clinically well-validated regression equation was published by Taylor et al. [110–112]

$$d_L = 16.7\frac{w}{h} + 0.027a - 0.940$$
$$d_R = 15.13\frac{w}{h} + 0.022a - 0.077 \tag{12.20}$$

where a [yrs] is the age of the patient. Taylor's equation estimates kidney depth with a mean absolute prediction error of 10–12 mm, which is similar to analogous equations derived by Inoue and colleagues [113] and Lythgoe and colleagues [107], the latter originally developed for children. However, only about 50 per cent of individual errors are below 1 cm, while 90 per cent of errors are lower than 2.5 cm [109].

If planar dynamic renal study is performed with a SPECT/CT camera, kidney depth can be measured in a single slice recorded by low-dose CT through the kidneys with a minimum radiation dose. Except in ectopic kidneys and inborn malformations, this is not necessary in children. Alternatively, scout image can be used as an approximate transmission map.

An attenuation map can be obtained also using radionuclide transmission sources (Figure 12.8). Cumbersome work with liquid flat phantom filled with the solution of ^{99m}Tc radiopharmaceutical can be avoided with a solid cobalt ^{57}Co flood source used for gamma camera quality control [114]. The radiation dose to the patient (additional to dynamic renal scintigraphy) is below 5 μSv. Additional time required for transmission measurement is just a couple of minutes. The flood phantom is placed directly on the lower detector below the patient table, and the two opposite projections are

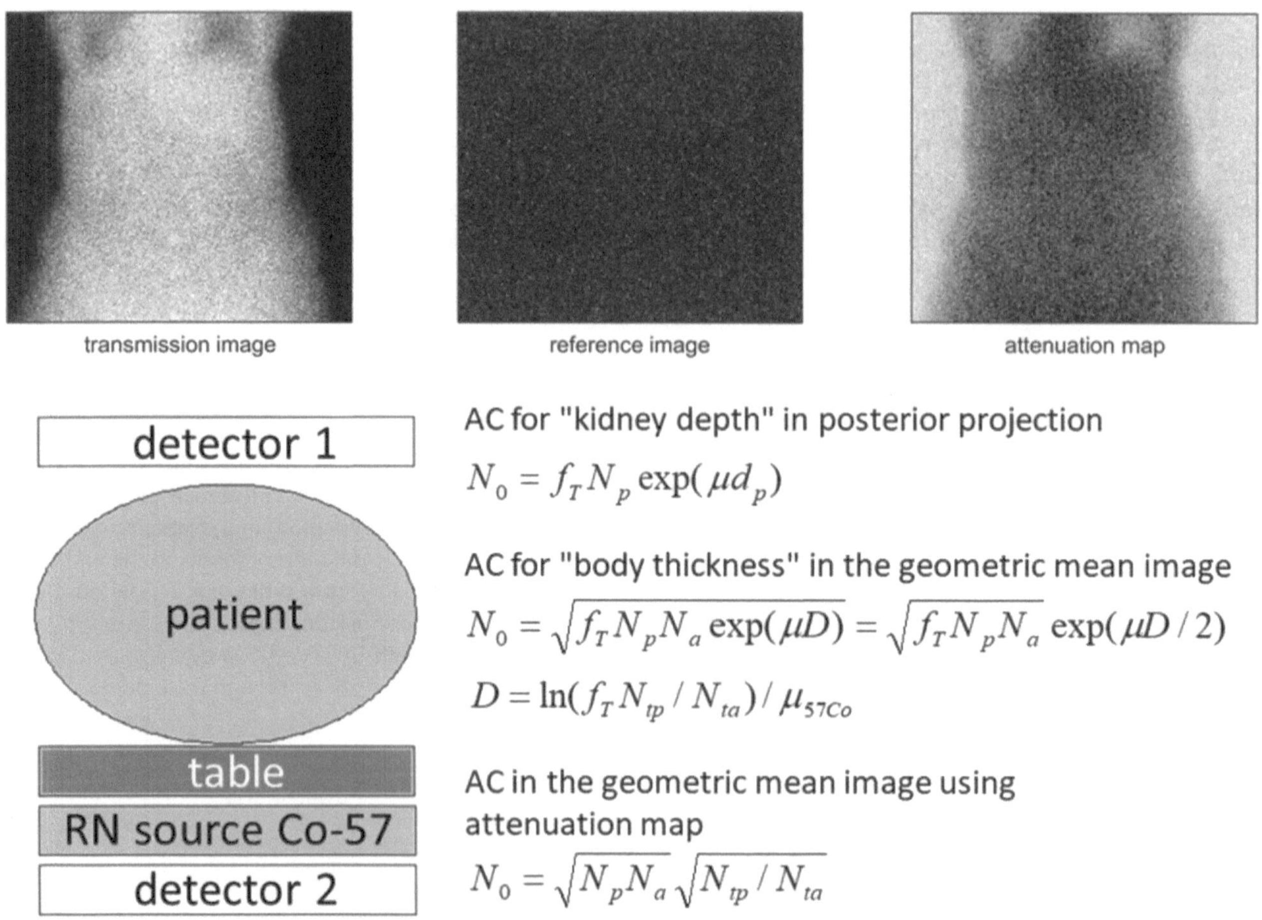

FIGURE 12.8 Transmission (camera detector 1) and reference (camera detector 2) images of ^{57}Co source used to calculate attenuation map (AC – attenuation correction, N_0 – true counts, N_p – posterior emission counts, N_a – anterior emission counts, f_T – table attenuation factor, d_p – individual kidney depth (kidney to posterior surface of the body), D – body thickness measured through the kidneys, μ – attenuation coefficient, N_{tp} – reference image counts, N_{ta} – transmission image counts).

recorded with the patient on the table (before injection of a radiopharmaceutical). The geometric mean image is then corrected pixel-wise by the attenuation map as

$$N = \sqrt{N_p N_a}\sqrt{N_{tp} / N_{ta}} \tag{12.21}$$

where N_p and N_a are pixel values in posterior and anterior emission projections of the patient, N_{tp} are non-attenuated reference counts from transmission source recorded by the lower detector (detector 2 in Figure 12.8) and N_{ta} are the counts from transmission source attenuated by the camera table and the patient's body (recorded by the upper detector in ^{99m}Tc energy window subsequently used for the examination). Although it is only an approximate solution that does not take into account several important factors (background, scatter, thickness of the flood source, sensitivity of the detector heads, position and width of the energy window, type of collimator, and geometry of measurement), recovery of realistic kidney phantom activity exceeds 80 per cent, which can be increased up to 100 per cent using simple corrections for the energy difference between ^{57}Co and ^{99m}Tc, different sensitivity of the posterior and anterior detectors, and for scatter.

Scatter correction is required only in the measurement of single-kidney or global renal function and can be performed by measurement or using broad-beam attenuation correction [115–119].

12.7.5 Relative Renal Function

Relative renal function is measured during the time interval of renal uptake before the tracer starts to leave the kidney (i.e. well before the peak of the kidney ROI curve). The measurement interval can be either fixed to 1–2 minutes (later and longer intervals may include the peak in some renograms), defined manually using the time-activity curve from the kidney ROI (between the end of vascular phase and the end of approximately linear uptake phase of the renogram), identified by deconvolution (corresponding to the length of the plateau of renal retention function), or by the Patlak-Rutland plot (corresponding to its linear segment). It is imperative that the uptake interval is visually checked with every kidney ROI curve for potential inclusion of the curve peak and appropriately modified if necessary.

The simplest and most commonly used technique for the measurement of relative renal function is the integral method. It integrates background-corrected time-activity curves from the left and right kidney ROIs over the interval of renal uptake. Integration has to be made over the same interval in both kidneys. If uptake intervals of the left and right kidneys are somewhat different, measurement should be performed in their overlap (between the later start and earlier end point). Relative renal function is then calculated as

$$LK\ \% = 100 S_L / (S_L + S_R) \tag{12.22}$$

where S_L and S_R are the sums or integrals of background-corrected time-activity curves of the left and right kidneys over the uptake interval (Figure 12.9).

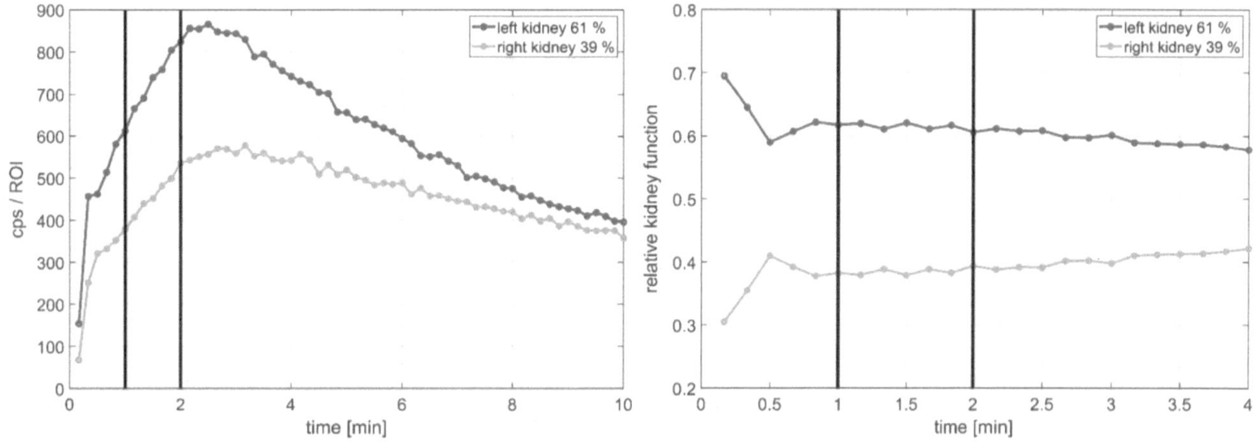

FIGURE 12.9 Measurement of relative renal function by integral method using the kidney ROI time-activity curves (left) between 1 and 2 minutes and point-wise values of relative renal function obtained from respective points of kidney curves (right).

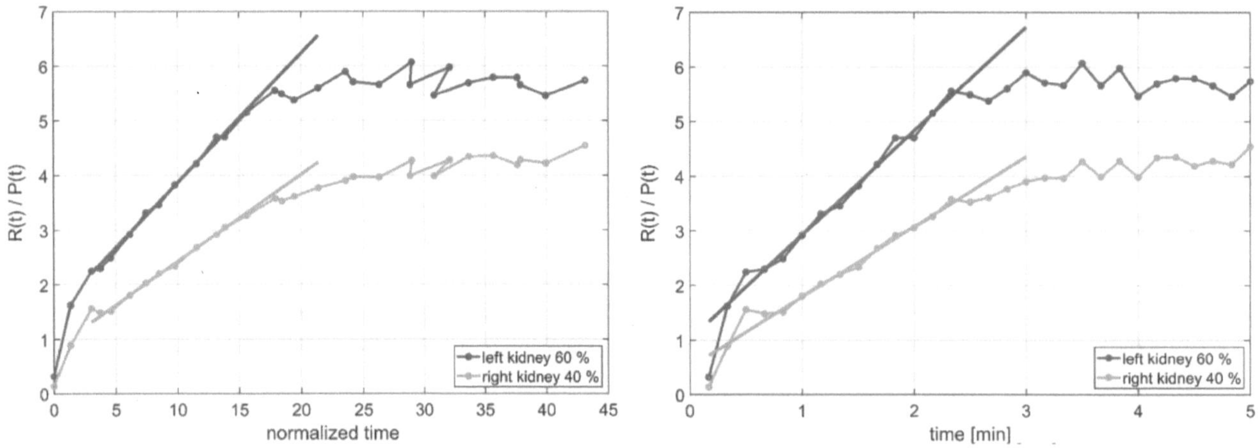

FIGURE 12.10 Patlak-Rutland (left) and normalized slope (right) plots of the left and right kidneys presented in Fig. 12.5. In both plots, relative kidney function is calculated using the slopes of straight lines fitted to the linear segment of the plot for each kidney (the slope is proportional to renal clearance). Normalized time in Patlak-Rutland plot is the ratio of cumulative integral of the heart ROI curve and the heart ROI curve itself.

The integral method can be improved by calculating relative renal function point-wise over the increasing part of the kidney curves or an even longer interval. In the interval of renal uptake, individual points of relative renal function should plot a horizontal segment that provides additional quality control. Absence of any slope demonstrates appropriate background subtraction; its length indicates the length of uptake interval, and the scatter of points around the horizontal line reflects reliability of measurement. The result of measurement then can be presented as the mean value and standard deviation of individual point values in the horizontal segment.

The accuracy of the integral method depends on accurate subtraction of both extravascular and intravascular (extrarenal and intrarenal) background. With different background ROIs (each with different contribution from intravascular and extravascular background), the values of relative renal function may vary substantially [87, 90, 100, 101, 120].

Measurement of relative renal function using the Patlak-Rutland plot (Figure 12.10) has been described above demonstrating its advantage of "automatic" subtraction of vascular background (both intrarenal and extrarenal). The method is sometimes avoided just because it presumably requires the heart in the field of view – a requirement that cannot be satisfied in some patients with medical preference to include the urinary bladder, or who are simply too tall. There is already enough evidence that in the calculation of relative renal function with the Patlak-Rutland plot, the heart ROI curve can be safely substituted by the liver or the spleen ROI curve that is available in all patients [15–17]. With the liver or the spleen ROI curve, the values of relative renal function are the same as those obtained with the heart ROI curve.

The method similar to the Patlak-Rutland plot that provides identical values of relative renal function is sometimes called as the "corrected-slope" or "normalized-slope" method [74, 121]. It plots the values $R(t)/P(t)$ directly against standard time in minutes (instead of "normalized time" $\int P(t)dt/P(t)$ as in the Patlak-Rutland plot). The rest is the same as with the Patlak-Rutland plot. The slope of the linear segment from each kidney is substituted into Eqn. 12.18 and relative renal function is calculated. Another approach producing identical results as the Patlak-Rutland plot is the "blood-pool compensation" method [88, 122], which simply plots renal counts $R(t)$ against the liver or the heart ROI counts $P(t)$. Linear segments of the plot of the left and right kidneys are back-extrapolated to $P(t) = 0$, and their intercepts with vertical $R(t)$ axis compared in a similar way as the slopes in the Eqn. 12.18. The method does not correct for extravascular background (as was originally expected) but accurately corrects for intravascular background [15, 16, 123].

Relative renal function can be derived also from the renal retention function where the level of the plateau is proportional to renal function [124]. It is good practice to measure relative renal function with several different methods. While good agreement of their results does not confirm the correctness, large differences between them indicate a problem that should be identified and commented in the report. Normal values of relative renal function are between 45–55 per cent.

12.7.6 Absolute Renal Function

Information on both relative and absolute (single kidney and global) renal function can best be obtained by combination of the measurement of plasma clearance (to assess global function) and static or dynamic renal scintigraphy (to assess relative function of the left and right kidneys). It is also possible to combine dynamic renal scintigraphy with subsequent blood sampling (using ^{99m}Tc-DTPA or a combination of ^{99m}Tc-MAG3 with ^{51}Cr-EDTA). Although these procedures provide the most accurate results, they are laborious and time-consuming. In practice, they are usually reserved for research and special clinical circumstances when high accuracy is vital.

In routine clinical practice, absolute renal function is estimated from scintigraphic data without blood sampling by regression and model-based methods (Figure 12.11). Regression methods are derived from comparisons of radiopharmaceutical plasma clearance (dependent variable) and (background and attenuation corrected) activity accumulated in the kidneys (independent variable). The first gamma-camera regression methods for effective renal plasma flow and glomerular filtration tracers were introduced by Schlegel and Hamway [125], and by Gates [126, 127]. Despite subsequent criticism, based on common experience, that they are well reproducible but not accurate [79, 128–131], these methods and their modifications are still part of commercial nuclear medicine software. Later, more advanced studies such as those by Inoue and colleagues with DTPA [113, 132], and by Taylor and colleagues with MAG3 [133] independently resulted in similar regression equations validated in several centres [112, 134, 135].

Model-based methods calculate single-kidney and global renal function using equations derived from physiological models. The simplest approach employs the normalized slope of the Patlak-Rutland plot [136]. It has been shown above that the slope of its linear segment Z / V_h equals to renal clearance as a fraction or a multiple of plasma volume in the heart ROI. Unknown volume V_h varies in individual patients, but normalization of the heart ROI to the same size in all patients (or, alternatively, normalizing the heart ROI curve to reflect count rate per pixel of the heart ROI) approximately unifies V_h to the same value V_{hn} in all patients and substantially reduces variability between individual examinations due to ROI size. Residual variability (due to variable pixel depth) is lower, and the values of the Patlak-Rutland slope, reflecting absolute kidney function (clearance Z), can be compared on both intra-patient and inter-patient bases. The values of normalized uptake index Z/V_{hn} can be converted to renal clearance by regression derived from simultaneous examinations of plasma and renal clearance.

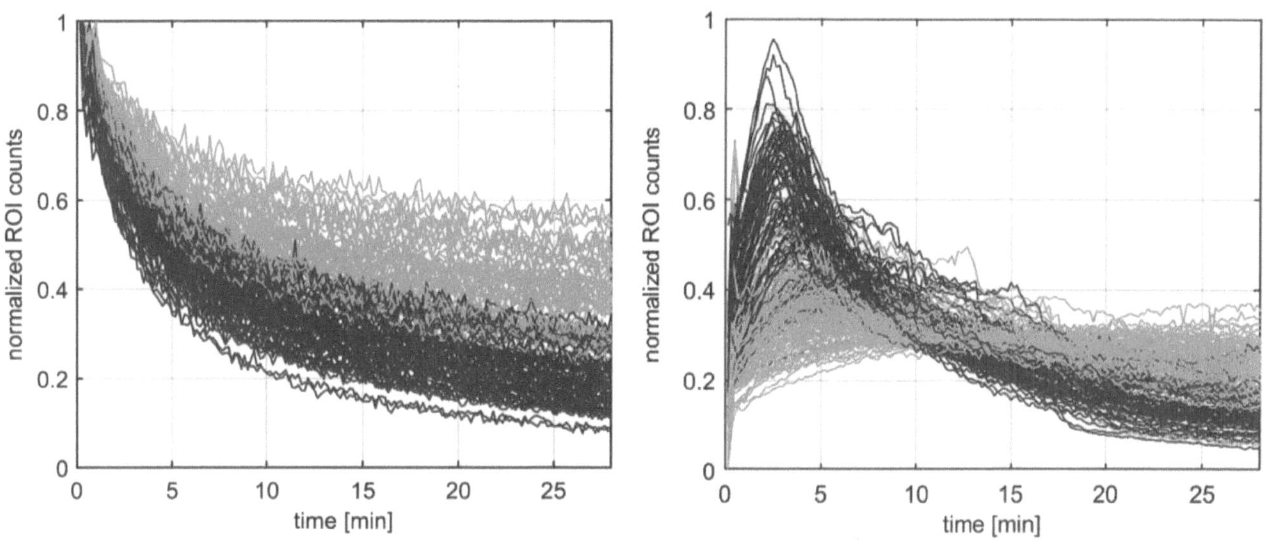

FIGURE 12.11 Simple demonstration that information on global and individual kidney function is included in data of dynamic renal scintigraphy (n = 107 dynamic renal studies). The heart ROI curves are presented on the left and single-kidney curves on the right. Reference plasma clearance of ^{99m}Tc-MAG3 was measured by blood sampling and assigned to each kidney using relative renal function. The heart ROI curves (normalized to common maximum) in the patients with reference plasma clearance above the median value (dark curves) decrease faster than in the patients with the reference clearance below the median (grey curves). Individual kidney ROI curves (normalized to unit area under the curve) increase faster in the patients with plasma clearance above the median.

Alternatively, with or without previous normalization of plasma volume in the heart ROI, V_h can be related to total plasma volume V_p as

$$V_h = \frac{P(t)}{c(t)} = \frac{P(0)}{c(0)} = \frac{P(0)}{D} V_p \tag{12.23}$$

where *P(t)* [cps] is the heart ROI curve, *P(0)* the initial value of *P(t)* back-extrapolated to time zero (for curve fitting it is important to use the curve segment corresponding to uptake interval), *c(t)* [cps/ml] plasma concentration of the tracer, and *D* [cps] injected activity. Subsequently, the slope of the Patlak-Rutland plot (renal clearance as fraction of plasma volume in the heart ROI) can be converted into renal clearance as a fraction of total plasma volume

$$\frac{Z}{V_p} = \frac{Z}{V_h}\frac{P(0)}{D} \tag{12.24}$$

Fractional renal clearance Z/V_p [1/min] was first derived by Rutland and colleagues as FUR, for fractional uptake rate [34]. It is a single-kidney clearance expressed as a fraction of total plasma volume cleared per unit time. Plasma volume V_n in millilitres can be estimated by regression [32] from individual body surface area *BSA* [m²]

$$V_p = 1645\,BSA \tag{12.25}$$

or, alternatively, FUR can be normalized directly to standard body surface area $BSA_n = 1.73$ m² as

$$\begin{aligned} Z_n &= \frac{Z}{V_p} V_{pn} \\ V_{pn} &= 1645\,BSA_n = 2846 \\ Z_n &= 2846 \frac{Z}{V_p} \end{aligned} \tag{12.26}$$

using standard (normal) plasma volume $V_{pn} = 2846$ ml to obtain normalized clearance Z_n in usual units of [ml/min/1.73 m²].

Absolute global renal function (of both kidneys) can be also calculated as fractional plasma clearance from the heart ROI curve [9, 86, 96, 137], using equations 12.5, 12.8 and 12.13

$$\begin{aligned} Z &= \frac{D}{\int_0^\infty c(t)dt} = V_p \frac{V_h c(0)}{\int_0^\infty P(t)dt} = V_p \frac{V_h P(0)}{V_h \int_0^\infty P(t)dt} = V_p \frac{P(0)}{\int_0^\infty P(t)dt} \\ \frac{Z}{V_p} &= \frac{P(0)}{\int_0^\infty P(t)dt} \end{aligned} \tag{12.27}$$

and normalized using Eqn. 12.26 as above. The value *P(0)* is obtained by back extrapolation of monoexponential fit to the heart ROI curve through the uptake interval to time zero. The total area under *P(t)* is obtained by numerical integration of the observed part of the curve and by extrapolating the monoexponential fit through the last segment of the curve. While fractional renal clearance (FUR) requires careful background subtraction and attenuation correction, Eqn. 12.27 provides estimates of fractional plasma clearance without corrections, probably due to approximate compensation of the underlying processes like diffusion, contribution of tissue background, attenuation, and scatter [138]. The true mechanism of compensation remains to be established.

The accuracy of model-based estimates of renal and plasma clearance by gamma camera without blood sampling is approaching the accuracy of single-sample ^{99m}Tc-MAG3 techniques by Russell and colleagues [131, 139] and Bubeck [60, 61]. Reference values of plasma clearance of MAG3 measured with 2 or more blood samples are estimated with the mean absolute error of prediction 10–20 ml/min. Model-based estimates without blood sampling cannot substitute blood-sampling methods but they overcome prediction equations based on plasma creatinine concentration and may represent a useful supplement to diagnostic information reported in dynamic renal scintigraphy.

12.7.7 Renal Transit Times

Renal transit times quantify periods of time required for the molecules of a radiopharmaceutical to transit through renal parenchyma (parenchymal transit time) and through the whole kidney including the renal pelvis (whole-kidney transit time). Transit time is a sensitive diagnostic index whose prolongation demonstrates impaired function of renal parenchyma and delayed outflow of urine from the kidney. As the only diagnostic parameter in radionuclide nephrourology currently available, parenchymal transit time is assumed to predict kidney function and thus facilitate timely decisions on kidney surgery including estimation of its potential effect in children with hydronephrosis [140, 141]. A comprehensive review on renal transit times has been published by the International Scientific Committee of Radionuclides in Nephrourology in 2008 [92].

An observed kidney ROI curve can be modelled by convolution of input function (the heart ROI curve) and renal retention function which is an analogy of residue or impulse retention function in tracer dilution studies [91]. The renal retention function is the curve that would be observed in the kidney ROI after injection of a small bolus of the radiopharmaceutical (unit impulse) directly into the renal artery [142]. It starts with a short spike corresponding to vascular transit followed by a horizontal plateau reflecting transit of injected activity through the kidney (Figure 12.7). After the shortest minimum transit time, activity starts to leave the kidney and the retention function decreases to zero. Less invasively, it can be obtained by deconvolution of the standard kidney ROI curve by input function [143–147]. Individual transit times (minimum, mean, and maximum transit time through the kidney) are then measured as the length of the retention curve plateau, area under the curve, and the time when the curve decreases to zero. A spectrum of transit times can be observed on output (by analogy with the impulse response function) and obtained by differentiating the cumulative distribution function of transit times that complements the residue function to one [91].

The renal retention function $H(t)$ quantifies the amount $q(t)$ of the tracer in the kidney at time t as a fraction of the tracer remaining in the organ after an impulse input $q(0)$

$$H(t) = \frac{q(t)}{q(0)} \tag{12.28}$$

Mean transit time MTT is then

$$MTT = \int_0^\infty H(t)\,dt \tag{12.29}$$

and the amount of tracer remaining in the organ, $Q(t)$

$$Q(t) = \int_0^\infty I(t-T)H(T)\,dT \tag{12.30}$$

where $I(t)$ is the actual rate of tracer input into the organ. If $I(t)$ and $Q(t)$ can be measured (as the corrected heart and kidney ROI curves), $H(t)$ and MTT can be determined by solving the convolution Eqn. 12.30. The solution of the equations above will only be valid if underlying assumptions of stationarity and linearity are satisfied. If they are violated – for example, if a diuretic is administered during a renogram so that the urine flow rate changes during examination, then the result of deconvolution becomes unpredictable [148].

Successful transit time measurement requires data acquisition starting before the tracer injection (to ensure that the peak of the heart ROI curve is included and can be defined as time zero), a sufficiently fast frame rate (10 s frames represent good trade-off between frequency and sufficient count rate), a good count rate, an appropriate smoothing, the heart in the field of view (as the source of input function), and a monotonically decreasing input curve, careful definition of ROIs, and a robust method for deconvolution [77, 145, 148]. Measurement of transit times requires strict adherence to theoretical assumptions and methodological details, the lack of which may reduce the validity of results [76, 102].

Transit times are measured in the ROI including the whole kidney (both renal parenchyma and pelvis) as the whole-kidney transit time, and as parenchymal transit time in the narrow ROI including renal parenchyma or cortex but carefully excluding all the hollow structures as renal calyces and pelvis. Whole-kidney transit time covers both renal transit and outflow while parenchymal transit time reflects exclusively the function of the renal cortex. The parenchymal transit time index (difference between the mean and minimum parenchymal transit time) has been introduced to improve detection of kidney obstruction [149].

Besides deconvolution, renal transit can be roughly estimated by several surrogate measures (visual examination of the images and curves, measurement of time to peak and half time of the kidney ROI curve or pelvic appearance time, and assessment of outflow indices such as renal output efficiency and normalized residual activity). Despite mutual correlations, each index has its specific advantages and disadvantages. Descriptive parameters of the ROI curves depend on the curve shape and have higher negative rather than positive predictive value, while model-based methods are more consistent with deconvolution approach [76, 92, 137, 150].

A simple and robust (though subjective and approximate) method to estimate the quality of the parenchymal transit was demonstrated by Schlotmann and colleagues [151, 152] in both animal experiments and in patients. Instead of numerical results, the method classifies estimated transit times into three categories: timely, delayed, and indeterminate. Classification is based on visual assessment of activity distribution in the kidney after the uptake interval. If the pelvis region remains photopenic and activity in renal parenchyma does not decrease or even grows, parenchymal transit time (reported as tissue transit time by the authors) is considered to be delayed. The method was successfully clinically validated in several pediatric departments [141, 153].

12.7.8 Outflow Indices

Outflow indices quantify drainage of the kidney. They are derived in both standard and diuretic renography that includes application of diuretics to increase urine flow and differentiate between renal obstruction and dilated, but unobstructed, renal pelvis that mimics obstruction.

The simplest and often used but poorly defined and least reliable index is the outflow half time (T/2 or T½), the time at which the count rate in the kidney decreases to a half of some reference value. It is affected by several renal functions (uptake, transit, outflow) and by the overall shape of the renogram. Its most serious disadvantage is lack of standardization [2, 92, 154]. Different users count it from different reference times – from the beginning of the study, from the peak of the kidney ROI curve, from the time of diuretic injection, or from the beginning of its effect, still others derive the half time from exponential decay fitted to the decreasing part of the renogram – and relate it to different reference values (curve maximum, renogram value before application of the diuretic, renogram value before emptying of the patient's bladder, etc.). Half time is also measured in different ROIs (whole kidney ROI or the ROI restricted to retained activity in the collecting system). There is general agreement that prompt clearance of the radiopharmaceutical from the renal collecting system excludes obstruction while prolonged voiding should never be the sole criterion determining its presence [1, 2].

Another approximate measure of renal outflow is the time to the peak of the kidney ROI curve. It is well defined but lacks clear physiological meaning (it is the time when inflow and outflow in the kidney ROI equal each other). Peak time is regularly reported and measured with relatively small variance between individual users and centres. Its measurement is easy unless the kidney curve is flat. Then background subtraction or smoothing may shift the peak substantially [102]. Normal peak time has a high negative predictive value – if it is short enough, both renal uptake and minimum transit time are all right. Normal values are below 5 minutes for both DTPA and MAG3 [1, 92].

Physiological, well defined, and more accurate outflow indices are normalized residual activity (NORA) and renal output efficiency (ROE). Normalized residual activity [155, 156] substituted intuitive parameters such as R20 / R2 (the ratio of the renogram value in the 20th minute to its value in the 2nd minute) and similar indices used previously [157]. NORA can be applied to various times indicated by subscript ($NORA_{20}$ = R20 / R2, $NORA_{30}$ = R30 / R2, etc.), while denominator remains R2. The most frequently used index is $NORA_{20}$. The value R2 is usually integrated over the second minute of the renogram between 1–2 minutes from time zero, the value R20 between 19–20 minutes from time zero. Even NORA thus suffers from some lack of standardization that does not, however, modify its results substantially. Time interval R2 is counted between 1–2 minutes or 1.5–2.5 minutes, R20 between 19–20, 20–21, or 19.5–20.5 minutes, and so forth. Reporting specific time intervals used for calculation of NORA is thus mandatory [92].

The most robust and informative indicator of renal outflow is renal output efficiency (*ROE*) or, for short, output efficiency (*OE*)

$$ROE(t) = \frac{ZOC(t) - R(t)}{ZOC(t)} \tag{12.31}$$

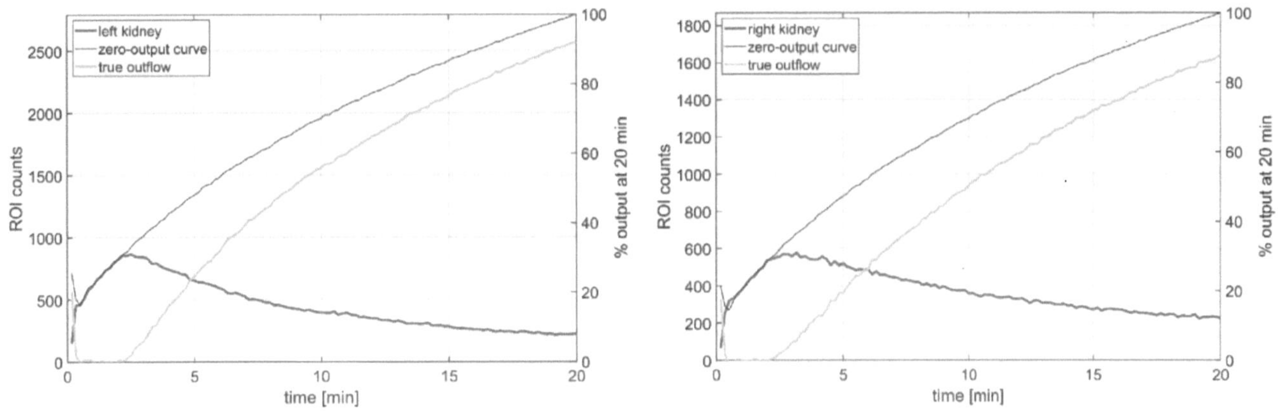

FIGURE 12.12 Renal output efficiency of the left and right kidneys demonstrated in Fig. 12.5 and Fig. 12.6. Light grey curve demonstrates true outflow from each kidney (output efficiency), reaching the values 51, 80, 88, and 92 per cent at 5, 10, 15, and 20 minutes on the left, and 41, 72, 83, and 88 per cent on the right, indicating normal outflow from both kidneys.

where *ZOC(t)* is "zero output curve" and *R(t)* background-corrected renogram. Zero-output curve is a theoretical renogram that would be observed if there was no outflow from the kidney [158, 159]. It is calculated as the integral of the curve from the heart ROI scaled to fit the uptake phase of the renogram (proportionality constant k and intercept h can be calculated by regression over the uptake interval or obtained equivalently from the Eqn. 12.16 and Patlak-Rutland plot)

$$ZOC(t) = k\int_0^t P(t)\,dt + hP(t) \tag{12.32}$$

Output efficiency is usually reported in percentage, multiplying the result of Eqn. 12.31 by 100, as ROE_{20}, the value at 20 minutes from time zero (peak of the heart ROI curve). However, both *ROE(t)* and *ZOC(t)* can be plotted separately against time to visualize the time course of renal outflow (Figure 12.12), which is especially useful in the assessment of the kidney reaction to diuretics. Both NORA and ROE can and should be also quantified in static post-void images.

The problem with both NORA and ROE is their residual dependence on global renal function that modifies the shape of the heart ROI curve (input function). This does not, however, invalidate the parameters and, if necessary, can be removed by appropriate corrections [137]. Normal values of ROE_{20} at 20 minutes are above 80 per cent (normally at least 80% of activity integrated by the kidney should be excreted by 20 minutes). Lower values indicate an outflow problem.

12.7.9 Reporting

As the minimum number of quantitative indices, the report on dynamic renal scintigraphy should contain the age, gender, weight, and height of the patient (both measured in the department), injected activity, relative renal function (preferably measured by more than one method to demonstrate stability of results), peak time of the kidney curve, NORA, and ROE. If a diuretic was applied, its amount and time of injection with respect to time zero (peak of the heart ROI curve) should be reported. The text should specify the methods used for the measurement (including background subtraction) and calculations. Normal values and their limits should be indicated. Time-activity curves for individual kidneys and the heart (if it is in the field of view) should be presented. The report should be completed by the images integrated over the uptake period (also demonstrating the ROIs), the images from the end of the study (demonstrating residual activity in the kidneys), and potentially recorded post-void images. A guidance document for structured reporting of diuresis renography has been published by Taylor and colleagues [160].

12.7.10 Database of Dynamic Renal Scintigraphy

To support standardization of analytical methods and data-processing software in dynamic renal scintigraphy, national and international audits, development of the new data-processing methods and their validation, publicly available database www.dynamicrenalstudy.org has been developed, including anonymized image data sets, basic laboratory and clinical information on the patients (both children and adults), dynamic studies obtained with physical phantom of the kidneys, and realistic Monte-Carlo simulated data with "known truth" [161, 162]. The database is suitable also for training and teaching purposes.

12.8 RADIONUCLIDE CYSTOGRAPHY

The aim of radionuclide cystography is to detect problems with emptying of the urinary bladder, especially vesico-ureteral reflux (abnormal flow of urine back from the bladder to the ureters and renal pelves). Radionuclide cystography also makes it possible to assess urine flow through a megaureter (a pathological condition of the ureter) and to measure residual volume of urine in the incompletely voided bladder, all representing potential causes of pyelonephritis with subsequent risk of the reduction of renal parenchyma and function [163].

The urinary bladder can be filled by the tracer either directly by a catheter introduced through the patient's urethra (ascending or direct radionuclide cystography) or indirectly (physiologically) at the end of dynamic renal scintigraphy (descending or indirect radionuclide cystography). Unlike similar radiological examinations (voiding cystourethrography, VCUG) scintigraphy does not provide accurate quantitative indices although potential reflux can be classified semi-quantitatively [164]. The radiation dose with radionuclide cystography (especially a direct one) is much lower than the dose from VCUG. Therefore, it is a good strategy to quantify reflux with VCUG and subsequently monitor the patient by radionuclide cystography with substantially lower radiation dose.

12.9 POSITRON EMISSION TOMOGRAPHY

Positron emission tomography PET/CT is routinely used for tumour imaging of the kidney and urogenital system, including prostate cancer. At the time of compilation of this text, application of PET/CT to non-oncological diseases of the kidneys is rare. However, its potential to examine kidney function is significant [165–167]. In contrast to conventional modalities discussed above, PET offers several advantages such as better spatial–temporal resolution, rapid three-dimensional imaging, absolute camera-based quantification, and anatomical co-registration with high-resolution multislice CT. New positron-emitting radiopharmaceuticals are being developed and experimentally validated to examine perfusion and function of the kidneys. An example is ^{68}Ga-EDTA for quantitative imaging of glomerular filtration. Other new radiopharmaceuticals are tested for imaging of amyloidosis of the kidneys, quantitative imaging of renal ischemia, assessment of kidney transplant nephropathy, obstructive renal disease, and imaging of angiotensin receptors. Clinical applications of non-oncological PET/CT imaging in nephrology and urology will depend on the value of clinical information provided by the new radiopharmaceuticals, their cost, and on both radioactive and non-radioactive risks especially in children.

REFERENCES

[1] A. T. Taylor, "Radionuclides in nephrourology, Part 1: Radiopharmaceuticals, quality control, and quantitative indices," *J Nucl Med,* vol. 55, no. 4, pp. 608–15, 2014, doi: 10.2967/jnumed.113.133447.

[2] A. T. Taylor, "Radionuclides in nephrourology, Part 2: Pitfalls and diagnostic applications," *J Nucl Med,* vol. 55, no. 5, pp. 786–98, 2014, doi: 10.2967/jnumed.113.133454.

[3] S. T. Treves, Harmon, W., Packard, A. B., and Kuruc, A., "Kidneys," in *Pediatric Nuclear Medicine / PET*, S. T. Treves, Ed. New York: Springer, 2007, pp. 239–85.

[4] S. T. Treves *et al.*, "Nuclear medicine in the first year of life," *J Nucl Med,* vol. 52, no. 6, pp. 905–25, 2011, doi: 10.2967/jnumed.110.084202.

[5] C. Lote, *Principles of Renal Physiology*, 5th ed. New York: Springer Science and Business Media, 2012.

[6] B. M. Koeppen, and Stanton, B. A., *Renal Physiology*, 6th ed. (Mosby's Physiology Series). Philadelphia: Elsevier, 2019.

[7] A. M. Peters, and Myers, M.J., *Physiological Measurements with Radionuclides in Clinical Practice.* New York: Oxford University Press., 1998.

[8] C. D. Greaves, Ed. *Mathematical Techniques in Nuclear Medicine* (IPEM Report). York: Institute of Physics and Engineering in Medicine, 2011.

[9] M. Goris, *Nuclear Medicine Applications and Their Mathematical Basis.* Singapore: World Scientific Publishing Co., 2011.

[10] A. M. Peters, "Quantification of renal haemodynamics with radionuclides," *Eur J Nucl Med,* vol. 18, no. 4, pp. 274–86, 1991, doi: 10.1007/BF00186653.

[11] A. M. Peters, "A unified approach to quantification by kinetic analysis in nuclear medicine," *J Nucl Med,* vol. 34, no. 4, pp. 706–13, 1993.

[12] C. S. Patlak, Blasberg, R. G., and Fenstermacher, J. D., "Graphical evaluation of blood-to-brain transfer constants from multiple-time uptake data," *J Cereb Blood Flow Metab,* vol. 3, no. 1, pp. 1–7, 1983, doi: 10.1038/jcbfm.1983.1.

[13] C. S. Patlak, Blasberg, R. G., and Fenstermacher, J. D., "Graphical evaluation of blood-to-brain transfer constants from multiple-time uptake data. Generalizations," *J. Cereb. Blood Flow Metab.,* vol. 5, no. 4, pp. 584–90, 1985.

[14] M. D. Rutland, "A comprehensive analysis of renal DTPA studies. I. Theory and normal values," *Nucl Med Commun,* vol. 6, no. 1, pp. 11–20, 1985, doi: 10.1097/00006231-198501000-00003.

[15] G. Conrad, "How robust are the new blood-pool compensation (BPC) and new modified Patlak-Rutland (MP-R) methods for determination split renal function in 99mTc-MAG3 scintigraphy?," *J. Nucl. Med.,* abstract vol. 57, 2, p. 542, 2016.

[16] M. Samal, Brink, A., Ptacnik, V., Jiskrova, H., Skibova, D., and Kubinyi, J., "Blood-pool compensation method revisited: Similarity and difference in comparison with Patlak-Rutland plot," *Eur J Nucl Med Mol Imaging,* abstract vol. 43, 1, p. S20, 2016.

[17] J. M. James, Lawson, R. S., Hussain, T., Read, N. A., Cunningham, L., Burges, R., and Yeung, G. J. M., "Renogram processing revisited," *Nucl Med Commun,* abstract vol. 22, no. 4, pp. 459–60, 2001.

[18] O. Schück, *Examination of Kidney Function.* Boston: Martinus Nijhoff, 1984.

[19] A. M. Peters, "The kinetic basis of glomerular filtration rate measurement and new concepts of indexation to body size," *Eur J Nucl Med Mol Imaging,* vol. 31, no. 1, pp. 137–49, 2004, doi: 10.1007/s00259-003-1341-8.

[20] E. Durand, "Measurements of renal function with radionuclide techniques in adults," in *Functional Imaging in Nephrourology*, A. Prigent and A. Piepsz, Ed. London and New York: Taylor & Francis, 2006, pp. 19–29.

[21] U. L. Henriksen and Henriksen, J. H., "The clearance concept with special reference to determination of glomerular filtration rate in patients with fluid retention," *Clin Physiol Funct Imaging,* vol. 35, no. 1, pp. 7–16, 2015, doi: 10.1111/cpf.12149.

[22] V. Ptáčník, Zogala, D., Skibová, D., Jiskrová, H., Trnka, J., Tesař, V., Ryšavá, R., and Šámal, M., "Assessment of renal function before contrast media injection: Right decisions based on inaccurate estimates," *Eur. Radiol.,* vol. 29, no. 6, pp. 3192–99, 2019.

[23] A. J. Bleyer, "A reciprocal graph to plot the reciprocal serum creatinine over time," *Am J Kidney Dis,* vol. 34, no. 3, pp. 576–78, 1999, doi: 10.1016/s0272-6386(99)70089-2.

[24] L. G. Hutton, Porter, C. A., Morgan, A. J., Bradley, K. M., and McGowan, D. R., "An investigation into the accuracy of using serum creatinine estimated glomerular filtration rate to predict measured glomerular filtration rate," *Nucl Med Commun,* vol. 40, no. 4, pp. 349–52, 2019.

[25] L. Silva da Selistre, Rech, D. L., Souza, de V., Iwaz, J., Lemoine, S., and Dubourg, L., "Diagnostic performance of creatinine-based equations for estimating glomerular filtration rate in adults 65 years and older," *JAMA Intern Med,* vol. 179, no. 6, pp. 796–804, 2019.

[26] G. B. Haycock, Schwartz, G. J., and Wisotsky, D. H., "Geometric method for measuring body surface area: A height-weight formula validated in infants, children, and adults," *J Pediatr,* vol. 93, no. 1, pp. 62–66, 1978, doi: 10.1016/s0022-3476(78)80601-5.

[27] G. M. Blake and Grewal, G. S., "An evaluation of the body surface area correction for 51Cr-EDTA measurements of glomerular filtration rate," *Nucl Med Commun,* vol. 26, no. 5, pp. 447–51, 2005, doi: 10.1097/00006231-200505000-00009.

[28] G. Redlarski, Palkowski, A., and Krawczuk, M., "Body surface area formulae: An alarming ambiguity," *Sci Rep,* vol. 6, 27966, 2016, doi: 10.1038/srep27966.

[29] C. C. Geddes, Woo, M. Y., and Brady, S., "Glomerular filtration rate – what is the rationale and justification of normalizing GFR for body surface area?" *Nephrol Dial Transplant,* vol. 23, no. 1, pp. 4–6, 2008.

[30] E. V. Newman, Bordley, J. III, Winternitz, J., "The interrelationships of glomerular filtration rate (mannitol clearance), extracellular fluid volume, surface area of the body, and plasma concentration of mannitol," *Johns Hopkins Hosp Bull,* vol. 75, pp. 253–68, 1944.

[31] L. Jødal, and Brøchner-Mortensen, J., "Simplified methods for assessment of renal function as the ratio of glomerular filtration rate to extracellular fluid volume," *Nucl Med Commun,* vol. 33, no. 12, pp. 1243–53, 2012.

[32] T. Dissmann, Gotzen, R., Neuber, K., Offermann, B., and Schwab, M., "Das Erythrocyten- und Plasmavolumen sowie die Relation zwischen Körperhämatokrit und venösen Hämatokrit in verschiedene Stadien der essentiellen und bei renovasculärer Hypertonie," *Klin. Wochenschr.,* vol. 49, no. 16, pp. 915–27, 1971.

[33] T. C. Pearson *et al.*, "Interpretation of measured red cell mass and plasma volume in adults: Expert Panel on Radionuclides of the International Council for Standardization in Haematology," *Br J Haematol,* vol. 89, no. 4, pp. 748–56, 1995, doi: 10.1111/j.1365-2141.1995.tb08411.x.

[34] M. Rutland, Que, L., and Hassan, I. M., ""FUR" – one size suits all," *Eur J Nucl Med,* vol. 27, no. 11, pp. 1708–13, 2000.

[35] A. J. White and Strydom, W. J., "Normalisation of glomerular filtration rate measurements," *Eur J Nucl Med,* vol. 18, no. 6, pp. 385–90, 1991, doi: 10.1007/BF02258428.

[36] H. McMeekin, Barnfield, M., Wickham, F., and Burniston, M., "99mTc DTPA vs. 51Cr EDTA for glomerular filtration rate measurement: Is there a systematic difference?," *Nucl Med Commun,* vol. 40, no. 12, pp. 1224–29, 2019.

[37] R. Agarwal *et al.*, "Assessment of iothalamate plasma clearance: Duration of study affects quality of GFR," *Clin J Am Soc Nephrol,* vol. 4, no. 1, pp. 77–85, 2009, doi: 10.2215/CJN.03720708.

[38] P. Delanaye, Ebert, N., Melsom, T., Gaspari, F., Mariat, C., Cavalier, E., Björk, J., Christensson, A., Nyman, U., Porrini, E., Remuzzi, G., Ruggenenti, P., Schaeffner, E., Soveri, I., Sterner, G., Eriksen, B. O., and Bäck, S.-E., "Iohexol plasma clearance for measuring glomerular filtration rate in clinical practice and research: A review. Part I: How to measure glomerular filtration rate with iohexol?," *Clin. Kidney J.,* vol. 9, no. 5, pp. 682–99, 2016.

[39] M. Rehling, M. L. Moller, B. Thamdrup, J. O. Lund, and J. Trap-Jensen, "Simultaneous measurement of renal clearance and plasma clearance of 99mTc-labelled diethylenetriaminepenta-acetate, 51Cr-labelled ethylenediaminetetra-acetate and inulin in man," *Clin Sci (Lond),* vol. 66, no. 5, pp. 613–19, 1984, doi: 10.1042/cs0660613.

[40] G. Bibbo, Munn, C., and Kirkwood, I., "Comparison of glomerular filtration rates determined using two- and single-blood sample methods with a three-blood sample technique for 2922 paediatric studies," *Nucl. Med. Commun.,* vol. 40, no. 12, pp. 1204–10, 2019.

[41] A. M. Peters, "A new single-sample, time-flexible, empirically-derived formula for measuring glomerular filtration rate from a single blood sample," *Nucl Med Commun,* vol. 40, no. 10, pp. 1029–35, 2019, doi: 10.1097/MNM.0000000000001077.

[42] J. S. Fleming, Persaud, L., and Zivanovic, M. A., "A general equation for estimating glomerular filtration rate from a single plasma sample," *Nucl Med Commun,* vol. 26, no. 8, pp. 743–48, 2005, doi: 10.1097/01.mnm.0000171783.18650.80.

[43] H. R. Ham, and Piepsz, A., "Estimation of glomerular filtration rate in infants and children using a single-plasma sample method," *J Nucl Med,* vol. 32, no. 6, pp. 1294–97, 1991.

[44] V. Ptacnik, Kubinyi, J., Jiskrova, H., Skibova, D., Chroustova, D., Rysava, R., Tesar, V., and Samal, M., "One late blood sample as a compromise between two- and multiple sample method and prediction equations," *Eur J Nucl Med Mol Imaging,* abstract vol. 41, 2, p. S357, 2014.

[45] W. S. Watson, "A simple method of estimating glomerular filtration rate," *Eur J Nucl Med,* letter vol. 19, no. 9, p. 827, 1992, doi: 10.1007/BF00182829.

[46] M. D. Blaufox *et al.*, "Report of the Radionuclides in Nephrourology Committee on renal clearance," *J Nucl Med,* vol. 37, no. 11, pp. 1883–90, 1996.

[47] M. Burniston, "Clinical guideline for the measurement of glomerular filtration rate (GFR) using plasma sampling," British Nuclear Medicine Society, guideline 2018, www.bnms.org.uk/page/BNMSClinicalGuidelines.

[48] J. S. Fleming, Zivanovic, M. A., Blake, G. M., Burniston, M., and Cosgriff, P. S., "Guidelines for the measurement of glomerular filtration rate using plasma sampling," *Nucl Med Commun,* vol. 25, no. 8, pp. 759–69, 2004.

[49] A. Piepsz, Colarinha, P., Gordon, I., Hahn, K., Olivier, P., Sixt, R., and Van Velzen, J. "Guidelines for glomerular filtration rate determination in children." https://eanm.org/publications/guidelines/gl_paed_gfrd.pdf

[50] A. Prigent, Piepsz, A., Ed. *Functional Imaging in Nephro-urology.* London and New York: Taylor & Francis, 2006.

[51] K. L. Zierler, "Circulation times and the theory of indicator-dilution methods for determining blood flow and volume," in *Handbook of Physiology*, vol. Section 2. Circulation. Volume 1., W. F. Hamilton, and P. Dow, Eds. Washington, DC: *Am. Physiol. Soc.,* 1962, pp. 585–615.

[52] J. S. Fleming, "An improved equation for correcting slope-intercept measurement of glomerular filtration rate for the single exponential approximation," *Nucl Med Commun,* vol. 28, no. 4, pp. 315–20, 2007.

[53] H. McMeekin, Wickham, F., Barnfield, M., and Burniston, M., "A systematic review of single-sample glomerular filtration rate measurement techniques and demonstration of equal accuracy to slope-intercept methods," *Nucl Med Commun,* vol. 37, no. 7, pp. 743–55, 2016, doi: 10.1097/MNM.0000000000000448.

[54] H. McMeekin, Wickham, F., Barnfield, M., and Burniston, M., "Effectiveness of quality control methods for glomerular filtration rate calculation," *Nucl Med Commun,* vol. 37, no. 7, pp. 756–66, 2016, doi: 10.1097/MNM.0000000000000520.

[55] A. B. Christensen and Growth, S., "Determination of 99mTc-DTPA clearance by a single plasma sample method," *Clin Physiol,* vol. 6, no. 6, pp. 579–88, 1986, doi: 10.1111/j.1475-097x.1986.tb00790.x.

[56] M. C. Gref and K. Karp, H., "Single-sample 99mTc-diethylenetriamine penta-acetate plasma clearance in advanced renal failure by the mean sojourn time approach," *Nucl Med Commun,* vol. 30, no. 3, pp. 202–5, 2009, doi: 10.1097/MNM.0b013e328315e0c6.

[57] C. A. Wesolowski, Puetter, R. C., Ling, L., and Babyn, P. S., "Tikhonov adaptively regularized gamma variate fitting to assess plasma clearance of inert renal markers," *J Pharmacokinet Pharmacodyn,* vol. 37, no. 5, pp. 435–74, 2010, doi: 10.1007/s10928-010-9167-z.

[58] C. A. Wesolowski *et al.*, "Validation of Tikhonov adaptively regularized gamma variate fitting with 24-h plasma clearance in cirrhotic patients with ascites," *Eur J Nucl Med Mol Imaging,* vol. 38, no. 12, pp. 2247–56, 2011, doi: 10.1007/s00259-011-1887-9.

[59] A. T. Taylor, Lipowska, M., and Cai, H., "99mTc(CO)3(NTA) and 131I-OIH: Comparable plasma clearances in patients with chronic kidney disease," *J Nucl Med,* vol. 54, no. 4, pp. 578–84, 2013, doi: 10.2967/jnumed.112.108357.

[60] B. Bubeck, Piepenburg, R., Grethe, U., Ehrig, B., and Hahn, K., "A new principle to normalize plasma concentration allowing single-sample clearance determination in both children and adults," *Eur. J. Nucl. Med. Mol. Imaging,* vol. 19, no. 7, pp. 511–16, 1992.

[61] B. Bubeck, "Renal clearance determination with one blood sample: Improved accuracy and universal applicability by a new calculation principle," *Semin Nucl Med,* vol. 23, no. 1, pp. 73–86, 1993, doi: 10.1016/s0001-2998(05)80064-9.

[62] K. Weyer, Nielsen, R., Petersen, S. V., Christensen, E .I., Rehling, M., and Birn, H., "Renal uptake of 99mTc-dimercaptosuccinic acid is dependent on normal proximal tubule receptor-mediated endocytosis," *J Nucl Med,* vol. 54, no. 1, pp. 159–65, 2013.

[63] J. S. Fleming, "A technique for analysis of geometric mean renography," *Nucl Med Commun,* vol. 27, no. 9, pp. 701–8, 2006, doi: 10.1097/01.mnm.0000230070.16200.71.

[64] J. S. Fleming, "A technique for the absolute measurement of activity using a gamma camera and computer," *Phys. Med. Biol.,* vol. 24, no. 1, pp. 176–80, 1979.

[65] S. R. Cherry, Sorenson, J. A., and Phelps, M. E., *Physics in Nuclear Medicine*, 3rd ed. Philadelphia: Saunders, An Imprint of Elsevier Science, 2003.

[66] J. S. Fleming, Cosgriff, P. S., Houston, A. S., Jarritt P. H., Skrypniuk, J. V., and Whalley, D. R., "UK audit of relative renal function measurement using DMSA scintigraphy," *Nucl Med Commun,* vol. 19, no. 10, pp. 989–97, 1998.

[67] E. Durand, and Prigent, A., "Can dimercaptosuccinic acid renal scintigraphy be used to assess global renal function?" *Eur J Nucl Med,* vol. 27, no. 6, pp. 727–30, 2000.

[68] C. De Wiele, Van Den Eeckhaut, A., Van Verweire, W., Van Haelst, J. P. Versijpt, J., and Dierckx, R. A., "Absolute 24 h quantification of 99mTc-DMSA uptake in patients with severely reduced kidney function: A comparison with 51Cr-EDTA clearance," *Nucl. Med. Commun.,* vol. 20, no. 9, pp. 829–32, 1999.

[69] H. M. Goodgold, Fletcher, J. W., and Steinhardt, G. F., "Quantitative technetium-99M dimercaptosuccinic acid renal scanning in children," *Urology,* vol. 47, no. 3, pp. 405–8, 1996, doi: 10.1016/s0090-4295(99)80461-0.

[70] J. M. Bland, and Altman, D. G., "Statistical methods for assessing agreement between two methods of clinical measurement," *The Lancet,* vol. 327, no. 8476, pp. 307–10, 1986.

[71] L. Kabasakal, Turoğlu H. T., Önsel, Ç., Özker, K., Uslu, İ., Atay, S., Cansiz, T., Sönmezoğlu, K., Altiok, E., Isitman, A. T., Kapicioğlu, T., and Urgancioğlu, I., "Clinical comparison of technetium-99m EC, technetium-99m MAG3 and iodine-131-OIH in renal disorders," *J Nucl Med,* vol. 36, no. 2, pp. 224–28, 1995.

[72] P. S. Cosgriff, "Quality assurance in renography: A review," *Nucl Med Commun,* vol. 19, no. 7, pp. 711–16, 1998, doi: 10.1097/00006231-199807000-00014.

[73] P. Cosgriff, Lawson, R. S., and Nimmon, C. C., "Towards standardization in gamma camera renography.," *Nucl Med Commun,* vol. 13, no. 8, pp. 580–85, 1992.

[74] A. Prigent, Cosgriff, P., Gates, G. F., Granerus, G., Fine, E. J., Itoh, K., Peters, M., Piepsz, A., Rehling, M., Rutland, and M., Taylor, A. Jr., "Consensus report on quality control of quantitative measurement of renal function obtained from the renogram: International consensus committee from the scientific committtee of radionuclides in nephrourology," *Semin Nucl Med,* vol. 29, no. 9, pp. 146–59, 1999.

[75] H. Bergmann, Mostbeck, A., Kletter, K., Nicoletti, R., and Oberladstätter, M., "Recommendations for carrying out nuclear medicine kidney function tests with a gamma camera and a computer.," *Acta Medica Austriaca*, vol. 43, Supplement 1, pp. 1–19, 1991.

[76] C. C. Nimmon, Fleming, J. S., and Samal, M., "Probable range for whole kidney mean transit time values determined by reexamination of UK audit studies," *Nucl Med Commun,* vol. 29, no. 11, pp. 1006–14, 2008, doi: 10.1097/MNM.0b013e32830f4adb.

[77] A. Kuruc, Caldicott, W. J., and Treves, S., "An improved deconvolution technique for the calculation of renal retention functions," *Comput Biomed Res,* vol. 15, no. 1, pp. 46–56, 1982, doi: 10.1016/0010-4809(82)90052-0.

[78] M. D. Blaufox *et al.*, "The SNMMI and EANM practice guideline for renal scintigraphy in adults," *Eur J Nucl Med Mol Imaging,* vol. 45, no. 12, pp. 2218–28, 2018, doi: 10.1007/s00259-018-4129-6.

[79] I. Gordon, Piepsz, A. Sixt, R. and M. Auspices of Paediatric Committee of European Association of Nuclear, "Guidelines for standard and diuretic renogram in children," *Eur J Nucl Med Mol Imaging,* vol. 38, no. 6, pp. 1175–88, 2011, doi: 10.1007/s00259-011-1811-3.

[80] E. V. Garcia, Folks, R. Pak, S., and Taylor, A., "Totally automatic definition of renal regions of interest from 99mTc-MAG3 renograms: Validation in patients with normal kidneys and in patients with suspected renal obstruction," *Nucl Med Commun,* vol. 31, no. 5, pp. 366–74, 2010, doi: 10.1097/MNM.0b013e3283362aa3.

[81] M. Samal and Nimmon, C., "Automatic definition of renal regions of interest," *Nucl Med Commun,* vol. 32, no. 5, pp. 419–20; author reply 420-1, 2011, doi: 10.1097/MNM.0b013e3283406dea.

[82] J. S. Fleming, "Quantitative measurement from gamma camera images," in *Mathematical Techniques in Nuclear Medicine*, vol. IPEM Report 100, C. D. Greaves Ed. York: Instittute of Physics and Engineering in Medicine, 2011, pp. 22–74.

[83] A. S. Houston, "Factor analysis," in *Mathematical Techniques in Nuclear Medicine*, vol. IPEM Report 100, C. D. Greaves Ed. York: Institute of Physics in Engineering and Medicine, 2011, pp. 153–171.

[84] M. Sámal, Nimmon, C. C., Britton, K. E., and Bergmann, H., "Relative renal uptake and transit time measurements using functional factor images and fuzzy regions of interest," *Eur J Nucl Med,* vol. 25, no. 1, pp. 48–54, 1998.

[85] H. Bergmann, Dworak E., König, B., Mostbeck, A., and Šámal, M., "Improved automatic separation of renal parenchyma and pelvis in dynamic renal scintigraphy using fuzzy regions of interest," *Eur. J. Nucl. Med.,* vol. 26, no. 8, pp. 837–843, 1999.

[86] A. M. Peters, Allison, H., and Ussov, W., "Measurement of the ratio of glomerular filtration rate to plasma volume from the technetium-99m diethylene triamine pentaacetic acid renogram: Comparison with glomerular filtration rate in relation to extracellular fluid volume," *Eur J Nucl Med,* vol. 21, no. 4, pp. 322–27, 1994, doi: 10.1007/bf00947967.

[87] A. M. Peters, Gordon, I., Evans, K., and Todd-Pokropek, A., "Background in 99mTcm DTPA renography evaluated by the impact of its components on individual kidney glomerular filtration rate," *Nucl Med Commun,* vol. 9, no. 8, pp. 545–52, 1988.

[88] M. D. Blaufox, "Renal background correction and measurement of split renal function: The challenge: Editorial Comment: EJNM-D-15-00322, M. Donald Blaufox, MD, PhD," *Eur J Nucl Med Mol Imaging,* vol. 43, no. 3, pp. 548–49, 2016, doi: 10.1007/s00259-015-3253-9.

[89] M. Šámal, Ptáčník, V., Jiskrová, H., Skibová, D., "Background subtraction in dynamic renal scintigraphy revisited," *Eur J Nucl Med Mol Imaging,* abstract vol. 44, 2, 2017.

[90] A. L. Martel, and Tindale, W. B., "Bacgkround subtraction in 99Tcm-DTPA renography using multiple background regions: A comparison of methods," *Nucl Med Commun,* vol. 15, no. 8, pp. 636–42, 1994.

[91] J. B. Bassingthwaighte and Holloway, Jr., G. A., "Estimation of blood flow with radioactive tracers," *Semin Nucl Med,* vol. 6, no. 2, pp. 141–61, 1976, doi: 10.1016/s0001-2998(76)80002-5.

[92] E. Durand *et al.*, "International Scientific Committee of Radionuclides in Nephrourology (ISCORN) consensus on renal transit time measurements," *Semin Nucl Med,* vol. 38, no. 1, pp. 82–102, 2008, doi: 10.1053/j.semnuclmed.2007.09.009.

[93] M. D. Rutland, "A single injection technique for subtraction of blood background in 131-I-hippuran renograms," *Br J Radiol,* vol. 52, no. 614, pp. 134–37, 1979.

[94] A. Gjedde, "Origins of the Patlak plot," *Nucl Med Commun,* vol. 16, no. 11, pp. 979–80, 1995, doi: 10.1097/00006231-199511000-00050.

[95] A. Gjedde, "Dark origins of the Patlak-Gjedde-Blasberg-Fenstermacher-Rutland-Rehling plot," *Nucl Med Commun,* vol. 18, no. 3, pp. 274–75, 1997.

[96] A. M. Peters, "Graphical analysis of dynamic data: The Patlak-Rutland plot," *Nucl Med Commun,* vol. 15, no. 9, pp. 669–72, 1994, doi: 10.1097/00006231-199409000-00001.

[97] M. D. Rutland, "Origin of the Patlak-Rutland plot," *Nucl Med Commun,* vol. 17, no. 5, p. 441, 1996, doi: 10.1097/00006231-199605000-00015.

[98] M. L. Goris, Daspit, S. G., McLaughlin, P., and Kriss, J. P., "Interpolative background subtraction," *J Nucl Med,* vol. 17, no. 8, pp. 744–47, 1976.

[99] G. W. Middleton, Thomson, W. H., Davies, I. H. and Morgan, A., "A multiple regression analysis for accurate background subtraction in 99Tcm-DTPA renography," *Nucl Med Commun,* vol. 10, no. 5, pp. 315–24, 1989, doi: 10.1097/00006231-198905000-00002.

[100] M. Caglar, Gedik, G. K., and Karabulut, E., "Differential renal function estimation by dynamic renal scintigraphy: Influence of background definition and radiopharmaceutical," *Nucl Med Commun,* vol. 29, no. 11, pp. 1002–5, 2008, doi: 10.1097/MNM.0b013e32830978af.

[101] B. K. Geist, Dobrozemsky, G., Samal, M., Schaffarich, M. P., Sinzinger, H., and Staudenherz, A., "WWSSF – A worldwide study on radioisotopic renal split function: Reproducibility of renal split function assessment in children," *Nucl Med Commun,* vol. 36, no. 12, pp. 1233–38, 2015, doi: 10.1097/MNM.0000000000000380.

[102] A. S. Houston, Whalley, D. R., Skrypniuk, J. V., Jarritt, P. H., Fleming, J. S., andCosgriff, P. S., "UK audit and analysis of quantitative parameters obtained from gamma camera renography," *Nucl Med Commun,* vol. 22, no. 5, pp. 559–66, 2001.

[103] L. Lezaic, Hodolic, M., Fettich, J., Grmek, M., and Milcinski, M., "Reproducibility of 99mTc-mercaptoacetyltriglycine renography: Population comparison," *Nucl Med Commun,* vol. 29, no. 8, pp. 695–704, 2008, doi: 10.1097/MNM.0b013e3283013d69.

[104] K. S. Nijran, Houston, A. S., Fleming, J. S., Jarritt, P. H., Heikkinen, J. O., Skrypniuk, J. V., and Institute of Physics and Engineering in Medicine Nuclear Medicine Software Quality Group, "UK audit of analysis of quantitative parameters from renography data generated using a physical phantom," *Nucl Med Commun,* vol. 35, no. 7, pp. 745–54, 2014.

[105] S. D. Bell, and Peters, A. M., "Extravascular chest wall technetium-99m diethylene triamine pentaacetic acid: Implications for the measurement of renal function during renography," *Eur. J. Nucl. Med.,* vol. 18, no. 2, pp. 87–90, 1991.

[106] D. C. Maneval, Magill, H. L., Cypess, A. M., and Rodman, J. H., "Measurement of skin-to-kidney distance in children: Implications for quantitative renography," *J Nucl Med,* vol. 31, no. 3, pp. 287–91, 1990.

[107] M. F. Lythgoe, Gradwell, M. J., Evans, K., and Gordon, I., "Estimation and relevance of depth correction in paediatric renal studies," *Eur J Nucl Med,* vol. 25, no. 2, pp. 115–19, 1998, doi: 10.1007/s002590050202.

[108] K. H. Tonnesen, Munck, O., Hald, T., Mogensen, P., and Wolf, H., "Influence on the radiorenogram of variation in skin to kidney distance and the clinical importance hereof," in *Proceedings of the International Symposium on Radionuclides in Nephrourology*, K. Winkel, M. D. Blaufox, and J.-L. Funck-Bretano, Eds. Stuttgart: Thieme, 1974, pp. 79–86.

[109] M. Samal, Kubinyi, J., Kotalova, D., Steyerova, P., Chroustova, D., Ptacnik, V., and Danes, J., "Prediction of kidney depth and body thickness in renal scintigrpahy," *J Nucl Med,* abstract vol. 50, 2, p. 1402, 2009.
[110] A. Taylor, "Formulas to estimate renal depth in adults (letter)," *J Nucl Med,* letter vol. 35, no. 12, pp. 2054–55, 1994.
[111] A. Taylor, Lewis, C., Giacometti, A., Hall, E. C., and Barefield, K. P., "Improved formulas for the estimation of renal depths in adults," *J Nucl Med,* vol. 34, no. 10, pp. 1766–69, 1993.
[112] A. Taylor, Jr. *et al.*, "Multicenter trial validation of a camera-based method to measure Tc-99m mercaptoacetyltriglycine, or Tc-99m MAG3, clearance," *Radiology,* vol. 204, no. 1, pp. 47–54, 1997, doi: 10.1148/radiology.204.1.9205222.
[113] Y. Inoue *et al.*, "Attenuation correction in evaluating renal function in children and adults by a camera-based method," *J Nucl Med,* vol. 41, no. 5, pp. 823–29, 2000.
[114] M. Samal, Sirova, V., Skibova, D., Trnka, J., Ptacnik, V., and Kubinyi, J., "Attenuation correction in planar scintigraphy using transmission measurement with flood source," *Eur J Nucl Med Mol Imaging,* abstract vol. 42, 1, p. S364, 2015.
[115] I. Buvat, Benali, H., Todd-Pokropek, A., and Di Paola, R., "Scatter correction in scintigraphy: The state of the art," *Eur J Nucl Med,* vol. 21, no. 7, pp. 675–94, 1994, doi: 10.1007/BF00285592.
[116] I. Buvat, Rodriguez-Villafuerte, M., Todd-Pokropek, A., Benali, H., and Di Paola, R., "Comparative assessment of nine scatter correction methods based on spectral analysis using Monte Carlo simulations," *J. Nucl. Med.,* vol. 36, no. 8, pp. 1476–88, 1995.
[117] K. F. Koral, and Zaidi, H., "Methods for planar image quantification," in *Quantitative Analysis in Nuclear Medicine Imaging*, H. Zaidi, Ed. New York: Springer Science & Business Media, 2004, pp. 414–34.
[118] H. Zaidi and K. F. Koral, "Scatter modelling and compensation in emission tomography," *Eur J Nucl Med Mol Imaging,* vol. 31, no. 5, pp. 761–82, 2004, doi: 10.1007/s00259-004-1495-z.
[119] M. Ljungberg, King, M. A., Hademenos, G. J., and Strand, S. E., "Comparison of four scatter correction methods using Monte Carlo simulated source distributions.," *J Nucl Med,* vol. 35, no. 1, pp. 143–51, 1994.
[120] A. Piepsz, A. Dobbeleir, and H. R. Ham, "Effect of background correction on separate technetium-99m-DTPA renal clearance," *J Nucl Med,* vol. 31, no. 4, pp. 430–35, 1990.
[121] M. Moonen, L. Jacobsson, G. Granerus, P. Friberg, and R. Volkmann, "Determination of split renal function from gamma camera renography: A study of three methods," *Nucl Med Commun,* vol. 15, no. 9, pp. 704–11, 1994, doi: 10.1097/00006231-199409000-00007.
[122] M. J. Wesolowski, Conrad, G. R., Šámal, M., Watson, G., Wanasundara, S. N., Babyn, P., and Wesolowski, C. A., "A simple method for determining split renal function from dynamic 99mTc-MAG3 scintigraphic data," *Eur J Nucl Med Mol Imaging,* vol. 43, no. 3, pp. 550–58, 2016.
[123] M. Šámal, Brink, A., Wesolowski, M. J., Conrad, G. R., Wesolowski, C. A., "Preliminary validation of blood-pool compensation method for assessment of relative renal uptake in children," *Eur J Nucl Med Mol Imaging* abstract vol. 42, 1, p. S771, 2015.
[124] R. Volkmann, Ekman, M., Carlsson, S. Jensen, G., Moonen, M., and Friberg, P., "Estimation of renal function by means of deconvolution analysis," *Nucl Med Commun,* abstract vol. 16, no. 5, p. 404, 1995.
[125] J. U. Schlegel and S. A. Hamway, "Individual renal plasma flow determination in 2 minutes," *J Urol,* vol. 116, no. 3, pp. 282–85, 1976, doi: 10.1016/s0022-5347(17)58783-2.
[126] G. F. Gates, "Glomerular filtration rate: Estimation from fractional renal accumulation of 99mTc-DTPA (stannous)," *AJR Am J Roentgenol,* vol. 138, no. 3, pp. 565–70, 1982, doi: 10.2214/ajr.138.3.565.
[127] G. F. Gates, "Split renal function testing using 99mTc-DTPA. A rapid technique for determining differential glomerular filtration," *Clin Nucl Med,* vol. 8, no. 9, pp. 400–07, 1983.
[128] J. S. Mulligan, Blue, P. W., and Hasbargen, J. A., "Methods for measuring GFR with technetium-99m-DTPA: An analysis of several common methods," *J Nucl Med,* vol. 31, no. 7, pp. 1211–19, 1990.
[129] N. Prasad, Barai, S., Gambhir, S., Parasar, D. S., Ora, M., Gupta, A., and Sharma, R. K., "Comparison of glomerular filtration rate estimated by plasma clearance method with modification of diet in renal disease prediction equation and Gates method," *Indian J Nephrol,* vol. 22, no. 2, pp. 103–7, 2012.
[130] C. D. Russell and E. V. Dubovsky, "Gates method for GFR measurement," *J Nucl Med,* vol. 27, no. 8, pp. 1373–74, 1986.
[131] C. D. Russell and E. V. Dubovsky, "Measurement of renal function with radionuclides," *J Nucl Med,* vol. 30, no. 12, pp. 2053–57, 1989.
[132] Y. Inoue, Ohtake, T., Homma, Y., Yoshikawa, K., Nishikawa, J., and Sasaki, Y., "Evaluation of glomerular filtration rate by camera-based method in both children and adults," *J Nucl Med,* vol. 39, no. 10, pp. 1784–88, 1998.
[133] A. Taylor, Jr. *et al.*, "Measuring technetium-99m-MAG3 clearance with an improved camera-based method," *J Nucl Med,* vol. 36, no. 9, pp. 1689–95, 1995.
[134] M. Bocher *et al.*, "Tc-99m mercaptoacetyltriglycine clearance: Comparison of camera-assisted methods," *Clin Nucl Med,* vol. 26, no. 9, pp. 745–50, 2001, doi: 10.1097/00003072-200109000-00001.
[135] F. P. Esteves, R. K. Halkar, M. M. Issa, S. Grant, and A. Taylor, "Comparison of camera-based 99mTc-MAG3 and 24-hour creatinine clearances for evaluation of kidney function," *AJR Am J Roentgenol,* vol. 187, no. 3, pp. W316-9, 2006, doi: 10.2214/AJR.05.1025.

[136] M. Rehling, M. L. Moller, J. O. Lund, K. B. Jensen, B. Thamdrup, and J. Trap-Jensen, "99mTc-DTPA gamma-camera renography: Normal values and rapid determination of single-kidney glomerular filtration rate," *Eur J Nucl Med,* vol. 11, no. 1, pp. 1–6, 1985, doi: 10.1007/BF00440952.

[137] C. C. Nimmon, M. Samal, and K. E. Britton, "Elimination of the influence of total renal function on renal output efficiency and normalized residual activity," *J Nucl Med,* vol. 45, no. 4, pp. 587–93, 2004.

[138] M. Samal, Ptacnik, V., Skibova, D., Jiskrova, H., and Kubinyi, J., "Simple model-based method for gamma-camera measurement of 99mTc-MAG3 plasma clearance," *J Nucl Med,* abstract vol. 54, 1, pp. 170P–171P, 2013.

[139] C. D. Russell, A. T. Taylor, and E. V. Dubovsky, "Measurement of renal function with technetium-99m-MAG3 in children and adults," *J Nucl Med,* vol. 37, no. 4, pp. 588–93, 1996.

[140] K. E. Britton, C. C. Nimmon, H. N. Whitfield, W. F. Hendry, and J. E. Wickham, "Obstructive nephropathy: successful evaluation with radionuclides," *Lancet,* vol. 1, no. 8122, pp. 905–7, 1979, doi: 10.1016/s0140-6736(79)91377-1.

[141] A. Piepsz, Tondeur, M., Nogarede C., Collier, F., Ismaili, K., Hall, M., Dobbeleir, A., and Ham, H., "Can serverely impaired cortical tranist predict which children with pelvi-ureteric junction stenosis detected antenatally might benefit from pyeloplasty?" *Nucl Med Commun,* vol. 32, no. 3, pp. 199–205, 2010.

[142] M. H. Farmelant, K. Bakos, and B. A. Burrows, "Physiological determinants of renal tubular passage times," *J Nucl Med,* vol. 10, no. 10, pp. 641–45, 1969.

[143] B. L. Diffey, Hall, F. M., and Corfield, J. R., "The 99mTc-DTPA dynamicrenal scan with deconvolution analysis," *J. Nucl. Med.,* vol. 17, no. 5, pp. 352–55, 1976.

[144] J. S. Fleming, and Goddard, B. A., "A technique for the deconvolution of the renogram.," *Phys Med Biol,* vol. 19, no. 4, pp. 546–49, 1974.

[145] D. Sutton, "Deconvolution of the renogram," in *Mathematical Techniques in Nuclear Medicine*, vol. IPEM Report 100, C. D. Greaves Ed. York: Institute of Physics and Engineering in Medicine, 2011, pp. 106–29.

[146] D. G. Sutton and V. Kempi, "Constrained least-squares restoration and renogram deconvolution: A comparison by simulation," *Phys Med Biol,* vol. 37, no. 1, pp. 53–67, 1992, doi: 10.1088/0031-9155/37/1/004.

[147] D. G. Sutton and V. Kempi, "Constrained least-squares restoration and renogram deconvolution: A comparison with other techniques," *Phys Med Biol,* vol. 38, no. 8, pp. 1043–50, 1993, doi: 10.1088/0031-9155/38/8/003.

[148] R. S. Lawson, "Application of mathematical methods in dynamic nuclear medicine studies," *Phys Med Biol,* vol. 44, no. 4, pp. R57–98, 1999, doi: 10.1088/0031-9155/44/4/028.

[149] K. E. Britton, Nawaz M. K., Whitfield, H. N., Nimmon, C. C., Carroll, M. J., Granowska, M., and Mlodkowska, E., "Obstructive nephropathy: Comparison between parenchymal transit time index and frusemide diuresis," *Br. J. Urol.,* vol. 59, no. 2, pp. 127–32, 1987.

[150] J. S. Fleming and P. M. Kemp, "A comparison of deconvolution and the Patlak-Rutland plot in renography analysis," *J Nucl Med,* vol. 40, no. 9, pp. 1503–7, 1999.

[151] A. Schlotmann, Clorius, J. H., Rohrschneider, W. K., Clorius, S. N., Amelung, F., and Becker, K., "Diuretic renography in hydronephrosis: Delayed tissue transit accompanies both functional decline and tissue reorganization," *J Nucl Med,* vol. 49, no. 7, pp. 1196–1203, 2008.

[152] A. Schlotmann, J. H. Clorius, and S. N. Clorius, "Diuretic renography in hydronephrosis: Renal tissue tracer transit predicts functional course and thereby need for surgery," *Eur J Nucl Med Mol Imaging,* vol. 36, no. 10, pp. 1665–73, 2009, doi: 10.1007/s00259-009-1138-5.

[153] A. I. Santos *et al.*, "Interobserver agreement on cortical tracer transit in 99mTc-MAG3 renography applied to congenital hydronephrosis," *Nucl Med Commun,* vol. 38, no. 2, pp. 124–28, 2017, doi: 10.1097/MNM.0000000000000620.

[154] L. P. Connolly, Zurakowski, D., Peters, C. A., Dicanzio, J., Ephraim, P., Paltiel, H. J., Share, J. C., and Treves, S. T., "Variability of diuresis renography interpretation due to method of post-diuretic renal pelvic clearance half-time determination," *J. Urol.,* vol. 164, no. 2, pp. 467–71, 2000.

[155] A. Piepsz, Tondeur, M., and Ham, H., "NORA: A simple and reliable parameter for estimating renal output with or without frusemide challenge," *Nucl Med Commun,* vol. 21, no. 4, pp. 317–23, 2000.

[156] A. Piepsz, Kuyvenhoven, J. D., Tondeur, M., and Ham, H., "Normalized residual acitivity: Usual values and robustness of the method," *J Nucl Med,* vol. 43, no. 1, pp. 33–38, 2002.

[157] P. Schmidlin, J. H. Clorius, E. M. Lubosch, H. Siems, and K. Dreikorn, "Evaluation of scintigraphic data of renal transplants," *Comput Biomed Res,* vol. 19, no. 4, pp. 330–9, 1986, doi: 10.1016/0010-4809(86)90046-7.

[158] K. E. Britton, Brown, N. J. G., *Clinical Renography*. London: Lloyd–Luke, 1971.

[159] T. Chaiwatanarat, Padhy, A. K., Bomanji, J. B., Nimmon, C. C., Sonmezoglu, K., and Britton, K. E., "Validation of renal output efficiency as an objective quantitative parameter in the evaluation of upper urinary tract obstruction," *J Nucl Med,* vol. 34, no. 5, pp. 845–48, 1993.

[160] A. T. Taylor *et al.*, "Guidance document for structured reporting of diuresis renography," *Semin Nucl Med,* vol. 42, no. 1, pp. 41–48, 2012, doi: 10.1053/j.semnuclmed.2010.12.006.

[161] G. Brolin, Gleisner, K. S., and Ljungberg, M., "Dynamic 99mTc-MAG3 renography: Images for quality control obtained by combining pharmacokinetic modelling, an anthropometric computer phantom and Monte Carlo simulated scintillation camera imaging," *Phys. Med. Biol.,* vol. 58, no. 10, pp. 3145–61, 2013.

[162] J. O. Heikkinen, "New automated physical phantom for renography," *J Nucl Med,* vol. 45, no. 3, pp. 495–99, 2004.

[163] S. T. Treves, and Grant, F. D., "Vesicoureteral reflux and radionuclide cystogrpahy," in *Pediatric Nuclear Medicine and Molecular Imaging*, S. T. Treves, Ed. New York: Springer, 2014, pp. 335–53.

[164] O. Ozdogan, Turkmen, M., Atasever, S., Arslan, G., Soylu, A., Kasap, B., Kavukcu, S., and Degirmeneci, B., "New quantitative parameters for evaluating radionuclide cystography and their value in understanding physiology of reflux," *J Nucl Med Technol,* vol. 37, no. 2, pp. 101–06, 2009.

[165] Z. Szabo, N. Alachkar, J. Xia, W. B. Mathews, and H. Rabb, "Molecular imaging of the kidneys," *Semin Nucl Med,* vol. 41, no. 1, pp. 20–28, 2011, doi: 10.1053/j.semnuclmed.2010.09.003.

[166] H. Pawelski, U. Schnockel, D. Kentrup, A. Grabner, M. Schafers, and S. Reuter, "SPECT- and PET-based approaches for noninvasive diagnosis of acute renal allograft rejection," *Biomed Res Int,* vol. 2014, p. 874785, 2014, doi: 10.1155/2014/874785.

[167] R. A. Werner *et al.*, "The next era of renal radionuclide imaging: Novel PET radiotracers," *Eur J Nucl Med Mol Imaging,* vol. 46, no. 9, pp. 1773–86, 2019, doi: 10.1007/s00259-019-04359-8.

13 Neuroimaging in Nuclear Medicine

Anne Larsson Strömvall and Susanna Jakobson Mo

CONTENTS

13.1 INTRODUCTION

13.1.1 Anatomy

The brain is made up of nerve cells (neurons), and supporting tissue (glia cells), and is protected by the thick cranial bone and three different layers of membranes, that is, the meninges. In addition, the blood- brain barrier (BBB) and the cerebrospinal fluid (CBF) that flows around the brain and spinal cord and fills the ventricles in the centre of the brain, are protective components. For simplicity, the brain is here divided into the cerebrum, the cerebellum, and the brainstem (i.e. the medulla oblongata, the pons, and the midbrain) (Figure 13.1).

A tough membrane made up of meningeal tissue, called the cerebellar tentorium, separates the cerebrum from the cerebellum. Therefore, one sometimes speaks of supra-tentorial and infra-tentorial structures or functions, referring to the parts of the brain lying above or below the tentorium, respectively. The surface of the two cerebral hemispheres, the cortex, is a 3–4 mm thick layer of mainly neuronal cell bodies – the so-called grey matter. The surface is folded into multiple furrows and ridges (i.e. sulci and gyri) to increase the surface area. In each hemisphere there are four lobes, roughly demarcated by certain of the larger sulci – the frontal, parietal, temporal, and occipital lobes (Figure 13.2).

The neurons have thin extensions, called axons, through which they interconnect via electrical impulses, causing the release of neurotransmitters. An isolating fatty covering called the myelin sheath often encases axons in order to enhance and speed up the transmission of the electrical impulses. The brain tissue beneath the cortex – that is, the subcortical

DOI: 10.1201/9780429489501-13

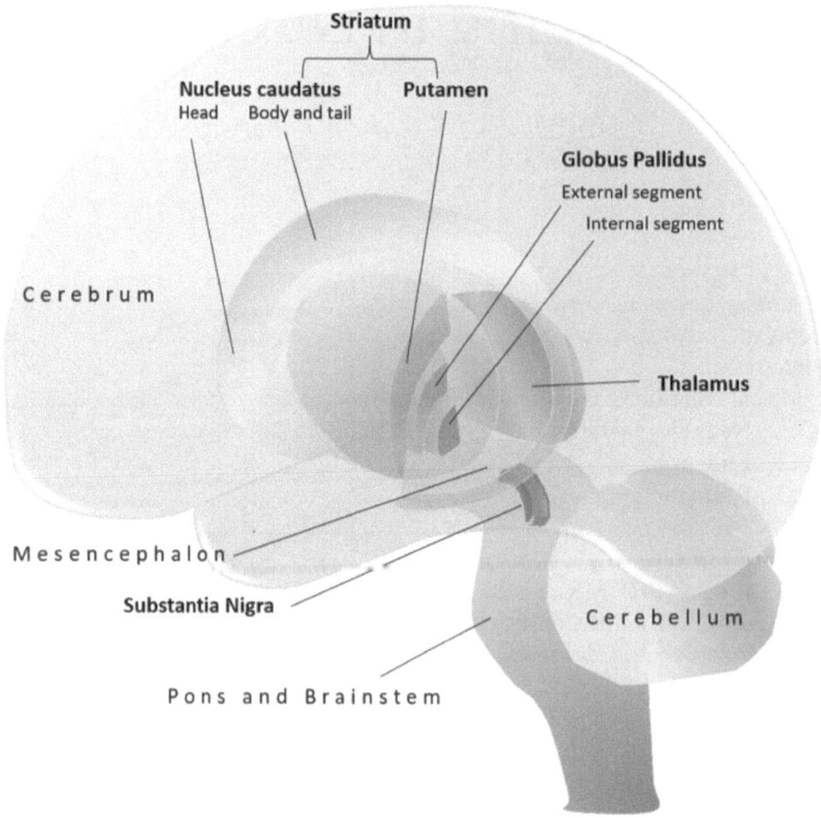

FIGURE 13.1 Anatomy – schematic.

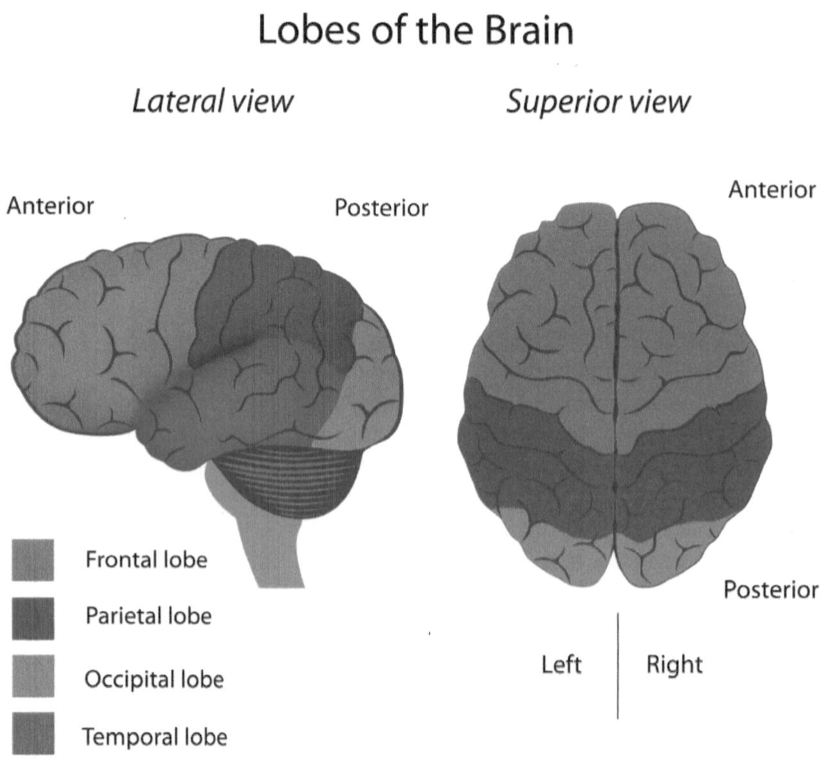

FIGURE 13.2 Anatomy – schematic map of the lobes in the brain.

white matter – is dominated by myelinated nerve fibres. Deep within the white matter of each cerebral hemisphere, there are defined areas, nuclei, of grey matter, for example, the basal ganglia. The basal ganglia include the striatum, which in turn is divided into the caudate nucleus and the putamen. A tiny, paired area of grey matter in the brainstem is called the substantia nigra, which is also considered part of the basal ganglia. The thalamus is another important subcortical nucleus lying medial to the striatum. The cerebellum also has two hemispheres and a middle part called the vermis. The cerebellum is heavily folded, and its proportion of white matter is smaller compared to the cerebrum.

13.1.2 Physiology

The brain is dependent on oxygen and glucose and a continuous supply of nutrients and oxygen is crucial for normal cerebral function. The oxygen and energy consumption are locally regulated according to the specific activity in the brain. The metabolism in the brain is aerobic, and the glucose turnover/metabolism is closely related to the oxygenated arterial cerebral blood flow. Imaging of cortical brain function/activity may either be done by the use of the sugar analogue Fluorodeoxyglucose (FDG) or indirectly by tracers for imaging brain perfusion. Both methods may visualize degenerative, epileptical, malignant, or ischemic processes in the brain. In PET and SPECT, a reference region is often used, that is. a region in which the uptake of a radiopharmacon is compared to the rest of the brain. The cerebellum or the occipital cortex are often used as reference regions, since ischemic and degenerative processes are less common in these regions.

The vessel walls of the brain are sealed by special tight connections between cells in the wall and by foot-like extensions of special kinds of glia cells called astrocytes. This blood-brain barrier (BBB) prevents large molecules (e.g. bacteria, proteins or blood cells) to pass through the wall of the vessel into the cerebrospinal fluid (CBF) and reach the brain tissue. The BBB allows for free movement of water, gases, and small molecules such as glucose. Also, the more fat soluble the molecule is, the easier it will pass through the BBB. In a normal state most parts of the brain are protected by the BBB.

Neurons communicate by neurotransmitters that are released from the so-called endplate or terminal upon an electrical impulse. The neurotransmitter enters the space between two endplates, the synaptical cleft (Figure 13.3).

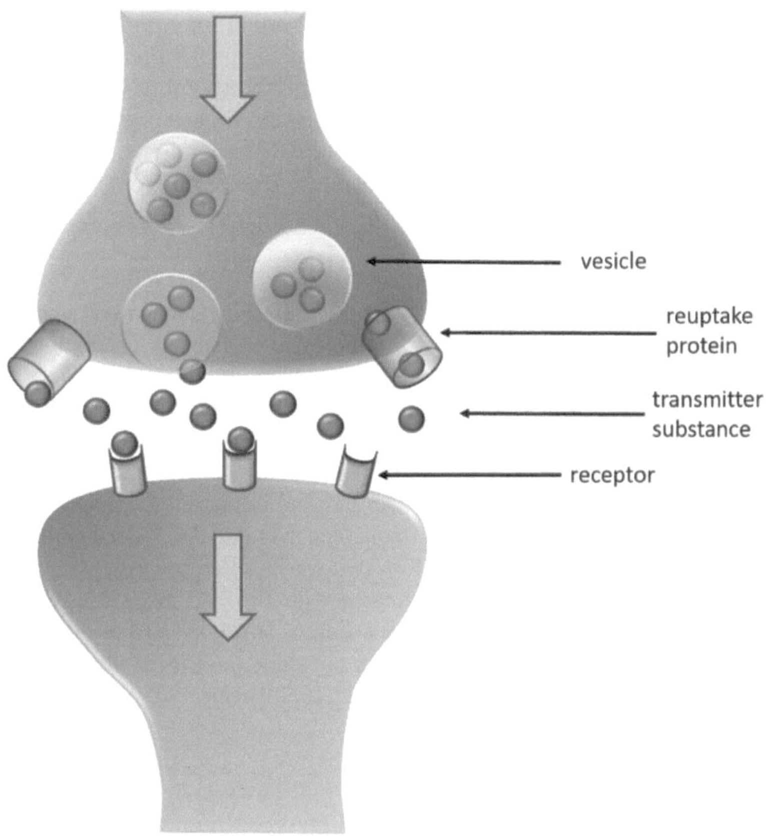

FIGURE 13.3 Schematic illustration of the synapse. Colour image available at www.routledge.com/9781138593312.

Like a key in a keyhole locking up a door, a neurotransmitter fits into special receptors, which in the receiving neuron triggers a specific process leading to action, a nerve impulse. There may be different kinds of receptors for each kind of neurotransmitter. Dopamine, for example, fits several kinds of dopamine receptors, called the D1, D2, D3, D4, and D5 receptors. The strength of the signal that is transmitted between two neurons depends on the amount of neurotransmitter in the synapse and the amount of time the neurotransmitter is allowed to act on the receptors. Therefore, there are specialized proteins or enzymes that degrade or recycle the released neurotransmitter in order to tune the signal. For example, neurotransmitters called monoamines (dopamine, serotonin, and noradrenaline) have their own transporter proteins (monoamine transporter proteins, MAPs) located at the nerve terminals. These reabsorb the neurotransmitter back into the nerve terminal. In this way, MAPs regulate the amount of available neurotransmitter in the synapse and thereby the response is tuned. In addition, some of the released neurotransmitter is recycled and may be re-used the next time. The dopamine transporter (DAT) is a well-known transporter protein, exclusively found on dopamine producing neurons. Apart from the MAPs, monoaminergic neurotransmission is regulated by enzymes called monoamine oxidase (MAO). The MAOs reduce the amount of available monoaminergic neurotransmitters in the synaptic cleft by decomposition.

13.1.3 Pathological Conditions and Clinical Implications

13.1.3.1 Neurodegenerative Disorders

Neurodegenerative disorders are common in the elderly population but may also occur in middle age. These disorders are neurodegenerative disorders causing progressive deterioration of neuronal function, typically leading to cognitive impairment, memory loss, and/or motoric dysfunction. Neurodegenerative diseases often share underlying mechanisms and features, and the distinction between these diseases may be difficult to assess clinically. The diagnosis is largely made by clinical examination. Functional imaging is often used to support the diagnosis.

In Alzheimer´s disease (AD), the most common form of dementia, amyloid plaques, mainly composed of clumps of β-amyloid (a part of a protein occurring in the synapse) and neurofibrillary tangles, are formed by aggregation of an intracellular protein type called tau. In AD the degenerative process causes damage to nerve-cells communicating with the neurotransmitter acetylcholine. In the brain, acetylcholine is important for cognitive functions, including memory. As the neurons die and the function in affected areas is decreased, the regional blood flow and glucose metabolism will be lower than in normally functioning areas. Typically, the parietal and temporal lobes are affected in Alzheimer´s disease, causing memory loss as well as, for example, difficulties in time and space orientation.

Frontotemporal degeneration (FTD) is a generic term for several different dementing disorders. These conditions are dominated by tau accumulation. As implicated by the term, the frontal and temporal lobes are mainly affected, causing personality changes and speech impairment, and may affect social behaviour. Memory loss may develop in later stages.

In another common neurodegenerative disorder, Parkinson´s disease (PD), clumps of protein called Lewy bodies (protein aggregates dominated by α-synuclein) accumulate inside neurons. Like in AD and FTD, neurons producing acetylcholine degenerate in PD. However, the hallmark of PD is loss of dopamine-producing nerve cells. In particular, dopamine-producing neurons located in the substantia nigra (see section 13.1.1) are affected. These cells normally produce lots of dopamine that is released in the striatum, regulating movements, but is also important in cognitive function. The typical movement symptoms of PD, Parkinsonism (i.e. slowness of movement, muscular stiffness/rigidity, balance impairment, and resting tremor) are largely associated with the loss of dopamine, and treatment with dopamine-like drugs (e.g. L-dopa) substitute for the loss of neurotransmitter and improve symptoms through action on the dopamine receptors. Parkinsonism may be seen in other conditions not associated with dopamine loss. Hereditary tremor, for example, is a relatively common neurological condition that may sometimes resemble early Parkinson's disease, but is not a neurodegenerative disease and should not be treated with anti-Parkinson drugs (i.e. dopaminergic agents). Parkinsonism is also seen in so-called atypical Parkinson syndromes (Multisystem atrophy, Progressive supranuclear palsy, and Corticobasal syndrome). These are far less common, but share the loss of dopamine in the striatum with PD. However, in these diseases, also the dopamine receptors are lost, and therefore dopaminergic treatment is usually not helpful.

Lewy-body dementia (LBD) is a neurodegenerative disorder that shares the features of both Alzheimer´s disease and Parkinson's disease with the formation of both amyloid plaques and Lewy body inclusions in the brain. Patients with LBD develop both cognitive impairment and Parkinsonism. In Lewy body dementia, there is both a dopamine deficit and a reduction of metabolism in the parietal and temporal lobes. In addition, often the occipital metabolism is affected.

13.1.3.2 Epilepsy

Epileptic seizures are effects of transient uncontrolled hyperactivity in cortical neurons somewhere in the brain. An epileptic seizure may start in, and be restricted to, a particular region in the brain, or spread to other parts or start within the interconnecting network between the hemispheres and instantly engage large parts of the brain. Depending on the start and development of epileptic activity, the symptoms may vary. An epileptic seizure may be the first symptom of a brain tumour, but can also be caused by acute damage to the brain such as infection, stroke, intoxication, or trauma. However, epileptic seizures may happen without obvious underlying or provoking factors. A diagnosis of epilepsy is made when a patient has had two spontaneous epileptic seizures without an acute underlying disease or condition. In some cases, the underlying cause may be due to damage to the brain in early life or malformations in the brain, or be due to a slowly growing brain tumour, but in many cases it is not possible to assess the cause of epilepsy. The diagnosis of epilepsy involves a thorough examination, including electroencelography (EEG) and MRI imaging. Epilepsy is treated by antiepileptical drugs, but sometimes neurosurgery is indicated. However, neurosurgical treatment in epilepsy warrants a very precise location of the epileptic focus and thorough presurgical mapping is important to avoid damage to important normally functioning areas in the brain. In situations where neurosurgical treatment is indicated, preoperative functional imaging may be of additional diagnostic value to pinpoint the origin of the epileptic activity. The diagnostic work-up in these situations is carried out by highly specialized multidisciplinary epilepsy teams or centres.

13.1.3.3 Brain Tumours

Tumours in the brain are considered "benign" if they are composed of non-cancerous cells or are not infiltrating the brain tissue. However, these tumours can still be life-threatening because of their mass effect within the restricted space in the skull, or if they grow in areas controlling vital functions. Malignant primary brain tumours are composed of cancerous cells that infiltrate the brain tissue. While malignant tumours in other parts of the body may metastasize to the brain, malignant brain tumours rarely metastasize to other parts of the body.

A common type of benign brain tumour is meningioma, arising from cells in one of the membranes (meninges) around the brain. There are several types of malignant primary brain tumours. These may arise from any cell type in the brain. For example, a type of malignant tumour occurring in children and called medulloblastoma, arises from nerve cells. However, the largest group of brain tumours in adults are called gliomas, arising from different kind of supporting cells in the brain, the glia cells. There are different types of gliomas. For example, a subtype of glioma called astrocytoma and glioblastoma multiforme arise from astroglia cells, and oligodendrogliomas arise from glia cells called oligodentrocytes. Gliomas may also be classified according to malignancy, as in the WHO-classification [1], which by characterization of specific features in the tumour cells separates them into four grades of malignancy; a WHO grade I tumour is benign and a grade IV tumour is highly malignant.

Brain tumours are treated with surgery and in many cases chemotherapy and radiotherapy are indicated. Imaging of brain tumours primarily rely on MRI. However, there are several implications for functional neuroimaging in the diagnostic work-up and during the course of treatment of gliomas [2].

13.2 TECHNOLOGY

13.2.1 SPECT AND SPECT/CT

Early SPECT systems were based on a single gamma camera head rotating around the patient's head, but dual-headed, triple-headed and even four-headed systems became available during the 1990s. These systems lead to an n-fold increase in sensitivity where n is the number of gamma camera detectors, enabling a higher image quality. The multi-headed systems were also useful for some dynamic applications, due to the more rapid image acquisition. Fast dynamic studies are, however, very difficult to perform with SPECT systems based on conventional setups with rotating detector heads.

Integrated SPECT/CT systems became commercially available in 1999, with the release of the Hawkeye (GE Healthcare). This low-dose, low-cost, and relatively slow CT, which in the beginning provided 10 mm CT slices, was a multi-purpose machine, but was also used for neuroimaging. SPECT/CT image fusion then became an option, as well as CT-based attenuation correction. The Hawkeye was later upgraded to a 4-slice system with 5.0 mm slices, and now some vendors can provide fully diagnostic integrated CTs that can help to improve the diagnostic workflow for neurodegenerative diseases.

Dedicated brain SPECT systems have been developed, both commercial systems and prototypes. Some early prototypes were the SPRINT II system [3] and the ASPECT system [4]. Both were ring-based and provided better spatial resolution and a higher sensitivity compared to conventional systems. An early commercial system was the

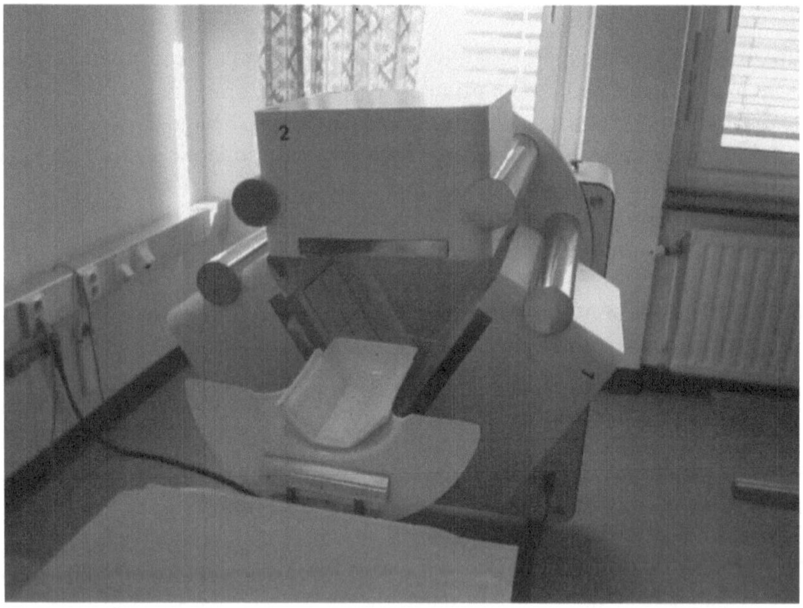

FIGURE 13.4 Neurocam gamma camera.

Neurocam (General Electric/CGR, France) which was a triple-headed gamma camera with a fixed rotation radius of 12.25 cm (Figure 13.4). Field of view (FOV) was 20.0 x 17.6 cm, and with low-energy high-resolution collimators, the resolution was 9.0 mm in air, in the centre of the FOV [5]. This system was built with a conventional approach, using three gamma camera heads rotating around the patient's head.

None of the above systems, however had the capability of providing rapid high-quality dynamic imaging which, for example, is of interest in neuroimaging research. In recent years, multiple pin-hole systems have gained in popularity for neuro imaging, due to the more favourable trade-off between resolution and sensitivity. A recent development is the prototype G-SPECT system (MILabs, The Netherlands), which is developed for large animals but with an intended future clinical use. It is based on multi-pinhole collimation with full angular coverage in a non-rotation gantry, with an inner diameter of 40 cm. The system consists of 9 large FOV cameras with 6 focusing pinholes each, pointing towards a central volume. The bed is automatically stepped to include all aspects of the part of the body studied in this volume. Acquisition and reconstruction protocols are relatively complicated, but results are impressive. Using 3.0 mm and 4.5 mm-diameter pinhole collimators, respectively, resolution is either 2.5 or 3.5 mm FWHM, and sensitivity 415 cps/MBq or 896 cps/MBq [6]. If attenuation correction is needed, however, the user has to co-register a CT or MRI scan for this purpose.

Another high-sensitivity SPECT system is the VERITON (Spectrum Dynamics, Florida) which also is available with a CT (the VERITON-CT), and is released for clinical use. The innovative design with 360 degrees of coverage, is based on 12 detector arms with tungsten parallel-hole collimators and cadmium zinc telluride (CZT) detectors. Contouring detection allows each arm to get as close to the patient as a few mm, which is positive for spatial resolution. VERITON is claimed to have a three times higher volumetric sensitivity compared with conventional SPECT, and to have a higher image contrast and resolution.

13.2.2 PET, PET/CT and PET/MR

Most PET brain studies are performed using conventional whole-body PET/CT equipment. The development of 3D PET, of crystal materials such as LSO and LYSO, new crystal cutting techniques, and SiPM detectors to replace photomultiplier (PM) tubes, has been advantageous for neuro PET as for PET in general. Novel PET scanners with a relatively large axial field of view are advantageous since fast dynamic studies with full-brain coverage and high sensitivity, can be performed. The spatial resolution is however limited to about 3 mm FWHM for whole-body PET scanners, provided that resolution recovery is included in the iterative reconstruction. Without resolution recovery, the resolution is usually in the order of 4–6 mm FWHM.

There are however specialized PET systems, developed for high-resolution brain imaging, with smaller-diameter detector rings and smaller crystal elements. A well-known example is the ECAT high-resolution research tomograph (HRRT) (CTI/Siemens) with a FOV diameter of 31.2 cm. The first two prototypes were developed in the 1990s and later developed to become a commercial product. This was the first commercially available camera with a double layer of LSO/LYSO scintillation crystals to provide depth-of-interaction (DOI) information. This was retrieved by measuring the decay time of crystal light, which is about 40 ns for LSO and 53 ns for LYSO. Spatial resolution varies between 2.3 mm and 3.4 mm in different parts of the FOV [7]. No CT is included, and attenuation correction is performed using a ^{137}Cs transmission source.

Neuroimaging was one of the motives for the development of the PET/MRI technique, with the first prototype presented in 1997 [8]. With integrated scanners, hybrid imaging has reached a new level with PET and MR images that are both spatially and temporally matched. Temporal matching can be of special interest in neuroimaging research. Functional MRI (fMRI) studies can then be performed in combination with dynamic PET scans which, for example, can lead to new insights in neurotransmitter release and activated brain networks [9]. PET/MRI can also provide a more efficient diagnostic procedure, since the patient can be spared from doing the PET and MRI separately.

Another approach to brain PET imaging has been to make the imaging device portable. With the Ambulatory Microdose Positron Emission Tomography (AM-PET) scanner, it is possible for the imaged subject to stand, or even move around during scanning, since the detectors are mounted on a helmet [10]. The prototype, which still is under development, allows functional studies during behavioural tasks, with a high sensitivity. It is based on twelve SiPM modules in a ring that fits around the head and covers a vertical FOV of approximately 5 cm. Although portable, the device still weighs 3.6 kg, and needs a supporting cord to reduce the weight on the subject's head and neck.

13.3 OPTIMIZATION OF ACQUISITION AND RECONSTRUCTION

In neuroimaging, acquisition times may be long, and to guarantee a high spatial resolution and reduce the risk of artefacts, it is important with an efficient fixation system for the head, to prevent motion. Some neurodegenerative diseases may also increase the risk of motion, which should be taken into consideration. For relatively short static scans of 10–30 min, it may be sufficient with supporting cushions in the head holder and a Velcro strap over the forehead, but for long dynamic research scans, it is not unusual with tightly fastened thermoplastic face masks (Figure 13.5). These masks are not suitable for all patients, as they may introduce claustrophobic reactions.

13.3.1 SPECT Acquisition

One of the most important optimization parameters for SPECT is the choice of collimator and, as for SPECT in general, the most common collimator used for neuroimaging is the parallel-hole collimator. Low-energy, high-resolution

FIGURE 13.5 Facemask.

(LEHR), or ultra-high resolution (LEUHR) collimators are commonly used for neuroimaging with ^{99m}Tc-based radiopharmaceuticals such as, for example, ^{99m}Tc-HMPAO. There are, however, several neuro tracers labelled with ^{123}I, such as for example ^{123}I FP-CIT used mostly for Parkinson diagnostics. The primary gamma energy (83% per decay) of ^{123}I is 159 keV, which nicely fits with the specifications for most low-energy collimators, but the isotope also emits small amounts of gamma photons with higher energies. The total amount of high-energy emissions is 3.0 per cent per decay, with the most prominent peak of 1.4 per cent at 529 keV. These data are obtained from the Lund/LBNL Nuclear Data Search website [11]. Since these high-energy photons have a much higher probability of passing the collimator septa and becoming detected after scattering, their contribution to the image is relatively important and can constitute up to 40 per cent of the detections in the main energy window for neuro imaging [12]. The use of medium-energy collimators has shown quantitative advantages for ^{123}I-imaging of other organs [13, 14], but for neuroimaging with its high-resolution requirements, it should be preferable to use low-energy high-resolution collimators and use the best available methods to correct for the septal penetration photons.

As has been described in Chapters 13 and 14, Volume I, spatial resolution degrades rapidly with increased distance from the collimator surface, and it is therefore important that the gamma camera detectors can rotate as closely as possible to the patient's head. A dedicated head-holder, careful patient positioning and, if possible, an automatic contour-detection system are therefore important factors for image quality. However, multi-purpose gamma cameras are rarely optimized for neuroimaging. The dead space outside the FOV can be several centimetres, and due to the patient's shoulders, it can be difficult to get a tight orbit around the patient's head while including the cerebellum. The frame outside the FOV is mainly due to the large size of the PM tubes. Novel machines based on CZT technology have no need for PM tubes, which is an advantage for neuroimaging. GE Healthcare claims to have reduced the frame size to 2.5 cm on the NM/CT 870 CZT scanner, as compared to the 7.5 cm on previous conventional machines.

Due to the relatively small size of the imaged object, the brain, converging collimators, as described in Chapter 14, Volume I, are relatively popular. The fan-beam collimator can be used to increase image resolution in the transaxial plane, and cone-beam collimators can be used to increase resolution in both the transaxial and axial plane. Both collimators will also increase sensitivity compared to a parallel-hole collimator, and they have the advantage that they can be mounted on conventional multi-purpose gamma cameras. Disadvantages with these approaches are distortions and sensitivity variations with the distance to the collimator, which increases the uncertainties in quantitative accuracy. Also, it is not a matter of course that console reconstruction software can include corrections for attenuation and scatter when these collimators are used.

A more exotic collimator that shows promising results and is well-suited for medium-sized organs as the brain is the slit-slat collimator [15, 16]. This collimator combines characteristics from the parallel-hole collimator and the pinhole collimator and can be used in a stationary system. An interesting application is a prototype SPECT/MRI system [17] where a stationary SPECT acquisition is essential due to the high magnetic field strength.

Apart from the collimator choice, there are other factors that need to be considered, such as the matrix size, pan and zoom factors, the number of projections and the total acquisition time. Matrix size in brain SPECT is usually 128x128 pixels, and the zoom factor should be chosen to set a pixel size that is small enough to not limit the spatial resolution, usually about 2–3 mm. For multi-purpose SPECT systems of conventional FOV dimensions and parallel-hole collimators, the zoom factor can be in the order of 1.5. It is however important to also consider what part of the FOV is zoomed, since the brain is not centred in the FOV. The pan factor needs to be adjusted for this purpose, so that the lower part of the FOV is used.

The choice of number of projections has been described in Chapter 15, Volume I, and recommendations can also be found in international guidelines. The uptake of the radiopharmaceutical in the brain can be relatively low, especially for ^{123}I-based substances where the injected activity can be less than 200 MBq. Scanning times are therefore usually relatively long, and the recommendation for ^{123}I-labelled dopamine transporter ligands, such as FP-CIT, is 30 min [18].

13.3.2 PET Acquisition

The overwhelming majority of PET cameras used today can only measure in 3D acquisition mode, which means that all coincidences in the FOV are detected, also the oblique planes. There are however older PET cameras, still in use, that can measure both in 2D mode, with septa inserted between the detector rings, and 3D mode without septa. The use of 3D mode will increase sensitivity at the expense of a higher fraction of scatter and random detections. With modern reconstruction techniques, the unwanted detections can be corrected for, and 3D mode is usually advantageous, but it depends of course on the accuracy of the corrections.

In a conventional PET/CT scanner, the axial FOV usually allows for full brain coverage, although with reduced sensitivity close to the edges. If the axial FOV is similar to the length of the brain, it is important to be extra careful when aligning the patient's head, relative to the FOV, to avoid sensitivity loss and truncation. The axial FOV is also a very important factor for the sensitivity of the PET scanner. On the latest developed commercial scanners, the FOV can be up to 26 cm, which is a clear increase from previous generations that are still in use. The higher sensitivity can be used to obtain a higher image quality with less noise, reduce scanning time and, therefore, increase patient throughput, or to reduce the radiation dose to the patient. Different local needs may lead to different optimization strategies, but the recommended number of detected events by international guidelines should always be considered. The transaxial FOV should be chosen to allow full head coverage, and the matrix size must be chosen so that the pixel-size does not limit spatial resolution. A 256x256 matrix is a typical choice.

For static protocols, the examination time rarely exceeds 30 min for PET neuroimaging. For dynamic imaging, protocols of 1.5 hours are, however, not unusual, especially for research purposes. Here, the dynamic protocol must take the radiopharmaceutical kinetics and physical half-life into account. The method of evaluation is also important, since a full parametric evaluation will need a higher temporal resolution during the first minutes compared to a Patlak or Logan evaluation, where parameters are more dependent on a later linear phase (see Chapter 5, Volume II). List-mode acquisition, where all detections are registered with time stamps, is crucial if time frames need to be adjusted retrospectively. Very short time frames, in the order of 20–30 s, are associated with quantitative uncertainties and noise-induced reconstruction bias [19] and should be carefully evaluated before they are applied.

In PET/CT, the CT protocol also needs to be considered. If the CT only is intended for attenuation correction, it is obvious that the CT should be in low-dose mode with a low tube current (typically up to 30 mAs). If no recent anatomical information is available, a diagnostic CT scan may, however, be preferred. In PET/MRI, the MRI sequences employed for different neurodegenerative diseases vary between hospitals, but typical examples have been listed in the work by Barthel and others [20]. Attenuation correction in brain PET/MRI still remains an area of research although quantitative accuracy has markedly improved during recent years.

13.3.3 Reconstruction

A high spatial resolution has always been of special interest in neuroimaging, both for visual interpretation and a high quantitative accuracy, and dedicated SPECT and PET cameras have been optimized with this parameter in mind. For multi-purpose equipment, it is important to choose appropriate reconstruction methods and, if applicable, also reconstruction filters with the highest possible critical frequencies, while maintaining noise at a reasonable level.

Attenuation and scatter are almost always corrected for in PET, and it is usually based on attenuation maps derived from CT, MRI, or transmission measurements. In SPECT, the effects of these interactions are less, and the corrections have not been considered as important if quantitative measurements are not performed. The corrections have however gained in popularity, also in neuroimaging, and have been proven to increase image quality and diagnostic accuracy. For ^{123}I-labelled radiopharmaceuticals, the septal penetration photons also need to be corrected for. Advanced model-based corrections are rarely available in scanner console reconstruction software, but window-based scatter corrections can also be used for the combined correction of scatter and septal penetration. Since the 159 keV photo-peak window is contaminated by both lower and higher energy photons, it should be appropriate with scatter windows on both sides of the peak, which means that the triple-energy-window (TEW) method [21], as described in Chapter 26, Volume I, can be an option.

Resolution recovery, to compensate for the detector response, has been popular in both SPECT and PET for many years, and it has been shown that resolution increases and contrast-to-noise improves. Small-lesion detectability can be markedly improved. Drawbacks with the technique are Gibb's ringing artefacts, changes in noise structure and a considerably slower convergence of the iterative reconstruction. This can increase the variability and therefore reduce the quantitative accuracy, especially for small structures [22]. It is therefore important to verify the advantages of this correction before implementation in clinic or in a research study.

13.4 METHODS FOR EVALUATION

13.4.1 Image Presentation

Visual interpretation of SPECT and PET brain images is still the most important factor for a correct diagnosis, and image presentation needs therefore to be optimized. A standardized orientation of the brain is usually used in radiology, with images reconstructed parallel to the anterior commissure – posterior commissure (AC-PC) line. It may however be difficult to make a perfect reorientation of PET and SPECT brain images, since resolution is relatively poor, and the anatomical landmarks can be difficult to find. With experience it should, however, be possible to find an approximate procedure for a standardized image orientation.

There is a variety in colour scales used at different sites and for different types of examinations, for example, linear grey scales, monochromatic scales as "hot iron" and multicolour scales such as "GE-colour," "Rainbow" or "Sokoloff." Some colour scales have different intervals between different colours, so that within the particular colour scale, one colour may represent a wide range of intensities / uptake levels in the image and another colour in the spectrum denotes a narrow intensity interval. Other colour scales have equally distributed intervals between the different colours, for example, the "step 10" colour scale, where there is a 10 per cent difference between each fixed colour level. It is important that the scale is easily interpreted and that the nuclear medicine physicians are experienced with it. For hybrid imaging, image fusion can also be used for adding anatomical information from CT or MRI. The anatomical information is usually presented in grey scale, whereas the PET or SPECT images are presented in a semi-transparent colour-scale, where the degree of transparency can be adjusted.

In visual interpretation of PET or SPECT images, it is important to standardize conditions in the image. This can be done by adjusting the colour intensity in the image so that the background or reference area is displayed within the lower third of the colour scale. For example, if using the GE colour scale in DAT SPECT imaging, the cortex and white matter should be displayed as light blue/turquoise in order to achieve the adequate contrast to the striatum uptake as exemplified in Figure 13.6. In other instances, it is more relevant to adjust the image according to highest intensity. For example, in FDG-PET imaging, the highest metabolism is normally seen in the basal ganglia or visual cortex/occipital lobe and those areas should be displayed with a colour intensity close to maximum in the chosen colour scale.

13.4.2 Quantitative Techniques

A visual interpretation is often done in combination with a quantitative or semi-quantitative technique to support the visual interpretation and improve the diagnostic accuracy, or to get numerical values to assess the specific uptake in a specific region. Basic semi-quantitative approaches rely on ROIs or VOIs that are drawn manually or are automatically applied to the structure of interest and to a reference region. A simple ratio or a specific uptake ratio can then be calculated from the average values in the respective regions. The standardized uptake value (SUV) is often used in PET, in which the uptake value is standardized to, for example, body weight and the amount of injected activity.

Due to the rigid structure of the brain, quantitative analyses where a patient's PET or SPECT uptake is compared to a healthy uptake, is relatively uncomplicated compared with studies of other parts of the body. Using geometrical transformations, a patient's brain volume can be matched to a template brain using either the SPECT or PET uptake pattern, or the corresponding anatomical information retrieved by the hybrid CT or MR scan. A variety of software exists, covering perfusion, metabolism and neurotransmitting imaging. The different software typically includes a normal

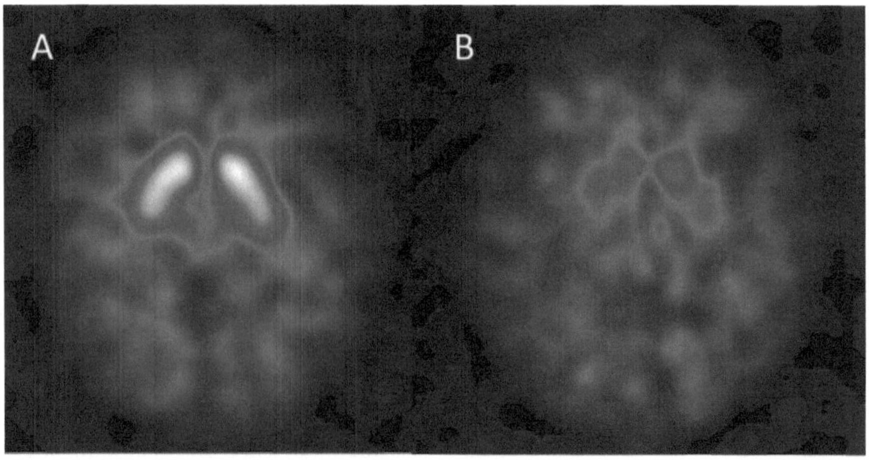

FIGURE 13.6 ^{123}I FP-Cit SPECT (DaTSCAN™) where (A) is healthy and (B) is a PD patient. Colour image available at www.routledge.com/9781138593312.

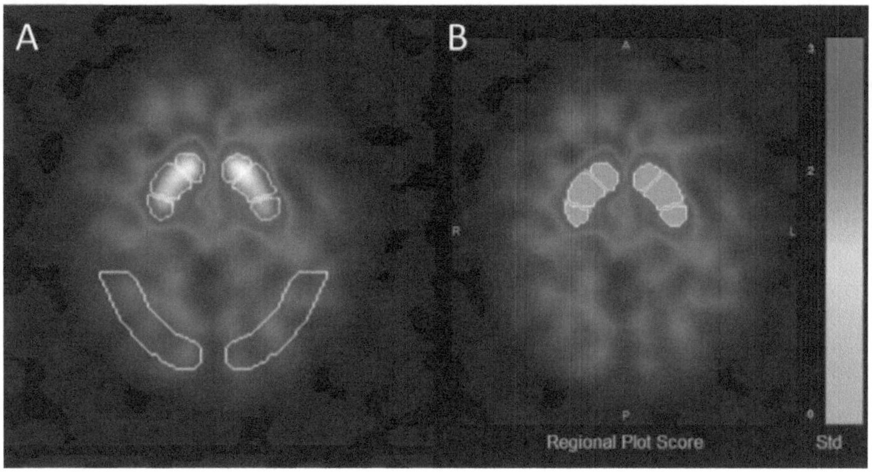

FIGURE 13.7 DaTQUANT™. ^{123}I FP-Cit SPECT of a healthy individual and the corresponding statistical map. Colour image available at www.routledge.com/9781138593312.

material, composed of a number of normal scans from healthy individuals, which sometimes are divided into different groups dependent on age and sex. The patient scan can then be compared with the normal material, resulting in a map of statistical differences. An example of such a map for a DAT SPECT study, retrieved from the software DaTQUANT (GE Healthcare) can be seen in Figure 13.7.

Three-dimensional stereotactic spatial normalization mapping (3D-SSP) [23] is a feasible quantitative method that has become frequently used and is advocated in the daily clinical routine of interpreting, for example, FDG and amyloid PET imaging. The technique is integrated in several commercially available software that automatically maps the uptake value in each voxel in the individual patient and calculates a Z-score relative to a built-in pooled normal material provided by the software. The result is displayed in different colours, which makes it easy and fast to detect significant deviations from normal uptake. However, when using such software with a built-in specific normal material, one should be aware that it may be necessary to use the same type of scanner, and, in particular that it is also important that reconstruction methods, with its filters and parameters, match.

13.5 APPLICATIONS

Preferentially, PET and SPECT images of the brain are merged with MR or CT images for best anatomical location and detection of structural lesions, such as infarction or tumours. SPECT or PET imaging of cortical brain function may be done either by using the sugar analogue Fluorodeoxyglucose or by tracers for imaging of brain perfusion. Both methods may tell us about degenerative, epileptical, malignant, or ischemic events in the brain.

By using SPECT or PET tracers simulating a neurotransmitter, very specific neuronal functions or functional deficits may be visualized, for example, different dopaminergic functions using agents acting on the different dopamine receptors in, for example, psychiatric illness or the dopamine transporter in, for example, neurodegenerative illness.

In pathological conditions, for example, trauma, ischemia, infection, inflammation or malignancy, the BBB will be more permeable locally or globally, and larger molecules can leak into the cerebrospinal fluid and spread in the brain tissue. This may facilitate the access of pharmacological substances for treatment and may be directly or indirectly utilized in brain imaging.

13.5.1 Imaging of Cerebral Blood Flow

Imaging of cerebral blood flow/cerebral perfusion is mainly indicated in the diagnosis of neurodegenerative diseases by reflecting decreased perfusion corresponding to areas of neuronal degeneration in the brain. Different areas may be affected in different neurodegenerative diseases, and the pattern indicates the specific type of dementia. For example, classical Alzheimer′s disease typically shows decreased blood flow in the temporal and parietal lobes, while other neurodegenerative diseases affect other brain areas. Preferentially, PET and SPECT images of the brain are merged with MR or CT images for best anatomical location and detection of structural lesions such as concomitant infarction or tumours.

Another indication for perfusion imaging is to determine the origin of an epileptic seizure, the epileptic focus. Perfusion imaging during a seizure (ictal imaging) may reveal the epileptic focus by a locally increased blood flow in the epileptogenic abnormally hyperactive neurons. The technique needs to be monitored by EEG (electroencephalography) and imaging in inter-ictal state (i.e. normal condition) is usually compared to the ictal image.

In PET, $^{15}O\ H_2O$ is used for imaging of perfusion. In SPECT ^{99m}Tc-HMPAO (Hexametylpropylenaminoxin, Ceretec), ^{99m}Tc bicisate (Neurolite) and ^{123}I Iofetamine (Perfusamine or SPECTamine) are examples of commercially available radiopharmaceuticals for perfusion imaging.

13.5.2 Imaging of Brain Metabolism

Imaging of cerebral metabolism is mainly indicated in early detection and differential diagnosis of dementing disorders by reflecting decreased sugar consumption corresponding to areas of neuronal degeneration in the brain. As in imaging of cerebral blood flow, the pattern of decreased glucose consumption, reflected by decreased ^{18}F-FDG-uptake, may indicate an early-stage dementia or be indicative of the type of neurodegenerative disease. For example, classical Alzheimer´s disease (AD) typically shows decreased metabolism in the temporal and parietal lobes, as seen in Figure 13.8A, while fronto-temporal dementia typically affects the frontal and temporal lobes as seen in Figure 13.8B. A normal ^{18}F-FDG-uptake can be seen in Figure 13.8C.

Lewy-body dementia (LDB) is a neurodegenerative disease sharing the features of AD and Parkinson´s disease and may be diagnosed by a combination of metabolic and dopaminergic imaging, as seen in Figure 13.9. Also, in atypical parkinsonian diseases, for example multisystem atrophy (MSA) and progressive supranuclear palsy (PSP), imaging of brain metabolism is used in combination with dopamine transporter imaging to determine the specific type of disease.

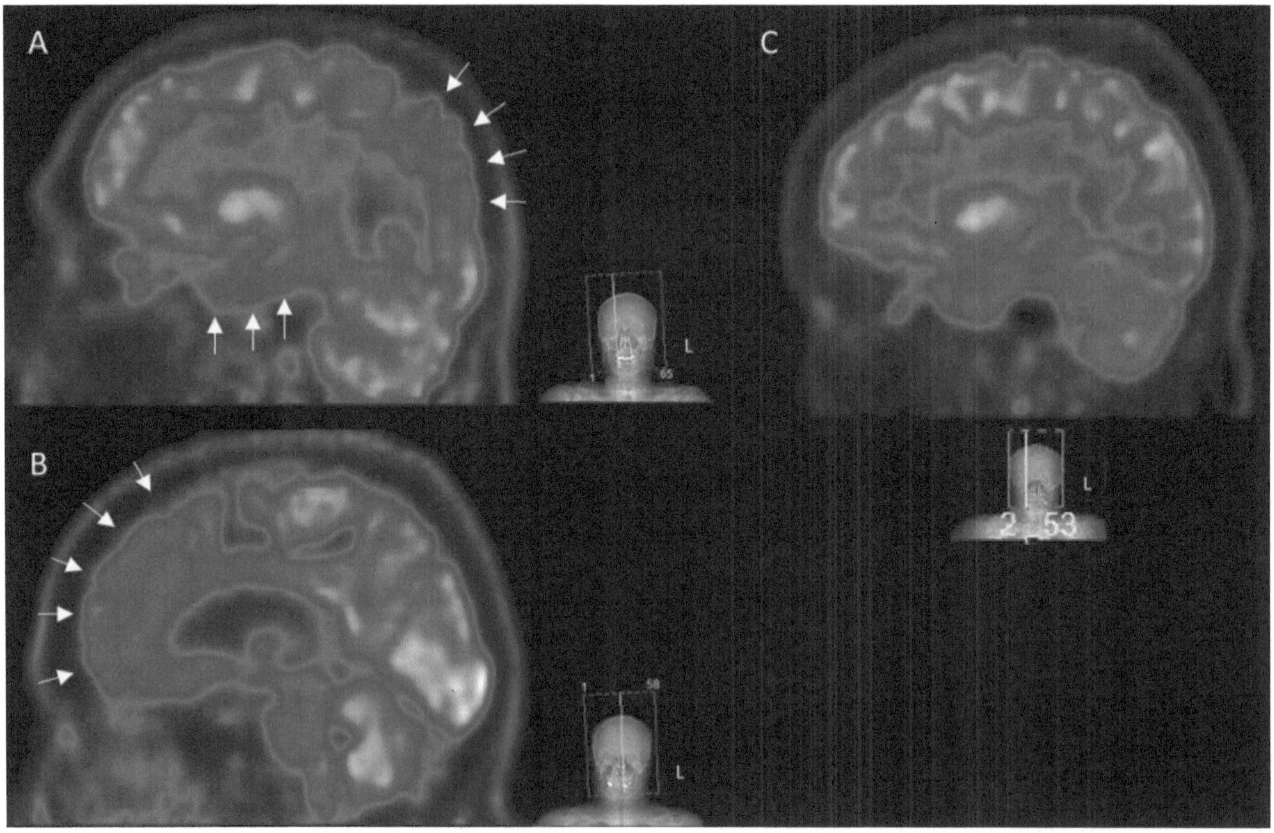

FIGURE 13.8 Images of ^{18}F-FDG PET. (A) Alzheimer's dementia, (B) Frontotemporal dementia and (C) normal ^{18}F-FDG uptake. White arrows indicate hypometabolic cortical areas. Colour image available at www.routledge.com/9781138593312.

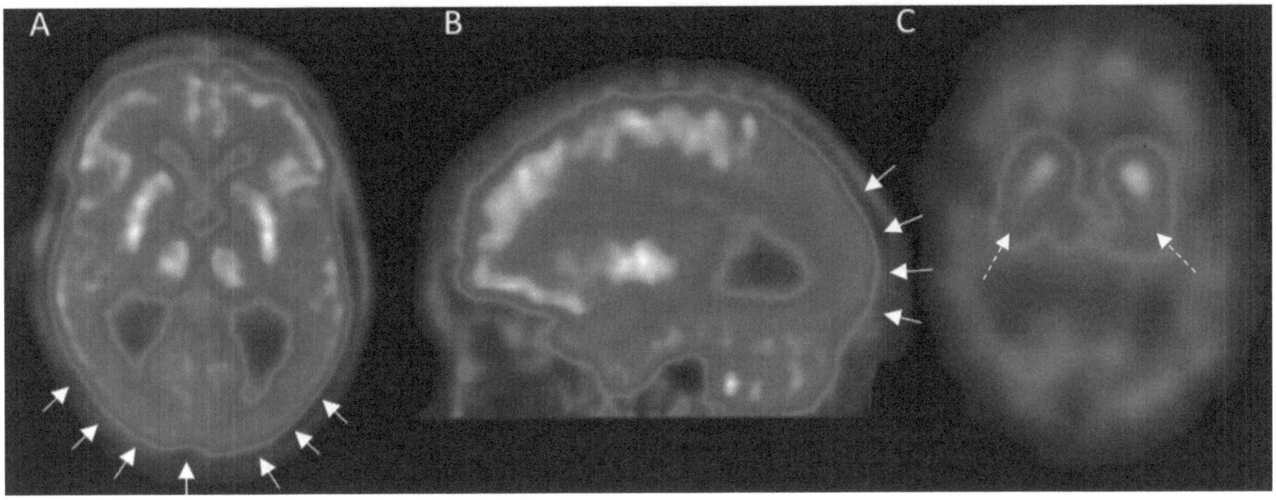

FIGURE 13.9 Lewy body dementia (A and B) ^{18}F-FDG PET and (C) ^{123}I FP-Cit SPECT). White solid arrows indicate hypometabolic cortical area. White dashed arrows indicate reduced DAT-activity. Colour image available at www.routledge.com/9781138593312.

In epilepsy, imaging of brain metabolism is indicated to detect an epileptic focus. It may also contribute with prognostic information before surgery. Imaging is done in inter-ictal state, monitored by EEG. Images may show decreased metabolism in the epileptogenic area due to neuronal dysfunction. However, the hypo metabolism is not always constrained to the ictal focus, but may be seen in other brain regions, which are secondarily affected by the primary focus. PET imaging with ^{18}F-FDG (Fluorodeoxyglucose) is used for imaging of brain metabolism. There are no available radiopharmaceuticals for imaging of metabolism with SPECT.

13.5.3 Imaging of Neurotransmitter Systems

Dopamine Transporter Imaging: The dopamine transporter (DAT) is a specialized protein for recycling of released dopamine, located on dopamine producing neurons. The DAT is found mainly at the nerve terminals in the striatum. Imaging of DAT reflects the integrity of dopaminergic neurons, and decreased DAT tracer uptake is typical for parkinsonian diseases. DAT imaging is generally not affected by medication with anti-Parkinson pharmaceuticals. However, drugs acting on the DAT, such as amphetamine, may interfere with the tracer uptake. The most common radiopharmaceuticals are ^{123}I-FP-Cit (Ioflupan) and ^{123}I-β-Cit (Iometapan), which both are commercially available for SPECT. For PET imaging, ^{11}C-PE2I and ^{18}F-FE-PE2I are the most DAT selective tracers.

Dopamine Receptor Imaging: Dopamine receptors are mainly postsynaptic and exist in different types in different parts of the brain. Imaging of the D2 and D3 receptors, primarily located in the striatum, is well established especially in brain research. Imaging of D2/D3 receptors may be useful in patients with parkinsonian symptoms. Decreased tracer uptake indicates an atypical parkinsonian syndrome, for example, MSA or PSP. PET imaging with ^{11}C-Raclopride is widely used in research, for example in memory research settings. For SPECT, ^{123}I-IBZM is the most common radiopharmaceutical. Both radioligands have affinity to the D2 and D3 receptors and example images can be seen in Figure 13.10.

The interpretation of D2/D3 imaging may be difficult, since high levels of dopamine in the synapse interfere with the binding of the tracer, causing decreased uptake of the tracer. Also, many different pharmacological substances and medications act on D2-receptors, thus competing with the tracer, which renders a false low uptake. Therefore, medication with any substance that mimics dopamine or interacts on dopaminergic receptors must be avoided before imaging to ensure best quality of the imaging result.

Imaging of the D1 receptor, which is located throughout in the brain, is less common but has implications in, for example, psychiatric research. The compound ^{11}CSCH 23390 is among the best documented tracers for D1 imaging with PET [24].

DOPA imaging: L-DOPA is a precursor for dopamine by enzymatic decarboxylation in the so-called catecholamine pathway, in which dopamine in its turn can be metabolized into noradrenaline and adrenaline. Imaging with ^{18}F-DOPA

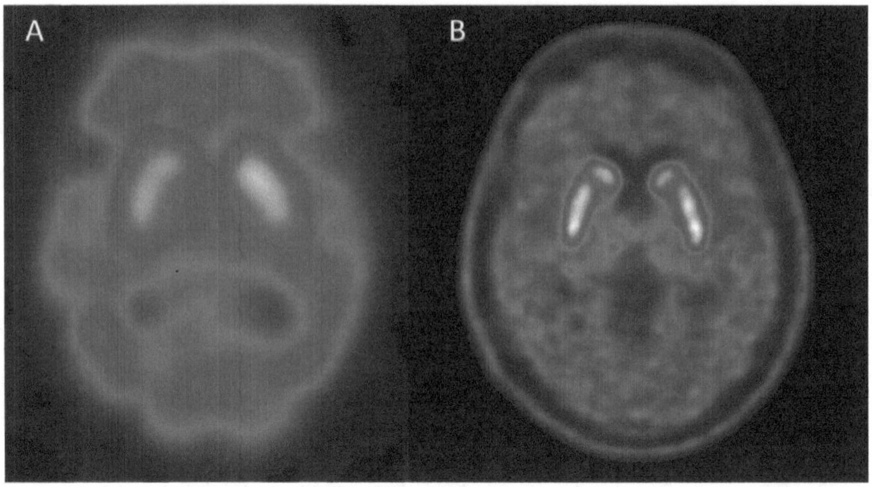

FIGURE 13.10 D2/D3 receptor imaging. (A) shows ^{123}I-IBZM SPECT and (B) ^{11}C-Raclopride PET is showing the binding potential (dynamic study). Colour image available at www.routledge.com/9781138593312.

or ^{11}C-DOPA reflects the activity in the catecholamine pathway and is used for imaging in parkinsonian syndromes to detect lack of dopamine synthesis.

13.5.4 Imaging of Other Cerebral Functions

Amyloid imaging: The formation of amyloid plaques (Aβ plaques) and neurofibrillary tangles is a hallmark of Alzheimer's disease. However, plaque formation may occur in healthy individuals and in other types of dementia. Therefore, the use of amyloid imaging warrants caution in patient selection and interpretation. Sometimes the clinical diagnosis of a dementing disorder is difficult if, for example, the onset or course or symptoms are not fully corroborant with established clinical criteria. However, the proper indications for amyloid imaging are still focus for research. The first and most well-documented amyloid tracer, the ^{11}C Pittsburg compound B (Pib), has been used for several years, mainly in research settings. Lately, radiopharmaceuticals labelled with ^{18}F for amyloid imaging for clinical practice (and research) have become commercially available: ^{18}F-florbetaben (NeuraCeq™), ^{18}F florbetapir (Amyvid) and ^{18}F-flutemetamol (Vizamyl). Up until now these are approved in Europe and the United States. The manufacturers have set up special recommendations for use and interpretation of each substance. For example, images should be interpreted only after a special training program provided by the manufacturer.

^{18}F-flutemetamol images should be displayed with Sokoloff or rainbow or equivalent colour scale. In contrast, grey-scale images should be used for ^{18}F-florbetaben and ^{18}F florbetapir. Guidelines for the use of these tracers are set up by the Society of Nuclear Medicine and Molecular Imaging (SNMMI) [25].

13.5.5 Brain Tumour Imaging

Although MRI is the principal mainstay for brain tumour imaging, there is a wide range of indications for functional imaging with PET as a complementary imaging method in gliomas. Both ^{18}F-FDG and in particular so-called amino acid PET radiotracers, for example, ^{11}C-MET, ^{18}F-FET and ^{18}F-FDOPA are useful for imaging of gliomas. The amino acid tracers reflect tissue growth and the uptake of the tracers is generally low in healthy brain tissue, which is an advantage in imaging of gliomas. ^{11}C-MET and ^{18}F-FET pass through the BBB. Since the glucose metabolism in normal brain is very high, the distinction between tumour tissue and normal brain may be difficult. Nevertheless, ^{18}F-FDG uptake may be higher than background in highly malignant, proliferative tissue, and ^{18}F-FDG PET still has its place in differing between highly malignant and slowly growing tumours as well as in detecting the best target for a tumour biopsy in malignant gliomas. Recently, comprehensive guidelines and standards for PET imaging in gliomas have been published by the European Association of Nuclear Medicine (EANM), the Society of Nuclear Medicine and Molecular Imaging, the European Association of Neurooncology, and the working group for Response Assessment in Neurooncology with PET [26]. In principle, PET imaging is indicated in both the primary diagnostic work-up, the planning of radiation therapy or surgical resection, detection of recurrent tumour growth, for differentiation between post-irradiation necrosis and tumour recurrence and for tumour treatment response monitoring.

REFERENCES

[1] D. R. Johnson, J. B. Guerin, C. Giannini, J. M. Morris, L. J. Eckel, and T. J. Kaufmann, "2016 Updates to the WHO Brain Tumor Classification System: What the Radiologist Needs to Know," *RadioGraphics,* vol. 37, no. 7, pp. 2164–80, 2017, doi: 10.1148/rg.2017170037.

[2] N. L. Albert *et al.*, "Response Assessment in Neuro-Oncology working group and European Association for Neuro-Oncology recommendations for the clinical use of PET imaging in gliomas," *Neuro Oncol,* vol. 18, no. 9, pp. 1199–208, 2016, doi: 10.1093/neuonc/now058.

[3] W. L. Rogers *et al.*, "SPRINT II: A second generation single photon ring tomograph," *IEEE Trans Med Imaging,* vol. 7, no. 4, pp. 291–97, 1988, doi: 10.1109/42.14511.

[4] S. Genna and A. P. Smith, "The development of ASPECT, an annular single crystal brain camera for high efficiency SPECT," *IEEE Trans Nucl Sci,* vol. 35, no. 1, pp. 654–658, 1988, doi: 10.1109/23.12806.

[5] K. Kouris, P. H. Jarritt, D. C. Costa, and P. J. Ell, "Physical assessment of the GE/CGR Neurocam and comparison with a single rotating gamma-camera," *Eur J Nucl Med,* vol. 19, no. 4, pp. 236–42, 1992, doi: 10.1007/bf00175135.

[6] Y. Chen, B. Vastenhouw, C. Wu, M. C. Goorden, and F. J. Beekman, "Optimized image acquisition for dopamine transporter imaging with ultra-high resolution clinical pinhole SPECT," *Phys Med Biol,* vol. 63, no. 22, p. 225002, 2018, doi: 10.1088/1361-6560/aae76c.

[7] H. W. de Jong, F. H. van Velden, R. W. Kloet, F. L. Buijs, R. Boellaard, and A. A. Lammertsma, "Performance evaluation of the ECAT HRRT: an LSO-LYSO double layer high resolution, high sensitivity scanner," *Phys Med Biol,* vol. 52, no. 5, pp. 1505–26, 2007, doi: 10.1088/0031-9155/52/5/019.

[8] Y. Shao *et al.*, "Simultaneous PET and MR imaging," *Phys Med Biol,* vol. 42, no. 10, pp. 1965–70, 1997, doi: 10.1088/0031-9155/42/10/010.

[9] D. L. Bailey *et al.*, "Combined PET/MRI: from Status Quo to Status Go. Summary Report of the Fifth International Workshop on PET/MR Imaging; February 15-19, 2016; Tubingen, Germany," *Mol Imaging Biol,* vol. 18, no. 5, pp. 637–50, 2016, doi: 10.1007/s11307-016-0993-2.

[10] S. Melroy *et al.*, "Development and Design of Next-Generation Head-Mounted Ambulatory Microdose Positron-Emission Tomography (AM-PET) System," *Sensors (Basel),* vol. 17, no. 5, 2017, doi: 10.3390/s17051164.

[11] S. Chu, L. Ekström, and R. Firestone. "The Lund/LBNL Nuclear Data Search Version 2.0." http://nucleardata.nuclear..lu.se/toi/.

[12] A. Larsson, M. Ljungberg, S. Jakobsson Mo, K. Riklund, and L. Johansson, "Correction for scatter and septal penetration in (123)I brain SPECT imaging – A Monte Carlo study," *Phys Med Biol,* vol. 51, pp. 5753–67, 2006.

[13] D. J. Macey, G. L. DeNardo, S. J. DeNardo, and H. H. Hines, "Comparison of Low- and Medium-Energy Collimators for SPECT Imaging with Iodine 123-Labeled Antibodies," *J Nucl Med,* vol. 27, pp. 1467–74, 1986.

[14] H. J. Verberne, C. Feenstra, W. M. de Jong, G. A. Somsen, B. L. van Eck-Smit, and E. Busemann Sokole, "Influence of collimator choice and simulated clinical conditions on 123I-MIBG heart/mediastinum ratios: A phantom study," *Eur J Nucl Med Mol Imaging,* vol. 32, no. 9, pp. 1100–7, 2005, doi: 10.1007/s00259-005-1810-3.

[15] K. Van Audenhaege, R. Van Holen, S. Vandenberghe, C. Vanhove, S. D. Metzler, and S. C. Moore, "Review of SPECT collimator selection, optimization, and fabrication for clinical and preclinical imaging," *Med Phys,* vol. 42, no. 8, pp. 4796–813, 2015, doi: 10.1118/1.4927061.

[16] R. Accorsi, J. R. Novak, A. S. Ayan, and S. D. Metzler, "Derivation and validation of a sensitivity formula for slit-slat collimation," *IEEE Trans Med Imaging,* vol. 27, no. 5, pp. 709–22, 2008, doi: 10.1109/tmi.2007.912395.

[17] K. Erlandsson, D. Salvado, A. Bousse, and B. F. Hutton, "Design optimization and evaluation of a human brain SPECT-MRI insert based on high-resolution detectors and slit-slat collimators," in *2013 IEEE Nuclear Science Symposium and Medical Imaging Conference (2013 NSS/MIC)*, 2013, pp. 1–4, doi: 10.1109/NSSMIC.2013.6829144.

[18] J. Darcourt *et al.*, "EANM procedure guidelines for brain neurotransmission SPECT using (123)I-labelled dopamine transporter ligands, version 2," *Eur J Nucl Med Mol Imaging,* vol. 37, no. 2, pp. 443–50, 2010, doi: 10.1007/s00259-009-1267-x.

[19] E. Wallstén, J. Axelsson, M. Karlsson, K. Riklund, and A. Larsson, "A Study of Dynamic PET Frame-Binning on the Reference Logan Binding Potential," *IEEE Transactions on Radiation and Plasma Medical Sciences,* vol. 1, no. 2, pp. 128–35, 2017, doi: 10.1109/TNS.2016.2639560.

[20] H. Barthel, M. L. Schroeter, K. T. Hoffmann, and O. Sabri, "PET/MR in dementia and other neurodegenerative diseases," *Semin Nucl Med,* vol. 45, no. 3, pp. 224–33, 2015, doi: 10.1053/j.semnuclmed.2014.12.003.

[21] K. Ogawa, Y. Harata, T. Ichihara, A. Kubo, and S. Hashimoto, "A practical method for position-dependent Compton-scatter correction in single photon emission CT," *IEEE Trans. Med. Imag.,* vol. 10, pp. 408–12, 1991, doi: 10.1109/42.97591.

[22] A. Rahmim, J. Qi, and V. Sossi, "Resolution modeling in PET imaging: theory, practice, benefits, and pitfalls," *Med Phys,* vol. 40, no. 6, p. 064301, 2013, doi: 10.1118/1.4800806.

[23] S. Minoshima, K. A. Frey, R. A. Koeppe, N. L. Foster, and D. E. Kuhl, "A diagnostic approach in Alzheimer's disease using three-dimensional stereotactic surface projections of fluorine-18-FDG PET," *J Nucl Med,* vol. 36, no. 7, pp. 1238–48, 1995.

[24] S. Cervenka, "PET radioligands for the dopamine D1-receptor: Application in psychiatric disorders," *Neurosci Lett,* vol. 691, pp. 26–34, 2019, doi: 10.1016/j.neulet.2018.03.007.

[25] S. Minoshima *et al.*, "SNMMI Procedure Standard/EANM Practice Guideline for Amyloid PET Imaging of the Brain 1.0," *J Nucl Med,* vol. 57, no. 8, pp. 1316–22, 2016, doi: 10.2967/jnumed.116.174615.

[26] I. Law *et al.*, "Joint EANM/EANO/RANO practice guidelines/SNMMI procedure standards for imaging of gliomas using PET with radiolabelled amino acids and [(18)F]FDG: version 1.0," *Eur J Nucl Med Mol Imaging,* vol. 46, no. 3, pp. 540–57, 2019, doi: 10.1007/s00259-018-4207-9.

14 Methodology and Clinical Implementation of Ventilation/Perfusion Tomography for Diagnosis and Follow-up of Pulmonary Embolism and Other Pulmonary Diseases

Clinical Use of Hybrid V/P SPECT-CT

Marika Bajc and Ari Lindqvist

CONTENTS

14.1 LUNG SPECT

Lung scintigraphy demonstrates patterns of ventilation and perfusion. It is an essential method to study pulmonary function. In a healthy individual there is a balance between regional perfusion and ventilation to achieve optimal gas exchange. Some pulmonary disease causes changes in both ventilation and perfusion; they are "matched." Some other diseases can only cause changes in perfusion or ventilation, that is, lead to mismatch that implies imbalance between perfusion and ventilation. The single most important application of lung scintigraphy is the evaluation of patients with suspected pulmonary embolism (PE). Ventilation/perfusion SPECT (V/P SPECT) is the recommended scintigraphic

DOI: 10.1201/9780429489501-14

technique for the diagnosis of pulmonary embolism, and many other disorders that affect lung function. It is noteworthy that there are no contraindications to V/P SPECT, and that even very sick and breathless patients can be studied.

14.2 PULMONARY EMBOLISM

Pulmonary embolism is a very common disease and is presented globally as the third most frequent cardiovascular syndrome, with about 250 000 patients diagnosed each year in the United States [1]. In spite of advanced technology, it remains a major diagnostic challenge because the clinical symptoms and signs that are frequently observed in PE are also a feature of other conditions. Accordingly, the initial clinical suspicion needs to be confirmed or negated by using a conclusive imaging test. Routinely computed tomography pulmonary angiography (CTPA) is suggested as the initial imaging study [2]. However, CTPA is overused in a great number of patients with low prevalence of PE [3, 4]. A PE located centrally in the pulmonary circulation can be detected by CTPA with a high positive predicted value. The positive predicted value decreases at segmental and subsegmental levels [5].The latest evidence show that the optimal test is V/P SPECT interpreted with holistic principles according to European Guideline [6].

Before performing imaging tests, it is recommended to estimate the clinical probability for PE [7]. Usually, a Wells score is applied. The measurement of D-dimer – a breakdown product of cross-linked fibrin clot – is widely used in the investigative workup of patients with suspected venous thromboembolism. However, D-dimer has a low specificity (40%) because a number of conditions, other than venous thromboembolism, may cause it to be elevated: For example, acute myocardial infarction, stroke, inflammation, active cancer, and pregnancy. The specificity declines even further with age and, in the elderly, may reach only 10 per cent [8]. Due to the low predictive value, a positive quantitative D-dimer test does not modify the pre-test probability. A negative quantitative D-dimer test combined with a low clinical probability is associated with a low risk of thromboembolic disease. At moderate to high pre-test clinical probability, D-dimer has no incremental value.

14.2.1 Basic Principles of PE Diagnosis with V/P SPECT

The lung circulation has a distinct architecture, where a single end-artery supplies each broncho-pulmonary segment and sub-segment. Emboli are usually multiple, occluding the arteries, causing segmental or sub-segmental perfusion defects within still-ventilated lung areas. The result is a so-called mismatch, as shown in Figure 14.1.

PE is often a recurring process giving rise to multiple emboli in various stages of resolution (Figure 14.2).

In clinical practice, it is essential to have a procedure that is both fast and conclusive to avoid the risks associated with an untreated disease. Therefore, it is recommended that imaging tests for PE diagnosis should be carried out as soon as possible, preferably within 24 hours of the onset of symptoms.

14.2.2 Imaging Protocols

14.2.2.1 V/P SPECT Acquisition

Administration of ventilation and perfusion agents should be performed with patients in a supine position in order to minimize gravitational gradients. During inhalation, activity over the lungs should be monitored to ensure adequate pulmonary deposition. The procedure starts with ventilation scintigraphy, which is usually based upon inhalation of a radio-aerosol (Table 14.1). Particles larger than 2 μm deposit mainly by impaction in large airways. Very fine particles smaller than 1 μm mainly move through the conductive airways and deposit in alveoli by diffusion. Technegas consisting of very small hydrophobic aerosolized particles (0.09 μm) diffuse more effectively through the central airways to periphery compared to liquid aerosols (like DTPA aerosols size, 1.4 -2 μm), minimizing hotspot formation in small airways. This is especially advantageous in patients with obstructive lung disease [9].

Perfusion tomography follows immediately after ventilation SPECT without changing the patient's position. Universally, the most-used agent for perfusion scintigraphy is technetium-labelled particles of macro-aggregates of human albumin (^{99m}Tc-MAA). After i.v. injection, the particles of size 15–100 μm are lodged in the pulmonary capillaries and in the precapillary arterioles in proportion to perfusion.

To achieve adequate imaging quality, low radiation exposure, and a short imaging time, relationships between activities, acquisition times, collimators, and matrices for SPECT imaging must be optimized. This problem was systematically analyzed by Palmer and colleagues in the context of a dual head gamma camera [10]. Doses of 25–30 MBq for ventilation studies and 120–160 MBq for perfusion studies were found optimal by using a general-purpose collimator,

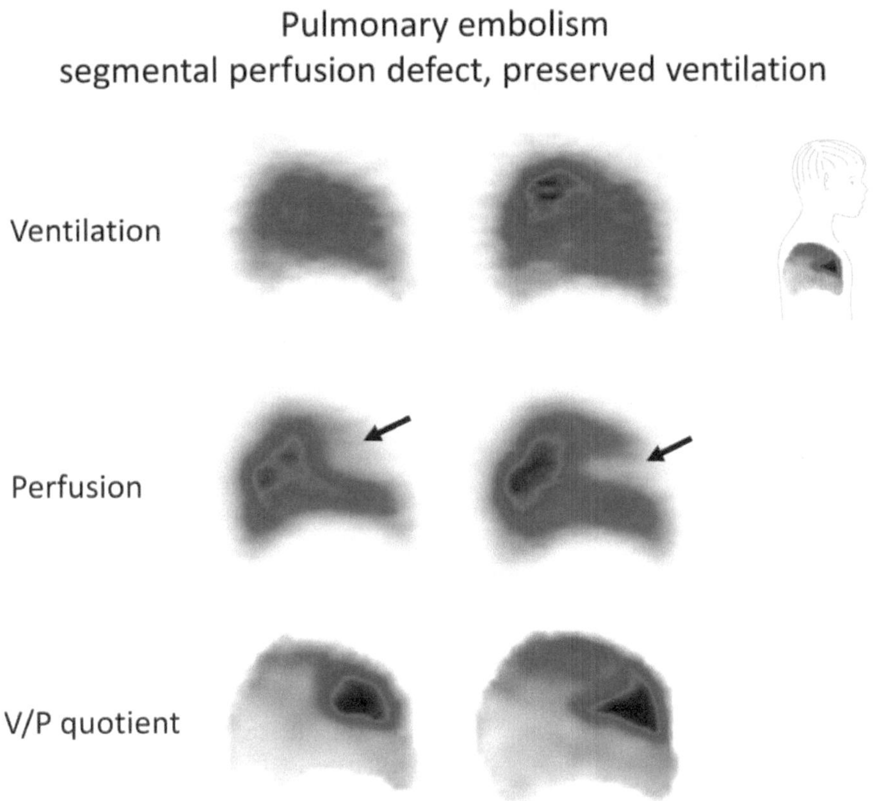

FIGURE 14.1 Patient with PE; segmental perfusion defect is seen on sagittal slices (PE, arrow) in area with preserved ventilation. Colour image available at www.routledge.com/9781138593312.

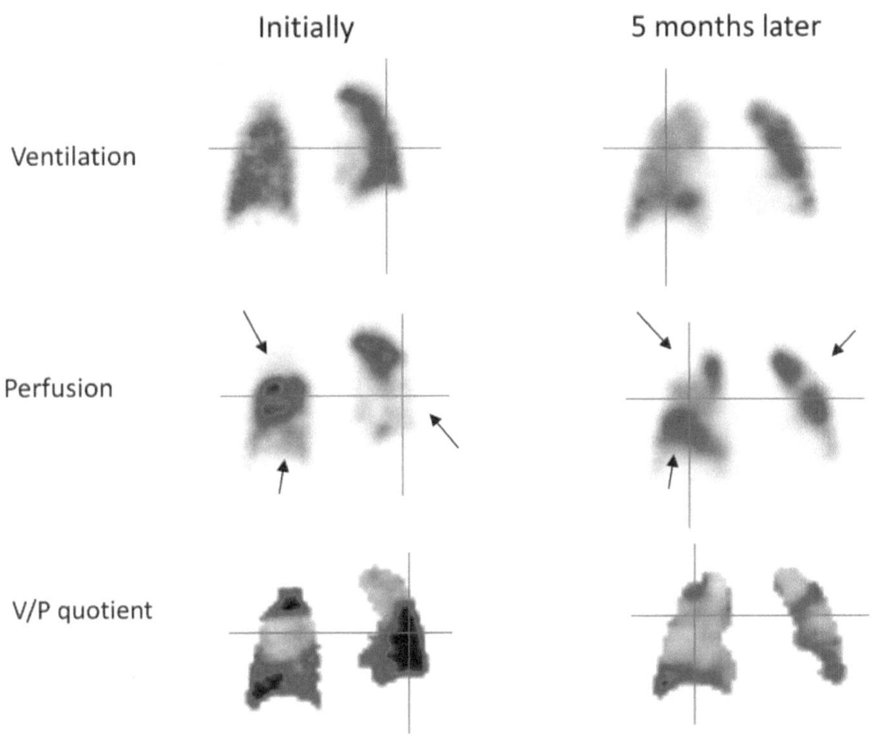

FIGURE 14.2 Patient with recurrent PE. Initially, segmental perfusion reductions are seen on coronal slices (PE, arrows). Follow-up scan five month later shows partially improved perfusion defects (arrows). Colour image available at www.routledge.com/9781138593312.

TABLE 14.1
Ventilation/Perfusion Protocol for Lung Tomography (V/P SPECT)

	Ventilation	Perfusion
Administration	inhalation	iv injection
Radiopharmaceutical & administered activity	Technegas ® 25-30 MBq to reach the lung	^{99m}Tc-MAA 120-160 MBq
Particle size	0.09 μm	15-100 μm
Time of Imaging	ca 11 minutes	ca 5 minutes
Acquisition protocol	general purpose collimator: 64 × 64 matrix, 60-64 steps for each head, 10 s/step	general purpose collimator: 64 × 64 matrix, 60-64 steps for each head, 5 s/step

Patient is in the supine position, during inhalation, I.V. injection and during the acquisition

and 64x64 matrix. Total acquisition time is about 20 minutes. If a matrix of 128x128 is used, higher doses and/or a longer acquisition time is required. This is not promoted, as it does not yield images of significantly higher quality [10]. To follow good medical practice, radiation exposure should be minimized to the lowest level consistent with satisfactory image quality [11].

For V/P SPECT it is essential to use iterative reconstruction. Recommended is OSEM (Ordered-Subset Expectation Maximization) with 4 iterations and 8 subsets. Standard software can be used for image presentation in coronal, sagittal, and transversal projections as well as for presentation of rotating 3-D images. Palmer and colleagues developed a way of calculating and displaying ventilation/perfusion quotient images. In this method, ventilation is normalized to perfusion counts. Then V/P quotient images are calculated. V/P quotient images facilitate diagnosis and quantification of PE extension. Using this protocol, attenuation correction is not needed [10].

For quality control and fast orientation, an overview of ventilation and perfusion in coronal and sagittal slices is useful. It is important to present the images so that ventilation and perfusion are carefully aligned to each other, as shown in Figure 14.3.

This is greatly facilitated by the one session protocol with the patient in an unchanged position. The option to triangulate between coronal, sagittal and transverse slices is valuable for identification of matching and non-matching ventilation and perfusion changes. Proper alignment is also a prerequisite for V/P quotient images. These facilitate the interpretation and quantification of PE extension and all ventilation and perfusion defects. However, quotient images are not a prerequisite for a high-quality V/P SPECT.

14.2.2.2 Reporting Findings

For V/P SPECT, interpretation criteria are as important as the imaging technique itself. According to European Guideline, all patterns of V and P and the number of defects need to be described, and clinical probability needs to be taken into account. This holistic principle for reporting gives also a clear answer: The defect (e.g. PE) either exists or it does not exist. This goal was not achieved with previous probabilistic reporting methods according to PIOPED or modified PIOPED [12, 13]. Large V/P SPECT studies have shown that interpretation of all patterns representing ventilation, together with perfusion, achieves this goal [14–18]. Conclusive reports can be given in 97–99 per cent of cases.

14.2.2.3 Criteria for Acute PE

Recommended criteria for reading V/P SPECT with respect to acute PE are the following:

PE:

- V/P mismatch of at least one segment or two sub-segments that conforms to the pulmonary vascular anatomy.

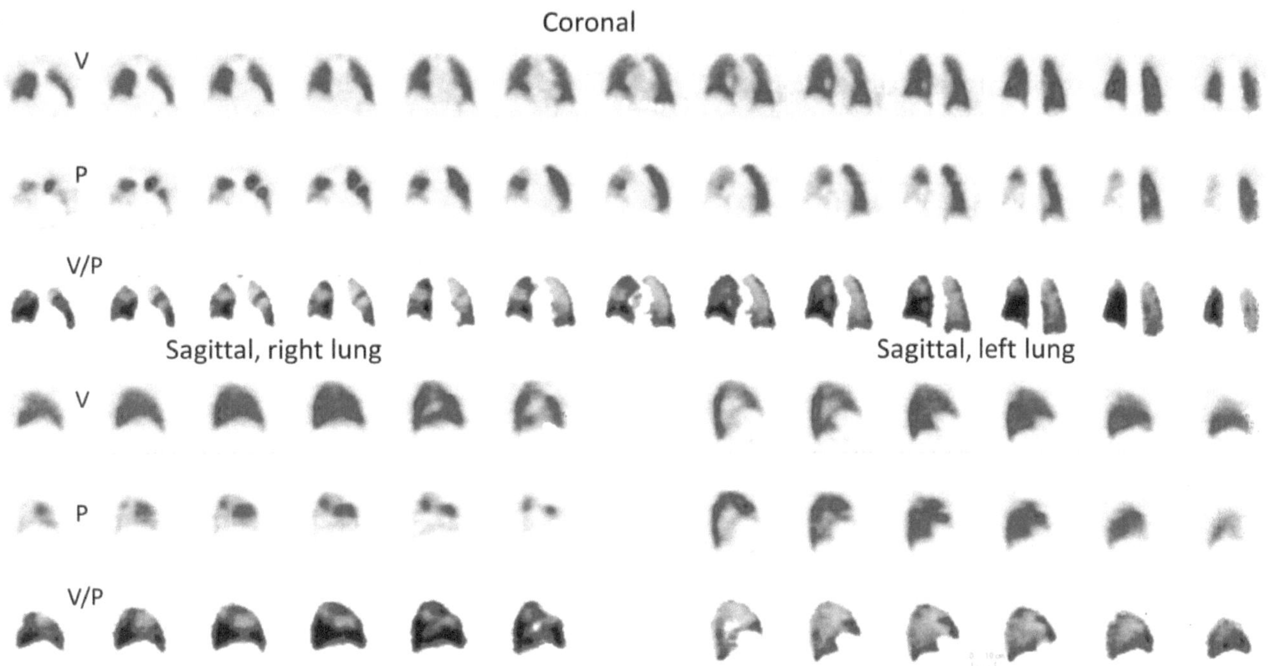

FIGURE 14.3 Overview of images on coronal and sagittal slices to check that ventilation and perfusion are carefully aligned to each other. Colour image available at www.routledge.com/9781138593312.

No PE:

- Normal perfusion pattern conforming to the anatomic boundaries of the lungs matched or reversed mismatch V/P defects of any size, shape or number in the absence of mismatch.
- Mismatch that does not have a lobar, segmental, or sub-segmental pattern.

Non-diagnostic for PE:

- Multiple V/P abnormalities that are not typical for a specific disease.

Applying these principles of interpretation, recent V/P SPECT studies amounting to over 5,000 cases report a negative predictive value of 97–99 per cent, sensitivity of 96–99 per cent, and specificity of 96–98 per cent for PE diagnosis. Rates of non-diagnostic findings are 1–4 per cent [19].

14.2.3 Quantification of V/P SPECT

An important step in the diagnostic procedure is to quantify the extent of embolism. V/P SPECT is particularly suitable for this because it has a greater sensitivity and lower radiation dose than CTPA. The number of segments and sub-segments indicating typical mismatch in PE are counted and expressed in per cent of the total lung parenchyma [20]. Furthermore, areas with ventilation abnormalities are recognized. This procedure allows estimation of the degree of lung malfunction. Quantification of V/P SPECT helps clinicians to evaluate home treatment of PE. Patients with a PE compromising up to 40 per cent of the lung perfusion can be safely treated at home if ventilation abnormalities engage not more than 20 per cent of the lung [21].

14.2.4 Comparison to CTPA

The other important method to visualize PE is computed tomographic angiography of the pulmonary artery (CTPA). Pulmonary emboli are visualized as so-called filling defects during the passage of intravenously injected iodinated contrast material through pulmonary arteries, which are otherwise homogenously contrast-filled. CTPA is easy to perform in a few minutes. According to the European Society of Cardiology Guidelines, CTPA needs to be embedded in decision strategies that are based on the assessment of clinical PE likelihood.

CTPA confirms the diagnosis of PE in clinically high-probability patients (a positive predictive value of > 95%). In cases of high pre-test probability of PE and a negative CTPA, current data on diagnostic accuracy are inconsistent. The positive predictive value decreases at segmental and subsegmental levels. When clinical probability of PE is low or intermediate, CTPA may overdiagnose PE, leading to a low negative predictive value.

CTPA is overused in a great number of patients with a low prevalence of PE. A PE located centrally in the pulmonary circulation can be detected by CTPA with a high positive predictive value.

CTPA has the potential to visualize additional pathologies other than PE such as pneumothorax, pneumonia, interstitial lung disease, pleural disease, aortic dissection, and pathologies of the spine and rib cage.

14.3 IMPORTANCE OF V/P SPECT IN DIAGNOSIS OF OTHER LUNG DISEASES

V/P SPECT gives the possibility to localize ventilation and perfusion impairment and estimate the total lung function.

14.3.1 Chronic Obstructive Pulmonary Disease (COPD)

In COPD ventilation is generally uneven. Inhaled aerosol deposits focally at sites of airway obstruction [22, 23]. The degree of unevenness of aerosol distribution correlates with lung function tests [24-26]. In healthy subjects, even distribution of Technegas with good peripheral penetration and without accumulation in large or small airways is observed. The grading of obstruction in ventilation SPECT has been standardized using Technegas ventilation SPECT [27] (Figure 14.4):

Mild airway obstruction (grade 1):

- slightly uneven distribution of Technegas with some deposition of Technegas in small and intermediate airways. Only minor areas with reduced peripheral penetration of Technegas are observed.

Moderate airway obstruction (grade 2):

- deposition of Technegas in intermediate and large airways and diminished peripheral penetration with a maximum accumulation of Technegas in the central half of the lung.

Severe airway obstruction (grade 3)

- central deposition of Technegas in large airways with severely impaired penetration of Technegas, and major areas with reduced or abolished function.

14.3.2 Pneumonia

Pneumonia (lung inflammation, often caused by a bacterial, viral or fungal infection) is a frequent finding in patients investigated for a suspected PE [28]. Blood biomarkers are not sufficient for diagnosing pneumonia, and unspecific clinical symptoms can lead to diagnostic problems [29]. In pneumonia, V/P SPECT shows ventilation defects that usually exceed perfusion defects known as reverse mismatch (reversed V/P mismatch) [30]. Preserved perfusion along the pleural border peripheral to a central matched defect recognized as the "stripe sign" is a specific pattern of pneumonia [31] (Figure 14.5).

Figure 14.5 shows a pneumonia in the posterior part of right lung together with a PE in the middle lobe. PE could not have been identified without SPECT and ventilation images. In some patients, V/P defects typical for pneumonia reduce the total lung function in the absence of any morphological CT changes. In these clinical scenarios, PE is frequently missed by CT [28]. This is important information because, in general, current clinical and nuclear medicine practices do not recognize nor use V/P SPECT as a potential imaging method to diagnose or manage pneumonia.

14.3.3 Left Heart Failure

Antigravitational perfusion distribution from posterior to anterior region indicates pulmonary congestion. The pattern was described already in 1966 [32] and studied later [33–35]. As ventilation is usually less affected, the typical pattern

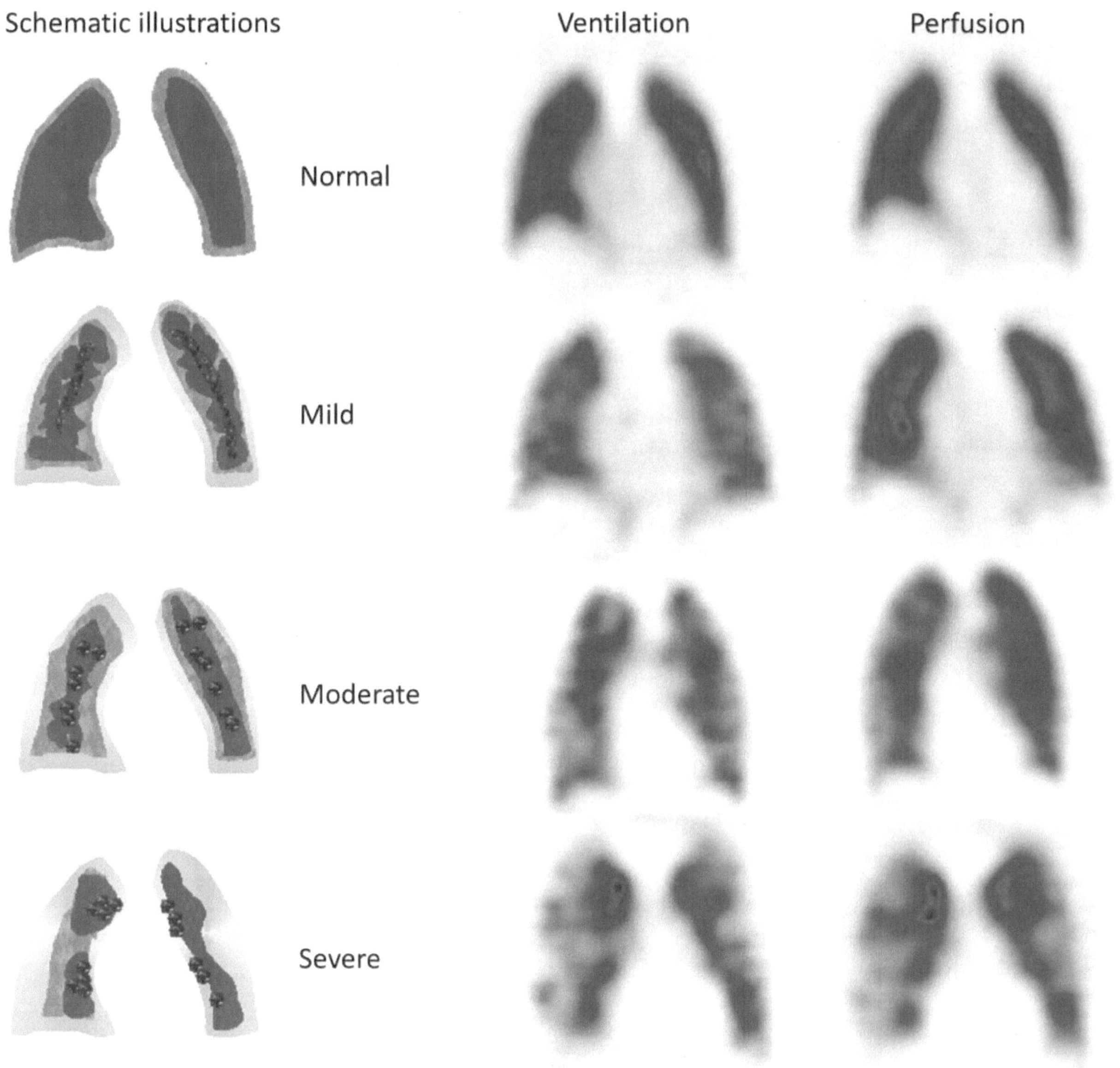

FIGURE 14.4 The grading of obstruction in ventilation SPECT images, schematic presentation, and corresponding coronal images of patients with different degree of COPD. Colour image available at www.routledge.com/9781138593312.

is anti-gravitational redistribution of perfusion and V/P mismatch in dorsal regions of the lung. This V/P mismatch has a non-segmental pattern (does not conform to pulmonary vascular architecture) and should not be misinterpreted as pulmonary embolism (Figure 14.6).

14.4 CLINICAL USE OF HYBRID V/P SPECT/CT

The hybrid SPECT/CT system is a dual imaging modality technique. Its clinical application is particularly relevant in oncological diseases as it leads to improved sensitivity and specificity, combining co-registration of anatomical and functional data. It may lead to improved staging and treatment monitoring. Nuclear medicine procedures have the ability to visualize early functional changes sooner than structural changes occur. An additional CT procedure may improve correction for photon attenuation and allow co-registration of morphology and function. However, SPECT/CT acquisition of the chest constitutes a challenge due to respiratory movements, which can cause image artefacts and thus decrease diagnostic accuracy.

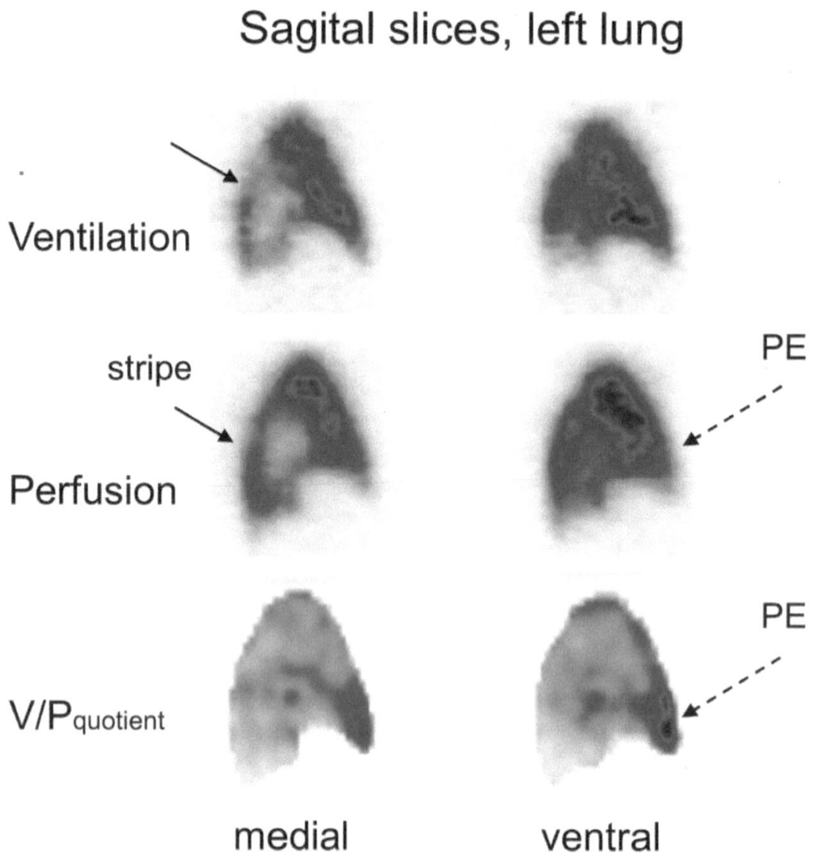

FIGURE 14.5 Patient with pneumonia and PE; reduction in ventilation is observed posteriorly on sagittal slices and stripe sign (solid arrows) is seen on perfusion image, Furthermore, perfusion defect is observed in lingula (PE, dashed arrows). Colour image available at www.routledge.com/9781138593312.

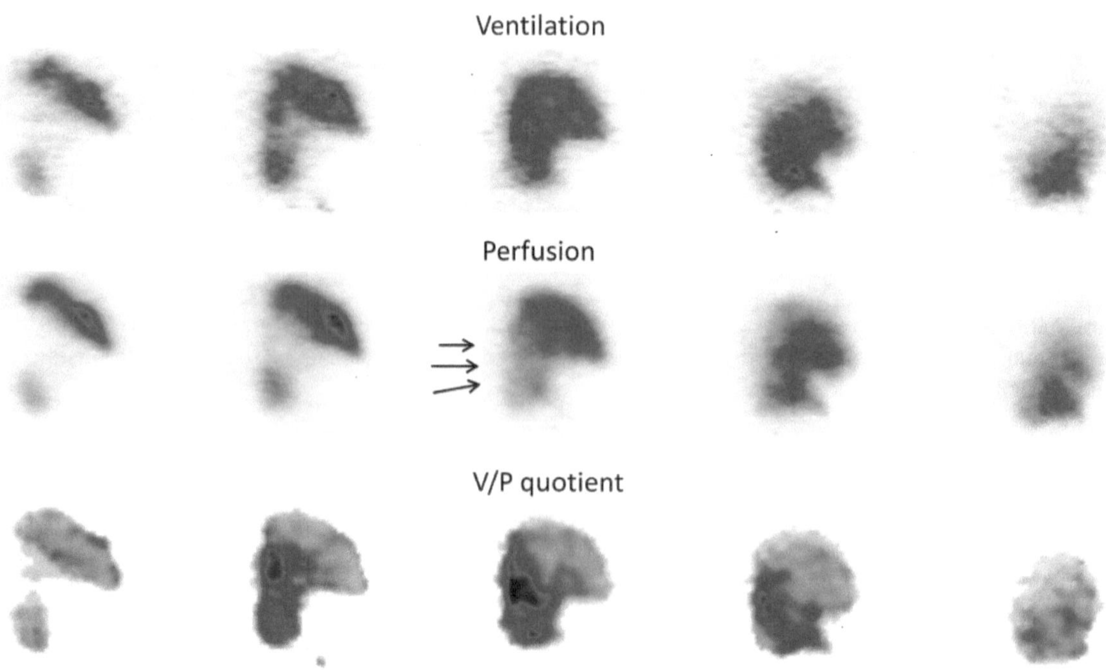

FIGURE 14.6 Patient with left heart failure; antigravitational redistribution of perfusion is seen on sagittal slices of the left lung (arrows). Ventilation is less affected, causing mismatch, however not of segmental character.

TABLE 14.2
V/P SPECT/CT Protocol CT Overview Image

	Tube potential	Tube current	Collimation/slice width	Pitch
Low dose CT	120 kV	20 mAs	16 x 1.5-0.5s rotation	0.813

The procedure starts with CT overview image and continues with diagnostic low-dose CT. Low-dose CT (120 kV, 20 mAs/slice, 16 × 1.5 collimator, 0.5-s rotation time, and pitch of 0.813) is not used for attenuation correction but to better to co-localize the morphological and functional changes visualized in either of the two modalities (Table 14.2). Thereafter follows the protocol for V/P SPECT as described above and according to the European Guideline. Low-dose CT delivers approximately 1 mSv when used for alignment and attenuation correction. However, as a diagnostic tool in this hybrid system, it delivers 2-3 mSv and V/P SPECT 2.1 mSv.

Some authors have recently recommended V/P SPECT/low-dose CT as a first line procedure in patients with suspected PE. This was based on their prospective study performing V/P SPECT and low-dose CT and making head-to-head comparisons with CTPA. In a total of 81 simultaneous studies, 38 per cent of patients had PE. They showed 97 per cent sensitivity and 88 per cent specificity when only V/P SPECT was used. However, adding low- dose CT the sensitivity was unchanged but specificity increased to 100 per cent. Interestingly 18 per cent of patients had false positive PE diagnosis when V/P SPECT alone was interpreted [15]. A reason for this may be that they were interpreting every mismatch as PE and not only mismatches that conform to segmental lung circulation as recommended by the European Guideline [19].

Accordingly, V/P SPECT is a primary tool in patients with suspected PE. Since 2003, more than 20,000 examinations have been performed. We recognize an ethical problem in advocating CT study in addition to V/P SPECT for every patient with suspected PE when in our experience prevalence of PE is about 30 per cent and might be as low as 10 per cent [36]. Following the good clinical practice recommendation to use a hybrid system for PE diagnosis is premature [11]. CT utilization has increased dramatically in the evaluation of patients with suspected PE without improving rate of PE or other clinically significant diagnoses [37]. Therefore, it is important to validate the V/P SPECT/CT system and to assess the benefits and risks [38]. We consider however, that the dual modality will have an impact in some groups of patients. SPECT/CT may provide adding value to COPD patients showing morphological changes in addition to functional defects, particularly in visualizing small tumours [39].

14.5 RADIATION DOSES

Following good clinical practice, it is important to minimize radiation exposure without sacrificing image quality and diagnostic accuracy. Based upon data from ICRP reports [40] the effective dose for V/P SPECT with the recommended protocol is 2 mSv. In a systematic review and meta-analysis of literature, radiation exposure was 2.12 mSv for V/P SPECT per correct diagnosis compared with 4.96 mSv for CTPA [41]. In clinical routine, radiation doses between 3.5 and 13.2 mSv have been reported for CTPA [42–44]. In general, higher effective doses (>5 mSv) are reported from automatically collected data [45, 46].

The most critical organ in CTPA is the female breast. Absorbed radiation doses to the female breast ranging from 8.6 mSv to 44 mSv have been reported [47–49]. Tube current modulation is able to decrease the breast dose from 51.5 mSv to 8.6 mSv [48] whilst shielding is less effective [50]. Absorbed radiation doses to the female breast from V/P SPECT is <1 mSv [47]. Fetal-absorbed doses for V/P SPECT and CTPA are similar and so small that they are unlikely to be clinically significant [47, 49, 51].

14.6 CONCLUSIONS

V/P SPECT is the recommended first choice imaging technique for diagnosis of acute PE and it is the gold standard for the diagnosis of chronic PE. An optimal combination of low nuclide activities and acquisition times for ventilation and perfusion, collimators, and imaging matrix yields a single session adequate V/P SPECT study in approximately 20 minutes of imaging time. Furthermore, full use should be made of display options, which are integrated in modern camera systems. Holistic interpretation strategy gives a clear report with respect to PE, its extension as well as other diagnoses based on ventilation/perfusion patterns typical for various diseases. V/P SPECT has the highest sensitivity and accuracy; it produces very few non-diagnostic reports, radiation doses caused by V/P SPECT are low, and it has neither contraindications nor complications. Low radiation doses are particularly important for women in the reproductive period and during pregnancy. The above-mentioned advantages of V/P SPECT for studying PE imply that it may be the most suitable technique, both for follow up in patients with PE as well as for research regarding its treatment and pathophysiology. Hybrid V/P SPECT-CT has added value in patients with COPD. However, further studies of V/P SPECT-CT are needed for validation in various categories of patients to define its value according to good clinical practice.

ACKNOWLEDGEMENT

The authors thank Kerstin Brauer, Skåne University Hospital, Sweden, for her excellent technical assistance.

REFERENCES

[1] M. D. Silverstein, J. A. Heit, D. N. Mohr, T. M. Petterson, W. M. O'Fallon, and L. J. Melton, 3rd, "Trends in the incidence of deep vein thrombosis and pulmonary embolism: A 25-year population-based study," *Arch Intern Med,* vol. 158, no. 6, pp. 585–93, 1998.

[2] British Thoracic Society, "Guidelines for the management of suspected acute pulmonary embolism," *Thorax,* vol. 58, no. 6, pp. 470–83, 2003.

[3] L. B. Feng, J. M. Pines, H. R. Yusuf, and S. D. Grosse, "U.S. trends in computed tomography use and diagnoses in emergency department visits by patients with symptoms suggestive of pulmonary embolism, 2001–2009," *Acad Emerg Med,* vol. 20, no. 10, pp. 1033–40, 2013, doi: 10.1111/acem.12221.

[4] B. D. Hutchinson, P. Navin, E. M. Marom, M. T. Truong, and J. F. Bruzzi, "Overdiagnosis of pulmonary embolism by pulmonary CT angiography," *American Journal of Roentgenology,* vol. 205, no. 2, pp. 271–77, 2015, doi: 10.2214/AJR.14.13938.

[5] P. Stein *et al.*, "Multidetector computed tomography for acute pulmonary embolism," *N Engl J Med,* vol. 354, no. 22, pp. 2317–27, 2006, doi: 10.1056/NEJMoa052367.

[6] M. Bajc *et al.*, "EANM guideline for ventilation/perfusion single-photon emission computed tomography (SPECT) for diagnosis of pulmonary embolism and beyond," *Eur J Nucl Med Mol Imaging,* 2019, doi: 10.1007/s00259-019-04450-0.

[7] R. A. Douma *et al.*, "Performance of 4 clinical decision rules in the diagnostic management of acute pulmonary embolism: A prospective cohort study," *Ann Intern Med,* vol. 154, no. 11, pp. 709–18, 2011, doi: 10.7326/0003-4819-154-11-201106070-00002.

[8] P. D. Stein *et al.*, "D-dimer for the exclusion of acute venous thrombosis and pulmonary embolism: A systematic review," *Ann Intern Med,* vol. 140, no. 8, pp. 589–602, 2004, doi: 140/8/589 [pii].

[9] J. Jogi, B. Jonson, M. Ekberg, and M. Bajc, "Ventilation-perfusion SPECT with 99mTc-DTPA versus Technegas: A head-to-head study in obstructive and nonobstructive disease," *J Nucl Med,* vol. 51, no. 5, pp. 735–41, 2010, doi: 10.2967/jnumed.109.073957.

[10] J. Palmer, U. Bitzen, B. Jonson, and M. Bajc, "Comprehensive ventilation/perfusion SPECT," *J Nucl Med,* vol. 42, no. 8, pp. 1288–94, 2001.

[11] M. Bajc, L. Maffioli, and M. Miniati, "Good clinical practice in pulmonary embolism diagnosis: Where do we stand today?" *Eur J Nucl Med Mol Imaging,* vol. 41, no. 2, pp. 333–36, 2014, doi: 10.1007/s00259-013-2612-7.

[12] P. D. Stein *et al.*, "Tracking the uptake of evidence: Two decades of hospital practice trends for diagnosing deep vein thrombosis and pulmonary embolism," *Arch Intern Med,* vol. 163, no. 10, pp. 1213–19, 2003.

[13] "Value of the ventilation/perfusion scan in acute pulmonary embolism. Results of the prospective investigation of pulmonary embolism diagnosis (PIOPED). The PIOPED Investigators," *JAMA*, vol. 263, no. 20, pp. 2753–59, 1990.

[14] M. Bajc, B. Olsson, J. Palmer, and B. Jonson, "Ventilation/Perfusion SPECT for diagnostics of pulmonary embolism in clinical practice," *J Intern Med.*, vol. 264, no. 4, pp. 379–87. 2008.

[15] H. Gutte *et al.*, "Detection of pulmonary embolism with combined ventilation-perfusion SPECT and low-dose CT: Head-to-head comparison with multidetector CT angiography," *Journal of Nuclear Medicine: Official publication, Society of Nuclear Medicine,* vol. 50, no. 12, pp. 1987–92, 2009.

[16] M. Leblanc, F. Leveillee, and E. Turcotte, "Prospective evaluation of the negative predictive value of V/Q SPECT using 99mTc-Technegas," *Nucl Med Commun,* vol. 28, no. 8, pp. 667–72, 2007.

[17] M. Lemb and H. Pohlabeln, "Pulmonary thromboembolism: a retrospective study on the examination of 991 patients by ventilation/perfusion SPECT using Technegas," *Nuklearmedizin,* vol. 40, no. 6, pp. 179–86, 2001, doi: 01060179 [pii].

[18] A. Le Duc-Pennec *et al.*, "Diagnostic accuracy of single-photon emission tomography ventilation/perfusion lung scan in the diagnosis of pulmonary embolism," *Chest,* vol. 141, no. 2, pp. 381–87, 2012, doi: 10.1378/chest.11-0090.

[19] M. Bajc *et al.*, "EANM guideline for ventilation/perfusion single-photon emission computed tomography (SPECT) for diagnosis of pulmonary embolism and beyond," *Eur J Nucl Med Mol Imaging,* 2019, doi: 10.1007/s00259-019-04450-0.
[20] C. G. Olsson *et al.*, "Outpatient tinzaparin therapy in pulmonary embolism quantified with ventilation/perfusion scintigraphy," *Med Sci Monit,* vol. 12, no. 2, pp. Pi9–13, 2006.
[21] J. E. Elf, J. Jogi, and M. Bajc, "Home treatment of patients with small to medium sized acute pulmonary embolism," *J Thromb Thrombolysis,* vol. 39, no. 2, pp. 166–72, 2015, doi: 10.1007/s11239-014-1097-y.
[22] B. N. Jobse, R. G. Rhem, C. A. McCurry, I. Q. Wang, and N. R. Labiris, "Imaging lung function in mice using SPECT/CT and per-voxel analysis," *PLoS One,* vol. 7, no. 8, p. e42187, 2012, doi: 10.1371/journal.pone.0042187.
[23] B. N. Jobse, R. G. Rhem, I. Q. Wang, W. B. Counter, M. R. Stampfli, and N. R. Labiris, "Detection of lung dysfunction using ventilation and perfusion SPECT in a mouse model of chronic cigarette smoke exposure," *J Nucl Med,* vol. 54, no. 4, pp. 616–23, 2013, doi: 10.2967/jnumed.112.111419.
[24] J. Jogi, M. Ekberg, B. Jonson, G. Bozovic, and M. Bajc, "Ventilation/perfusion SPECT in chronic obstructive pulmonary disease: an evaluation by reference to symptoms, spirometric lung function and emphysema, as assessed with HRCT," *Eur J Nucl Med Mol Imaging,* vol. 38, no. 7, pp. 1344–52, 2011, doi: 10.1007/s00259-011-1757-5.
[25] P. Norberg *et al.*, "Does quantitative lung SPECT detect lung abnormalities earlier than lung function tests? Results of a pilot study," *EJNMMI Res,* vol. 4, no. 1, p. 39, 2014, doi: 10.1186/s13550-014-0039-1.
[26] M. Bajc *et al.*, "Identifying the heterogeneity of COPD by V/P SPECT: A new tool for improving the diagnosis of parenchymal defects and grading the severity of small airways disease," *Int J Chron Obstruct Pulmon Dis,* vol. 12, pp. 1579–87, 2017, doi: 10.2147/copd.S131847.
[27] M. Bajc, H. Markstad, L. Jarenback, E. Tufvesson, L. Bjermer, and J. Jogi, "Grading obstructive lung disease using tomographic pulmonary scintigraphy in patients with chronic obstructive pulmonary disease (COPD) and long-term smokers," *Ann Nucl Med,* vol. 29, no. 1, pp. 91–99, 2015, doi: 10.1007/s12149-014-0913-y.
[28] A. Begic, E. Opankovic, V. Cukic, A. Lindqvist, M. Miniati, and M. Bajc, "Ancillary findings assessed by ventilation/perfusion tomography. Impact and clinical outcome in patients with suspected pulmonary embolism," *Nuklearmedizin,* vol. 54, no. 5, pp. 223–30, 2015, doi: 10.3413/Nukmed-0748-15-06.
[29] P. M. Scherer and D. L. Chen, "Imaging Pulmonary Inflammation," *J Nucl Med: Official publication, Society of Nuclear Medicine,* vol. 57, no. 11, pp. 1764–70, 2016, doi: 10.2967/jnumed.115.157438.
[30] D. J. Li, I. Stewart, K. A. Miles, and E. P. Wraight, "Scintigraphic appearances in patients with pulmonary infection and lung scintigrams of intermediate or low probability for pulmonary embolism," *Clin Nucl Med,* vol. 19, no. 12, pp. 1091–93, 1994, doi: 10.1097/00003072-199419120-00011.
[31] H. D. Sostman and A. Gottschalk, "The stripe sign: A new sign for diagnosis of nonembolic defects on pulmonary perfusion scintigraphy," *Radiology,* vol. 142, no. 3, pp. 737–41, 1982, doi: 10.1148/radiology.142.3.7063693.
[32] W. F. Friedman and E. Braunwald, "Alterations in regional pulmonary blood flow in mitral valve disease studied by radioisotope scanning. A simple nontraumatic technique for estimation of left atrial pressure," *Circulation,* vol. 34, no. 3, pp. 363–76, 1966, doi: 10.1161/01.cir.34.3.363.
[33] J. Jogi, M. Al-Mashat, G. Radegran, M. Bajc, and H. Arheden, "Diagnosing and grading heart failure with tomographic perfusion lung scintigraphy: Validation with right heart catheterization," *ESC Heart Fail,* vol. 5, no. 5, pp. 902–10, 2018, doi: 10.1002/ehf2.12317.
[34] J. Jogi, J. Palmer, B. Jonson, and M. Bajc, "Heart failure diagnostics based on ventilation/perfusion single photon emission computed tomography pattern and quantitative perfusion gradients," *Nucl Med Commun,* vol. 29, no. 8, pp. 666–73, 2008, doi: 10.1097/MNM.0b013e328302cd26.
[35] M. Pistolesi *et al.*, "Factors affecting regional pulmonary blood flow in chronic ischemic heart disease," *J Thorac Imaging,* vol. 3, no. 3, pp. 65–72, 1988.
[36] M. D. Mamlouk *et al.*, "Pulmonary embolism at CT angiography: Implications for appropriateness, cost, and radiation exposure in 2003 patients," *Radiology,* vol. 256, no. 2, pp. 625–32, 2010, doi: 10.1148/radiol.10091624.
[37] A. S. Coco and D. T. O'Gurek, "Increased emergency department computed tomography use for common chest symptoms without clear patient benefits," *J Am Board Fam Med,* vol. 25, no. 1, pp. 33–41, 2012, doi: 10.3122/jabfm.2012.01.110039.
[38] A. J. Einstein, K. W. Moser, R. C. Thompson, M. D. Cerqueira, and M. J. Henzlova, "Radiation dose to patients from cardiac diagnostic imaging," *Circulation,* vol. 116, no. 11, pp. 1290–305, 2007, doi: 10.1161/circulationaha.107.688101.
[39] J. Jogi, H. Markstad, E. Tufvesson, L. Bjermer, and M. Bajc, "The added value of hybrid ventilation/perfusion SPECT/CT in patients with stable COPD or apparently healthy smokers. Cancer-suspected CT findings in the lungs are common when hybrid imaging is used," *Int J Chron Obstruct Pulmon Dis,* vol. 10, pp. 25–30, 2015, doi: 10.2147/copd.S73423.
[40] ICRP, "Radiation dose to patients from radiopharmaceuticals, publication 53," ICRP, Oxford and New York, 53, 1988.
[41] J. J. Phillips, J. Straiton, and R. T. Staff, "Planar and SPECT ventilation/perfusion imaging and computed tomography for the diagnosis of pulmonary embolism: A systematic review and meta-analysis of the literature, and cost and dose comparison," *Eur J Radiol,* vol. 84, no. 7, pp. 1392–400, 2015, doi: 10.1016/j.ejrad.2015.03.013.
[42] A. M. Bucher *et al.*, "Systematic comparison of reduced tube current protocols for high-pitch and standard-pitch pulmonary CT angiography in a large single-center population," *Acad Radiol,* vol. 23, no. 5, pp. 619–27, 2016, doi: 10.1016/j.acra.2016.01.003.

[43] T. De Zordo *et al.*, "Comparison of image quality and radiation dose of different pulmonary CTA protocols on a 128-slice CT: High-pitch dual source CT, dual energy CT and conventional spiral CT," *Eur Radiol,* vol. 22, no. 2, pp. 279–86, 2012, doi: 10.1007/s00330-011-2251-y.

[44] I. A. Elbakri and I. D. C. Kirkpatrick, "Survey of clinical doses from computed tomography examinations in the Canadian province of Manitoba," *Radiat Prot Dosimetry,* vol. 157, no. 4, pp. 525–35, 2013, doi: 10.1093/rpd/nct168.

[45] C. R. Liang, P. X. H. Chen, J. Kapur, M. K. L. Ong, S. T. Quek, and S. C. Kapur, "Establishment of institutional diagnostic reference level for computed tomography with automated dose-tracking software," *J Med Radiat Sci,* vol. 64, no. 2, pp. 82–89, 2017, doi: 10.1002/jmrs.210.

[46] R. Smith-Bindman *et al.*, "Predictors of CT radiation dose and their effect on patient care: A comprehensive analysis using automated data," *Radiology,* vol. 282, no. 1, pp. 182–93, 2017, doi: 10.1148/radiol.2016151391.

[47] S. A. Astani, L. C. Davis, B. A. Harkness, M. P. Supanich, and I. Dalal, "Detection of pulmonary embolism during pregnancy: Comparing radiation doses of CTPA and pulmonary scintigraphy," *Nucl Med Commun,* vol. 35, no. 7, pp. 704–11, 2014, doi: 10.1097/mnm.0000000000000114.

[48] A. Sabarudin, Z. Mustafa, K. M. Nassir, H. A. Hamid, and Z. Sun, "Radiation dose reduction in thoracic and abdomen-pelvic CT using tube current modulation: A phantom study," *J Appl Clin Med Phys,* vol. 16, no. 1, p. 5135, 2014, doi: 10.1120/jacmp.v16i1.5135.

[49] J. Isidoro, P. Gil, G. Costa, J. Pedroso de Lima, C. Alves, and N. C. Ferreira, "Radiation dose comparison between V/P-SPECT and CT-angiography in the diagnosis of pulmonary embolism," *Phys Med,* vol. 41, pp. 93–96, 2017, doi: 10.1016/j.ejmp.2017.04.026.

[50] M. P. Revel *et al.*, "Breast dose reduction options during thoracic CT: Influence of breast thickness," *AJR Am J Roentgenol,* vol. 204, no. 4, pp. W421-8, 2015, doi: 10.2214/ajr.14.13255.

[51] T. Grüning *et al.*, "Diagnosing venous thromboembolism in pregnancy," *Br J Radiol,* vol. 89, no. 1062, p. 20160021, 2016, doi: 10.1259/bjr.20160021.

15 Myocardial Perfusion Imaging

Elin Trägårdh, David Minarik and Mark Lubberink

CONTENTS

15.1 INTRODUCTION

Coronary artery disease (CAD) – or ischemic heart disease (IHD) – is a disease where the blood flow to the heart muscle is reduced, most often due to buildup of plaque (atherosclerosis) in the coronary arteries. IHD can cause chest pain due to ischemia, often related to activity, or to myocardial infarction if the coronary arteries suddenly become blocked due to a plaque rupture. IHD is the leading cause of death for men and women worldwide, and it causes substantial disability and loss of productivity. The World Health Organization (WHO) predicts that the number of deaths due to IHD will increase worldwide. The cardiovascular mortality rate in Northern, Southern and Western Europe is relatively low compared with Central and Eastern Europe. Cardiovascular mortality rates for women are lower than those for men in all European countries [1]. Temporal trends suggest a decrease in the annual cardiovascular mortality rate [2]. However, the prevalence of a history of diagnosed IHD does not appear to have decreased, suggesting that the prognosis of those with IHD is improving. Improved sensitivity of diagnostic tools may additionally contribute to the contemporary high prevalence of diagnosed IHD.

The diagnosis of IHD involves clinical evaluation, including identifying risk factors and specific cardiac investigations such as stress testing or coronary artery imaging. These investigations may be used to confirm the diagnosis in patients with suspected IHD, assist in stratifying risk associated with IHD, and to evaluate efficacy of treatment. Many of the investigations used for diagnosis also offer prognostic information. Methods of detecting the presence and assessing the extent of IHD have become increasingly important in order to guide therapy.

Ischemia imaging has been regarded most appropriate in patients with intermediate pre-test probability (15-85%) of hemodynamically significant IHD. In asymptomatic patients or those with low or high pre-test probability, these tests are generally not recommended. The arsenal of imaging modalities to detect and manage patients with IHD is large: Coronary computed tomography angiography (CCTA), myocardial perfusion scintigraphy (MPS), positron emission tomography (PET), stress echocardiography, and cardiac magnetic resonance imaging (CMR). CCTA is the preferred test in patients with a lower range of clinical likelihood of IHD, no previous diagnosis of IHD, and characteristics associated with a high likelihood of good image quality. The non-invasive functional tests (MPS, PET, stress echocardiography, and CMR) for ischemia typically have better rule-in power, and may be preferred in patients

DOI: 10.1201/9780429489501-15

at the higher end of the range of clinical likelihood of IHD, depending on local expertise and availability [3]. In new guidelines from the European Society of Cardiology from 2019 [3], the terms IHD and CAD have been changed to chronic coronary syndrome, as opposed to acute coronary syndrome.

15.2 PATHOPHYSIOLOGY AND INDICATIONS

Myocardial ischemia occurs when the supply of oxygen to the myocytes is lower than the demand (Figure 15.1). The inotropic and chronotrophic states determine the myocardial demand. For example, during physical exercise, the increasing heart rate and blood pressure lead to an increased demand for oxygen. In normal conditions, the arterioles can dilate during exercise, and perfusion can increase from a resting value of 1 mL/min/g tissue to more than 3 mL/min/g. IHD is often caused by atherosclerotic stenosis in the coronary arteries, which increase the resistance in the vessels and limits the blood flow. The heart will compensate for this at rest by dilation of the arterioles, thereby preventing a flow decrease. This, however, means that part of the flow vasodilator reserve is already being used at rest, and the arterioles cannot dilate further during stress. Thus, there is a lower supply of oxygen to the myocytes at stress, but not at rest, which is why patients who have stress-induced ischemia often have normal perfusion at rest.

The use of radionuclide imaging, such as MPS and PET, is often part of the imaging of IHD. Stress and rest myocardial perfusion imaging can be performed to detect, localize, and quantify the degree of ischemia and infarction, as well as quantity the systolic function. Quantification of absolute myocardial perfusion can be done with PET. The results of an MPS or PET are related to the prognostic outcome of cardiac events (myocardial infarction and death). Therefore, MPS and PET are used to diagnose IHD and to guide treatment in patients with known IHD. They can also be used to determine the culprit lesion in order to only treat significant stenosis. MPS is a well-established technique and one of the most commonly used cardiac imaging modalities. There are international guidelines on when to use [3, 4], how to perform [5], and how to report [6, 7] MPS. PET is a newer nuclear medicine imaging technique. Its use in cardiology is

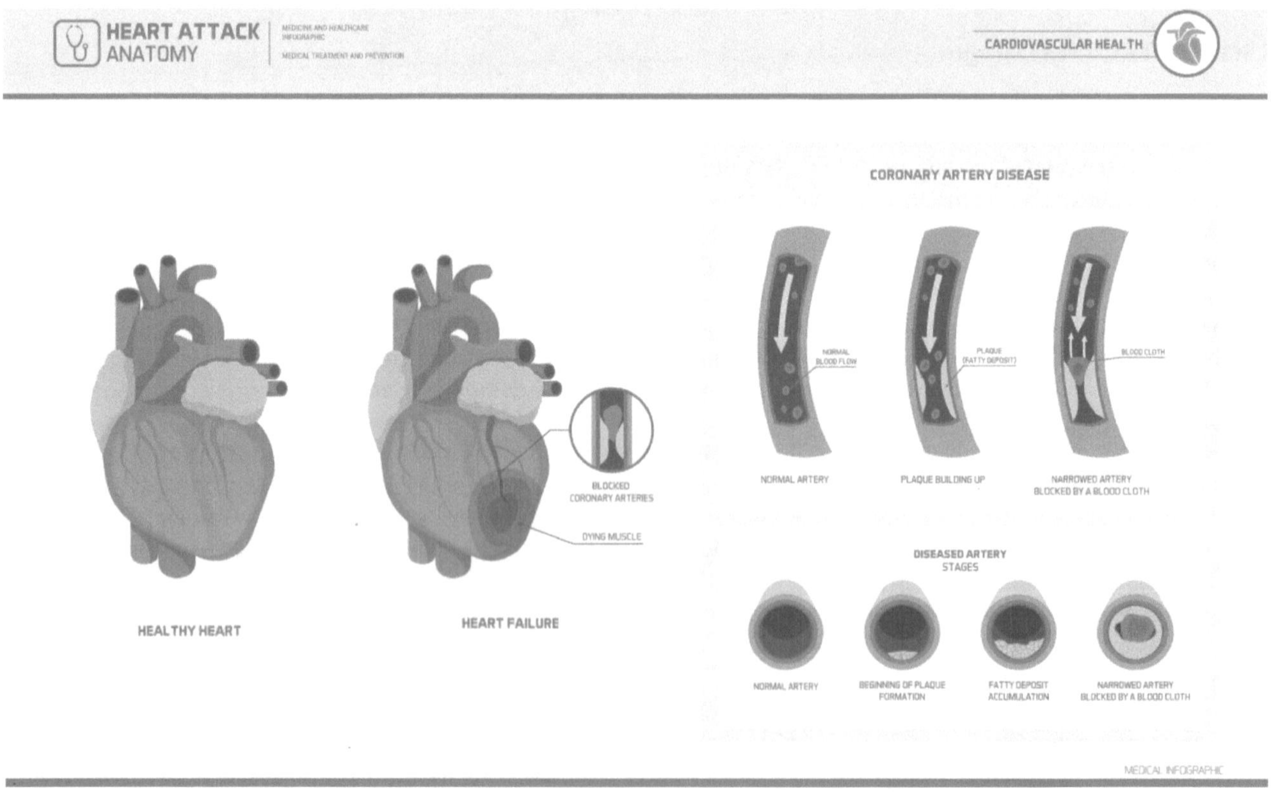

FIGURE 15.1 The build-up of an atherosclerotic plaque in a coronary artery can lead to ischemia if the blood supply is decreased. If the artery is completely blocked, for example if a plaque ruptures, a myocardial infarction occurs unless the blood flow is restored quickly. Colour image available at www.routledge.com/9781138593312.

increasing because of advantages compared with MPS, such as better image resolution and the possibility for quantification of absolute myocardial perfusion. However, the limited availability of the equipment (especially of the perfusion tracer) and higher costs are influencing the choice of method.

15.3 PROCEDURE

Images at stress and at rest should be obtained. It is preferred that the stress study be performed first, and if the results are normal, the rest study can be omitted. MPS studies can be performed as a 1-day or a 2-day protocol. If a 1-day protocol is used, in order to minimize the contribution of the residual activity from the first study to the second study, the second study should be performed at least 2 hours after the first and with a 3-fold increase in administered activity.

Stress test. In order to detect myocardial ischemia, a maximum increase in myocardial perfusion must be achieved. For MPS, the stress test may be undertaken by exercise, pharmacologically, or in combination. For PET, the stress test is performed with pharmaceuticals, due to the short half-life of PET perfusion pharmaceuticals.

Exercise test. A physical exercise is preferred whenever possible, since it provides additional information regarding heart performance. Different exercise modalities are available. Bicycle ergometry or treadmill exercise is normally used. The exercise stress test is performed stepwise starting around 25–50 W. Provided the stress test is not stopped because of limiting symptoms, a workload with a heart rate of at least 85 per cent of age-adjusted maximal predicted heart rate (220-age) should be achieved, otherwise the sensitivity for detecting ischemia is decreased. The radiopharmaceutical is injected at the highest stress level, which is maintained for 1–2 minutes. There are many absolute contraindications for a maximal exercise – for example acute coronary syndrome, acute pulmonary embolism, symptomatic severe aortic stenosis, acute myocarditis, and severe pulmonary hypertension, as well as relative contraindications such as left bundle branch block, ventricular paced rhythm, resting systolic or diastolic blood pressure >200/100 mmHg, recent stroke or inadequately controlled congestive heart disease. The electrocardiogram must be monitored continuously during the exercise test and for at least a few minutes of recovery, and the blood pressure should be controlled at least every 2–3 minutes during exercise.

Vasodilators. Adenosine, regadenoson or dipyramidole are used when an exercise test is not possible. Myocardium supplied by a significantly stenotic coronary artery has a reduced perfusion reserve compared with non-stenotic arteries, leading to heterogeneity of perfusion during vasodilation. This results in heterogeneous uptake of the radiopharmaceutical. Caffeine-containing beverages, foods and some medications (e.g. pain relievers) must be discontinued at least 12 h before vasodilator stress. An infusion pump is necessary for adenosine and dipyramidole administration at a constant infusion rate during 4–6 min, whereas regadenoson is administered intravenously by manual injection. Electrocardiographic and blood pressure monitoring should be carried out as with exercise testing. Low-level exercises can be performed in conjunction with a vasodilator stress test, since it reduces the side effects (flushing, dizziness, nausea, headache) and improves image quality due to lower extra-cardiac activity.

Catecholamines (dobutamine). Dobutamine is indicated in patients who cannot exercise and who have contraindications for vasodilators (for example high-degree atrioventricular block, severe chronic obstructive bronchospastic pulmonary disease). Dobutamine induces an increase in myocardial oxygen demand, increasing heart rate and blood pressure. Due to this effect, it causes coronary vasodilation similar to an exercise test. In areas supplied by significantly stenosed coronary arteries, the flow reserve is reduced. Dobutamine is infused incrementally until at least 85 per cent of the age-predicted maximal heart rate is reached. Contraindications are the same as for an exercise test, as well as β-blockers that cannot be discontinued before the stress test.

15.3.1 RADIOPHARMACEUTICALS

^{99m}Tc-labelled radiopharmaceuticals are normally used for MPS, whereas ^{201}Tl-labelled radiopharmaceuticals are nowadays not commonly used. Rubidium-82, ^{13}N-ammonia, ^{82}Rb-chloride and ^{15}O-water are PET-tracers.

- ^{99m}Tc-2-methoxyisobutylisonitrile (^{99m}Tc-sestamibi)
- ^{99m}Tc-1,2-bis[bis(2-ethoxyethyl) phosphino] ethane (^{99m}Tc-tetrofosmin)
- ^{201}Tl-chloride (Thallium-201)
- ^{82}Rb-chloride (Rubidium-82)
- ^{13}N-ammonia
- ^{15}O-water

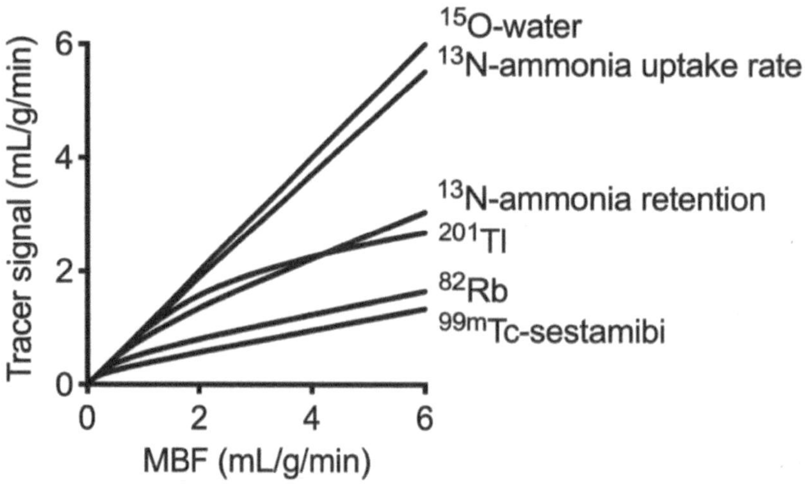

FIGURE 15.2 Relationship between uptake rate and MBF for the different radiopharmaceuticals. Only ^{15}O-water has a linear relationship between MBF and tracer signal. Adopted from [8].

Figure 15.2 shows the relationship between tracer signal (uptake rate) and myocardial blood flow (MBF) for each of these tracers. The fact that ^{15}O-water is freely diffusible and has a 100% extraction results in a linear relationship between MBF and signal, whereas all other tracers show a so-called "roll-off" effect due to a reduction of tracer extraction with increasing flow, resulting in a signal increasing less than proportionally with MBF. This has implications for sensitivity in detecting MBF defects, which will be discussed below. All tracers except ^{15}O-water are retained in the myocardial wall, which means that a scan performed several minutes after administration will give a qualitative image of the distribution of MBF. With ^{15}O-water, on the other hand, this is not possible, since both its uptake and its clearance are proportional to MBF. Hence, a fully quantitative analysis is necessary. Both ^{99m}Tc and ^{82}Rb are generator-produced, whereas both ^{13}N-ammonia and ^{15}O-water require the presence of an on-site cyclotron because of the 10 and 2 min half-lifes, respectively, of these tracers. This, and in case of ^{15}O-water the requirement of quantification, has been an obstacle for clinical introduction of ^{15}O-water and ^{13}N-ammonia. The availability of approved generators (at least in the USA) has made ^{82}Rb the most used PET tracer for myocardial perfusion imaging. However, small, dedicated cyclotrons with bed-side synthesis and administration units, together with automated software for quantification, are currently being introduced, which is expected to increase the clinical use of ^{15}O-water and ^{13}N-ammonia.

15.3.2 Image Acquisitions

MPS: Image acquisition with ^{99m}Tc-labelled radiopharmaceuticals starts 0,5-1 h after the injection, where the shorter time can be used when the patient has performed an exercise test. If image acquisition starts too early, the risk increases of high extra-cardiac uptake in the intestines, which can lead to image artefacts. If a standard system with a dual-head camera and a low-energy high-resolution collimator (LEHR) is used, the collimators are normally placed in a 90° setting. Standard procedure is that images are acquired with 60–120 projections over an 180° arc, starting from the 45° right anterior oblique position. Imaging time per projection needs to be optimized to the administered activity but is typically around 15 s when 600 MBq are administered. Preferably, a 64 x 64 or a 128 x 128 matrix should be used with a pixel size between 4 and 6 mm.

Other systems exist, such as dedicated heart systems with solid state detectors. These systems allow a reduction of the administered activity and/or the acquisition time. There are also cardiac specific collimators available, used on conventional SPECT systems, in which the holes in the collimator are converging with a changing focal point. The focal point at the centre of the collimator is short, changing to larger focal points (near parallel) at the edges of the collimator. This allows imaging the heart at the centre focal point without truncating extra-cardiac activity. One example is the SMARTZOOM IQ.SPECT system where, instead of a centre-of-rotation at the mechanical centre, a cardio-centric acquisition is used, that is, the heart is always at the centre-of-rotation, allowing for an approximately fourfold increase in sensitivity compared with LEHR collimators. Typically, 34 projections are acquired in an 180° arc, with the same starting point as for the standard system. As with the solid-state systems, this allows for a reduction in administered activity and/or acquisition time.

Usually, the images are gated to the electrocardiogram in order to calculate cardiac volumes and left ventricular (LV) ejection fraction (EF) (see below). The EF is a measurement of how much blood the left ventricle pumps out with each contraction, expressed as a percentage or a fraction.

Images are normally acquired with the patient in the supine position and, if available, a low-dose CT scan for attenuation correction. If no CT is available, it is possible as an adjunct to the supine position, to acquire images in the prone position, in order to better evaluate if defects seen on the images are due to attenuation or reduced perfusion. Attenuation defects when using a parallel hole collimator can typically be seen in non-corrected images in the inferior wall of the left ventricle in obese men, and in the anterior-septal part of the ventricle in women with large breasts (Figure 15.3). The attenuation pattern differs when using other types of collimators and the dedicated solid-state systems.

PET: Image acquisition with PET tracers, aiming to quantify MBF, is usually done using a dynamic scan beginning simultaneously with the start of the injection of the tracer. Dynamic scan durations are typically 4 min for ^{82}Rb and ^{15}O-water, and 5–10 min for ^{13}N-ammonia. In case of ^{82}Rb and ^{13}N-ammonia, this is then immediately followed by a 3–5 min gated static scan to obtain qualitative MBF images as well as cardiac volumes and ejection. Frame durations during the dynamic acquisition are typically increasing from 5 s during the first pass to 30–60 s for the last minute of the scan.

Gated list mode acquisition allows for combined acquisition of the dynamic and gated scan, with frame durations defined using post-processing. This is generally preferable because it allows for use of the total scan duration in the quantitative analysis, and it is essential for ^{15}O-water, where no static uptake image is available. Using list mode, a gated reconstruction of the first pass of the tracer through the heart can then be used to measure ejection fractions.

A fast-controlled bolus using an automated injection pump (for example, 1 mL/s during 5 s followed by a 35 mL saline flush at 2 mL/s) is preferred for quantification. When adenosine is used as stress agent, it is important that the adenosine infusion is continued through the entire dynamic ^{15}O-water or ^{82}Rb scan, since the short biological half-life of adenosine results in a rapid decrease in MBF after infusion is stopped, which affects quantitative analysis. Exercise stress imaging is not possible for quantitative measurement of MBF. For ^{15}O-water and ^{82}Rb, rest and stress imaging can be performed during the same session, with 10–15 min between administrations to allow for decay of the remaining activity from the rest scan. This allows for a complete rest-stress protocol to be performed within 25 min. For ^{13}N-ammonia, an interval of at least 50 min between stress and rest administrations is recommended because of the longer radioactive half-life. However, a short protocol where the amount of radioactivity administered during the stress scan is

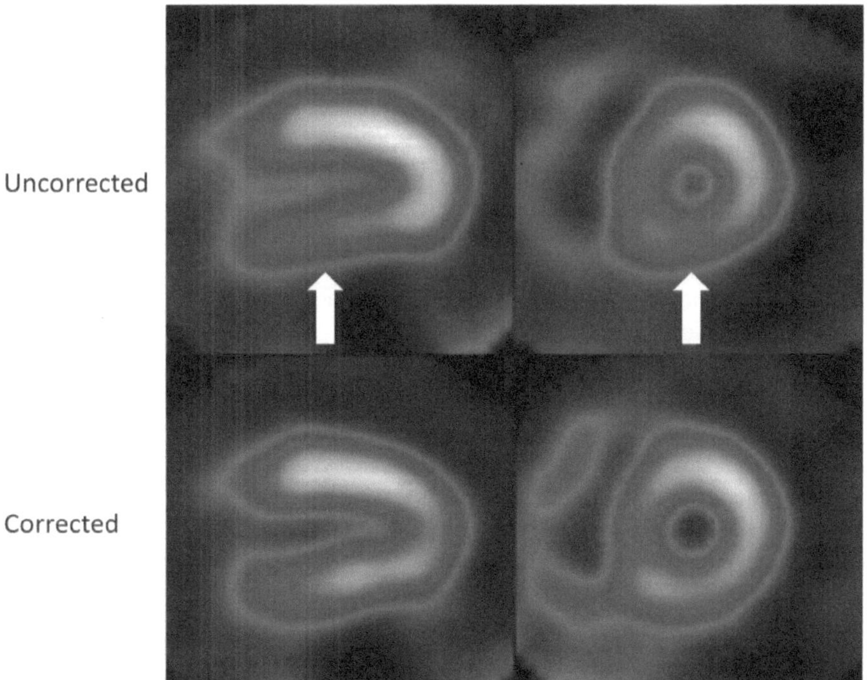

FIGURE 15.3 Attenuation defect in the inferior myocardial wall (white arrow) in an obese male. Colour image available at www.routledge.com/9781138593312.

TABLE 15.1
Properties of PET Tracers

	^{15}O-water	^{13}N-ammonia	^{82}Rb
Half-life (min)	2.03	9.97	1.25
Production	cyclotron	cyclotron	generator
Administered activity (MBq)	2x400	2×550 370+740 (short protocol)	2×1850 (2D) 2×550 (3D)
Radiation dose (mSv)	0.8	2.0	1.4-4.6
Static uptake images	no	yes	yes
Protocol duration (rest+ stress)	< 30 min	> 60 min < 30 min (short protocol)	< 30 min
Gating	First pass blood volume images	Late uptake images	Late uptake images
Quantification	+++	++	+

much higher than that during the rest scan, may be used. This, however, could affect quantitative accuracy if remaining activity from the rest scan is not properly accounted for. Table 15.1 summarizes the properties of PET tracers for myocardial perfusion imaging.

For computation of parametric perfusion images based on ^{15}O-water, showing MBF at the voxel level, it has been demonstrated that 3D imaging using a state-of-the-art PET scanner results in considerable higher quality images than 2D imaging, because of much higher noise-equivalent count rates (see Chapter 18 in Volume I) during the first pass of the tracer. If qualitative imaging is combined with quantitative assessment at regional levels only, choice of scanner is less critical but amounts of injected activity must be adapted to the acquisition method (2D/3D), see Table 15.1. As for SPECT, a 128x128 matrix with 4 mm voxels is sufficient for cardiac PET.

15.3.3 Reconstruction and Reorientation

Tomographic images must be reconstructed from the acquired projection data. Historically the reconstruction algorithm mostly used was the filtered back projection (FBP) algorithm but has now mostly been replaced with statistical reconstruction methods such as the ordered subset expectation maximization (OSEM) algorithm [9] or variants thereof. The OSEM algorithm and other iterative methods have been adapted for the specific heart cameras and collimators [10, 11]. An algorithm developed for the SMARTZOOM collimator is derived from an objective function based on a modified chi-squared distribution instead of the Poisson distribution, which the OSEM algorithm is derived from, and optimized with a gradient descent algorithm instead of the expectation maximization algorithm [12]. To facilitate an easier review of the images by the physician, the reconstructed transaxial images are then re-orientated with a new z-axis parallel to long axis of the left ventricle yielding images in the short axis (SA) plane of the LV (Figure 15.4). Coronal and sagittal views then represent horizontal and vertical planes with one axis parallel to the long axis of the LV, also called horizontal (HLA) and vertical (VLA) long axis images (Figure 15.5).

15.3.3.1 Cardiac Volumes and Ejection Fraction

Electrocardiogram (ECG) gating of the image acquisition allows for quantitative assessment of LV function. A 3-lead ECG is normally used and patients with a fairly regular heart rhythm can be studied. The fundamental principle of ECG gating is that one cardiac cycle is divided into 8 or 16 different imaging frames with the same duration. Data from each imaging frame is collected over multiple cardiac cycles and saved separately in 8 or 16 different bins. When all data in each bin are added, every image represents a specific phase of the cardiac cycle and can be viewed as a cine-loop. End-systolic and end-diastolic volumes (ESV and EDV) are calculated and from this the LV EF is calculated.

15.4 EVALUATION OF IMAGES

Images are evaluated visually, semi-quantitatively (MPS) and quantitatively (PET). Myocardial infarction is diagnosed when reduced radiotracer uptake (and preferable reduced wall motion) is present both at rest and stress. Myocardial

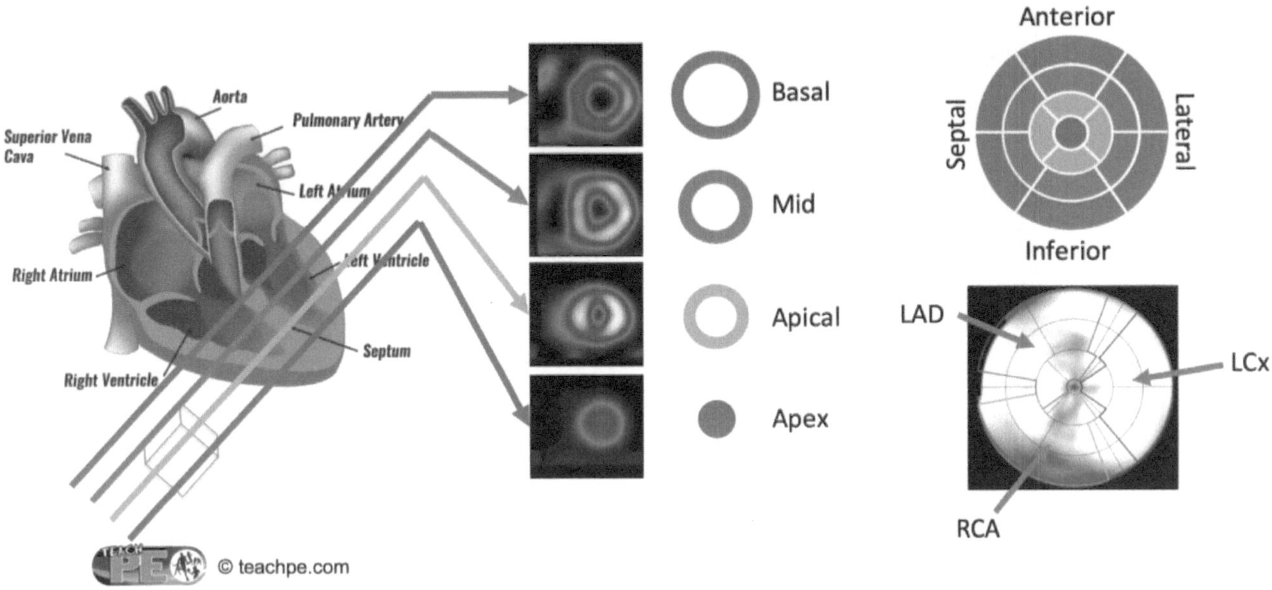

FIGURE 15.4 The figure to the left shows different SA slices through the left ventricle. The slices can also be visualized as a so-called polar plot (right images), divided into 17 segments (see below, "evaluation of images"). The figure in the bottom right corner shows the corresponding coronary artery territories. LAD = left anterior descending artery, LCx = left circumflex artery, RCA: right coronary artery. Colour image available at www.routledge.com/9781138593312.

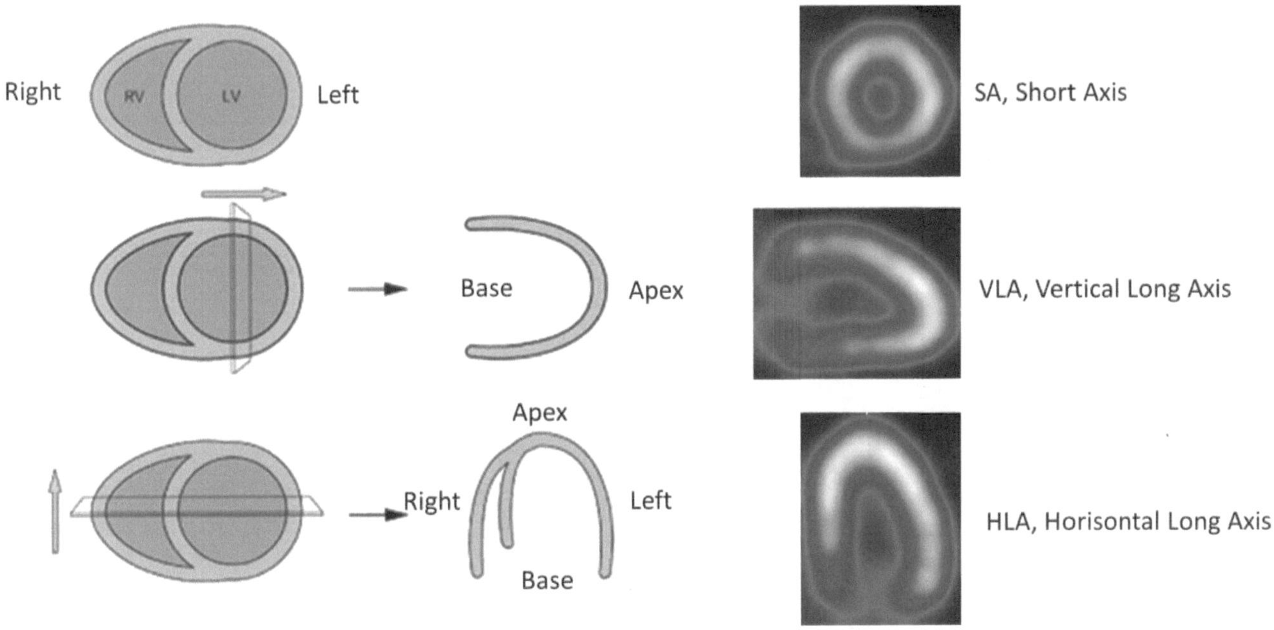

FIGURE 15.5 Images are shown as multiple slices throughout the left ventricle in the SA, VLA and HLA planes. The figure shows how the VLA and HLA planes are created. Colour image available at www.routledge.com/9781138593312.

ischemia is diagnosed when the radiotracer activity is reduced during the stress examination, but not during the rest examination. The location, severity and extent of ischemia and infarction are noted and related to coronary artery territories. For this purpose, the radiotracer activity is related to the 17 segments of the heart, established by the American Heart Association. Figure 15.6 shows an example of a patient with ischemia, examined with MPS.

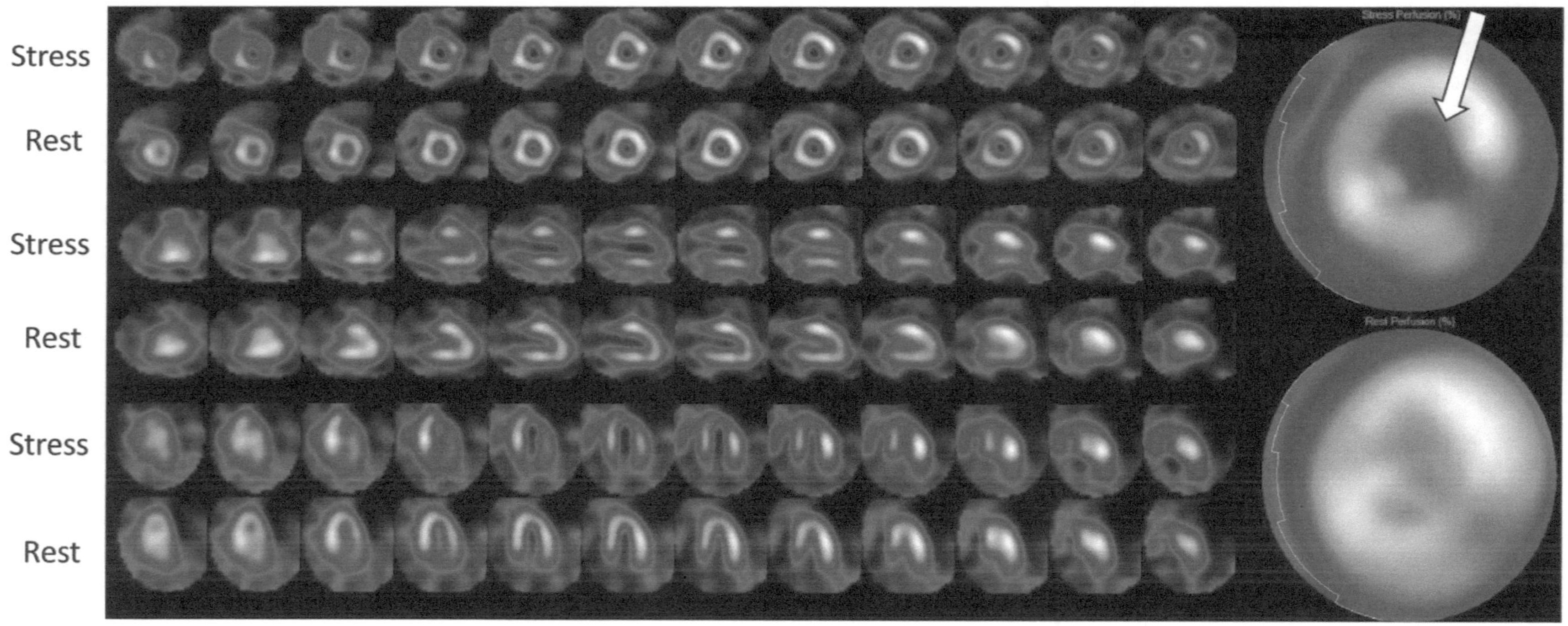

FIGURE 15.6 A patient with ischemia in the apex (white arrow shows reduced perfusion in the stress image). The figure shows SA slices in stress and rest (upper two rows), HLA slices (middle two rows) and VLA slices (lower two rows) of the left ventricle. The figures to the right are so called polar plot images, explained in Figure 15.4. The upper polar plot represents the stress study and the lower the rest study. Colour image available at www.routledge.com/9781138593312.

15.4.1 Normal Databases

In order to analyse the images, the perfusion images from a patient are compared to a normal database derived from patients with a low pre-test probability of IHD. It has previously been shown that there are major differences between United States and Japanese normal databases, particularly in the apex and in the anterior wall in women and in the inferior wall in men [13]. Also, different gamma camera types and reconstruction protocols yield differences in normal databases. In an ideal situation, every clinic should create its own normal databases, specific for gender, camera type, radiopharmaceutical, reconstruction, and attenuation correction system used.

15.4.2 Scoring Parameters

Software for the semi-quantitative assessment of perfusion compare the radiotracer uptake in the 17 segments of the images obtained from the patient with a normal database. Scores corresponding to the perfusion is obtained in each segment and the final overall score considers both the extent and the severity of ischemia or infarction in relation to the 17 segments. A score of 0 is normal perfusion. Mild and moderate perfusion deficit is indicated by 1 and 2 points. A score of 3 means a significant perfusion impairment and a score of 4 means no perfusion. The scoring is performed in both the stress and the rest images. The total score in the stress image is called summed stress score (SSS) and for the rest study summed rest score (SRS). The difference between the SSS and the SRS is called summed difference score (SDS). From this, the percentage of ischemia in the left ventricle can be calculated. Studies have shown that patients with an ischemic burden of less than 10 per cent should receive medical treatment only, whereas patients with ischemia of more than 10 per cent of the left ventricle benefit from revascularization through percutaneous coronary intervention (PCI) or coronary artery bypass grafting (CABG) [14].

15.4.3 Quantification of Myocardial Perfusion

Absolute MBF can be calculated using compartment modelling based on the dynamic PET scan (for details see Chapter 5 in Volume II), either at a regional level or for individual voxels, resulting in parametric images showing MBF at the voxel level [8]. The outcome measure of compartment modelling is K_1, the uptake rate of the tracer into myocardial tissue in mL per gram tissue per minute. For ^{15}O-water, K_1 (but also the clearance rate k_2) is identical to MBF, but for other tracers the relations in Figure 15.2 have to be used to convert K_1 into MBF. It has been shown that stress-MBF based on ^{15}O-water PET, using a cut-off value of 2.3 mL/g/min as threshold for normal MBF, has a higher accuracy than SPECT, CT coronary angiography, or PET coronary flow reserve in the detection of IHD using fractional flow reserve as reference standard [15]. Quantitative assessment of MBF allows for detection of balanced ischemia and microvascular disease, which cannot be seen with visual image analysis alone. An example of this is shown in Figure 15.7.

As can be seen in Figure 15.2, the relation between tracer signal and MBF levels out for MBF values above 1.5 mL/g/min for ^{82}Rb, much like for ^{99m}Tc-sestamibi. This implies that regional subtle decreases in stress MBF around 2 mL/g/min are likely harder to find, both visually and quantitatively, with SPECT and ^{82}Rb-PET than with ^{15}O-water-PET or ^{13}N-ammonia-PET. Examples of this have been shown for SPECT versus ^{15}O-water [16]. It should be noted that quantitative measurement of MBF using dynamic SPECT is also possible. However, in addition to the technical challenges associated with quantitative SPECT imaging, this does not overcome the limitations associated with the poor extraction of ^{99m}Tc-sestamibi at high flows.

15.4.3.1 Phase Analysis

Left ventricular mechanical dyssynchrony (LVMD) is defined by a difference in the timing of mechanical contraction between different segments of the left ventricle. LVMD can lead to heart failure [17]. Although echocardiography is the most used modality to assess LVMD, it is also possible to use gated MPS or PET. It is hoped that assessing LVMD can help in patient selection for cardiac resynchronization therapy (CRT), but so far it is not incorporated in the guidelines for CRT. LVMD by MPS or PET is calculated by creating images of the LV that correspond to sequential time points in the cardiac cycle and determine the variability in the timing in which different LV segments contract. Usually, the histogram phase bandwidth (BW), which is the time range during which 95 per cent of the left ventricle is initiating contraction, and the phase standard deviation (SD), which is the degree of heterogeneity in the onset of mechanical contraction, are calculated [18]. The parameters are obtained from standard software packages, but are normally not included in the routine clinical report of the MPS or PET examinations, although it contains prognostic information in patients with heart failure as well as coronary artery disease [19, 20].

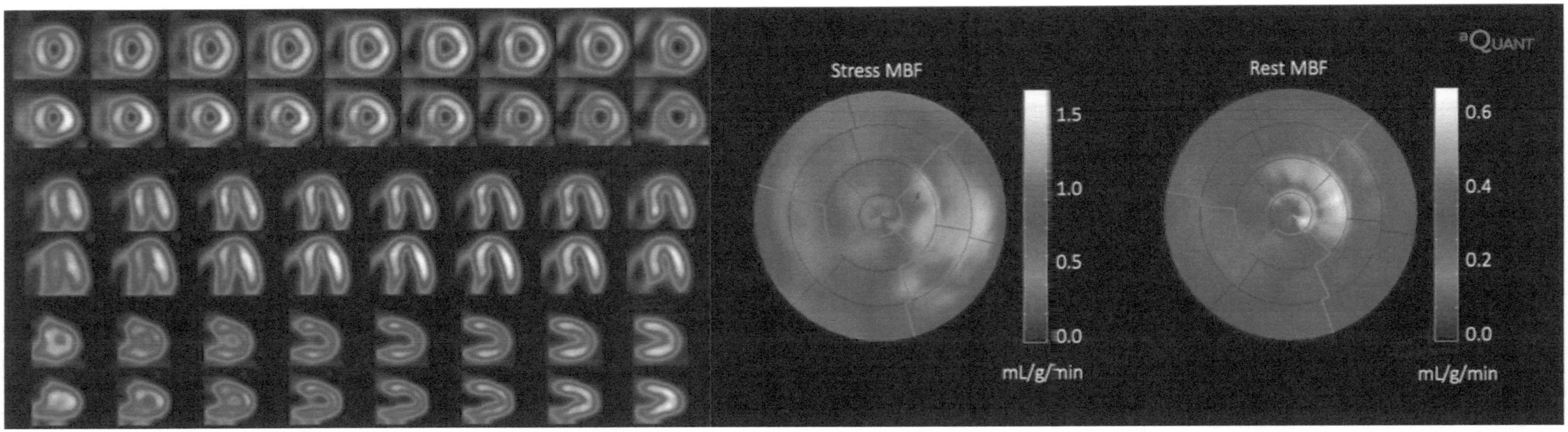

FIGURE 15.7 An example of a visually normal scan of a 65 y.o. female with angina and multiple risk factors (left), with associated polar maps (right) showing balanced ischemia with stress MBF below 1.5 mL/g/min in the entire left ventricular wall. Colour scales of stress and rest short and long axis images match those in the respective polar maps. Colour image available at www.routledge.com/9781138593312.

Source: Images courtesy Dr. Jens Sörensen, Uppsala University Hospital.

15.5 FUTURE OUTLOOK

Testing for ischemia by non-invasive imaging is not expected to decrease, but rather the opposite since an exercise test is no longer generally recommended for diagnosing ischemia [3]. In European guidelines, CCTA is recommended in patients with lower pre-test probability of IHD. The pre-test probability is estimated from symptoms, sex and age, and has been revised based on new populatoin studies and found to be lower than previously. CCTA could thus be expected to increase. However, the guidelines stress the importance of taking local expertise and availability into account when selecting the test for diagnosing IHD, why any transition towards one test or another will take time. No clear recommendations regarding when to use MPS, PET, stress echocardiography or CMR are given. Since the prevalence of chronic coronary syndrome is increasing, non-invasive imaging to evaluate new symptoms is expected to rise. As the availability of PET/CT scanners and suitable PET radiopharmaceuticals increase, the use of PET in this population is also expected to rise, especially when it is important to quantify the perfusion. For MPS, the use of cardiac-specific systems, as opposed to general gamma cameras, are expected to increase, at least in departments with a high number of examinations performed annually.

15.6 AVAILABLE GUIDELINES

Several European and US guidelines regarding how to perform myocardial perfusion studies as well as management of acute and chronic coronary syndromes exist. Here are some key guidelines listed:

- EANM procedural guidelines for radionuclide myocardial perfusion imaging with SPECT and SPECT/CT: 2015 revision
- EANM procedural guidelines for PET/CT quantitative myocardial perfusion imaging
- EANM procedural guidelines for myocardial perfusion scintigraphy using cardiac-centred gamma cameras
- ESC guidelines for the diagnosis and management of chronic coronary syndromes

REFERENCES

[1] WHO, "Cardiovascular diseases, deaths per 100 000. Data by country." http://apps.who.int/gho/data/node.main.A865CARDIOVASCULAR?lang=en.

[2] V. L. Roger *et al.*, "Heart disease and stroke statistics – 2012 update: A report from the American Heart Association," *Circulation,* vol. 125, no. 1, pp. e2-e220, 2012, doi: 10.1161/CIR.0b013e31823ac046.

[3] J. Knuuti *et al.*, "2019 ESC guidelines for the diagnosis and management of chronic coronary syndromes," *Eur Heart J,* 2019, doi: 10.1093/eurheartj/ehz425.

[4] R. C. Hendel *et al.*, "ACCF/ASNC/ACR/AHA/ASE/SCCT/SCMR/SNM 2009; *Appropriate Use Criteria for Cardiac Radionuclide Imaging/* Report of the American College of Cardiology Foundation Appropriate Use Criteria Task Force, the American Society of Nuclear Cardiology, the American College of Radiology, the American Heart Association, the American Society of Echocardiography, the Society of Cardiovascular Computed Tomography, the Society for Cardiovascular Magnetic Resonance, and the Society of Nuclear Medicine," *Circulation,* vol. 119, no. 22, pp. e561–87, 2009, doi: 10.1161/CIRCULATIONAHA.109.192519.

[5] H. J. Verberne *et al.*, "EANM procedural guidelines for radionuclide myocardial perfusion imaging with SPECT and SPECT/CT: 2015 revision," *Eur J Nucl Med Mol Imaging,* vol. 42, no. 12, pp. 1929–40, 2015, doi: 10.1007/s00259-015-3139-x.

[6] E. Tragardh *et al.*, "Reporting nuclear cardiology: A joint position paper by the European Association of Nuclear Medicine (EANM) and the European Association of Cardiovascular Imaging (EACVI)," *Eur Heart J Cardiovasc Imaging,* vol. 16, no. 3, pp. 272–79, 2015, doi: 10.1093/ehjci/jeu304.

[7] R. C. Hendel *et al.*, "American Society of Nuclear Cardiology Consensus Statement: Reporting of radionuclide myocardial perfusion imaging studies," *J Nucl Cardiol,* vol. 13, no. 6, pp. e152-56, 2006, doi: S1071-3581(06)00606-4.

[8] R. Sciagra *et al.*, "EANM procedural guidelines for PET/CT quantitative myocardial perfusion imaging," *Eur J Nucl Med Mol Imaging,* 2020, doi: 10.1007/s00259-020-05046-9.

[9] H. M. Hudson and R. S. Larkin, "Accelerated image reconstruction using ordered subsets of projection data," *IEEE Trans Med Imaging,* vol. 13, no. 4, pp. 601–9, 1994, doi: 10.1109/42.363108.

[10] M. Bocher, I. M. Blevis, L. Tsukerman, Y. Shrem, G. Kovalski, and L. Volokh, "A fast cardiac gamma camera with dynamic SPECT capabilities: Design, system validation and future potential," *Eur J Nucl Med Mol Imaging,* vol. 37, no. 10, pp. 1887–902, 2010, doi: 10.1007/s00259-010-1488-z.

[11] S. S. Gambhir *et al.*, "A novel high-sensitivity rapid-acquisition single-photon cardiac imaging camera," *J Nucl Med,* vol. 50, no. 4, pp. 635–43, 2009, doi: 10.2967/jnumed.108.060020.

[12] H. Vija, "Introduction to xSPECT* technology: Evolving multi-modal SPECT to become context-based and quantitative," *Siemens Medical Solutions USA Molecular Imaging,* 2013.

[13] K. Nakajima *et al.*, "The importance of population-specific normal database for quantification of myocardial ischemia: Comparison between Japanese 360 and 180-degree databases and a US database," *J Nucl Cardiol,* vol. 16, no. 3, pp. 422–30, 2009, doi: 10.1007/s12350-009-9049-1.

[14] R. Hachamovitch *et al.*, "Impact of ischaemia and scar on the therapeutic benefit derived from myocardial revascularization vs. medical therapy among patients undergoing stress-rest myocardial perfusion scintigraphy," *Eur Heart J,* vol. 32, no. 8, pp. 1012–24, 2011, doi: 10.1093/eurheartj/ehq500.

[15] I. Danad *et al.*, "Comparison of coronary CT Angiography, SPECT, PET, and Hybrid imaging for diagnosis of ischemic heart disease determined by fractional flow reserve," *JAMA Cardiol,* vol. 2, no. 10, pp. 1100–7, 2017, doi: 10.1001/jamacardio.2017.24712648688.

[16] W. J. Stuijfzand *et al.*, "Value of hybrid imaging with PET/CT to guide percutaneous revascularization of chronic total coronary occlusion," *Curr Cardiovasc Imaging Rep,* vol. 8, no. 7, p. 26, 2015, doi: 10.1007/s12410-015-9340-2.

[17] M. Fudim, F. Dalgaard, M. Fathallah, A. E. Iskandrian, and S. Borges-Neto, "Mechanical dyssynchrony: How do we measure it, what it means, and what we can do about it," *J Nucl Cardiol,* 2019, doi: 10.1007/s12350-019-01758-0.

[18] J. Chen *et al.*, "Onset of left ventricular mechanical contraction as determined by phase analysis of ECG-gated myocardial perfusion SPECT imaging: Development of a diagnostic tool for assessment of cardiac mechanical dyssynchrony," *J Nucl Cardiol,* vol. 12, no. 6, pp. 687–95, 2005, doi: 10.1016/j.nuclcard.2005.06.088.

[19] M. Fudim *et al.*, "The prognostic value of diastolic and systolic mechanical left ventricular dyssynchrony among patients with coronary artery disease and heart failure," *J Nucl Cardiol*, 2019, doi: 10.1007/s12350-019-01843-4.

[20] M. Fudim *et al.*, "The prognostic value of diastolic and systolic mechanical left ventricular dyssynchrony among patients with coronary heart disease," *JACC Cardiovasc Imaging,* vol. 12, no. 7 Pt 1, pp. 1215–26, 2019, doi: S1936-878X(18)30467-4 [pii]10.1016/j.jcmg.2018.05.018.

16 Infection and Inflammation

Erik H. J. G. Aarntzen and Andor W. J. M. Glaudemans

CONTENTS

Inflammation is part of the biological response of body tissues to potentially harmful stimuli, such as infections causing microorganisms (bacteria, fungi, viruses, and parasites). As such, inflammation aims to eliminate the cause, prevent further danger, and start tissue repair. The complex dynamics and contributions of different immune cell populations serves the delicate balance of tissue homeostasis. During our lifespan, the immune system faces the tremendous task of counteracting the continuous invasion of microorganisms over large surface areas. As too little or too severe, or inappropriate timing or duration may result in progressive tissue destruction, the immune system is a key component in a wide variety of clinical conditions.

In this chapter, we will discuss how clinical imaging can play a role in the assessment of inflammation, in particular when occurring with an infection caused by invading microorganisms. Figure 16.1 provides a framework that illustrates the different phases, dynamics, and immune cell populations involved in an ensuing inflammatory response – highlighting current targets for imaging.

DOI: 10.1201/9780429489501-16

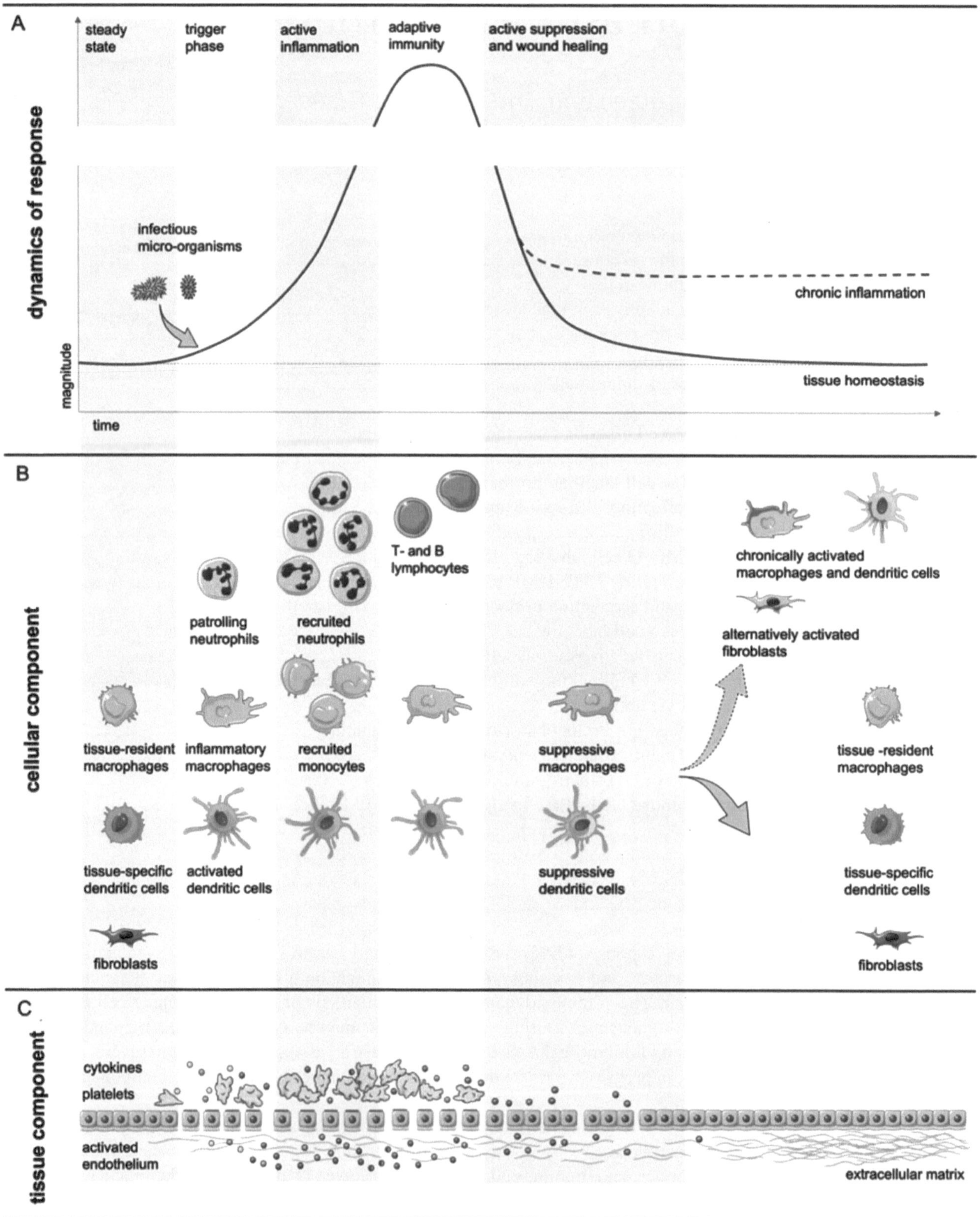

FIGURE 16.1 Inflamopmatory response at a glance. Colour image available at www.routledge.com/9781138593312.

16.1 INFLAMMATORY RESPONSES AT A GLANCE

16.1.1 Triggering an Immune Response

Although inflammatory responses can be triggered by physical or chemical stimuli, with in general overlapping characteristics, this chapter focuses on the response on pathogenic stimuli such as bacteria, viruses, and parasites. Pathogenic microorganisms elicit an immunological response once they have crossed the epithelial barriers – for example, skin, mucosal lining of the gastro-intestinal tract, or the respiratory system. At these large surface areas, cells of the innate immune system, such as neutrophils and tissue resident macrophages (e.g. Kupffer cells, Langerhans cells, alveolar macrophages) are actively surveilling [1-3]. Although these cells lack the specificity of the adaptive immune system, they express pattern-recognition receptors that recognize classes of molecules present on pathogens. For example, toll-like receptors (TLRs) recognize molecular patterns that are not found in normal vertebrates – for example, lipopolysaccharide (LPS), a component of bacterial cell wall that is recognized by TLR-4. Mannose receptors are expressed on macrophages to recognize sugar molecules present on most bacteria and some viruses. Scavenger receptors bind negatively charged cell-wall components from gram-positive bacteria, such as lipoteichoic acid.

These innate immune cells have two tasks; (1) constrain spread of the pathogen and (2) initiate an inflammatory cascade. Ligation of most of the cell-surface receptors leads to phagocytosis – for example, surrounding the pathogen with cell membrane and subsequent internalization in a membrane-bound vesicle called phagosomes, for subsequent killing in the intra-cellular milieu [4]. Furthermore, macrophages and neutrophils have membrane-bound granules, called lysosomes, that contain toxic products – for example, nitric oxide (NO), oxygen radicals or hydrogen peroxide – which can be released to destroy pathogens.

In particular, macrophages release soluble factors, for example, prostaglandins, leukotrienes, and platelet-activation factor, or chemokines, which facilitate the recruitment of other immune cells [5–7]. For example, tumour-necrosis factor α (TNFα) is an activator of endothelial cells; interleukin-8 (CXCL8) is involved in the recruitment of neutrophils to the site of infection, and interleukin-1β (IL-1β) and interleukin-6 (IL-6) induce systemic acute-phase responses in the liver. Antibody-antigen complexes and surface molecules on pathogens can induce the activation of other, non-eukaryotic, components of the immune system; the complement system, and platelets [8, 9]. The complement system consists of plasma proteins that react to mark pathogens for phagocytosis, a process called opsonization, and help exaggerate immune response.

16.1.2 Recruitment of Immune Cells

The production of inflammatory cytokines facilitates the recruitment of additional immune cells, dilatation of local small blood vessels, and increased expression of adhesion molecules of endothelial cells [5, 6, 9]. The blood vessels also become more permeable, which allows proteins from the plasma and fluids to leak into the interstitial tissue. Altogether, these effects result in tissue changes that represent the classical signs of inflammation: heat (calor), pain (dolor), redness (rubor) and swelling (edema).

16.1.3 Involvement of the Adaptive Immune System

In addition to direct recognition and killing by cells of our innate immune system, T- and B-cells can be mobilized when antigen-specific immune responses are required. Viruses, as well as some bacteria, survive and replicate in self-host cells and thereby escape from recognition by innate immune cells. Although macrophages and tissue dendritic cells are mainly phagocytic, they can be activated to express co-stimulatory molecules and major histocompatibility complexes (MHC) containing antigens derived from the infectious microorganism. In response to chemokines like CCL21, these antigen-presenting cells home to draining lymph nodes, bridging innate and adaptive immunity [10, 11].

Naïve T- and B-cells circulate from the bloodstream and enter lymphoid tissues to screen for their cognate antigen [12]. Naïve T- and B-cells that recognize their specific antigen, in the proper context of co-stimulation and inflammatory cytokines, will clonally expand to generate large numbers of antigen-specific effector cells. The armament of the adaptive immune system includes CD8+ cytotoxic T-cells, capable of killing virus-infected cells, CD4+ helper T-cells (Th-cells); grossly, Th1-cells activate macrophages and Th2-cells activate B-cells, as well as long-lived memory T cell populations [13, 14].

16.1.4 Systemic Inflammatory Responses

Macrophage derived pro-inflammatory cytokines, mainly TNFα, IL-6 and IL-1β, act not only on immune cells with professional antigen presenting capacity, but also systemically. The liver responds by producing acute-phase proteins, stimulating the activation of complement system, and fibrinogen. In case of a systemic inflammatory response, called sepsis, there is widespread leakage from blood vessels, leading to edema, decreased blood volume, collapse of vessels and disseminated intravascular coagulation [15]. This eventually may lead to organ failure and even death.

16.2 TRANSLATION TO TARGETS FOR IMAGING

16.2.1 The Unique Features of Inflammation

Some features of inflammation have major consequences for the requirements of in vivo imaging: The hematogenous spreading microorganisms and highly mobile immune cell subsets require a wide, that is, total-body, field of view. Imaging modalities with limited field of view or penetration depth, such as ultrasound, fluorescence, and bioluminescence are, therefore, less suitable. Next, inflammatory processes span a broad range of magnitudes. Viral or parasitic infections go relatively silent, whereas gram-positive cocci result in fulminant and purulent inflammation that can rapidly end in sepsis and death. The sensitivity of the imaging system should allow coverage of the whole spectrum, rendering highly sensitive techniques such as PET more attractive than other techniques, although its use can be limited by exposure limits and costs. Lastly, the complexity of an ongoing immune response is enormous, involving many soluble factors and cell subsets, challenging the specificity of the imaging system. Therefore, one should realize that labelling specific subsets of cells or imaging a single receptor may provide only fractional information.

16.2.2 Choice of Target

Closely related to the imaging modality of choice is the question of choosing suitable targets for imaging inflammation (Figure 16.2). Directly imaging the presence, numbers, and dynamics of pathogens would allow study of the interaction of pathogens with the immune system [16] and possibly guide antibiotic treatments. Several pathogens have been visualized using substrates for virus or bacteria-specific enzymes [17, 18], radiolabelled antibiotics [19, 20], pathogen specific antibodies, and antibody fragments, or via targeting bacterial products. However, most studies are performed in preclinical models [21] and are beyond the scope of this chapter.

Several studies have investigated radiolabelled cytokines and chemokines (reviewed in [21]) to image the presence of cells expressing specific receptors – for example, ^{99m}Tc-labelled interleukin-2, to visualize activated lymphocytes [22], or ^{99m}Tc-labelled interleukin-8 to image neutrophil recruitment [22, 23]. Radiolabelled antibodies against cell type, specific cell surface molecules, have also been designed – for example, anti-CD4 and anti-CD3 antibodies [24] or anti-CD56 to target NK-cells [25], but these are not available for routine clinical use. Nevertheless, ^{111}In-labelled polyclonal IgG has been applied in the clinic for infection imaging [26]. Several disadvantages exist for antibody-based approaches: Often the expression of cell surface molecules is dynamic; molecules are internalized and sometimes re-expressed; and the expression is often not specific but part of physiological processes as well. Moreover, radiolabelled antibodies have slow kinetics, that is, require days to accumulate in tissues, whereas many inflammatory steps go much faster. High background in liver, spleen, bone marrow, and blood circulation, due to unspecific uptake by phagocytic cells, is another drawback.

The increased trafficking of innate immune cells, mainly neutrophils and monocytes, towards sites of infection inevitably represents a "point-of-no-return" in ensuing inflammatory responses. These cells are abundantly present in the circulation and for this reason, ex vivo radiolabelled autologous leukocytes are an established tool to visualize infection using scintigraphy, which will be discussed in this chapter.

Changes in tissue characteristics can often be visualized by clinical imaging modalities using intravenous contrast, for example, contrast-enhanced computed tomography (CT) using iodine-contrast or dynamic contrast enhanced (DCE) magnetic resonance imaging (MR) using gadolinium-compounds. Enhanced perfusion and permeability are assessed by more rapid and more abundant passage of intravenous contrast than under physiological conditions. Although these imaging modalities have many favourable characteristics, the structural changes to the tissue follow functional changes in cell behaviour and occur relatively late during the process. The lower sensitivity probably does not allow visualization of the small changes in tissue microstructure at earlier time points. Moreover, given the diverse nature of inflammatory responses, changes in tissue structure do not provide information on the type of inflammation.

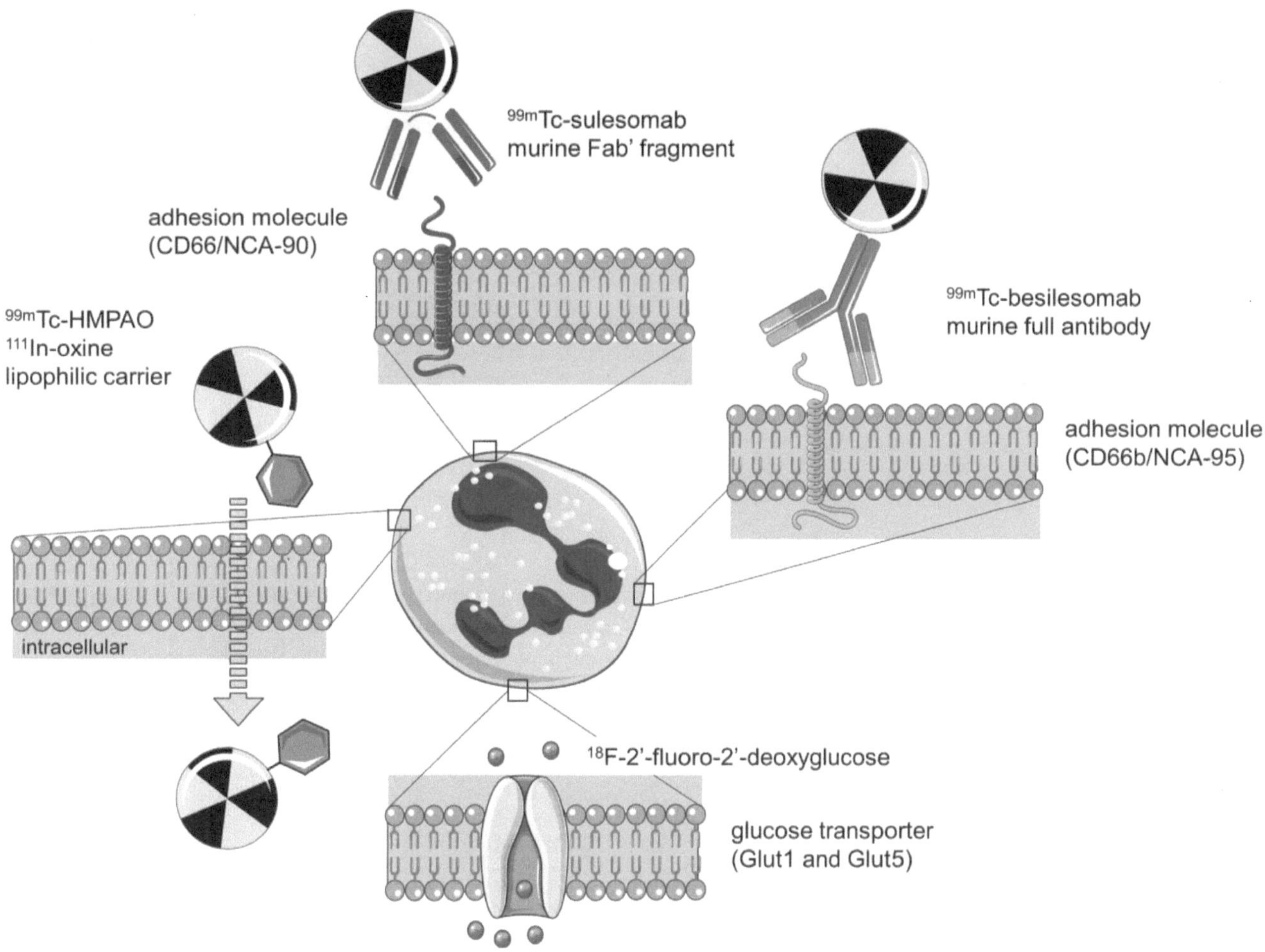

FIGURE 16.2 Clinically available labelling techniques for leukocyte imaging. Colour image available at www.routledge.com/9781138593312.

The metabolic switch upon activation of immune cells [27, 28] has a few characteristics that make it more attractive for imaging than other targets. In general, metabolism involves smaller molecules as substrate that rapidly diffuse over the cells and tissues, and thus are less dependent on tissue structure and perfusion; and metabolic substrates are often abundantly present. Imaging metabolic processes would result in the most minimal perturbation of the system, other than in vivo or ex vivo labelling functional receptors or cells. Moreover, the sensitivity of clinical PET imaging systems, has greatly improved, enabling the measurement of relatively small amounts of substrate.

16.3 EX VIVO CELL LABELLING: PASSIVE DIFFUSION OVER THE CELL MEMBRANE.

Lipophilic agents like oxine, hexamethyl-propylene amine oxime (HMPAO), pyruvaldehyde-bis-N4-methylthiosemicarbazone (PTSM), and tropolone passively diffuse over the cell membrane, complexes fall apart in the cytoplasma by reduction, and the radionuclide dissociates and binds to intracellular proteins. Of these, oxine and HMPAO are mostly used for labelling general leukocyte populations in infection and inflammation [29–31].

16.3.1 ^{111}In-oxine Labelling

With first reports in the mid-1970s, ^{111}In-oxine was approved in 1995 for human use to image infection. ^{111}In forms is an uncharged pseudo-octahedral N_3O_3 complex with three molecules of 8-hydroxyquinoline (oxine). Its neutral and lipophilic characteristics enable it to diffuse over the bilayer cell membrane. Intracellularly, ^{111}In-oxine binds to cytoplasmic proteins that chelate indium more strongly than 8-hydroxyquinoline, which is subsequently released from the cell.

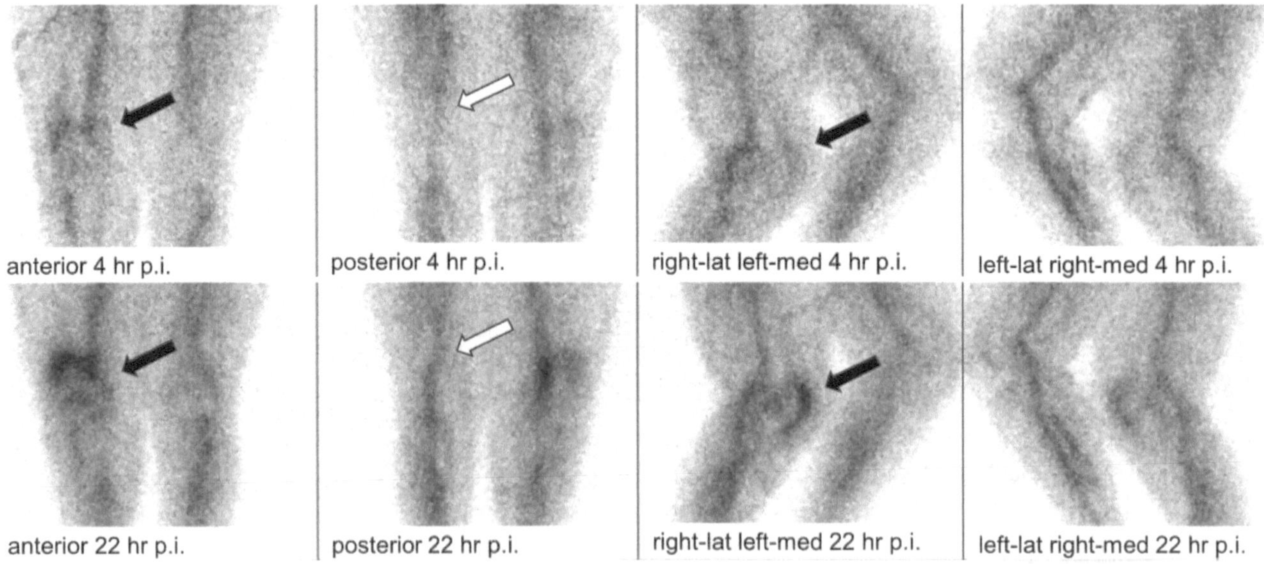

FIGURE 16.3 Example ^{111}In-labelled polyclonal IgG scintigraphy. Patient was scanned under suspicion of infected knee prothesis right-sided. Please note the accumulation of activity (black arrow) in the right prosthetic knee joint between 4 hr and 22 hr post-injection (p.i.). In addition, note the prolonged circulation time of immunoglobulines, visible in the large blood vessels of the lower extremities at both time points (white arrow).

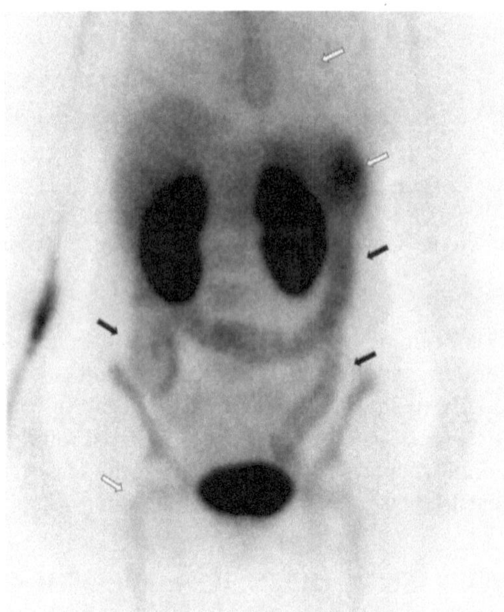

FIGURE 16.4 Example ^{99m}Tc-CXCL8 scintigraphy. Patients with active inflammatory bowel disease. Please note the diffuse uptake in the ascending, transverse, and descending colon (black arrow). Also note the physiological distribution of in vivo labelled neutrophils in spleen, bone marrow, and retention in the lung (white arrows).

Cell-labelling efficiency with ^{111}In-oxine is in general 65–85 per cent, with high cell viability after labelling (>75%). Potential disadvantages of this technique include efflux of ^{111}In from the cell (>60% after 48 hours) [32, 33]. The low-energy Auger electrons emitted by ^{111}In may cause damage to the cell [34, 35].

When compared to ^{99m}Tc-HMPAO, which will be discussed further, there are some advantages and disadvantages in using ^{111}In-oxine. The main advantage of ^{111}In-oxine over ^{99m}Tc-HMPAO is the generally higher labelling efficiency and less efflux of radioactivity from the cell. The costs of ^{111}In-oxine are generally lower than those of ^{99m}Tc-HMPAO, although in some countries HMPAO is available at low cost as a generic drug (i.e. exametazime). On the other hand, ^{111}In-oxine has to be ordered in advance, whereas ^{99m}Tc is readily available from portable generators, which favours its use in more urgent clinical indications. The most important disadvantage, however, is the radiation exposure of ^{111}In-oxine to labelled cells, critical organs (spleen), and the whole body, which is substantially higher than that from ^{99m}Tc-HMPAO. Furthermore, imaging time points are 24 and 48 hours after injection, making this a time-consuming procedure for the patient.

16.3.1.1 ^{111}In-oxine Cell Labelling Protocol

The ^{111}In-oxine complex is unstable in presence of (even traces of serum) proteins. As a consequence, extra care should be taken to remove remains of blood serum prior to radiolabelling. The recommended procedure for radiolabelling cells with ^{111}In-oxine is provided in the European Association for Nuclear Medicine (EANM) guideline [36]. Radiolabelling of a general leukocyte population is fairly simple, as it requires ex vivo cell isolation by ficoll density gradient via centrifugation, incubation with the tracer, and several washing steps. A similar protocol can be applied to radiolabel cells with other oxine compounds such as ^{89}Zr-oxine [37, 38]or ^{67}Ga-oxine [39].

16.3.1.2 Factors Affecting ^{111}In-oxine Uptake

The efficiency of ^{111}In-oxine labelling depends on the cell type, whereby cells with higher cell volume tend to take up a higher proportion of administered activity than those with a lower volume.

Uptake is influenced by a number of factors such as: (a) incubation temperature, (b) incubation volume and cell concentration, and (c) duration of incubation. The labelling efficiency of general leukocyte populations is in the range of 50–80 per cent, but it requires optimization of the following conditions for specific immune cell populations. Incubation of cells at 37°C in certain cases may modestly increase the uptake of ^{111}In-oxine compared to room temperature incubation. Higher cell concentrations and longer incubation times may augment the relative uptake of ^{111}In-oxine at the cost of possible loss of cell viability.

16.3.2 ^{99M}Tc-HMPAO Labelling

^{99m}Tc-HMPAO kit preparations have been commercially available since 1988. Upon reconstitution of the HMPAO kit (containing both D-and L-isomers) with ^{99m}Tc-pertechnetate from a fresh generator, eluate (a lipophilic complex) is formed. The lipophilic complex is transformed into free ^{99m}Tc-pertechnetate and a hydrophilic ^{99m}Tc-HMPAO complex in aqueous solution over time. Only ^{99m}Tc-HMPAO, prepared within 20 minutes, should therefore be used for cell labelling to avoid low labelling efficiencies. Intracellularly, the lipophilic ^{99m}Tc-HMPAO complex is reduced into a hydrophilic complex and, in addition, ^{99m}Tc-HMPAO binds to non-diffusible proteins. Release of ^{99m}Tc-HMPAO from the labelled cells after reinjection into the subject is often observed resulting in accumulation of radioactivity in the gastrointestinal and urinary tracts.

16.3.2.1 ^{99m}Tc-HMPAO Cell Labelling Protocol

^{99m}Tc-HMPAO is serum compatible and can therefore be performed in the patient's own blood serum. The recommended procedure for radiolabelling stem cells with ^{99m}Tc-HMPAO is provided in the EANM guideline [40].

16.3.2.2 Factors Affecting ^{99m}Tc-HMPAO Uptake

Similar to ^{111}In-oxine labelling, the efficiency of ^{99m}Tc-HMPAO labelling depends on the cell type, whereby cells with higher cell volume tend to take up a higher proportion of administered activity than those with a lower volume.

The typical percentage uptake depends very much on cell type and conditions, and labelling efficiency is between 40–80 per cent for general leukocyte populations. Uptake can be optimized under the same considerations as for ^{111}In-oxine, however considering the short shelf-life of ^{99m}Tc-HMPAO incubation time should be limited to a maximum of 30 minutes.

16.3.3 Patient Preparation and Acquisition Protocols for Ex Vivo Cell Labelling

Patients are not required to undergo any particular preparation, but conditions that interfere with an appropriate yield of $2x10^8$ cells should receive attention – for example, an extra syringe of whole blood should be drawn.

A large-field-of-view gamma camera with a medium energy collimator should be used and, in case of ^{111}In-oxine labelled leukocytes, both photopeaks at 173 and 245 keV (±10%) should be acquired.

Images of the chest, abdomen and pelvis should be obtained for at least 100,000 counts but images over the peripheral skeleton may be acquired for time (Signore et al. EJNMMI 2018). As ^{111}In-WBC scintigraphy is preferable in low-grade infection, delayed and late imaging are usually sufficient and not hampered by efflux of ^{111}In from the leukocytes. In abdominal infections or inflammatory bowel disease, differentiation between mucosal uptake (will decrease over time with stool passage), submucosal uptake (remains stable) or abscess (increase over time) require imaging at multiple time points, for example, 3–4 hours and 20–24 hours post-injection for ^{99m}Tc-HMPAO and, for example, 24 and 48 hours for ^{111}In-oxine. Although images are affected by low statistics due to the limitations in the injected activity for dosimetry reasons, the low background activity will nevertheless allow sufficient contrast, especially using SPECT.

16.3.4 Indications for Ex Vivo Cell Labelling

The increased influx of leukocytes, which basically involves monocytes and neutrophils, represents an early phase of an ensuing infection. It requires elevated concentrations of chemokines and cytokines as well as tissue changes to allow chemotaxis of these immune cells into infected tissue, which are produced by activated resident macrophages. This can be interpreted as active infection, where the causative microorganism is not yet cleared, and the immune response is in the upper slope of the curve (Figure 16.1). This is in contrast to inflammation induced by mechanical tissue injury or wound healing; under these circumstances the influx of monocytes and neutrophils is less prominent. For this reason, ex vivo cell-labelling techniques are considered to have a higher specificity than, for example, ^{18}F-FDG PET/CT, to discriminate infection from inflammation [41].

Commonly practiced indications for ^{111}In-oxine and/or ^{99m}Tc-HMPAO cell labelling include

- Osteomyelitis of the appendicular skeleton [42];
- Diabetic foot infection [43, 44];
- Infected prosthetic joint [45] and vascular graft infection [46].

Less commonly practiced indications for ^{111}In-oxine and/or ^{99m}Tc-HMPAO cell labelling follow:

- Inflammatory bowel disease and intra-abdominal infection; in these cases ^{111}In-oxine is preferred due to less bowel excretion;
- Endocarditis and cardiovascular implantable electronic device [47];
- Fever of unknown origin; and
- Neurological infections.

16.3.5 Image Interpretation of Ex Vivo Radiolabelled Cells

Images acquired at different time points should be displayed with the same intensity scale and in absolute counts acquired with a time-decay corrected protocol. A framework for image interpretation is provided by the EANM guidelines [36, 40, 48]. In general, up to 60 per cent of the injected radioactivity is taken up by the liver, spleen, and bone marrow immediately after injection. A transient accumulation in the lungs within 30 minutes after injection is considered normal. The remaining radioactivity in the circulation is cleared exponentially, whereas the clearance of radiolabelled leukocytes from spleen and liver is very slow and negligible. Little activity is expected to be excreted in urine and feces (the latter is expected more for ^{99m}Tc-HMPAO labelled cells).

Caution should be taken if radioactivity is retained in the lung beyond 30 minutes, but nevertheless is cleared within 3 hours, or when radioactivity in the liver is higher than in the spleen at any time point during the acquisition. Both conditions likely indicate cell damage, and the procedure should be considered to non-diagnostic. Normal biodistributions of radiopharmaceuticals has been summarized in Table 16.1 [48].

With ^{111}In-oxine leukocytes, any bowel activity is abnormal; swallowed neutrophils from upper and lower respiratory tract could be considered in some cases of mild bowel activity.

Images are classified as negative for infection, when there is no uptake or there is a clear decrease of activity from delayed to late images in the regions of interest, and classified as positive, when a clear increase is seen with time of

TABLE 16.1
Normal Biodistribution of Ex Vivo Labelled Leukocytes. The Table has been Remade from Data and Information Published in reference [48]

Location	^{99m}Tc-HMPAO-leukocytes early (30′–1h)	^{99m}Tc-HMPAO-leukocytes delayed (3–4 h)	^{99m}Tc-HMPAO-leukocytes late (20–24 h)	^{111}In-oxine leukocytes (3 h/24 h)	^{99m}Tc-anti-granulocyte mAbs (3 h/24h)
Blod/heart	+++	+	−	±	±
Lung	+	−	−	−/−	−/−
Liver	++	++	++	++/++	+++/+++
Spleen	+++	+++	+++	+++/+++	+/++
Kidneys	+	+	+	−/−	+/+
Bladder	−	+	+	−/−	+/+
Bowel	−	+	++	−/−[a]	−/−
Bone Marrow	+	++	++	++/++	+++/+++

uptake intensity or size in the region of interest. In all unequivocal cases a semi-quantitative evaluation can aid to discriminate infection from non-specific uptake. Regions of interest (ROIs) are drawn over the area of interest and copied to a presumed normal (e.g. contra-lateral) reference tissue. When the L/R ratio increases with time by at least 10 per cent, it indicates infection.

16.3.5.1 In Vivo Cell Labelling: Targeting Neutrophil Related Receptors

In vivo labelling of immune cell populations has attractive features, mainly because it circumvents the handling of blood products, which requires additional high-level facilities and qualified personnel. As discussed in the introduction, radiolabelled small peptides, such the cytokines interleukin-2 and interleukin-8, have been investigated for infection imaging. However, their stability as well as in vivo biological activity has hampered implementation in the clinic. In addition, antibodies for lineage-specific cell surface markers have been used in research settings, but the generally high background of radiolabelled antibody images in relation to the relative paucity of, for example, CD4 or CD56, positive cell populations in the circulation restricted its use. As a potential solution to Fc-mediated non-specific uptake, antibody fragments termed minibodies have now been engineered and employed in clinical studies targeting CD8 positive T-cells [49]. Polyclonal ^{111}In-labelled immunoglobulins initially were developed to target Fc-receptor expression on myeloid cell populations present in infection [50, 51], but their use is limited nowadays. Similarly, radiolabelled citrate [52] has been replaced by ^{18}F-FDG PET imaging in most institutes.

16.3.6 ^{99m}Tc-Anti-Granulocyte Antibodies

Monoclonal antibodies (mAbs) are engineered immunoglobulins designed to bind specific sequences of amino acid, termed antigens, with high affinity. Besilesomab is a full-size anti-granulocyte mAb produced in murine cells and designed to bind NCA-95, which is found on the surface of activated neutrophilic granulocytes. Sulesomab is a fragment antigen-binding mAb designed to target an antigen called NCA-90, also present on the surface of neutrophilic granulocytes.

Both mAbs can be radiolabelled with technetium-99m (^{99m}Tc) for in vivo labelling of neutrophils. As large numbers of neutrophils accumulate at the site of an infection, scintigraphic imaging can delineate the site of infection. Anti-granulocyte mAbs localize infectious foci by two pathways: (a) in vivo targeting of activated neutrophilic granulocytes, and (b) non-specific extravasation and retention due to the locally enhanced vascular permeability and retention [48, 53, 54].

16.4 PATIENT PREPARATION AND ACQUISITION PROTOCOLS FOR IN VIVO CELL LABELLING

In order to obtain images of best quality and to reduce the radiation exposure of the bladder, patients should be encouraged to drink sufficient amounts of water and to empty their bladder prior to and after the scintigraphic examination. An interval of at least 2 days is recommended between previous procedure with 99mTc-labelled agents and administration of ^{99m}Tc-labelled anti-granulocyte mAbs.

Please note that only ^{99m}Tc-Besilesomab is authorized for use in the European Union, whereas ^{99m}Tc-Sulesomab in adults has been temporarily withdrawn from use in the European Union.

16.4.1 Indications for ^{99m}Tc-anti-granulocyte Antibodies

Commonly practiced indications for 99mTc-anti-granulocyte antibodies include

- Peripheral bone osteomyelitis;
- Infected joint prosthesis and other orthopedic hardware; and
- Diabetic foot infection.

16.4.2 Image Interpretation of ^{99m}Tc-anti-granulocyte Antibodies

Comparing full size and fragmented radiolabelled anti-granulocyte mAbs, differences in physiological uptake should be taken into account. Both radiolabelled anti-granulocyte mAbs bind to circulating neutrophils but with substantial biodistribution differences. Besilesomab accumulates more than Sulesomab in normal bone marrow but binds more efficiently to neutrophils in blood and at sites of infection. It has slower plasma clearance than Sulesomab, thus mimicking more the biodistribution of radiolabelled autologous white blood cells [53]. Therefore, 99mTc-Besilesomab shows intense uptake in bone marrow, spleen uptake is higher than liver uptake already 1–4 h after injection, whereas both kidneys are shown only slightly. Non-specific bowel activity is regularly seen 20–24 h after injection due to the beginning of radiolabel instability. In contrast to ^{99m}Tc-Besilesomab, with ^{99m}Tc-Sulesomab, there is much less bone marrow uptake, the liver uptake is higher than the spleen uptake, very intense accumulation is seen in the kidneys due to predominantly renal excretion, and non-specific bowel activity is already seen 4–6 h after injection due to enzymatic liver degradation of the compound [48].

Scintigraphic planar (segmental and whole-body) acquisition with [^{99m}Tc]Tc-Besilesomab should be performed at 2–4 h and 16–24 h after radiopharmaceutical injection because a significant increase in sensitivity and specificity will be achieved with late 24 h images due to higher target-to-background ratios. Scintigraphic imaging with ^{99m}Tc-Sulesomab should be performed 1 h and 4-6 h after radiopharmaceutical injection.

In equivocal cases at both visual and semi-quantitative analysis, bone-marrow imaging using ^{99m}Tc-labelled sulfur colloids can be considered. Radiolabelled colloids and anti-granulocyte mAbs accumulate in healthy and displaced bone marrow, whereas radiolabelled colloids do not accumulate in infection sites.

False positive accumulation due to non-specific inflammation anti-granulocyte mAbs might occur within 3 months after surgery; artefacts related to the attenuation over-correction in patients with metallic devices; lesions of size lower than spatial resolution of method causing false negative results; false results in patients with diseases involving neutrophil defects and in patients with haematological malignancies; active substances may inhibit inflammation or affect the haematopoietic system (such as antibiotics and corticosteroids), which may lead to false negative results [48, 53, 54].

16.5 THE METABOLIC SWITCH IN IMMUNE CELLS: PET IMAGING OF GLUCOSE METABOLISM

Glucose fuels immune cell metabolism through two integrated pathways. The first is glycolysis, which is the conversion of glucose to pyruvate in the cytoplasm. Pyruvate can enter the tricarboxylic acid (TCA) cycle, which generates NADH in the cytoplasm to donate electrons to fuel the second process that takes place in the mitochondria, called oxidative phosphorylation [28, 55, 56]. Remarkably, immune cells preferentially use glycolysis to generate ATP, which is less efficient, even when oxygen is widely available, a process known as "aerobic glycolysis" or the "Warburg effect." They switch to this less-efficient pathway of ATP-production and by doing so, cells are able to meet their high demands for biosynthetic precursors (e.g. lipid, proteins, and nucleic acids that are by-products of glycolysis), required for effector functions. The extent to which this metabolic switch occurs might vary, but it is generally present in all immune cell populations relevant to inflammation.

Neutrophils: In steady state, neutrophil rely on glycolysis rather than on oxidative phosphorylation reflected by the low number of mitochondria in neutrophils. This dependency on glycolysis is even further increased once activated. The main reason for neutrophils to stick to glycolysis is their need to produce H_2O_2 in activated status, which is their main microbicial product. Animal studies evaluating [^{18}F]-FDG-uptake in models of pneumonia [57] or sepsis-induced lung injury [58] as well as clinical studies [59] confirm increased ^{18}F-FDG-uptake in neutrophil-infiltrated lung tissue by PET/CT. Validation by autoradiography in these studies demonstrated increased ^{3}H-DG uptake was dominated by neutrophils. Enhanced glucose uptake occurs upon priming of neutrophils but is not required for executing the effector functions itself, such as the respiratory burst or degranulation [60]. This suggests that ^{18}F-FDG PET/CT might represent a biomarker for *early* inflammatory processes, reflecting neutrophil activation and priming. The strong correlation of increased ^{18}F-FDG-uptake in the lung, with both the number of neutrophils and their activation status, supports this notion [59].

Lymphocytes: Multiple in vitro studies have shown increased glucose metabolism in different stages of functional differentiation of T cells [61]. When a naïve T cell encounters its cognate antigen in the proper stimulatory context, it undergoes a transcriptional program that is characterized by proliferation, rapid growth, and induction of specialized effector functions belonging to effector T cells. This reprogramming requires metabolic adaption from a catabolic metabolism to an anabolic metabolism because, unlike quiescent states, nutrients will not be used for homeostasis but will be incorporated in biosynthetic precursors required for daughter cells and effector functions. This demand for biosynthetic precursors drives *effector* T cells to increase their glycolysis, even in the presence of sufficient levels of oxygen. To the contrary, T cells destined to become *memory* cells must maintain catabolic metabolism, underlying their longevity and may postpone their terminal differentiation, so they rely mainly on mitochondrial fatty acid oxidation.

Macrophages: Macrophages are primarily glycolytic [62], as demonstrated by their relatively low numbers of mitochondria. In parallel with neutrophils, inflammatory macrophages require radical oxygen species (ROS) for bacterial killing in phagolysosomes. Indeed, inflammatory macrophages show increased ^{18}F-FDG-uptake as demonstrated in multiple clinical studies on atherosclerosis [63] and in alveolar macrophages in fibrotic lung disease [64].

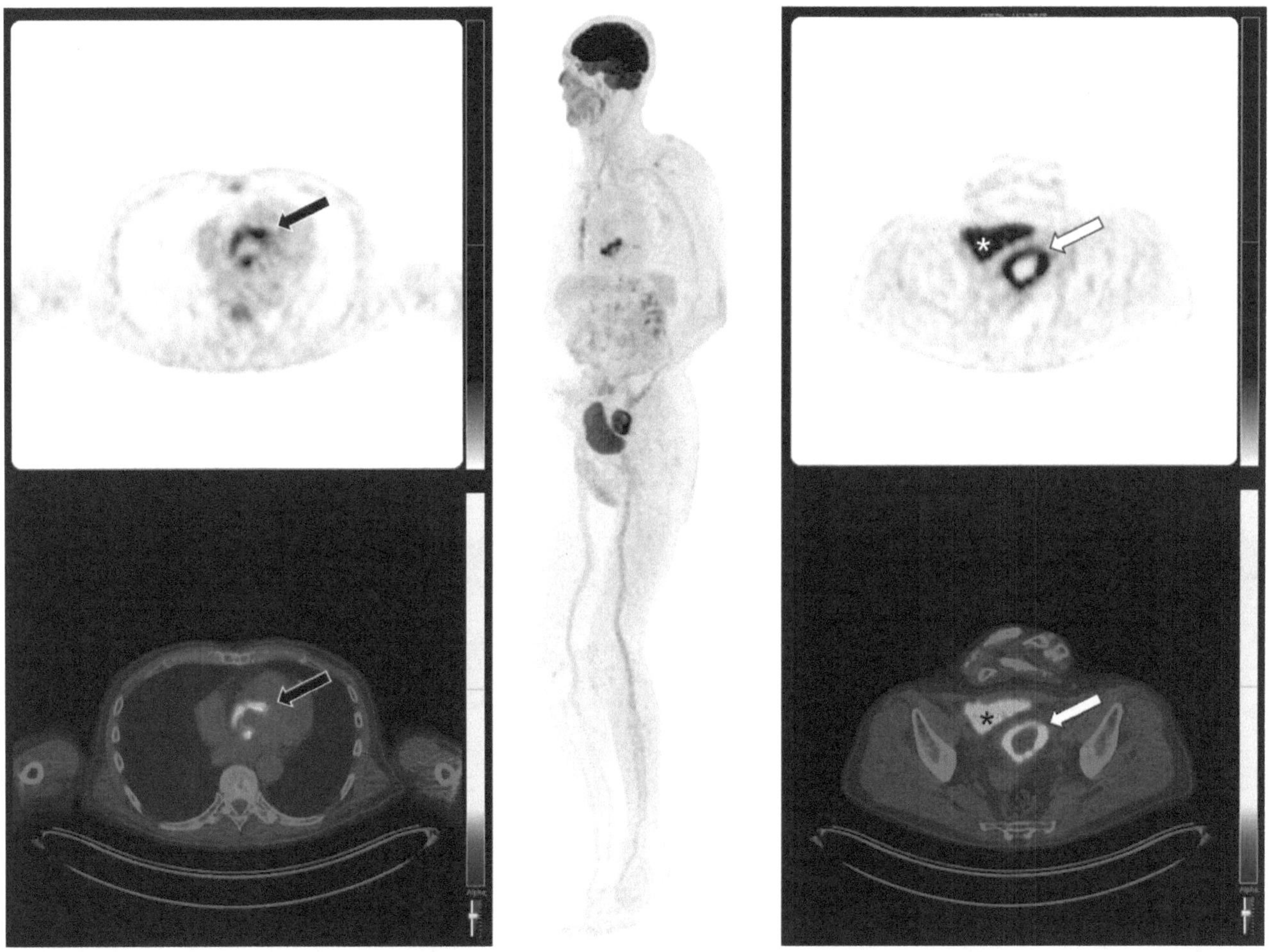

FIGURE 16.5 Example ^{18}F-FDG PET/CT. Microbiologically proven prosthetic aortic valve endocarditis associated with abscess upon perforated diverticulitis. Note the intense ^{18}F-FDG uptake in the granulomatous rim of the encapsulated perforation (white arrow) and irregular and increased ^{18}F-FDG uptake around the aortic valve (black arrow). Asterisk (*) denotes excretion of ^{18}F-FDG in the urine bladder. Colour image available at www.routledge.com/9781138593312.

16.5.1 [18F]FDG PET

^{18}F-fluorodeoxyglucose (^{18}F-2'-fluoro-2'-deoxyglucose or ^{18}F-FDG) is an analog of glucose and is taken up by living cells via cell membrane glucose transporters and subsequently phosphorylated with hexokinase inside most cells. Immune cells involved in infection and inflammation, especially activated neutrophils and the monocyte/macrophage lineages, express high levels of glucose transporters GLUT1 and GLUT3, and hexokinase activity [65].

16.5.2 Indications

The major indications for ^{18}F-FDG PET/CT in infection and inflammation show some overlap with ex vivo cell labelling and include [66]

- Sarcoidosis [67]
- Peripheral bone osteomyelitis (non-postoperative, non–diabetic foot) [45]
- Suspected spinal infection (spondylodiscitis or vertebral osteomyelitis, non-postoperative) [68],
- Evaluation of fever of unknown origin (FUO) [69, 70],
- Evaluation of metastatic infection and of high-risk patients with bacteremia [71, 72],
- Primary evaluation of vasculitis [73],
- Evaluation of potentially infected liver and kidney cysts in polycystic disease,
- AIDS-associated opportunistic infections, associated tumours, and Castleman disease,

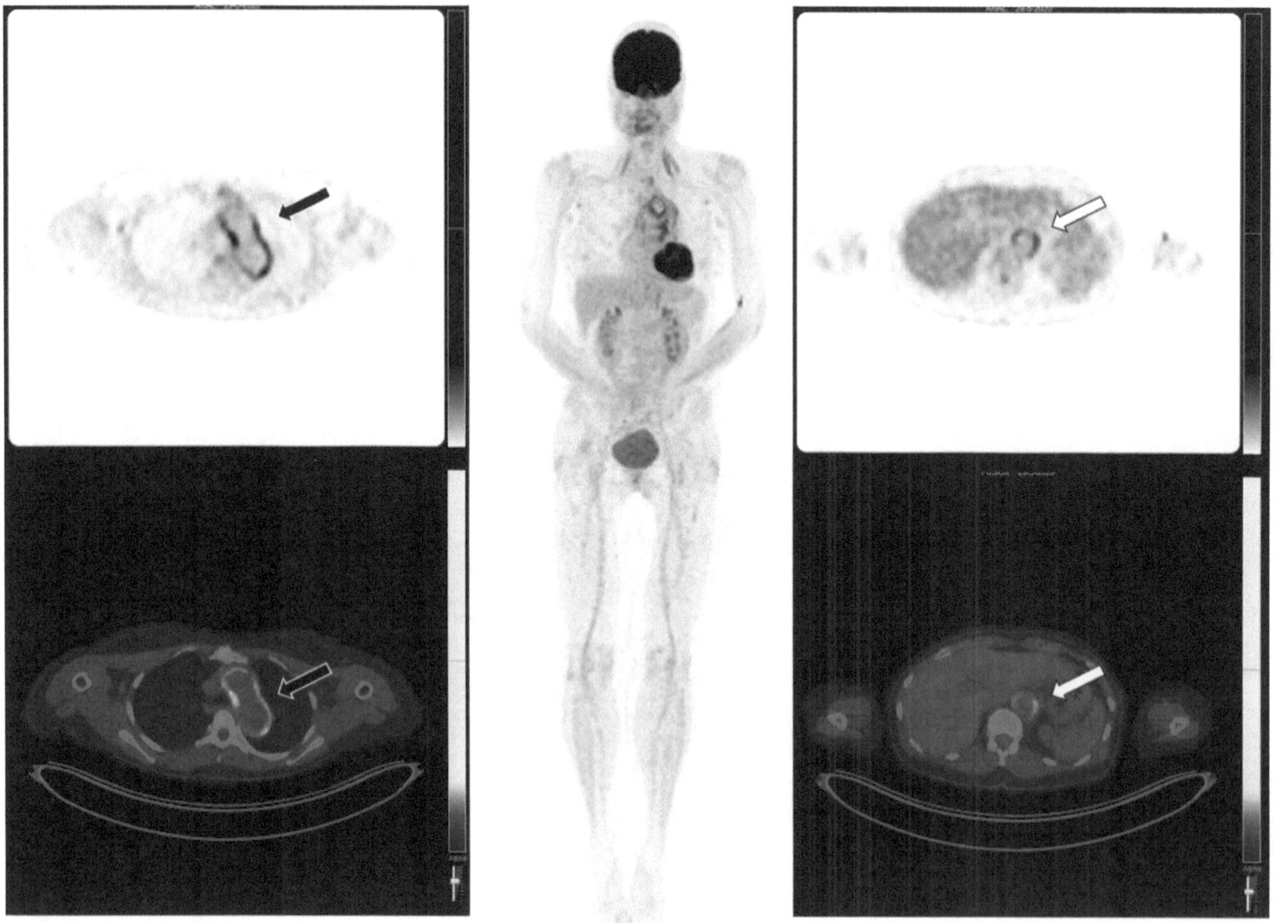

FIGURE 16.6 Example ^{18}F-FDG PET/CT. Exacerbation Takayashu inflammatory aortitis. Note the irregular and increased ^{18}F-FDG uptake in predominantly the ascending aorta and arch (black arrow) but involving the descending aorta as well (white arrow). Colour image available at www.routledge.com/9781138593312.

- Assessment of metabolic activity in tuberculosis lesions [74-76]
- Diabetic foot infections [44],
- Joint prosthetic infections [45],
- Vascular prosthetic infections [41], and
- Endocarditis and CIED infection [47].

16.5.3 Image Interpretation

Physiologic ^{18}F-FDG accumulation is high in the brain, heart, kidneys, and urinary tract at 60 min after injection [66]. The brain has a high uptake of ^{18}F-FDG (approximately 7% of injected activity). The myocardium in a typical fasting state primarily uses free fatty acids but, after a glucose load, switches to glucose. In the fasting state, ^{18}F-FDG uptake in the myocardium should be low, but this is variable. Unlike glucose, ^{18}F-FDG is excreted by the kidneys into the urine and accumulates in the urinary tract. ^{18}F-FDG may also be seen in muscles, depending on recent muscle activity (specific muscle groups involved in the exercise), and insulin (generalized muscle uptake). Uptake in the gastrointestinal tract varies from patient to patient and may be increased in patients taking metformin. Uptake is common in the secondary lymphoid tissue of the Waldeyer ring and in the lymphoid tissue of the terminal ileum and cecum. Physiologic thymic uptake may be present, especially in children and young adults. Uptake in brown fat may be observed mainly in young patients and when the ambient temperature is low. No physiologic uptake is noted in the bone itself (unless free ^{18}F-fluoride is present as a contaminant) but, especially in infected or inflamed patients, bone marrow uptake can be

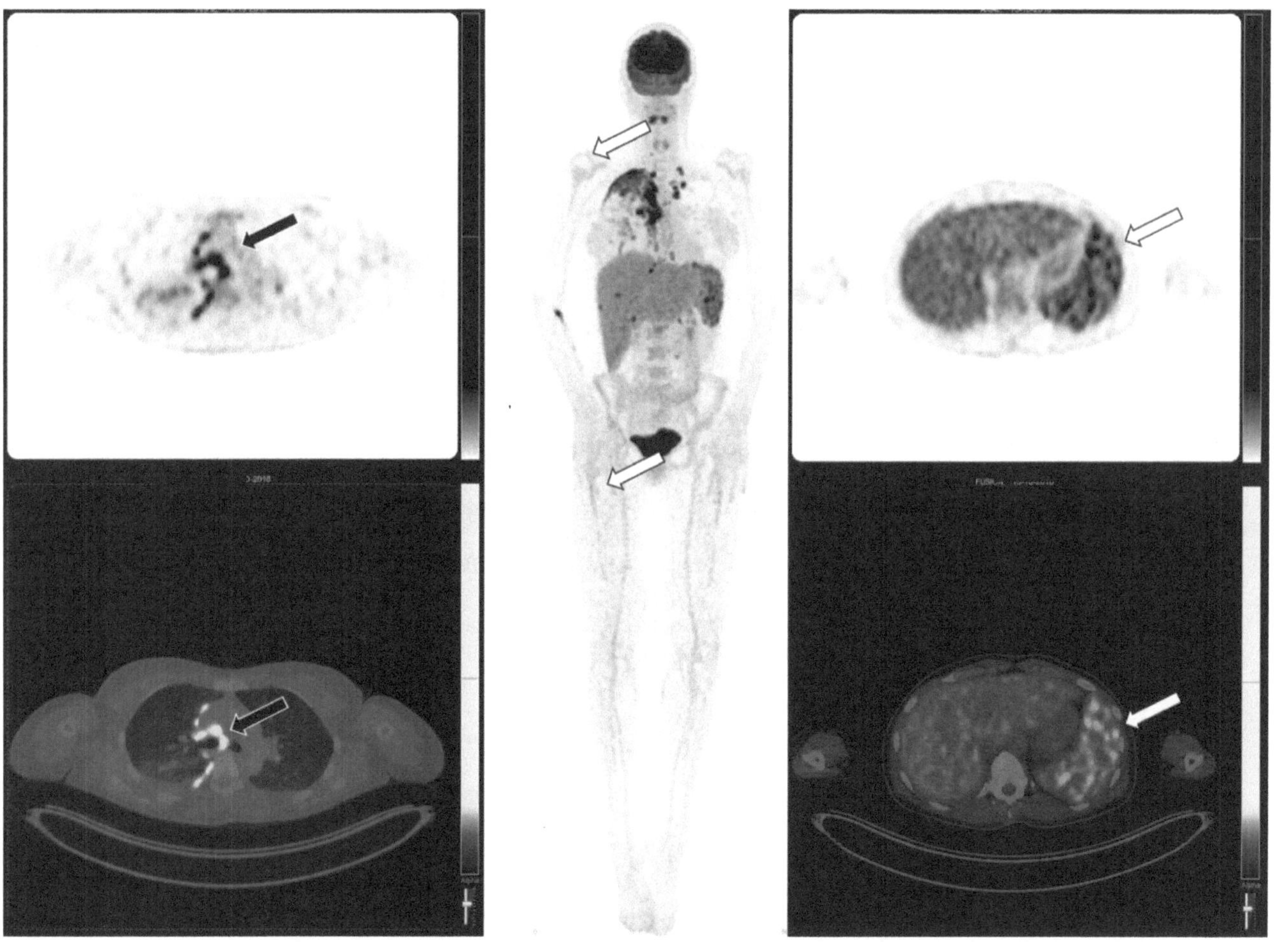

FIGURE 16.7 Example ^{18}F-FDG PET/CT. Non-tuberculosis mycobacterial infection of the right upper and lower lobe of the lung, pleura, and draining hilar and mediastinal lymph nodes (black arrow). Systemic inflammatory response with diffuse uptake in the spleen and increased uptake in the bone marrow proximal humeri and femora (white arrow). Colour image available at www.routledge.com/9781138593312.

noted to a variable level due to increased hematopoiesis. This is also true in patients with hematopoietic regeneration, such as after chemotherapy, either spontaneously or after administration of hematopoietic growth factors (e.g., granulocyte–macrophage colony-stimulating factor) [77].

There are no general criteria published for all inflammatory and infectious disorders. Most research articles on the subject have defined interpretation criteria for the purposes of the study. Some authors have reported specific interpretation criteria that can be used, although no definitive consensus has been agreed on.

- Sarcoidosis: Sarcoidosis can mimic malignancies and especially lymphoma. It has been reported that a high parenchymal lung uptake was predictive of disease activity, especially if the mediastinum and hilum SUV_{max} was low. Conversely, the absence of metabolic activity in the lung parenchyma was related to low-activity disease. For cardiac sarcoidosis, correct patient preparation, scan acquisition, and image interpretation is described in a recently published guideline [78].
- Vascular graft infection: Because physiologic uptake is often visible in vascular prostheses, patterns of interpretation have been discussed. It is now felt that linear, diffuse, and homogeneous uptake is not likely to represent infection, whereas focal or heterogeneous uptake with projection over the vessel on CT is highly suggestive of infection. Recently, a European guideline on vascular graft infections was published, including diagnostic methods [46].
- Vasculitis: Criteria for the diagnosis of active giant cell arteritis are commonly based on visual comparison of uptake in the aorta with that in the liver or brain [79, 80]. Recently, a European guideline was published in which the correct acquisition protocols and interpretation criteria are described [73].

The use of ^{18}F-FDG-PET/CT is now widely adapted for infection and inflammation imaging; initiatives to standardize and harmonize studies are growing. The advantage of the use of ^{18}F-FDG-PET/CT is the patients' comfort (especially when compared to the ex vivo cell labelling procedures) with procedures finished within 2 hours, the possibility to perform a contrast enhanced CT scan in the same procedure, and the wide availability of this tracer. The major disadvantage is that ^{18}F-FDG-PET/CT is not specific for differentiating between infection and inflammation, which leaves a role for tracers that target surface markers of immune cell populations that are primarily involved in infections as opposed to sterile inflammation.

REFERENCES

[1] T. A. Wynn, A. Chawla, and J. W. Pollard, "Macrophage biology in development, homeostasis and disease," *Nature,* vol. 496, no. 7446, pp. 445–55, Apr 25 2013, doi: 10.1038/nature12034.

[2] D. M. Mosser, "The many faces of macrophage activation," *J Leukoc Biol,* vol. 73, no. 2, pp. 209–12, Feb 2003, doi: 10.1189/jlb.0602325.

[3] P. Friedl and B. Weigelin, "Interstitial leukocyte migration and immune function," *Nat Immunol,* vol. 9, no. 9, pp. 960–69, Sep 2008, doi: 10.1038/ni.f.212.

[4] S. Gordon, "Phagocytosis: An immunobiologic process," *Immunity,* vol. 44, no. 3, pp. 463–75, Mar 15 2016, doi: 10.1016/j.immuni.2016.02.026.

[5] C. Shi and E. G. Pamer, "Monocyte recruitment during infection and inflammation," *Nat Rev Immunol,* vol. 11, no. 11, pp. 762–74, Oct 10 2011, doi: 10.1038/nri3070.

[6] A. Abtin *et al.*, "Perivascular macrophages mediate neutrophil recruitment during bacterial skin infection," *Nat Immunol,* vol. 15, no. 1, pp. 45–53, Jan 2014, doi: 10.1038/ni.2769.

[7] M. F. Bachmann, M. Kopf, and B. J. Marsland, "Chemokines: more than just road signs," *Nat Rev Immunol,* vol. 6, no. 2, pp. 159–64, Feb 2006, doi: 10.1038/nri1776.

[8] C. N. Jenne and P. Kubes, "Platelets in inflammation and infection," *Platelets,* vol. 26, no. 4, pp. 286–92, 2015, doi: 10.3109/09537104.2015.1010441.

[9] J. S. Pober and W. C. Sessa, "Evolving functions of endothelial cells in inflammation," *Nat Rev Immunol,* vol. 7, no. 10, pp. 803–15, Oct 2007, doi: 10.1038/nri2171.

[10] F. Geissmann, M. G. Manz, S. Jung, M. H. Sieweke, M. Merad, and K. Ley, "Development of monocytes, macrophages, and dendritic cells," *Science,* vol. 327, no. 5966, pp. 656–61, Feb 5 2010, doi: 10.1126/science.1178331.

[11] V. Angeli and G. J. Randolph, "Inflammation, lymphatic function, and dendritic cell migration," *Lymphat Res Biol,* vol. 4, no. 4, pp. 217–28, 2006, doi: 10.1089/lrb.2006.4406.

[12] M. Lee, J. N. Mandl, R. N. Germain, and A. J. Yates, "The race for the prize: T-cell trafficking strategies for optimal surveillance," *Blood,* vol. 120, no. 7, pp. 1432–38, Aug 16 2012, doi: 10.1182/blood-2012-04-424655.

[13] X. Fan and A. Y. Rudensky, "Hallmarks of Tissue-Resident Lymphocytes," *Cell,* vol. 164, no. 6, pp. 1198–1211, Mar 10 2016, doi: 10.1016/j.cell.2016.02.048.

[14] E. L. Pearce and H. Shen, "Making sense of inflammation, epigenetics, and memory CD8+ T-cell differentiation in the context of infection," *Immunol Rev,* vol. 211, pp. 197–202, Jun 2006, doi: 10.1111/j.0105-2896.2006.00399.x.

[15] R. S. Hotchkiss, L. L. Moldawer, S. M. Opal, K. Reinhart, I. R. Turnbull, and J. L. Vincent, "Sepsis and septic shock," *Nat Rev Dis Primers,* vol. 2, p. 16045, Jun 30 2016, doi: 10.1038/nrdp.2016.45.

[16] J. L. Coombes and E. A. Robey, "Dynamic imaging of host–pathogen interactions in vivo," *Nat Rev Immunol,* vol. 10, no. 5, pp. 353–64, May 2010, doi: 10.1038/nri2746.

[17] L. A. Diaz, Jr. *et al.*, "Imaging of musculoskeletal bacterial infections by [124I]FIAU-PET/CT," *PLoS One,* vol. 2, no. 10, p. e1007, Oct 10 2007, doi: 10.1371/journal.pone.0001007.

[18] M. Pullambhatla, J. Tessier, G. Beck, B. Jedynak, J. U. Wurthner, and M. G. Pomper, "[(125)I]FIAU imaging in a preclinical model of lung infection: quantification of bacterial load," *Am J Nucl Med Mol Imaging,* vol. 2, no. 3, pp. 260–70, 2012. [Online]. Available: www.ncbi.nlm.nih.gov/pubmed/23133816.

[19] G. Ferro-Flores, B. E. Ocampo-Garcia, and L. Melendez-Alafort, "Development of specific radiopharmaceuticals for infection imaging by targeting infectious micro-organisms," *Curr Pharm Des,* vol. 18, no. 8, pp. 1098–106, 2012, doi: 10.2174/138161212799315821.

[20] D. A. Dorward, C. D. Lucas, A. G. Rossi, C. Haslett, and K. Dhaliwal, "Imaging inflammation: molecular strategies to visualize key components of the inflammatory cascade, from initiation to resolution," *Pharmacol Ther,* vol. 135, no. 2, pp. 182–99, Aug 2012, doi: 10.1016/j.pharmthera.2012.05.006.

[21] G. Malviya, A. Signore, B. Lagana, and R. A. Dierckx, "Radiolabelled peptides and monoclonal antibodies for therapy decision making in inflammatory diseases," *Curr Pharm Des,* vol. 14, no. 24, pp. 2401–14, 2008, doi: 10.2174/138161208785777414.

[22] A. Annovazzi *et al.*, "99mTc-interleukin-2 and (99m)Tc-HMPAO granulocyte scintigraphy in patients with inactive Crohn's disease," *Eur J Nucl Med Mol Imaging,* vol. 30, no. 3, pp. 374–82, Mar 2003, doi: 10.1007/s00259-002-1069-x.

[23] H. J. Rennen *et al.*, "99mTc-labeled interleukin-8 for scintigraphic detection of pulmonary infections," *Chest,* vol. 126, no. 6, pp. 1954–61, Dec 2004, doi: 10.1378/chest.126.6.1954.

[24] G. Malviya, F. Galli, I. Sonni, and A. Signore, "Imaging T-lymphocytes in inflammatory diseases: a nuclear medicine approach," *Q J Nucl Med Mol Imaging,* vol. 58, no. 3, pp. 237–57, Sep 2014. [Online]. Available: www.ncbi.nlm.nih.gov/pubmed/25265246.

[25] F. Galli, S. Histed, and O. Aras, "NK cell imaging by in vitro and in vivo labelling approaches," *Q J Nucl Med Mol Imaging,* vol. 58, no. 3, pp. 276–83, Sep 2014. [Online]. Available: www.ncbi.nlm.nih.gov/pubmed/25265248.

[26] E. M. de Kleijn, W. J. Oyen, F. H. Corstens, and J. W. van der Meer, "Utility of indium-111-labeled polyclonal immunoglobulin G scintigraphy in fever of unknown origin. The Netherlands FUO Imaging Group," *J Nucl Med,* vol. 38, no. 3, pp. 484–89, Mar 1997. [Online]. Available: www.ncbi.nlm.nih.gov/pubmed/9074544.

[27] G. D. Norata *et al.*, "The Cellular and Molecular Basis of Translational Immunometabolism," *Immunity,* vol. 43, no. 3, pp. 421–34, Sep 15 2015, doi: 10.1016/j.immuni.2015.08.023.

[28] E. M. Palsson-McDermott and L. A. O'Neill, "The Warburg effect then and now: from cancer to inflammatory diseases," *Bioessays,* vol. 35, no. 11, pp. 965–73, Nov 2013, doi: 10.1002/bies.201300084.

[29] A. F. Tarantal, C. C. Lee, D. L. Kukis, and S. R. Cherry, "Radiolabeling human peripheral blood stem cells for positron emission tomography (PET) imaging in young rhesus monkeys," *PLoS One,* vol. 8, no. 10, p. e77148, 2013, doi: 10.1371/journal.pone.0077148.

[30] M. E. Roddie *et al.*, "Inflammation: imaging with Tc-99m HMPAO-labeled leukocytes," *Radiology,* vol. 166, no. 3, pp. 767–72, Mar 1988, doi: 10.1148/radiology.166.3.3340775.

[31] M. L. Thakur, J. P. Lavender, R. N. Arnot, D. J. Silvester, and A. W. Segal, "Indium-111-labeled autologous leukocytes in man," *J Nucl Med,* vol. 18, no. 10, pp. 1014–21, Oct 1977. [Online]. Available: www.ncbi.nlm.nih.gov/pubmed/409745.

[32] B. B. Chin, Y. Nakamoto, J. W. Bulte, M. F. Pittenger, R. Wahl, and D. L. Kraitchman, "111In oxine labelled mesenchymal stem cell SPECT after intravenous administration in myocardial infarction," *Nucl Med Commun,* vol. 24, no. 11, pp. 1149–54, Nov 2003, doi: 10.1097/00006231-200311000-00005.

[33] A. Aicher *et al.*, "Assessment of the tissue distribution of transplanted human endothelial progenitor cells by radioactive labeling," *Circulation,* vol. 107, no. 16, pp. 2134–39, Apr 29 2003, doi: 10.1161/01.CIR.0000062649.63838.C9.

[34] F. J. Gildehaus *et al.*, "Impact of indium-111 oxine labelling on viability of human mesenchymal stem cells in vitro, and 3D cell-tracking using SPECT/CT in vivo," *Mol Imaging Biol,* vol. 13, no. 6, pp. 1204–14, Dec 2011, doi: 10.1007/s11307-010-0439-1.

[35] B. Nowak *et al.*, "Indium-111 oxine labelling affects the cellular integrity of haematopoietic progenitor cells," *Eur J Nucl Med Mol Imaging,* vol. 34, no. 5, pp. 715–21, May 2007, doi: 10.1007/s00259-006-0275-3.

[36] M. Roca, E. F. de Vries, F. Jamar, O. Israel, and A. Signore, "Guidelines for the labelling of leucocytes with (111)In-oxine. Inflammation/Infection Taskgroup of the European Association of Nuclear Medicine," *Eur J Nucl Med Mol Imaging,* vol. 37, no. 4, pp. 835–41, Apr 2010, doi: 10.1007/s00259-010-1393-5.

[37] N. Sato, H. Wu, K. O. Asiedu, L. P. Szajek, G. L. Griffiths, and P. L. Choyke, "(89)Zr-Oxine complex PET cell imaging in monitoring cell-based therapies," *Radiology,* vol. 275, no. 2, pp. 490–500, May 2015, doi: 10.1148/radiol.15142849.

[38] N. Sato *et al.*, "In vivo tracking of adoptively transferred natural killer cells in rhesus macaques using (89)Zirconium-Oxine cell labeling and PET imaging," *Clin Cancer Res,* vol. 26, no. 11, pp. 2573–81, Jun 1 2020, doi: 10.1158/1078-0432.CCR-19-2897.

[39] G. Karanikas, M. Rodrigues, S. Granegger, and H. Sinzinger, "Platelet labeling with 67Ga chelates," *Appl Radiat Isot,* vol. 50, no. 3, pp. 505–11, Mar 1999, doi: 10.1016/s0969-8043(98)00084-0.

[40] E. F. de Vries, M. Roca, F. Jamar, O. Israel, and A. Signore, "Guidelines for the labelling of leucocytes with (99m)Tc-HMPAO. Inflammation/Infection Taskgroup of the European Association of Nuclear Medicine," *Eur J Nucl Med Mol Imaging,* vol. 37, no. 4, pp. 842–48, Apr 2010, doi: 10.1007/s00259-010-1394-4.

[41] E. I. Reinders Folmer *et al.*, "Diagnostic imaging in vascular graft infection: A systematic review and meta-analysis," *Eur J Vasc Endovasc Surg,* vol. 56, no. 5, pp. 719–29, Nov 2018, doi: 10.1016/j.ejvs.2018.07.010.

[42] A. Glaudemans *et al.*, "Consensus document for the diagnosis of peripheral bone infection in adults: A joint paper by the EANM, EBJIS, and ESR (with ESCMID endorsement)," *Eur J Nucl Med Mol Imaging,* vol. 46, no. 4, pp. 957–70, Apr 2019, doi: 10.1007/s00259-019-4262-x.

[43] C. Lauri *et al.*, "Detection of osteomyelitis in the diabetic foot by imaging techniques: A systematic review and meta-analysis comparing MRI, white blood cell scintigraphy, and FDG-PET," *Diabetes Care,* vol. 40, no. 8, pp. 1111–20, Aug 2017, doi: 10.2337/dc17s0532.

[44] C. Lauri *et al.*, "Comparison of white blood cell scintigraphy, FDG PET/CT and MRI in suspected diabetic foot infection: Results of a large retrospective multicenter study," *J Clin Med,* vol. 9, no. 6, May 30 2020, doi: 10.3390/jcm9061645.

[45] A. Signore *et al.*, "Consensus document for the diagnosis of prosthetic joint infections: A joint paper by the EANM, EBJIS, and ESR (with ESCMID endorsement)," *Eur J Nucl Med Mol Imaging,* vol. 46, no. 4, pp. 971–88, Apr 2019, doi: 10.1007/s00259-019-4263-9.

[46] N. Chakfe *et al.*, "Editor's Choice–European Society for Vascular Surgery (ESVS) 2020 Clinical Practice Guidelines on the Management of Vascular Graft and Endograft Infections," *Eur J Vasc Endovasc Surg,* vol. 59, no. 3, pp. 339–84, Mar 2020, doi: 10.1016/j.ejvs.2019.10.016.

[47] P. A. Erba *et al.*, "Recommendations on nuclear and multimodality imaging in IE and CIED infections," *Eur J Nucl Med Mol Imaging,* vol. 45, no. 10, pp. 1795–1815, Sep 2018, doi: 10.1007/s00259-018-4025-0.

[48] A. Signore, F. Jamar, O. Israel, J. Buscombe, J. Martin-Comin, and E. Lazzeri, "Clinical indications, image acquisition and data interpretation for white blood cells and anti-granulocyte monoclonal antibody scintigraphy: An EANM procedural guideline," *Eur J Nucl Med Mol Imaging,* vol. 45, no. 10, pp. 1816–31, September 01 2018, doi: 10.1007/s00259-018-4052-x.

[49] N. Pandit-Taskar *et al.*, "First-in-human imaging with (89)Zr-Df-IAB22M2C anti-CD8 minibody in patients with solid malignancies: Preliminary pharmacokinetics, biodistribution, and lesion targeting," *J Nucl Med,* Oct 4 2019, doi: 10.2967/jnumed.119.229781.

[50] W. J. Oyen, J. R. van Horn, R. A. Claessens, T. J. Slooff, J. W. van der Meer, and F. H. Corstens, "Diagnosis of bone, joint, and joint prosthesis infections with In-111-labeled nonspecific human immunoglobulin G scintigraphy," *Radiology,* vol. 182, no. 1, pp. 195–99, Jan 1992, doi: 10.1148/radiology.182.1.1727281.

[51] W. J. Oyen, R. A. Claessens, J. R. van Horn, J. W. van der Meer, and F. H. Corstens, "Scintigraphic detection of bone and joint infections with indium-111-labeled nonspecific polyclonal human immunoglobulin G," *J Nucl Med,* vol. 31, no. 4, pp. 403–12, Apr 1990. [Online]. Available: www.ncbi.nlm.nih.gov/pubmed/2324817.

[52] C. Love and C. J. Palestro, "Altered biodistribution and incidental findings on gallium and labeled leukocyte/bone marrow scans," *Semin Nucl Med,* vol. 40, no. 4, pp. 271–82, Jul 2010, doi: 10.1053/j.semnuclmed.2010.03.004.

[53] S. Gratz, P. Reize, B. Kemke, W. U. Kampen, M. Luster, and H. Hoffken, "Targeting osteomyelitis with complete [99mTc] besilesomab and fragmented [99mTc]sulesomab antibodies: Kinetic evaluations," *Q J Nucl Med Mol Imaging,* vol. 60, no. 4, pp. 413–23, Dec 2016. [Online]. Available: www.ncbi.nlm.nih.gov/pubmed/25325395.

[54] J. Meller, T. Liersch, M. M. Oezerden, C. O. Sahlmann, and B. Meller, "Targeting NCA-95 and other granulocyte antigens and receptors with radiolabeled monoclonal antibodies (Mabs)," *Q J Nucl Med Mol Imaging,* vol. 54, no. 6, pp. 582–98, Dec 2010. [Online]. Available: www.ncbi.nlm.nih.gov/pubmed/21221067.

[55] E. L. Pearce and E. J. Pearce, "Metabolic pathways in immune cell activation and quiescence," *Immunity,* vol. 38, no. 4, pp. 633–43, Apr 18 2013, doi: 10.1016/j.immuni.2013.04.005.

[56] E. L. Pearce, M. C. Poffenberger, C. H. Chang, and R. G. Jones, "Fueling immunity: insights into metabolism and lymphocyte function," *Science,* vol. 342, no. 6155, p. 1242454, Oct 11 2013, doi: 10.1126/science.1242454.

[57] H. A. Jones, R. J. Clark, C. G. Rhodes, J. B. Schofield, T. Krausz, and C. Haslett, "In vivo measurement of neutrophil activity in experimental lung inflammation," *American Journal of Respiratory and Critical Care Medicine,* vol. 149, no. 6, pp. 1635–39, Jun 1994, doi: 10.1164/ajrccm.149.6.7516252.

[58] L. Hansson *et al.*, "Glucose utilisation in the lungs of septic rats," *Eur J Nucl Med,* vol. 26, no. 10, pp. 1340–44, Oct 1999. [Online]. Available: www.ncbi.nlm.nih.gov/pubmed/10541834.

[59] D. L. Chen, D. B. Rosenbluth, M. A. Mintun, and D. P. Schuster, "FDG-PET imaging of pulmonary inflammation in healthy volunteers after airway instillation of endotoxin," (in English), *J Appl Physiol (1985),* Evaluation Studies Research Support, N.I.H., Extramural Research Support, Non-U.S. Gov't vol. 100, no. 5, pp. 1602–9, May 2006, doi: 10.1152/japplphysiol.01429.2005.

[60] H. A. Jones, K. A. Cadwallader, J. F. White, M. Uddin, A. M. Peters, and E. R. Chilvers, "Dissociation between respiratory burst activity and deoxyglucose uptake in human neutrophil granulocytes: Implications for interpretation of (18)F-FDG PET images," (in English), *Journal of Nuclear Medicine: Official publication, Society of Nuclear Medicine,* Research Support, Non-U.S. Gov't vol. 43, no. 5, pp. 652–57, May 2002. [Online]. Available: www.ncbi.nlm.nih.gov/pubmed/11994530.

[61] C. J. Fox, P. S. Hammerman, and C. B. Thompson, "Fuel feeds function: energy metabolism and the T-cell response," *Nat Rev Immunol,* vol. 5, no. 11, pp. 844–52, Nov 2005, doi: 10.1038/nri1710.

[62] R. Kubota, K. Kubota, S. Yamada, M. Tada, T. Ido, and N. Tamahashi, "Microautoradiographic study for the differentiation of intratumoral macrophages, granulation tissues and cancer cells by the dynamics of fluorine-18-fluorodeoxyglucose uptake," *Journal of Nuclear Medicine: Official publication, Society of Nuclear Medicine,* vol. 35, no. 1, pp. 104–12, Jan 1994. [Online]. Available: www.ncbi.nlm.nih.gov/pubmed/8271030.

[63] E. J. Folco *et al.*, "Hypoxia but not inflammation augments glucose uptake in human macrophages: Implications for imaging atherosclerosis with 18fluorine-labeled 2-deoxy-D-glucose positron emission tomography," *J Am Coll Cardiol,* vol. 58, no. 6, pp. 603–14, Aug 2 2011, doi: 10.1016/j.jacc.2011.03.044.

[64] S. El-Chemaly *et al.*, "Glucose transporter-1 distribution in fibrotic lung disease: association with [(1)(8)F]-2-fluoro-2-deoxyglucose-PET scan uptake, inflammation, and neovascularization," *Chest,* vol. 143, no. 6, pp. 1685–91, Jun 2013, doi: 10.1378/chest.12-1359.

[65] T. Mochizuki *et al.*, "FDG uptake and glucose transporter subtype expressions in experimental tumor and inflammation models," *J Nucl Med,* vol. 42, no. 10, pp. 1551–55, Oct 2001. [Online]. Available: www.ncbi.nlm.nih.gov/pubmed/11585872.

[66] F. Jamar *et al.*, "EANM/SNMMI guideline for 18F-FDG use in inflammation and infection," *J Nucl Med,* vol. 54, no. 4, pp. 647–58, Apr 2013, doi: 10.2967/jnumed.112.112524.

[67] G. Treglia, S. Annunziata, D. Sobic-Saranovic, F. Bertagna, C. Caldarella, and L. Giovanella, "The role of 18F-FDG-PET and PET/CT in patients with sarcoidosis: an updated evidence-based review," *Acad Radiol,* vol. 21, no. 5, pp. 675–84, May 2014, doi: 10.1016/j.acra.2014.01.008.

[68] E. Lazzeri *et al.*, "Joint EANM/ESNR and ESCMID-endorsed consensus document for the diagnosis of spine infection (spondylodiscitis) in adults," *Eur J Nucl Med Mol Imaging,* vol. 46, no. 12, pp. 2464–87, Nov 2019, doi: 10.1007/s00259-019-04393-6.

[69] C. M. Mulders-Manders, I. J. Kouijzer, M. J. Janssen, W. J. Oyen, A. Simon, and C. P. Bleeker-Rovers, "Optimal use of [18F]FDG-PET/CT in patients with fever or inflammation of unknown origin," *Q J Nucl Med Mol Imaging,* Jul 1 2019, doi: 10.23736/S1824-4785.19.03129-7.

[70] I. J. E. Kouijzer, C. M. Mulders-Manders, C. P. Bleeker-Rovers, and W. J. G. Oyen, "Fever of Unknown Origin: The Value of FDG-PET/CT," *Semin Nucl Med,* vol. 48, no. 2, pp. 100–7, Mar 2018, doi: 10.1053/j.semnuclmed.2017.11.004.

[71] F. J. Vos *et al.*, "Metastatic infectious disease and clinical outcome in Staphylococcus aureus and Streptococcus species bacteremia," *Medicine (Baltimore),* vol. 91, no. 2, pp. 86–94, Mar 2012, doi: 10.1097/MD.0b013e31824d7ed2.

[72] F. J. Vos *et al.*, "18F-FDG PET/CT for detection of metastatic infection in gram-positive bacteremia," *J Nucl Med,* vol. 51, no. 8, pp. 1234–40, Aug 2010, doi: 10.2967/jnumed.109.072371.

[73] R. Slart *et al.*, "FDG-PET/CT(A) imaging in large vessel vasculitis and polymyalgia rheumatica: joint procedural recommendation of the EANM, SNMMI, and the PET Interest Group (PIG), and endorsed by the ASNC," *Eur J Nucl Med Mol Imaging,* vol. 45, no. 7, pp. 1250–69, Jul 2018, doi: 10.1007/s00259-018-3973-8.

[74] M. Soussan *et al.*, "Patterns of pulmonary tuberculosis on FDG-PET/CT," *Eur J Radiol,* vol. 81, no. 10, pp. 2872–76, Oct 2012, doi: 10.1016/j.ejrad.2011.09.002.

[75] M. Sathekge, A. Maes, M. Kgomo, A. Stoltz, and C. Van de Wiele, "Use of 18F-FDG PET to predict response to first-line tuberculostatics in HIV-associated tuberculosis," *J Nucl Med,* vol. 52, no. 6, pp. 880–85, Jun 2011, doi: 10.2967/jnumed.110.083709.

[76] H. Esmail *et al.*, "Characterization of progressive HIV-associated tuberculosis using 2-deoxy-2-[(18)F]fluoro-D-glucose positron emission and computed tomography," *Nat Med,* vol. 22, no. 10, pp. 1090–93, Oct 2016, doi: 10.1038/nm.4161.

[77] F. J. Vos, C. P. Bleeker-Rovers, C. E. Delsing, B. J. Kullberg, and W. J. Oyen, "Bone-marrow uptake of (18) F-FDG during fever," *Lancet Infect Dis,* vol. 10, no. 8, pp. 509–10; author reply 510–1, Aug 2010, doi: 10.1016/S1473-3099(10)70137-7.

[78] R. Slart *et al.*, "A joint procedural position statement on imaging in cardiac sarcoidosis: From the Cardiovascular and Inflammation & Infection Committees of the European Association of Nuclear Medicine, the European Association of

Cardiovascular Imaging, and the American Society of Nuclear Cardiology," *J Nucl Cardiol,* vol. 25, no. 1, pp. 298–319, Feb 2018, doi: 10.1007/s12350-017-1043-4.

[79] J. Meller *et al.*, "Early diagnosis and follow-up of aortitis with [(18)F]FDG PET and MRI," *Eur J Nucl Med Mol Imaging,* vol. 30, no. 5, pp. 730–36, May 2003, doi: 10.1007/s00259-003-1144-y.

[80] H. Hautzel, O. Sander, A. Heinzel, M. Schneider, and H. W. Muller, "Assessment of large-vessel involvement in giant cell arteritis with 18F-FDG PET: Introducing an ROC-analysis-based cutoff ratio," *J Nucl Med,* vol. 49, no. 7, pp. 1107–13, Jul 2008, doi: 10.2967/jnumed.108.051920.

17 Special Considerations in Pediatric Nuclear Medicine

Sofie Lindskov Hansen, Søren Holm, Liselotte Højgaard and Lise Borgwardt

CONTENTS

17.1 INTRODUCTION

The development of nuclear medicine was fuelled by physicists, radiochemists, and physicians working with Hevesy's tracer principle in the 1940s and 1950s. It was recognized as a potential medical specialty already in the 1940s, when the effect of radioactive Iodine therapy was described in the *Journal of the American Medical Association* [1]. It has since then proved to be an indispensable tool for diagnosing and treating patients with a variety of diseases [2][3], the advantages being that the examinations are minimally invasive, sensitive, and highly specific. These qualities make diagnostic nuclear medicine procedures well suited for pediatric patients [4].

Pediatric nuclear medicine requires special attention as one cannot regard children as just small adults. Many things have to be taken into consideration and, to the physicist, the most important is the scaling of the administered activity according to, for example, body weight. This is important in all cases of nuclear medicine, but in pediatric nuclear medicine it is of special importance, since children are at greater risk than adults of developing secondary cancers following exposure to ionizing radiation. One reason is their longer life expectancy [5], and another is their morphology. The energy imparted from the emitted photons is distributed in smaller organs located closer together, due to the smaller physical size of the child. Finally, children are more radiosensitive, as they have a larger proportion of dividing cells undergoing DNA replication [6]. It is estimated that the life-time cancer risk following exposure in early childhood is approximately 3 times higher than the corresponding risk of an adult given the same radiation dose [7]. For this reason, it is important to be particularly careful and thorough when designing scan protocols for children. This is true for both the nuclear medicine examinations and the CT examination that may accompany the nuclear medicine examination. When dealing with pediatric nuclear medicine, it is always important to consider the fact that even small changes in tracer uptake can be pathological in children, contrary to adult patients. For this reason, nuclear medicine specialists with special training in pediatric nuclear medicine are needed when interpreting pediatric nuclear medicine scans.

Cancer in children accounts for less than 1 per cent of all new cancer cases [8], yet in the United States it is the second most common cause of death in children and adolescents, after trauma [9]. Early and accurate diagnosis plays a pivotal role in the outcome and survival of the child, which is why pediatric nuclear medicine is an important tool, as it provides a non-invasive method for diagnosing, staging, evaluating treatment, and finding recurrent disease.

DOI: 10.1201/9780429489501-17

Because the survival of children with cancer has increased markedly over the past decades [10], to almost 90 per cent on average, the importance of keeping the exposure to ionizing radiation to an absolute minimum has increased, as the risk of developing secondary cancer scales with the absorbed dose [11]. The ALARA principle states that radiation must be kept As Low As Reasonably Achievable, and all examinations should adhere to this principle. For this reason, it is always important to consider other non-radiating methods such as ultrasound or MRI. A physicist working at a nuclear medicine department where pediatric examinations are performed must be aware of exactly what clinical questions the pediatrician seeks to answer and should tailor the protocols to make sure that the clinical questions can be adequately addressed at the lowest possible exposure to ionizing radiation. Hence, it is important to understand how the parameters pertaining to the actual acquisition as well as the post-processing affect the image quality.

This chapter will introduce the most common issues concerning the investigations of pediatric patients and the radiopharmaceuticals used in these examinations.

17.2 COMMON PEDIATRIC EXAMINATIONS

The most common examinations in a department with pediatric nuclear medicine vary a lot depending on the clinicians who are being assisted, the patient populations, and thereby the scans being requested. The most dominating groups of diagnoses are children with renal diseases and children with oncological diseases. Childhood renal diseases are, first of all, congenital renal malformations and the many complications associated with this.

Nuclear medicine examinations may be performed on blood samples or using imaging equipment. The modalities in nuclear medicine include planar imaging with a gamma camera, SPECT or SPECT/CT scans and PET, PET/CT, or PET/MRI scans. Planar imaging with a gamma camera yields two-dimensional images, whereas both SPECT and PET yield three-dimensional images of the patient. Common to these examinations is the use of radiopharmaceuticals. The actual image acquisition, however, differs greatly between the modalities. Both planar imaging and SPECT imaging are based on single photon emissions, whereas PET is based on the 511 keV photon pair emitted as a result of an annihilation event between a positron and an electron.

When planar imaging is performed, the distribution of the relevant radiopharmaceutical is imaged by one or two gamma cameras at a fixed angular position. The patient is placed above and/or under the gamma camera, and if an area larger than the field of view of the camera is required, then the patient is moved on the scanner bed along the camera or vice versa. A two-dimensional projection of the distribution of the tracer is imaged. If information about the rate of change of the distribution is needed for the diagnostic evaluation, this may be obtained in a dynamic acquisition at one fixed camera position.

A SPECT scan utilizes the same type of gamma camera as the ones used in planar imaging, but in order to obtain three-dimensional images, two or sometimes three gamma camera heads rotate around the patient in order to acquire the spatial information required for the three-dimensional reconstruction. Typically, 120 projection images are acquired at 3-inch increments [12].

In a PET scanner the patient is placed inside a detector ring, and coincidence counts resulting from two annihilation photons are detected. There are thus no moving camera heads involved, and the patient is moved on the scanner bed through the stationary gantry. The arms of the patient are most often placed above the head to avoid the attenuation from the arms. Depending on the specific type of scanner, 15–25 cm of the patient can be imaged per bed position. A pediatric whole-body PET FDG examination can image the whole body from the fingertips to the toes but can also be performed from head to mid-thigh or from ears to mid-thigh depending on the diagnostic question to be answered. Usually, it takes around 3 minutes to image one bed position, and the whole examination takes around 15–30 minutes in total. As PET scanners with longer axial fields of view become available, the scan time can be reduced accordingly. Some PET systems scan bed positions consecutively, while some newer scanners move the patient through the bore continuously. New digital PET detector-systems with time-of-flight (TOF) have increased sensitivity compared to images obtained on PET scanners with an analogue detector system. Digital PET thus offers images of higher resolution or, alternatively, images of comparable resolution but acquired at a lower administered activity, or at reduced scan-time, which is very attractive for pediatric examinations.

It is for the nuclear medicine physician to decide which one of the possible types of investigations is relevant for the given indication, following national and international guidelines of relevance.

The examinations relevant for the pediatric population differ from the examinations most often performed on adult patients. The most common types of pediatric nuclear medicine examinations are renal clearance with glomerular filtration rate (GFR) examinations, renography, bone scintigraphy, metaiodobenzylguanidine (MIBG) scintigraphy, and PET FDG. These examinations will be mentioned briefly below.

A GFR examination is based on initial injection of a radiopharmaceutical, usually ^{99m}Tc labelled diethylene-triamine-pentaacetate (DTPA), that is exclusively excreted by the kidneys. ^{99m}Tc has a half-life of 6.01 h. After the initial injection subsequent blood samples yield information about the rate at which the radiopharmaceutical disappears from the blood/is excreted by the kidneys. The kidneys' excretory function can now be assessed from the data obtained from the blood samples. Specifically, the data from the blood samples yield information about the plasma clearance, C_p using the formula:

$$C_p = \frac{D}{\int_0^\infty P(t)\,dt} \tag{17.1}$$

where D is the injected dose, and P is the plasma concentration and t is time [13]. (It is currently debated how to evaluate the integral in the denominator; the discussion is beyond the scope of this chapter.)

Possible indications include suspicion of kidney disease, assessment of disease progression, evaluation of treatment, and monitoring kidney function when the child is treated with nephrotoxic pharmaceuticals.

A renography is a dynamic imaging procedure of the kidneys, showing excretion from the kidney to the bladder via the ureters and thereby allowing visualization of the renal function. A renography is performed by placing the child lying on his/her back above a gamma camera. The gamma camera images the distribution of the radiotracer over time. The radiotracer typically used is mercaptoacetyltriglycine (MAG3) or diethylene-triamine-pentaacetate (DTPA) labelled with [^{99m}Tc]. The diagnosis is decided based on the images of the radiotracer distribution in the kidneys and a time-activity curve over the kidneys and the bladder. An empty bladder/dry diaper at the onset of the examination is important, as the tracer is excreted in the urine, meaning that a large amount of activity will accumulate in the bladder/diaper. This area of increased activity will deteriorate the image quality in the vicinity of this hot spot. The length of the examination is typically 20 minutes.

Bone scintigraphies are used to assess focal pathological processes (osteomyelitis, cancer, etc.) in bones and/or joints as well as traumas and consequences of fractures. The radiotracer hydroxydiphosphonate (HDP) labelled with [^{99m}Tc] is injected intravenously, and its distribution reflects the metabolism in the bones. A planar image of the whole body is usually obtained first. Based on the two-dimensional images, the relevant area for a more detailed SPECT/CT scan can be identified, and a three-dimensional image of the area of interest is obtained. The length of the examination is 20–40 minutes depending on the necessary coverage of the body.

An MIBG scintigraphy is used for diagnosing and assessing treatment of tumours in the immature nerve cells, called neuroblastomas in children – the most common extracranial solid malignant tumour in children [2]. The radioactive isotope in MIBG for diagnostic imaging is ^{123}I, which has a half-life of 13.22 h. The radiotracer is injected one day prior to image acquisition. A planar image yielding a two-dimensional image of the whole body is recorded first. MIBG uptake can be seen in the upper and lower extremities in case of bone marrow infiltration. It is important to perform a whole-body scan in order to ensure optimal conditions for the interpretation. The planar scintigraphy is then used to decide the area over which a SPECT/CT scan should be obtained. The thorax and abdomen are always included and, depending on the planar scintigraphy, supplemented with other areas. The length of the examination is around 150 minutes including SPECT/CT depending on the height of the child.

PET [^{18}F]-FDG is useful for detection, grading, and evaluating response to therapy for many cancer diseases, infections, and rheumatological diseases. The radioactive nuclide is ^{18}F built into the glucose analogue fluorodeoxyglucose. ^{18}F has a half-life of 109.8 minutes, and the pediatric patient is injected 60 minutes prior to image acquisition. It is important that the patient has been fasting for 4–6 hours before the examination. The child is placed with arms above the head, and the whole body can be imaged in approximately 10–30 minutes, depending on the size of the child and the scanner generation.

Childhood cancers differ from adult cancers with respect to clinical behaviour, response to therapy, and prognosis, and some cancers are predominantly or exclusively pediatric cancers. These include neuroblastoma, Wilms tumour, rhabdomyosarcoma, and retinoblastoma. Leukemia, lymphomas, tumours of the central nervous system as well as primary bone malignancies and sarcomas are more frequent in children relative to adults [14].

17.3 IMPORTANT CONSIDERATIONS SCANNING PEDIATRIC PATIENTS

When scanning children, it is especially important to ensure a good and calm atmosphere, and make sure that the child is as relaxed as possible before initiating the procedure. It is extremely important that the child and parents are well informed and feel comfortable with the situation.

The relation between administered activity, examination time, and possible motion artefacts must be considered in order to find the best possible compromise between scan time and possible sedation and image quality. First and foremost, one must keep the radiation exposure as low as reasonably achievable. Here the word reasonably becomes important, as the examination time is also of importance, and the lower the administered dose, the longer the time required to achieve an image of diagnostic quality. The longer the examination lasts, the more counts the cameras will detect, and the better the image quality. However, the possibility of motion artefacts will increase with increased examination time, and if the child is under general anesthesia (GA), a short examination time is also desirable, since anesthesia in itself may well involve a risk that is comparable to the risk associated with exposure to small amounts of radiation.

The position of the child in the scanner has to be as symmetric and as comfortable as possible. A symmetric position makes the reading and analysis of the images easier, and a comfortable position is good for the patient and results in fewer motion artefacts, but one must take into account attenuation from, for example, arms, and place the child so that as many photons as possible reach the detector. Velcro straps and cushions can be used to immobilize the child, and if it does not interfere with the scan procedure, a movie can be projected onto the ceiling during the scan. Figure 17.1 illustrates some of the techniques used to immobilize and calm children before and during an examination.

All personnel present before, during, and after the examination of a child, should be aware of their appearance, and their communication with the child, parents, or other staff involved in the examination. The role of the physicist is seldom to communicate directly with the child or parents, but all communication with the physician or tech-staff should be professional, calm, and friendly, as to avoid any misinterpretations by the child or parents.

It is possible to limit the use of sedation by ensuring a quiet environment, giving the child small gifts after the procedure and cooperating with child and parents in order to make them as comfortable as possible. A dedicated children's ward with toys, colourful yet calm wall decorations and staff dedicated to pediatric examinations are advantageous if available. In smaller children it should be attempted to fit the imaging procedure with the child's sleep routine in order to avoid sedation [10]. However, in some instances sedation is necessary. This is especially true for smaller children, having motion artefact sensitive procedures performed (e.g. SPECT and PET). It is important to be aware, that some types of sedatives affect the cerebral blood flow and metabolism, and, for example, for FDG PET examinations, the sedative should therefore be given 40 minutes after tracer injection [10].

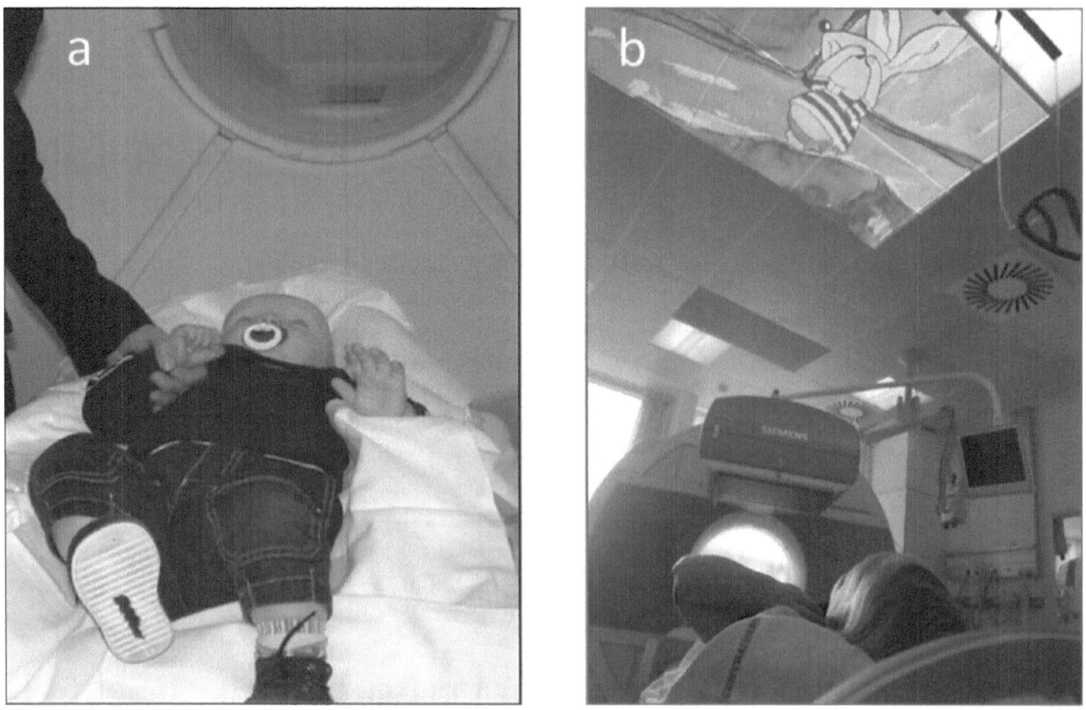

FIGURE 17.1 Left: A toddler getting ready for an examination. Cushions are used to immobilize the child. The presence of the parents often ensures a calm child. Right: An older child watching a cartoon while getting ready for an examination. Colour image available at www.routledge.com/9781138593312.

17.4 RADIOPHARMACEUTICALS AND DOSAGES

When talking about dosages in nuclear medicine, it must be considered that most of the examinations are done on hybrid scanners, meaning that the radiation dose to the patients stems from the administered radiopharmaceutical and the CT scan. This section deals with the dosages associated with the nuclear medicine part of the examinations, and the dosages associated with CT scans will be addressed in a later section.

The degree of detriment to biological tissue caused by radiation depends, not only on the type of radiation, but also on the type of tissue irradiated. The same equivalent dose delivered to two different tissues will result in different overall risks of radiation damage. For example, an equivalent dose of 1 mSv absorbed in the brain will not cause the same risk as 1 mSv absorbed in the lung or breast tissue. The tissue weighting factors reported in the International Commission on Radiological Protection (ICRP) 103 for radiation-protection purposes assign a 12 times higher (risk) value to the lungs and breast tissue than to the brain [7]. For this reason, girls in particular are at greater risk of developing secondary cancers as a result of exposure to ionizing radiation. In order to take the different tissue sensitivities into account, the effective dose is calculated. The effective dose is measured in Sievert (Sv) and is used to represent the stochastic health risk of an examination. It takes into account both the type of radiation and the radio-sensitivity of the organ or tissue irradiated.

Calculations of radiation dose to a patient require information about the anatomical and physiological characteristics of the exposed individual. For this reason, the ICRP published a report (publication 89, 2002) [15] with updated information on individual variation resulting from differences in age. Five postnatal age groups are defined: newborn, 1, 5, 10, and 15 years. Dosimetry calculations may be performed based on the anatomical and physiological parameters presented for the appropriate age group. This information is particularly important when assessing doses to body organs and tissues from internally deposited radionuclides, as is the case in nuclear medicine.

Below exemplary cases are mentioned, where the administered activity is correlated with an effective dose:

> A 1 year old child weighing 10 kg undergoing a bone scintigraphy using a [^{99m}Tc]-labelled phosphate or phosphonate will be injected with an activity of 94.9 MBq. The administered activity is determined using the EANM recommendations from "The new EANM pediatric dosage card" accessed via the PedDose App.

The resulting effective dose can be calculated from tables in the ICRP 128 [16]

$$1.8 * 10^{-2}\ \text{mSv/MBq} * 94.9\ \text{MBq} = 1.71\text{mSv}$$

In comparison, a standard adult man subjected to the same activity would receive an effective dose of 4.9 10^{-3} mSv/MBq * 94.9 MBq = 0.47 mSv.

For a PET FDG, a 5-year-old child weighing 20 kg will be injected with an activity of 3 MBq/kg * 20 kg = 60 MBq FDG. From the ICRP 128, a conversion factor of $5.6*10^{-2}$ mSv/MBq yields an effective dose of 3.36 mSv.

In comparison an adult subjected to the same activity would have received an effective dose of 1.14 mSv.

17.5 IMPORTANT CONSIDERATIONS REGARDING IMAGE QUALITY

In pediatric nuclear medicine, hybrid scans are preferred when both modalities are needed for diagnostics, since the two scans can be performed sequentially on one system. In pediatric nuclear medicine this is very important, since the child and parents only need to come to the hospital for one day and will only experience one stressful procedure, while they effectively get two scans. If the child is sedated or in general anesthesia only one sedation/general anesthesia needs to be performed. That is, either a PET or SPECT scan combined with a CT scan can be performed as a "one-stop-shop-scan." The combination of a PET and MR scanner yields a similar advantage.

PET/CT scanners are designed with a fixed gantry opening, and the patient-detector distance is thus fixed. The spatial resolution is best in the centre of the gantry but varies little across the gantry opening with newer scanners tending to have less variation due to better reconstruction methods and time-of-flight capabilities. Still, it is very important to centre the child in the gantry. For gamma cameras the spatial resolution depends on the distance between the source and detector as the photon flux follows the inverse square law. Thus, the gamma cameras are free to move so they can get as close to the patient as possible. An illustration of how the width of the point-spread function of a gamma camera depends on the distance between the detector and the point source is shown in Figure 17.2. This implies that, in a SPECT scan, the resolution is better at the surface of the patient than in the centre of the patient, as the resolution degrades with the source-detector distance.

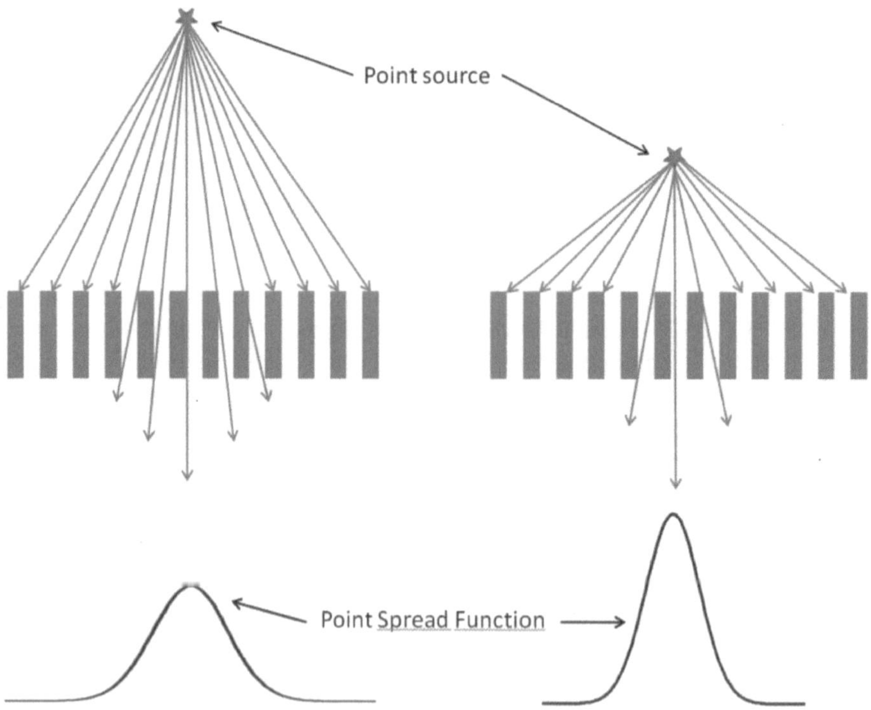

FIGURE 17.2 Illustration of how the point spread function (PSF) depends on the distance between the source and the detector.

17.6 PLANAR IMAGING – LIMITATIONS

Planar imaging yields a two-dimensional projection which, in some instances, such as renographies, may yield sufficient information for a diagnosis. If this is the case, this type of examination is preferred over a SPECT due to the ease of the investigation, and shorter examination time compared to a SPECT scan. However, there are obvious drawbacks of having data projected in a two-dimensional space. It makes the identification of pathological uptake "hiding behind" normal uptake very difficult. As an example, in the case of neuroblastoma, a tumour mass is often found in close proximity to the spine – a paravertebral element. In these cases, a three-dimensional image of the patient is required to evaluate the possible presence of tumour mass not visible on the planar images, and a SPECT/CT scan of the thorax and abdomen region is subsequently performed. This is illustrated in Figure 17.3, where the tumour mass is hardly visible on the two-dimensional planar image (lower left corner), but clearly visible on the relevant slices of the SPECT/CT.

17.7 COMPARING PET/CT, PET/MRI AND SPECT/CT

A typical PET scan has higher resolution than a typical SPECT scan due to the fact that the PET signal is not filtered by a collimator in order to ensure the directional information. Whether to choose a PET or a SPECT scan depends on the underlying clinical question, and the Society of Pediatric Nuclear Medicine has very well-defined guidelines [17] suggesting which procedures are most appropriate in a given clinical situation. The availability of relevant tracer nuclides and equipment will naturally play a role in the choice of procedure.

For musculoskeletal tumours, FDG-PET is comparable to scintigraphy in detecting the primary tumour but is superior for the detection of lymph-node involvement. Bone scintigraphy with [^{99m}Tc]MDP has low specificity for the assessment of response as there is increased uptake of the tracer in areas of tissue repair [18]. Consequently, FDG PET/CT is the gold standard for musculoskeletal tumours in children (and adults) for staging, monitoring, and finding recurrent disease. Bone scintigraphy is no longer used in the detection of musculoskeletal tumours. PET/MR is the most optimal modality for muscoloskeletal tumour evaluation since it also gives information on the primary tumour with the soft-tissue discrimination from MRI and the staging ability from PET, though at present a CT of the thorax also needs to be performed in order to get the highest diagnostic quality when evaluating the lungs.

Neuroblastoma can be imaged with the PET tracers [^{18}F]-FDG or 6-[^{18}F]-FDOPA or with the SPECT tracer [^{123}I]-MIBG. For the primary diagnosis MIBG and MRI are used. But in the past decade 6-[^{18}F]-FDOPA has slowly taken a

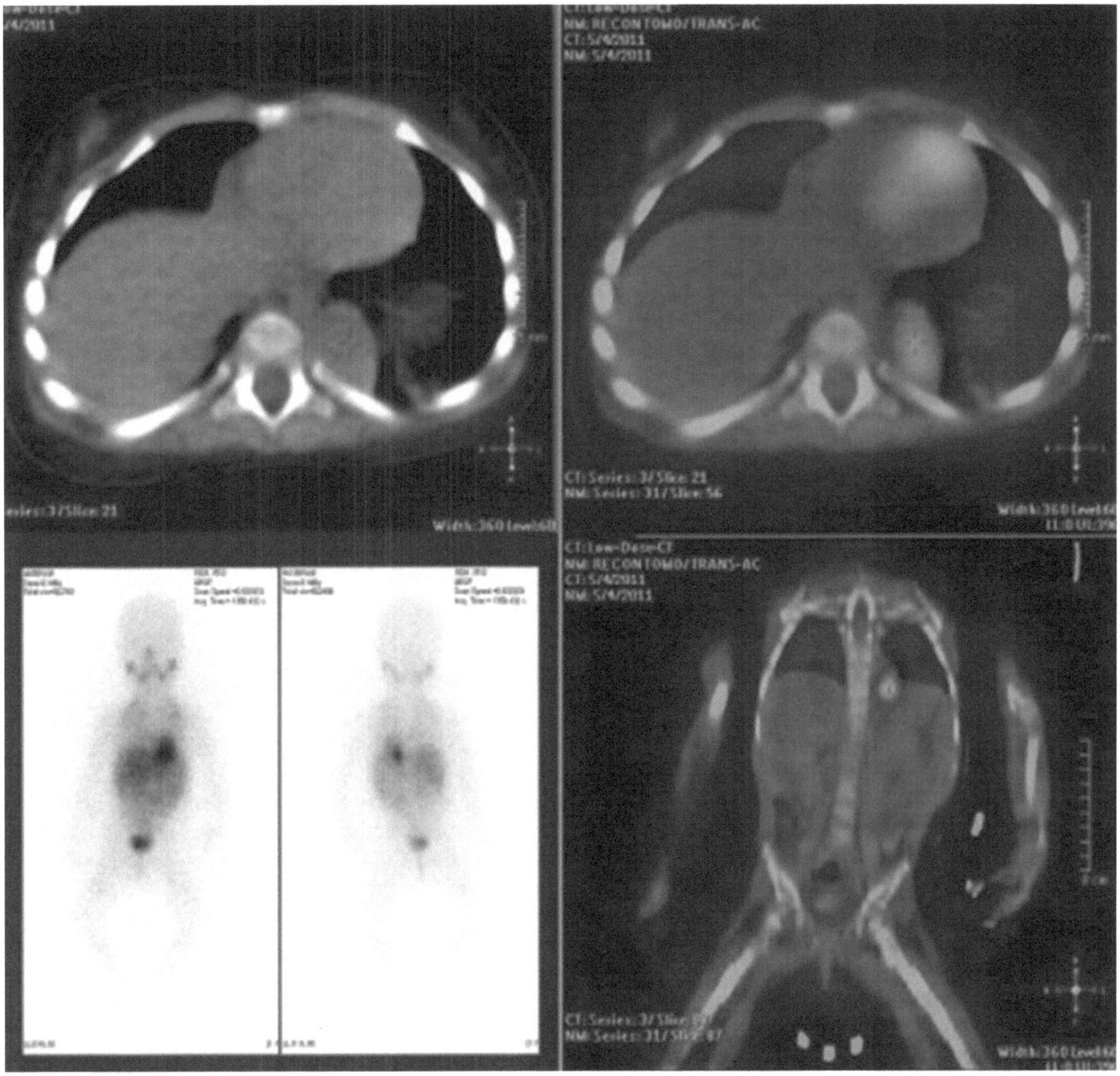

FIGURE 17.3 This is a 3-year-old girl with neurological symptoms with suspicion of neuroblastoma, a malignant tumour. It is very important to diagnose the tumour fast, not only to cure the cancer disease, but also because the severity of her neurological symptoms will increase, also chronically, with increased time to diagnosis. The tumour is hidden behind the heart on the planar images; the trained eye may be suspicious on the posterior images and perform a supplementary SPECT/LowDoseCT (LDCT), but it could be overlooked. SPECT/LDCT of the thorax and abdomen is therefore standard for a MIBG-scintigraphy. An illustration of how a tumour mass may be hard to see in planar images (lower left), while it is easily assessed in the relevant slices of the three-dimensional SPECT/LDCT scan, (lower right and upper left and right). Colour image available at www.routledge.com/9781138593312.

place in the diagnostic work-up of neuroblastoma. The difficulties in changing to the FDOPA PET tracer has been that the treatment protocols are based on the [^{123}I]-MIBG-scans, making treatment evaluation based on FDOPA scans difficult. Therefore the new PET tracer [18F]-meta-fluorobenzylguanidine ([^{18}F]-MFBG), a PET analogue of MIBG, seems very promising, as it allows for single-day, high resolution quantitative imaging [19].

From a technical point of view, there are important differences between PET/CT and SPECT/CT scanners, and there are different advantages and disadvantages of the two modalities. In most cases there is no choice between PET and

SPECT tracers for the same type of examination. However, in general, a SPECT examination takes more time than a PET examination of the same area due to the collimators in front of the gamma camera hampering the efficiency of the signal detection, and the resolution is higher in PET than SPECT scans, further PET is quantitative.

In PET/CT the CT scan may be replaced by a Magnetic Resonance Imaging (MRI) scan. However, PET/MRI scanners are most often only available in highly specialized departments. By omitting the CT scan and replacing it by an MRI scan, the radiation exposure can be significantly decreased, as there is no radiation dose associated with an MRI scan, which is especially important in pediatrics, as mentioned earlier [10]. Accordingly, the possibility to replace the CT scan by an MRI scan without compromising the diagnostic quality is a subject of great interest. Aside from the benefit of the decreased radiation exposure, MRI images display a high degree of soft tissue discrimination [20] [21] and evaluation of the bone marrow, which is advantageous in pediatric cancers such as sarcoma and many other pediatric diseases. But MRI has difficulties in evaluating bone metastases and lung metastases. Some studies indicate that lesion detection rates with PET/MR imaging are equivalent to that with PET/CT, and thus conclude that pediatric oncologic PET/MR is feasible, and show satisfactory performance for PET quantification with SUV's similar to those of PET/CT [22]. But it is still being discussed. The main argument for not always performing PET/MR instead of PET/CT, is the need to perform supplementary CT of the thorax in order to sufficiently diagnose lung metastasis in many pediatric oncological diseases, as the MRI-sequences are still not precise enough in diagnosing lung metastasis. Another issue is the increased need for sedation. In the PET/MRI-scanner, children typically need to be sedated until the age of 6–8 y, whereas in the PET/CT the age when sedation is needed is until approximately 4 years of age. PET/MR images are of great clinical relevance, and the setup with one-stop-shop scan and the lower radiation dose is very important in pediatrics. The use of the PET/MRI-scanning modality will increase as the method develops, when the length of the MRI-sequences is shortened, the sequences are less noisy, and preparation of the child and the parents for the procedure is developed, including standard procedures with playful pre-scanning training.

17.8 THE IMPORTANCE OF CT IN NUCLEAR MEDICINE EXAMINATIONS

Since the invention of CT in the 1970s, its use has rapidly increased due to the advantageous characteristics of the modality, such as excellent diagnostic accuracy, availability, and short acquisition time [6]. It is, however, associated with a higher radiation exposure than most other imaging modalities [6], and in 2009 it was estimated that CT scans are responsible for approximately 50 per cent of the collective effective dose associated with medical procedures [23]. Since then, a lot of technical improvements have been developed, and the radiation dose per scan is lower, but it is still an important issue to consider. As mentioned, children have a much larger lifetime risk per unit dose of radiation than adults. However, the use of hybrid imaging with CT in nuclear medicine has proven to contribute positively to the risk-benefit balance, as it serves as an added diagnostic modality that provides detailed anatomical information and has a high sensitivity for lesion detection and increases the diagnostic accuracy of many examinations in nuclear medicine [12]. The anatomical information in the CT images can sometimes aid the interpreting physician in distinguishing normal from abnormal physiological uptake [12]. Especially small sites of avidity are more easily assessed when the anatomical information from the CT scan is available. Furthermore, the information in the CT image can be used for attenuation correction, and thus improve the image quality of the nuclear medicine images by providing information about the inherent attenuation experienced by the photons emitted within the patient.

As the nuclear medicine images do not inherently contain any information regarding attenuation, an underestimation of the activity accumulated in the central part of the body and a relative overestimation of the activity closer to the surface of the body will result. (This is illustrated in Figure 17.4 and is especially pronounced in Figure 17.4 (a)).

In order to correct for both overestimation and underestimation of the activity due to attenuation, a map of the linear attenuation coefficients, μ, is required. Computed Tomography images represent tissue attenuation and can thus provide the information needed to correct for attenuation. Attenuation correction becomes increasingly important with the size of the attenuating object. For this reason, attenuation correction plays a smaller role in younger children than it does in adults. This is also visible when comparing sub-figures (a) and (b) with sub-figures (c) and (d) in Figure 17.4. However, as is visible from 17.4 (c) and (d), attenuation correction does play a significant role in depicting the distribution of activity accurately throughout the body.

The advantages of hybrid imaging and combining nuclear medicine modalities with a CT scanner are therefore both the added anatomical information *and* the improved depiction of the distribution of activity. However, in order to assure accuracy of the attenuation correction, it is important to keep in mind that the CT images represent the attenuation in terms of Hounsfield Units, and the images are acquired as a transmission map using a polychromatic spectrum of energies with a typical effective energy of 70 keV. The linear attenuation coefficient depends on material as well as photon

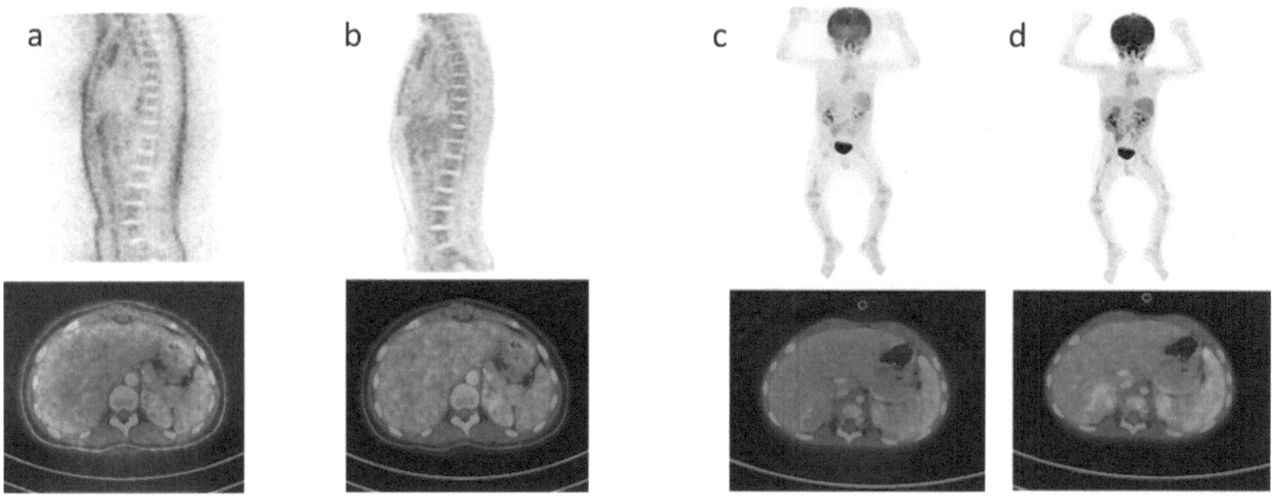

FIGURE 17.4 Illustration of the effect of attenuation correction. The top row shows PET images, and the bottom row shows fused PET and CT images. (a) Non-attenuation-corrected FDG PET images of a 16-year-old. The top image is the sagittal view of the PET image showing clear overestimation of the activity on the surface of the body. The bottom is an axial view fused with the corresponding diagnostic CT slice. (b) Attenuation-corrected FDG PET image of the same patient. (c) Non-attenuation-corrected FDG PET image of a 1.5-year-old child. The top shows the PET image, and the bottom shows an axial slice fused with the corresponding diagnostic CT slice. (d) The same images as in (c), corrected for attenuation. Colour image available at www.routledge.com/9781138593312.

energy, and the attenuation data measured at an effective energy of 70 keV must be converted to 140 keV for ^{99m}Tc or 511 keV in the case of annihilation photons in PET scans. The reason why the energies used in CT are much lower than the energies used in NM is that the type of interaction that gives rise to the highest degree of tissue differentiation is the photoelectric effect, and this effect dominates at lower energies. Furthermore, a CT scan involves a polychromatic spectrum, as the formation of the x-ray photons is governed by deceleration of electrons in the target material in the x-ray tube.

The conversion from Hounsfield Units to attenuation coefficients at a given energy is done using a bilinear model with simplifying assumptions regarding the composition of tissues [24][25].

An example of a bilinear conversion curve relating Hounsfield units to ^{99m}Tc is shown in Figure 17.5. Before the advent of hybrid scanners, such as PET/CT and SPECT/CT, the attenuation correction was performed using a radionuclide of a comparable energy to the injected tracer emission energy as a transmission source rotated slowly around the patient in a transmission scan. This was a lengthy procedure that only contributed to attenuation correction but did not add any other value to the image. In pediatric nuclear medicine the lengthy procedure is a problem. This is why the acquisition time of a CT scan in less than a minute, reducing the total imaging time significantly is of great importance. Additionally, the accurate anatomical image can be fused with the PET or SPECT images to yield a final image that contains both anatomical and physiological information. Both aspects make the combination of the nuclear medicine modality with a CT scanner favourable for pediatric patients. The shorter examination time again makes it less likely for the child to move during the procedure or having to be sedated for shorter time. Regarding the fused images, it has been shown that combining the nuclear medicine images with a CT image, adds diagnostic information [12]. It must be stressed, that the PET or SPECT scan must be acquired with the same patient positioning as the CT scan in order for the attenuation correction to be calculated and applied appropriately.

The CT component of the examination is controlled by several variable parameters. These parameters include kVp, mA, collimation, rotation speed, pitch, and the use of IV contrast. This means that the resulting image quality can be tailored very specifically to meet the requirements as dictated by the clinical need. For example, if the role of the CT scan is merely to correct for attenuation in the nuclear medicine image, the dose can be kept very low, as the required resolution is as low as the resolution of the nuclear medicine image. This class of CT images is often referred to as Attenuation Correction CTs (ACCT). If the CT image is to be used for attenuation correction as well as anatomical localization, the image should be acquired at an intermediate dose that allows for some degree of tissue differentiation. This class of CT images if often referred to as Low-dose CT (LDCT). Finally, CT images of diagnostic quality with very good resolution, good contrast, and tissue differentiation can be acquired by increasing the dose and using an iodinated

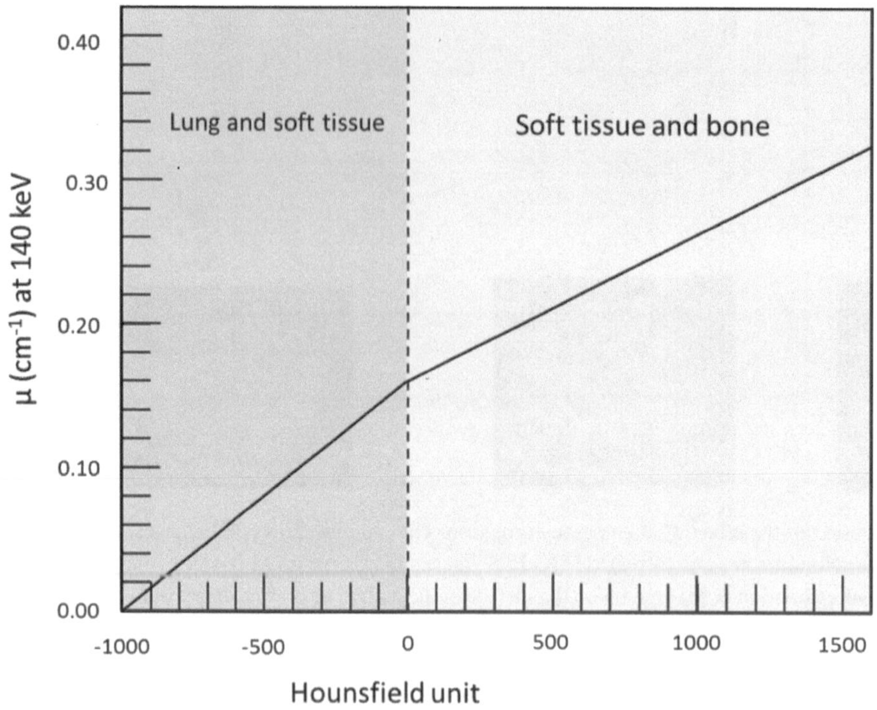

FIGURE 17.5 Illustration of the bilinear conversion curve used to convert the measured CT number obtained with an average energy of 70 keV to the corresponding linear attenuation coefficient for 140 keV photons.

contrast agent. These scans are referred to as Diagnostic CTs. Hence, it is very important to differentiate the indication for the CT in order to differentiate the CT dose given to the child. The goal of the CT scan can be grouped into three categories, from lowest to highest dose:

- CT for attenuation correction only (ACCT),
- CT for anatomical localization (LDCT),
- CT for additional diagnosis (DCT).

Diagnostic CTs are typically acquired using an iodine based intravenous contrast in order to opacify vascular structures; however, some centres chose to use IV contrast for low-dose CT examinations as well. This is currently debated, and the clinical value remains to be evaluated. Some studies indicate that the use of iodine-based contrast agents increase the absorbed radiation dose by up to 27 per cent in highly perfused organs [26].

17.9 CT DOSES AND REDUCTION TECHNIQUES

The availability combined with the ease of acquisition and good image quality, makes it tempting to order a PET/CT or SPECT/CT examination whenever a diagnostic challenge is encountered. However, one must always consider the benefit-risk ratio, and only perform a nuclear medicine examination when this is indicated. In this case, the referring physician must make very clear what clinical questions need to be answered so that the nuclear medicine physician and radiologist can assess the justification and choose the appropriate protocol.

CT scans are becoming increasingly popular in pediatric patients due to its accuracy and speed of acquisition. Although the risk associated with the delivered dose may be small for any one person, the increasing exposure to radiation on the population may become a public-health issue in the future [6].

In order to keep harvesting all the benefits of PET/CT or SPECT/CT scans, it is imperative to look at the possibility of obtaining images of good diagnostic quality, but at lower radiation doses.

This section will give an overview of typical CT doses administered to pediatric patients for a number of representative types of examinations relevant in nuclear medicine as well as discuss the current state of dose reduction techniques offered.

Doses from a CT examination vary greatly depending on the image quality requested by the physician and, naturally, the coverage of the CT scan. In pediatric nuclear medicine, the CT scan range can, when indicated, cover the whole body. This means that the Dose Length Product (DLP) is usually high compared to CT scans covering only parts of the body. The CT Dose Index (CTDI) measured in mGy and corresponding estimated effective doses (mSv) for whole-body CT scans performed at Rigshospitalet, Copenhagen, in 2020 are mentioned below. The conversion from CTDI to effective dose is done using the Monte Carlo–based program CT-expo [27].

A whole-body CT scan of a child between 2 and 5 years old, where the CT scan is of diagnostic image quality will result in CTDI values of around 2–3 mGy. This will result in effective doses in the order of 3–5 mSv. The same scan procedure performed on 5–10 year old children will result in CTDI values of around 3–5 mGy corresponding to effective doses of approximately 4–7 mSv. Children aged 15 and older will usually have a CTDI of around 5–7 mGy resulting in effective doses on the order of 7–10 mSv.

Low-dose CT scans of children younger than 15 years should give rise to CTDI values less than 1 mGy, and the corresponding effective doses can be kept around or below 1 mSv.

Finally, the dose associated with CT scans for attenuation correction only is so low, that effective doses <1 mSv will suffice [28].

Radiation-induced carcinogenesis is accepted as a stochastic process, meaning that the probability of developing a radiation-induced cancer increases with increasing radiation dose, yet the severity of the effect of the radiation is not influenced by the dose [29].

As the frequency of CT examinations increases, the number of radiation-induced cancers is likely to increase as well. In order to minimize the detrimental effects of the examinations, dose reduction techniques that will not compromise the diagnostic quality of the CT images are sought after.

Most commercial scanners allow the user to choose between filtered back projection (FBP) and iterative reconstruction methods. Furthermore, if the iterative reconstruction method is chosen, the "strength" can be set by the user. Iterative reconstruction methods have the potential to reduce the dose for a given image quality or to reduce the image noise for a given level of radiation exposure [30]. Although iterative reconstruction techniques increase the image quality when measured quantitatively, the images may appear smoother or glossier, which is disturbing to radiologists familiar with the texture known from FBP-reconstructed images. Consequently, vendors and scientists have embarked upon new ways of generating something that looks like a high-dose FBP image from low-dose images, and one promising path is to use deep-learning algorithms to up-convert low-dose images to high dose quality without changing the noise-texture of the images. One such commercially available deep-learning approach was recently presented. It is called True Fidelity and is developed by GE Healthcare. Their Deep Learning Image Reconstruction (DLIR) features a deep neural network that has been trained on high-quality FBP data sets in order to learn how to differentiate noise from signals, and thus to suppress noise without impacting anatomical and pathological structures [31].

The importance of dose reduction is evident, especially in children, and adherence to the ALARA principle is a goal for all pediatric imaging facilities worldwide. Consequently, information and experience regarding dose reduction techniques must be shared freely and willingly between facilities. The Image Gently Alliance is a coalition whose goal it is to provide safe pediatric imaging of high-quality worldwide by raising awareness [32].

REFERENCES

[1] S. M. Seidlin, L. D. Marinelli, and E. Oshry, "Radioactive iodine therapy: Effect on functioning metastases of adenocarcinoma of the thyroid," *CA. Cancer J. Clin.*, vol. 40, no. 5, pp. 299–317, Sep. 1990, doi: 10.3322/canjclin.40.5.299.

[2] A. Loft, P.-L. Khong, L. Borgwardt, A. K. Berthelsen, and L. Højgaard, "Pediatric whole-body oncology," in *Molecular Anatomic Imaging*, 3rd ed., G. K. von Schulthess, Ed. 2016, pp. 653–60.

[3] L. Borgwardt and L. Højgaard, "Pediatric inflammation and other nonneoplastic diseases," in *Molecular Anatomic Imaging*, 3rd ed., G. K. von Schulthess, Ed. 2016, pp. 660–68.

[4] S. T. Treves, "Introduction," in *Pediatric Nuclear Medicine/PET*, 3rd ed., S. T. Treves, Ed. Springer, 2007, pp. 1–16.

[5] "The ALARA concept in pediatric CT: Myth or reality? 1," doi: 10.1148/radiol.2231012100.

[6] D. J. Brenner, E. J. Hall, and D. Phil, "Computed tomography: An increasing source of radiation exposure," 2007.

[7] "ICRP 103." [Online]. Available: www.icrp.org/publication.asp?id=ICRP Publication 103. [Accessed: 23-Jun-2020].

[8] "Children's cancer statistics | Cancer Research UK." [Online]. Available: www.cancerresearchuk.org/health-professional/cancer-statistics/childrens-cancers#heading-Zero. [Accessed: 24-Feb-2020].

[9] R. M. Cunningham, M. A. Walton, and P. M. Carter, "The major causes of death in children and adolescents in the United States," *N. Engl. J. Med.*, vol. 379, no. 25, pp. 2468–75, Dec. 2018, doi: 10.1056/NEJMsr1804754.

[10] L. Borgwardt, A. K. Berthelsen, A. Loft, and L. Højgaard, "Special considerations in pediatric patients," in *Molecular Anatomic Imaging*, 3rd ed., G. k. von Schulthes, Ed. Wolters Kluwer, 2016, pp. 647–53.

[11] International Commission on Radiological Protection., *Low-dose Extrapolation of Radiation-related Cancer Risk*. Published for the International Commission on Radiological Protection by Elsevier, 2006.

[12] J. A. Patton and T. G. Turkington, "SPECT/CT physical principles and attenuation correction," *J. Nucl. Med. Technol.*, vol. 36, no. 1, pp. 1–10, Mar. 2008, doi: 10.2967/jnmt.107.046839.

[13] K. Bouchelouche, M. Rehling, and J. Frøkiær, "Urogenitalsystemet," in *Klinisk Nuklearmedicin*, 2011, pp. 92–119.

[14] L. P. Conolly, L. A. Drubach, and T. S. Treves, "Applications of nuclear medicine on pediatric oncology," *Clin. Nucl. Med.*, vol. 27(2), pp. 117–25, 2002, doi: 10.1097/00003072-200202000-00009.

[15] "ICRP 89." [Online]. Available: https://journals.sagepub.com/doi/pdf/10.1177/ANIB_32_3-4. [Accessed: 17-Jun-2020].

[16] "ICRP 128," 2015.

[17] "EANM Pediatric guidelines." [Online]. Available: www.eanm.org/publications/guidelines/paediatrics/. [Accessed: 16-Jun-2020].

[18] J. Stauss, K. Hahn, M. Mann, and D. De Palma, "Guidelines for paediatric bone scanning with 99mTc-labelled radiopharmaceuticals and 18F-fluoride," *Eur. J. Nucl. Med. Mol. Imaging*, vol. 37, no. 8, pp. 1621–28, Aug. 2010, doi: 10.1007/s00259-010-1492-3.

[19] N. Pandit-Taskar *et al.*, "Biodistribution and dosimetry of 18 F-meta-fluorobenzylguanidine: A first-in-human PET/CT imaging study of patients with neuroendocrine malignancies," *J. Nucl. Med.*, vol. 59, no. 1, pp. 147–53, Jan. 2018, doi: 10.2967/jnumed.117.193169.

[20] A. Kjær *et al.*, "PET/MRI in cancer patients: First experiences and vision from Copenhagen," *Magnetic Resonance Materials in Physics, Biology and Medicine*, vol. 26, no. 1. Springer, pp. 37–47, 25-Feb-2013, doi: 10.1007/s10334-012-0357-0.

[21] F. W. Hirsch *et al.*, "PET/MR in children. Initial clinical experience in paediatric oncology using an integrated PET/MR scanner," *Pediatr. Radiol.*, vol. 43, no. 7, pp. 860–75, Jul. 2013, doi: 10.1007/s00247-012-2570-4.

[22] J. F. Schäfer *et al.*, "Simultaneous whole-body PET/MR imaging in comparison to PET/CT in pediatric oncology: Initial results," *Radiology*, vol. 273, no. 1. Radiological Society of North America, pp. 220–31, 01-Oct-2014, doi: 10.1148/radiol.14131732.

[23] F. A. Mettler *et al.*, "Radiologic and nuclear medicine studies in the United States and worldwide: Frequency, radiation dose, and comparison with other radiation sources – 1950-2007 1," doi: 10.1148/radiol.2532082010.

[24] P. E. Kinahan, B. H. Hasegawa, and T. Beyer, "X-ray-based attenuation correction for positron emission tomography/computed tomography scanners," *Semin. Nucl. Med.*, vol. 33, no. 3, pp. 166–79, Jul. 2003, doi: 10.1053/snuc.2003.127307.

[25] K. J. LaCroix, B. M. W. Tsui, B. H. Hasegawa, and J. K. Brown, "Investigation of the use of X-ray CT images for attenuation compensation in SPECT," *IEEE Trans. Nucl. Sci.*, vol. 41, no. 6, pp. 2793–99, 1994, doi: 10.1109/23.340649.

[26] P. Sahbaee, E. Abadi, W. P. Segars, D. Marin, R. C. Nelson, and E. Samei, "The effect of contrast material on radiation dose at CT: Part II. A systematic evaluation across 58 patient models1," *Radiology*, vol. 283, no. 3, pp. 749–57, Jun. 2017, doi: 10.1148/radiol.2017152852.

[27] "CT-Expo – a novel program for dose evaluation in CT|INIS." [Online]. Available: https://inis.iaea.org/search/search.aspx?orig_q=RN:34039421. [Accessed: 23-Jun-2020].

[28] F. H. Fahey, M. R. Palmer, K. J. Strauss, R. E. Zimmerman, R. D. Badawi, and S. T. Treves, "Dosimetry and adequacy of CT-based attenuation correction for pediatric PET: Phantom study," *Radiology*, vol. 243, no. 1, pp. 96–104, Apr. 2007, doi: 10.1148/radiol.2431060696.

[29] Y. Nagayama *et al.*, "Radiation dose reduction at pediatric CT: Use of low tube voltage and iterative reconstruction," *RadioGraphics*, vol. 38, no. 5, pp. 1421–40, Sep. 2018, doi: 10.1148/rg.2018180041.

[30] A. C. Silva, H. J. Lawder, A. Hara, J. Kujak, and W. Pavlicek, "Innovations in CT dose reduction strategy: Application of the adaptive statistical iterative reconstruction algorithm," *Am. J. Roentgenol.*, vol. 194, no. 1, pp. 191–99, Jan. 2010, doi: 10.2214/AJR.09.2953.

[31] J. Hsieh, E. Liu, B. Nett, J. Tang, J.-B. Thibault, and S. Sahney, "A new era of image reconstruction: TrueFidelity technical white paper on deep learning image reconstruction," 2019.

[32] "Pediatric radiology & imaging | Radiation safety – Image gently." [Online]. Available: www.imagegently.org/. [Accessed: 09-Jun-2020].

18 Antibody-based Radionuclide Imaging

Steffie M. B. Peters, Erik H. J. G. Aarntzen and Sandra Heskamp

CONTENTS

18.1 INTRODUCTION

In the past years, PET and SPECT imaging with radiollabeled antibodies (immunoPET/SPECT) have shown great potential, especially in the field of oncology, but also beyond. In this chapter, we will focus on antibody imaging in cancer patients, as there is limited clinical experience in other diseases. However, it must be noted that the general principles discussed here can be translated to other disease settings as well. We will first provide a general overview of the different applications of antibody-based radionuclide imaging in oncology and provide some basic background about antibody structure and function, and the most commonly used antibody radiolabelling strategies. Subsequently, we will discuss practical aspects to consider when designing an antibody–radionuclide imaging study. Finally, we will address different aspects that should be taken into account when interpreting immunoPET/SPECT scans, and we will give a few examples of how this imaging approach has been applied in research and clinical practice.

18.2 APPLICATIONS OF ANTIBODY-BASED RADIONUCLIDE IMAGING

Molecular imaging with radiolabelled antibodies can play a significant role in early drug development and molecular characterization of tumours for individualized anti-cancer treatment. In early drug development, immunoPET/SPECT could provide information about the pharmacokinetics of monoclonal antibodies [1–3]. An important safety issue in drug

DOI: 10.1201/9780429489501-18

development is the maldistribution of antibodies, resulting in an adverse balance between safety (from effects in non-target tissues) and efficacy (on-target effects in target tissues). When applied in early drug development, immunoPET/SPECT can help to select antibodies which showed favourable pharmacokinetics. Furthermore, once a candidate antibody has been selected, immunoPET/SPECT can be used as a companion diagnostic to select specific patients that are most likely to benefit from the selected treatment [4, 5].

In clinical research and daily practice, immunoPET/SPECT has several potential applications [4, 6]. It can be used to non-invasively assess the expression levels of tumour-associated antigens. In this regard, imaging has several advantages over conventional approaches such as immunohistochemistry on tumour biopsies. First of all, it allows measurement of target expression of whole tumour lesions and their metastases, thereby avoiding misinterpretation due to tumour heterogeneity or sampling errors. Furthermore, it allows longitudinal monitoring of target expression, which could be of clinical relevance since target expression can change in time during disease progression or treatment. In certain cases, it can be used as a diagnostic tool if pathological findings and conventional imaging such as CT or [^{18}F]FDG PET are inconclusive. This is, for example, the case for antibodies that are highly specific for a tumour-associated antigen, such as HER2, CAIX, or PSMA. These antigens are expressed on breast tumours, renal cell cancer, and prostate cancer, respectively [7–9]. Next to target expression, in vivo imaging also takes into account target accessibility after systemic administration. Factors such as vascular permeability, interstitial fluid pressure, blood flow, and vessel density affect the uptake of monoclonal antibodies in a tumour (see section 18.6.2). If target accessibility is low, the therapeutic agent might not reach the tumour cells despite adequate expression of the target [10–12]. Therefore, the antitumour efficacy of the antibody is likely to be limited in this specific tumour. In research, immunoPET/SPECT can be used to visualize the expression and accessibility of target antigens in tumours and normal tissues, which can potentially play a role in patient selection and early-response monitoring of targeted therapies [13–15]. Furthermore, immunoPET/SPECT can be used for accurate dose planning for individualized radioimmunotherapy with ^{177}Lu- or ^{90}Y-labeled antibodies, such as for [^{90}Y]Y-DTPA-ibritumomab [^{177}Lu]Lu-DOTA-J519, or [^{177}Lu]Lu-DOTA-girentuximab [3, 8, 16–19].

18.3 BASIC STRUCTURE, IN VIVO BEHAVIOUR, AND RADIOLABELLING STRATEGIES OF ANTIBODIES

18.3.1 Basic Structure of Antibodies

Antibodies are heavy (~150 kDa) and Y-shaped globular plasma proteins, predominantly referred to as immunoglobulins (Ig). They consist of two regions, the antigen-binding fragment (Fab region) and the fragment crystallizable (Fc) region (Figure 18.1). The variable region of the Fab part is responsible for binding to specific antigens, while the Fc region interacts with immune cells and proteins. The Fc region also determines the antibody isotype: IgG, IgA, IgM, IgE, or IgD, each of which differs in function and distribution in the human body. For imaging and therapy, IgG is the most commonly used antibody isotype. Within IgG there are four subclasses: IgG1, IgG2, IgG3, and IgG4, each of which has a distinct role in the immune response. While the subtypes share 90 per cent similarity, small structural differences can lead to substantial changes in their in vivo behaviour. For example, the serum half-life of IgG1, IgG2, and IgG4 is approximately 21 days, while for IgG3 this is only 7 days [20, 21].

18.3.2 Long Circulation Time

In order to accurately interpret PET/SPECT scans of patients administered with radiolabelled antibodies, it is essential to understand the in vivo behaviour of antibodies. Also, it is important to understand the in vivo fate of the radionuclide in case the antibody is catabolized or the radionuclide becomes detached from the antibody. The latter will be discussed in section 18.4.1. For molecular imaging, radiolabelled antibodies are generally administered intravenously, resulting in rapid systemic distribution via the bloodstream. Circulating radiolabelled antibodies are taken up by different cells followed by degradation or recycling. Radiolabelled antibodies can be rescued from intracellular catabolism through their interaction with the neonatal Fc-receptor (FcRn), which is expressed by endothelial cells lining blood vessels, hepatocytes, and several immune cells such as macrophages, monocytes, and dendritic cells [21, 22]. By binding to FcRn, the antibody is protected from lysosomal degradation, and it is released back in the extracellular space or blood stream. This is the key mechanism responsible for the long serum-half-life of radiolabelled antibodies [21, 23].

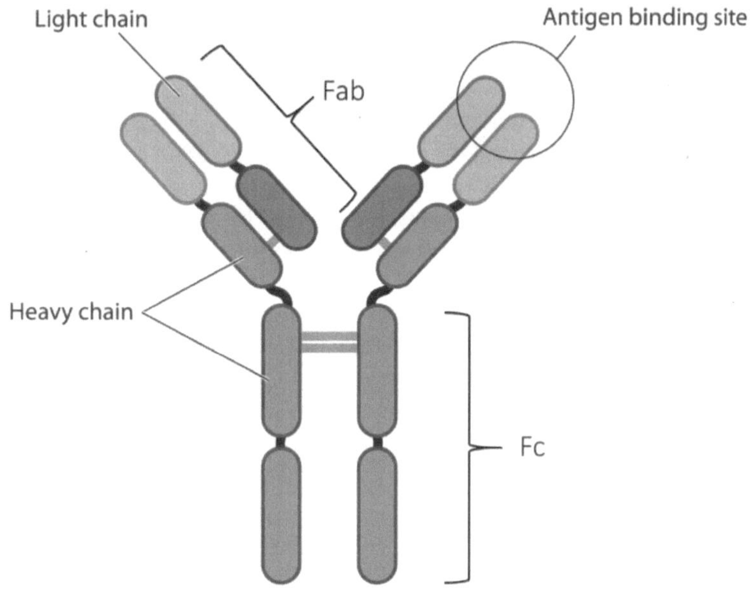

FIGURE 18.1 Basic structure of an immunoglobulin G (IgG).

18.3.3 Slow Tumour Penetration

In order to bind its receptor in the tumour, the radiolabelled antibody has to transport from the circulation, into the interstitial space of tissue. As antibodies have a high molecular weight and polarity, transport across the vessel wall is slow. On the other hand, tumour vasculature is often poorly organized, which aids the transport of the antibody into the tumour. In the human body, several tissues, including spleen, bone marrow, and liver, have naturally fenestrated capillaries resulting in high concentrations of radiolabelled antibody being driven into these tissues [21].

Within the tumour, the distribution of antibodies depends on diffusion and convection, the latter occurring when the antibody is carried along with the fluid that moves into the tumour. Also here, the large size of the antibody is a limiting factor and therefore transport within the tumour occurs extremely slow. The intratumoural distribution of the antibody is further hampered by the high interstitial fluid pressure, which presents strong physical resistance [21, 24].

18.3.4 Elimination from the Human Body

Systematic elimination of antibodies occurs mainly via catabolism. Unlike small molecules, the large size of full-length antibodies limits glomerular filtration and renal excretion. Biliary excretion also accounts for only a very small fraction of antibody elimination. Most of the radiolabelled antibodies are eliminated through intracellular catabolism. This is a non-specific process that can occur via two mechanisms: (1) non-specific endocytosis in cells, and (2) proteolysis in liver and reticuloendothelial system (RES). Non-specific endocytosis occurs in endothelial cells lining the blood vessels. Due to the large surface area of all endothelial cells in the human body, this process efficiently eliminates antibodies, and it occurs mostly in organs rich in capillary beds such as skin, muscles, and the gastrointestinal tract [25, 26]. It should be noted however, that approximately two thirds of the antibodies that are internalized via this pathway are protected against proteolytic degradation via FcRn-mediated recycling, as described previously. The remaining antibodies will be degraded to amino acids in the lysosomes. In addition to this process, antibodies are also capable of binding the Fc-gamma receptors (FcγR), which are expressed by Kuppfer cells in liver, monocytes and/or macrophages of the RES. Upon binding to FcγR, the antibody will be internalized and degraded in the lysosomes [21, 27].

18.4 PROPERTIES AND RADIOLABELLING APPROACHES OF COMMONLY USED RADIONUCLIDES FOR ANTIBODY IMAGING

Antibodies can be modified in several ways to allow radiolabelling for PET or SPECT imaging. Table 18.1 describes the different radionuclides suitable for antibody-based imaging. The choice for a suitable radionuclide depends on several factors, including matching the physical half-life of the radionuclide to the serum half-life of the mAb. This ensures that the radioactivity can be detected long enough so it can be imaged once the drug has reached its target, while preventing too much exposure to harmful radiation after the imaging timepoint. The serum half-life of mAbs or antibody-related therapeutics depends to a large part on the structure and size of the mAb and can vary from 30 minutes to 30 days [28] but is generally a couple of days. Therefore, the physical half-life of radionuclides should be in this range as well in order to be suitable for antibody imaging with a good signal-to-noise ratio.

Another important characteristic of radionuclides is the type of radiation. Positron-emitting radionuclides can be used for imaging with PET, while gamma-emitting radionuclides are suitable for imaging on a gamma camera or SPECT system. Generally, imaging with PET yields images with higher resolution. This is mainly a result of the fact that in PET there is no need for the use of a collimator for localization, leading to higher sensitivity. In addition to type of radiation, the emitted energy, the yield and, for positron emitting radionuclides, the positron range, are other factors that also affect the imaging quality.

In the next paragraphs, we will briefly describe the most common approaches to labelling antibodies with radionuclides, and we will discuss the in vivo fate of the radionuclide in case it gets detached from the antibody.

18.4.1 Iodine Isotopes

Radiolabelling of antibodies with iodine (^{123}I, ^{124}I, ^{125}I, ^{131}I) is generally performed via a direct electrophilic radioiodination of the tyrosine residues in the antibodies. This technique has been applied widely because of several favourable factors including the ease of radiolabelling, low costs, and availability of these isotopes. However, a major drawback is the fact that the radioiodine easily detaches from the antibody. When a radioiodinated antibody is taken up by the cell, it will be degraded in the lysosomes, resulting in the release of monoiodotyrosine, which is subsequently broken down to free iodine. This process occurs in tumour cells upon receptor-mediated uptake, resulting in reduced radioactivity concentration in the tumour. It also occurs in other cells that catabolize antibodies. Free iodine is released in the circulation and is rapidly taken up by tissues that express the sodium-iodine (Na^+/I^-) transporter, such as the thyroid gland and stomach (Table 18.1). In order to prevent the release of radioiodine from the cell, several radioiodinated prosthetic groups have

TABLE 18.1
Physical Characteristics of Radionuclides Used in Antibody-based Imaging [30]

Radionuclide	Emission	Half-life	Energy [keV]	Yield [%]	Positron Range [mm]	Main sites of accumulation of free radionuclide
^{64}Cu	β^+	12.7 h	653 (max) 278 (mean)	17.5	2.5 (max) 0.7 (mean	Liver
^{86}Y	β^+	14.7 h	1221 (max) 535 (mean)	31.9	5.6 (max) 1.9 (mean)	Bone
^{89}Zr	β^+	3.3 d	902 (max) 398 (mean)	22.7	3.8 (max) 1.3 (mean)	Bone
^{124}I	β^+	4.2 d	2138 (max) 0.975 (mean)	22.7	10 (max) 4.4 (mean)	Thyroid, stomach
^{99m}Tc	γ	6.0 h	140.5	90.0	n/a	Thyroid, salivary glands, stomach, colon
^{123}I	γ	13.2 h	159		n/a	Thyroid, stomach
^{111}In	γ	2.8 d	173 247	90.5 94.0	n/a	Liver, kidneys
^{177}Lu	γ	6.6 d	113 208	6.2 10.4	n/a	Bone
^{131}I	γ	8.0 d	364	81	n/a	Thyroid, stomach

been developed that help trap the radionuclide in the lysosomes. These strategies can result in enhanced tumour uptake and reduced uptake in thyroid gland and stomach. However, the radiochemistry of this approach is more challenging and, so far, results are mainly based on preclinical research. Therefore, the application in clinical practice is not yet clear [21].

18.4.2 Radiometals

In contrast to the direct labelling reaction of radioiodine, radiometals require a chelating moiety to couple the radionuclide to the antibody. Different radiometals require different chelators to allow stable and efficient labelling of the radionuclide. Here we will discuss the general principle and discuss some of the most frequently used radiometals for antibody imaging.

In general, the first step to label radiometals is to couple the chelating moiety to the antibody. In most cases, the chelator is randomly coupled to the free amine groups in the antibodies. However, new site-specific approaches are emerging that allow tight control of the localization and number of chelating moieties coupled to the antibody. This has a clear advantage compared with random conjugation, as the resulting product is more uniform, and the risk of altering the affinity and pharmacokinetics of the antibody is reduced. After conjugation of the chelator, the antibody-chelator complex can be labelled with the radionuclide of interest. In specific situations – for example if the radiolabelling requires conditions that damage the antibody – it is also possible to first label the chelator with the radionuclide and, in a second step, couple the chelator-radionuclide complex to the antibody. However, in most cases this is less practical or efficient.

The most commonly used radiometals for antibody imaging are ^{111}In, ^{89}Zr, and ^{64}Cu (Table 18.1). ^{111}In has a half-life of 2.8 days which matches the pharmacokinetics of antibodies. One of the most commonly used metal chelators that allow stable trapping of ^{111}In is DTPA. If ^{111}In-labeled antibody is taken up in the cell, the whole complex will be degraded in the lysosomes. However, in contrast to radioiodine, the ^{111}In-chelator-metabolite complex will not be transported out of the cell but will be trapped in the lysosomes and residualize in the cells, although there are reports that small amounts of ^{111}In-radiometabolites can be found in blood, urine, and feces as well. In general, trapping of ^{111}In-metabolites in the cell results in increased retention of the radionuclide in tumour, but also in non-target tissues. The chelation of ^{111}In by DTPA is very stable. However, if ^{111}In is released, it will be cleared via the kidneys or transchelate to iron carrying proteins in blood, such as transferrin and ferritin, leading to accumulation in the liver [21].

^{64}Cu is a positron-emitting radionuclide with a half-life of 12.7 hours. Although this is quite short for antibody imaging, several studies have used ^{64}Cu-labeled antibodies. This is possible when the injected activity dose is high and the time interval between injection and imaging is relatively short. A commonly used chelator for ^{64}Cu is DOTA. ^{64}Cu-labeled antibodies are rapidly catabolized upon internalization by the cell. The radiocopper-chelator-metabolite-complex is trapped within the lysosome, similar to ^{111}In. A drawback of ^{64}Cu-labeled antibodies is the high radioactivity concentration in the liver. This is because ^{64}Cu is very susceptible to transchelation to copper-binding proteins, which are abundantly present in the liver. New chelators, more resistant towards this transchelation issue, are currently being evaluated in preclinical and clinical studies to reduce liver retention [24].

Recently, ^{89}Zr gained a lot of attention as a new radiometal for immunoPET imaging. It emits low-energy positrons and has a physical half-life of 3.3 days. Labelling of antibodies with ^{89}Zr is mostly performed using the chelator DFO. Similar to ^{111}In and ^{64}Cu, ^{89}Zr is a residualizing radionuclide as well. However, ^{89}Zr is an osteophilic (bone-seeking) cation, thus, release of ^{89}Zr from the chelator results in accumulation in the mineralized constituents of the bone. This has especially been reported in mouse models when imaging was applied at several days post-injection of the antibody. From current clinical studies, this is less evident. However, current research is ongoing to find a more optimal chelator for chelation of ^{89}Zr – an alternative that shows less bone uptake [29].

18.5 PRACTICAL ASPECTS

18.5.1 Acquisition Protocols and Timing of Imaging

Acquisition protocols for antibody-based radionuclide imaging vary widely and are mainly dependent on the radionuclide of choice, the imaging system, the target region, and the pharmacokinetics of the mAb. (For elaborated information on the choice of many different acquisition parameters, please refer to Volume I, Chapter 15: "Image Acquisition Protocols"). One important practical aspect in antibody-based imaging is that full-sized antibodies require several days to reach the target tissue with sufficient target-to-blood ratios. This means that the patient will have to revisit the imaging facility several days after intravenous injection of the radiolabelled antibody. Several studies have compared immunoPET/SPECT scans acquired at different time points after injection, ranging from a few hours after injection up to 7 days, and found 5–7 days to be the optimum. At this time point, tracer accumulation in the tumour and clearance from circulation is sufficient to visualize lesions with good tumour-to-background ratios.

18.5.2 The Importance of Antibody Dosing

In order to obtain high tumour-normal tissue contrast images, it is important to achieve high targeting of the radiolabelled antibody in the tumour. How much antibody accumulates in the tumour depends on several factors, including target expression levels in the tumour, target accessibility for antibodies from the circulation, and the circulation time of the antibody in blood. In practice, antibody doses between 2–50 mg are used in clinical immunoPET/SPECT studies. However, the optimal dose may differ between antibodies and mainly depends on the expression levels of the antigen in tumour and normal tissues. At one hand, a low tracer dose is warranted in order to prevent saturation of the receptor of interest on the tumour cells. If the antibody dose is too high, all targets will be saturated, resulting in high concentrations of radiolabelled antibody in blood and low tumour-normal tissue contrast. On the other hand, most targets are not restricted to the tumour. For example, PD-L1 is also highly expressed on normal tissues, such as immune cells in the spleen. As a consequence, at low antibody doses, most of the injected tracer will accumulate in the spleen, resulting in rapid blood clearance and minimal targeting to PD-L1 positive cells in the tumour. In this case, the spleen is a so-called *sink organ* for radiolabelled anti-PD-L1 antibodies. By increasing the antibody dose, spleen uptake can be saturated, resulting in prolonged circulation time of the antibody and increased targeting to PD-L1 positive cells in the tumour. A comparable phenomenon has been described for HER2. Breast cancer cells highly over-express HER2 and by a phenomenon called "shedding," tumour cells can release HER2 domains in the circulation. Upon injection of a low dose of radiolabelled trastuzumab (anti-HER2 antibody), most of the injected antibody readily binds circulating HER2 which is subsequently cleared via the liver, resulting in short circulation time and high liver uptake. By increasing the antibody dose, circulating HER2 can be saturated, thereby restoring the circulation time, reducing hepatic uptake, and increasing the uptake in HER2-positive tumours. These two examples illustrate that for every tracer, the optimal antibody dose should be determined.

18.5.3 Radiation Safety

Evidently, when using radionuclides for imaging, the patient will be exposed to radiation. However, the final effective dose that the patient receives depends on many different factors. Each radionuclide disposes a different dose per injected activity, dependent on type of radiation and emission energy. The total amount of injected activity is furthermore determined by the imaging characteristics of the radionuclide. The amount of activity should be sufficient for imaging with an acceptable signal-to-noise ratio at the relevant timepoint. The mAb that the radionuclide is coupled to, mainly determines the location in the body at which the radionuclide will deliver its radiation and, thereby, which organs and structures are mainly affected by the radiation. As an example, imaging using the PET radiotracer ^{89}Zr with a typical injected activity of ±37 MBq leads to an effective dose to the patient of around 22 mSv. Although it might seem counterintuitive, using the SPECT radiotracer ^{111}In typically leads to the same effective dose of 22 mSv to the patient, despite the lower dose-conversion factor. This is due to the fact that a higher injected activity is required (~100MBq) due to the lower sensitivity of the SPECT scanner.

Radiation safety for proxies is another issue to take into account when working with radiolabelled antibodies. Patients might need to follow special guidelines in daily life directly after injection. Whether this is necessary is mainly dependent on the external dose rate of the patient after injection and the risk of causing a radioactive contamination, which in its turn depends on the specific radionuclide and the amount of injected activity. For example, when injecting ^{89}Zr-labeled antibodies, it is recommended that patients follow a specific set of rules for around a week, including limiting the time of close contact to proxies, extra attention for toilet hygiene and informing other healthcare institutions about the received activity when visiting. These precautions ensure that other people are not exposed to high radiation doses. The radiation safety precautions are generally developed by the imaging facility itself, in close collaboration with the radiation safety officer.

18.5.4 Production of Radiolabelled Antibodies under Good Manufacturing Practice (GMP) Guidelines

Production of radiolabelled antibodies for immunoPET/SPECT imaging requires dedicated laboratory facilities and skilled personnel, which are not available at every hospital. When a radiolabelled antibody is to be used in a clinical trial, the radiolabelling should be performed under GMP conditions, and the radiolabelling process should be carefully validated. An investigational medicinal product dossier (IMPD) needs to be compiled that specifies the manufacturing process and control of all critical steps such as purification, quality control methods, vialing, storage, stability, and validation [31, 32]. The need for dedicated facilities and personnel may hamper the clinical application of radiolabelled antibodies for immunoPET/SPECT imaging. However, because of the relatively long half-life of the used radionuclides, radiolabelled antibodies can be produced at a central radiopharmacy and shipped to other hospitals, depending on stability of the radiolabelled antibody. At these remote sites, the shipped radiolabelled antibody can be administered to a patient, followed by PET or SPECT imaging.

18.6 IMAGE INTERPRETATION

While interpreting immunoSPECT/PET scans, it is important to have basic knowledge about the in vivo distribution of antibodies and the fate of the radionuclide used. This is important to distinguish target-mediated uptake from physiological antibody uptake or even free radionuclides which can detach from the antibody. More details on this can be found in section 18.3 and 18.4, below we will discuss the consequences of this for the correct image interpretation.

18.6.1 Physiological Uptake in Normal Tissues

In the previous section, we already discussed that antibodies have a long circulation time, therefore high background concentrations are observed in all well-perfused tissues, especially in scans acquired at early time points after injection. In addition to this, high concentrations of radiolabelled antibodies can be found in tissues that have naturally fenestrated capillaries, such as spleen, bone marrow, and liver. Furthermore, the Kuppfer cells in the liver are responsible for the degradation of a large fraction of the antibodies – therefore high radioactivity concentrations are found in the liver. Another factor contributing to high liver uptake is the presence of circulation antigens, such as described for HER2. When attempting to visualize tumour lesions, uptake in organs such as liver, spleen, and bone marrow should be taken into account.

18.6.2 Factors that Influence Tumour Uptake

In many studies, radiolabelled antibodies are used to determine the expression levels of the target of interest in a tumour. Unfortunately, the reality is more complicated and how much antibody will accumulate in the tumour is a sum of many factors, including receptor expression, enhanced permeability and retention (EPR), affinity, and internalization of the radiolabelled antibody, and subsequent residualization of the radionuclide [33]. We will shortly address these issues.

In theory, one could expect that if an antibody is highly specific for the receptor of interest, it will accumulate in high concentrations in receptor-positive tumours while the uptake in receptor-negative tumours is low or neglectable. However, preclinical studies have demonstrated that there is never a perfect correlation between tracer uptake and receptor expression. In clinical practice this is more difficult to assess, as we usually only have an archival biopsy of a small fraction of the tumour and, therefore, it is difficult to directly correlate imaging findings to immunohistochemical analysis of tumour tissue. There are many explanations for the discrepancy between receptor expression and antibody uptake. First of all, the antigen can be present, but its in vivo accessibility for the antibody is hampered. Poor tumour vascularization, low vascular permeability, and high intratumoural interstitial fluid pressure can prevent the antibody from reaching its target in tissue. How well a tumour is vascularized, and how accessible it is for antibodies, depends on different factors, including tumour type, size, location, and previous treatment (e.g. antiangiogenic therapy) [10]. Therefore, not all tumour areas that are positive for the target of interest necessarily demonstrate uptake of the radiolabelled antibody. The opposite can also occur due to the EPR effect. Antibodies can selectively leak out from tumour vessels and not from vessels in healthy tissue. This effect is caused by the abnormal tumour vessel architecture, including defective lining of the endothelial cells with wide fenestrations and ineffective lymphatic drainage. This causes accumulations of antibodies in tumour tissue, even in the absence of the target of interest [34].

Also, characteristics of the antibody itself affect the accumulation and penetration in the tumour. High target affinity results in strong binding between the antibody and antigen of interest. However, it also limits the penetration in the tumours, caused by a phenomenon called the *antigen binding barrier*. Antibodies that enter the tumour are in close proximity to antigens for which they have a very high affinity. Upon entering the tumour, these antibodies will bind their respective antigens, which prevents further transport of the antibody into the tumour. This can result in a heterogeneous antibody distribution in the tumour, despite the fact that the antigen itself can be homogeneously distributed.

By modifying the affinity of the antibody or by increasing the antibody dose, a more homogeneous distribution within the tumour can be obtained [21, 35].

Finally, both the type of radioisotope and the antigen itself also influence how much of the radioactivity is retained in the tumour. Certain tumour-associated targets are internalized upon binding of the radiolabelled antibody (e.g. PSMA, HER2, CAIX). Following internalization, the radiolabelled antibody can be degraded and, in case of a residualizing radionuclide (such as the radiometals ^{89}Zr, ^{64}Cu, and ^{111}In), the radionuclide-chelator complex is trapped inside the cell, while non-residualizing radionuclides (e.g. radioiodine) are excreted from the cell. Trapping of the radionuclide in the cells leads to increased tumour-to-background contrast [21].

18.6.3 Accumulation of Detached Radionuclides in Healthy Tissue

Depending on the radionuclide and radiolabelling procedure, there is always a risk that a fraction of the radionuclide gets detached from the antibody. This can occur because of transchelation – for example, in the case of ^{64}Cu, leading to high hepatic uptake. But it can also occur upon degradation of the antibody – for example, in the case of radioiodine, which is released from the tumour and cell and taken up by cells expressing Na^+/I^- transporters. Another example is ^{89}Zr. A fraction of this radionuclide can be released from its chelator DFO, which can result in bone accumulation. The amount of radionuclide that is released, and if or how this will influence the scan, depends on many factors, including the type of radionuclide, radiolabelling method used, timing of imaging, target of interest, and location of the tumour lesion [21].

18.6.4 Examples of Antibody-radionuclide Imaging from Clinical Practice

In this section, we will discuss a couple of examples of how antibody-radionuclide imaging can be used in clinical trials or in daily practice. Rather than providing a complete overview, this section is meant to illustrate the potential role of immunoPET, as these have been described in detail in several excellent reviews [5, 36–38]. Because of the favourable characteristics of ^{89}Zr – such as the 78.4 h half-life, which perfectly matches the pharmacokinetics of antibodies, the relatively low-energy positrons, which provide high-resolution PET images, and the residualizing properties, which improve tumour retention and enhance tumour-to-tumour tissues ratios – we will focus the next section on ^{89}Zr-labelelled antibodies. Table 18.2 provides an overview of ^{89}Zr-labelled antibodies that have been used in clinical trials.

The first clinical immunoPET study using a ^{89}Zr-labeled antibody study was reported by Börjesson and colleagues. They showed that primary head and neck squamous cell carcinomas (HNSCC) could be detected by PET imaging using ^{89}Zr-labeled chimeric anti-CD44v6 antibody U36 [1]. Since then, several studies have successfully used immunoPET for tumour imaging. One prime example is [^{89}Zr]Zr-DFO-trastuzumab. Dijkers and colleagues have shown that [^{89}Zr]Zr-DFO-trastuzumab PET can detect metastatic liver, lung, bone, and even brain lesions in patients with HER2-positive breast cancer [2]. Furthermore, [^{89}Zr]Zr-DFO-trastuzumab immunoPET supported clinical decision making in patients where HER2 status could not be determined by standard work up. The addition of a [^{89}Zr]Zr-DFO-trastuzumab PET scan to conventional work up changed treatment approach, increased the physicians' confidence, and improved disease understanding in a subset of patients [9, 39]. In line with this, [^{89}Zr]Zr-DFO-girentuximab PET demonstrated to be of additional value in solving diagnostic dilemmas in patients suspicious for clear cell renal cell carcinoma [8]. Figure 18.2 shows a typical example of a [^{89}Zr]Zr-DFO-girentuximab PET scan

Another potential application for antibody imaging is patient selection for targeted therapy. In past years, only a few studies have been published evaluating the predictive value of immunoPET for treatment response. However, results are promising. For example, patients with a positive [^{89}Zr]Zr-trastuzumab PET scan were more likely to respond to treatment with the antibody drug conjugate trastuzumab-emtansine (TDM1), compared with patients with a negative scan.[14] Similar results have been reported for antibodies blocking immune checkpoints in cancer. Bensch and colleagues showed that clinical responses in patients treated with atezolizumab better correlated with pre-treatment [^{89}Zr]Zr-DFO-atezolizumab PET than with immunohistochemistry or RNA-sequencing-based predictive biomarkers. Although these studies were performed in small groups of patients, and larger clinical trials are warranted to further investigate the potential imaging, their results illustrate the potential role of imaging in patient selection for targeted therapy.

Antibody-radionuclide imaging can also be used for accurate pre-treatment dose planning for individualized radioimmunotherapy with ^{177}Lu- or ^{90}Y-labaled antibodies. A proof-of-concept of such an approach was demonstrated by Rizvi and colleagues, who showed that a pre-treatment [^{89}Zr]Zr-DFO-ibritumomab PET scan could be used to

TABLE 18.2
Overview of ^{89}Zr-labeled Antibodies for ImmunoPET Imaging of Tumours

Tracer	Target	Patient population	Timing between injection and imaging	Antibody dose (mg)	Radioactivity dose (MBq)	Reference
[^{89}Zr]Zr-DFO-trastuzumab	HER2	Breast cancer	4-5 days	50	37	[2, 9, 14, 39]
[^{89}Zr]Zr-DFO J591	PSMA	Prostate cancer	6 – 8 days	20-25	180 – 200	[3, 7]
[^{89}Zr]Zr-DFO-girentuximab	CAIX	Renal cell cancer	4-5 days	5 mg	37 MBq	[8, 40]
[^{89}Zr]Zr-DFO-atezolizumab	PD-L1	Bladder, lung, and breast cancer	4-7 days	11 mg	37 MBq	[41]
[^{89}Zr]Zr-DFO-nivolumab	PD-1	Lung cancer	7 days	2 mg	37 MBq	[42]
[^{89}Zr]Zr-DFO-ibritumomab	CD20	Non-Hodgkin's Lymphoma	3 – 7 days	2 mg	48 – 88 MBq	[19, 43]
[^{89}Zr]Zr-DFO-bevacizumab	VEGF	Lung, breast, mCCRCC, neuroendocrine	4-7 days	5 mg	37 MBq	[44-47]
[^{89}Zr]Zr-DFO-pertuzumab	HER2	Breast cancer	5-8 days	50 mg	74 MBq	[48]
[^{89}Zr] Zr-DFO-U36	CD44v6	Head and neck squamous cell carcinoma	1-6 days	10 mg	75 MBq	[1]
[^{89}Zr]Zr-DFO-rituximab	CD20	B-cell lymphose, rheumathoid arthritis, Graves' orbitopathy, orbital inflammatory disease, neurolymphomatosis, multiple sclerosis	3 – 6 days	10 mg (preloaded with a therapeutic rituximab dose)	18 – 111 MBq •	[49-54]

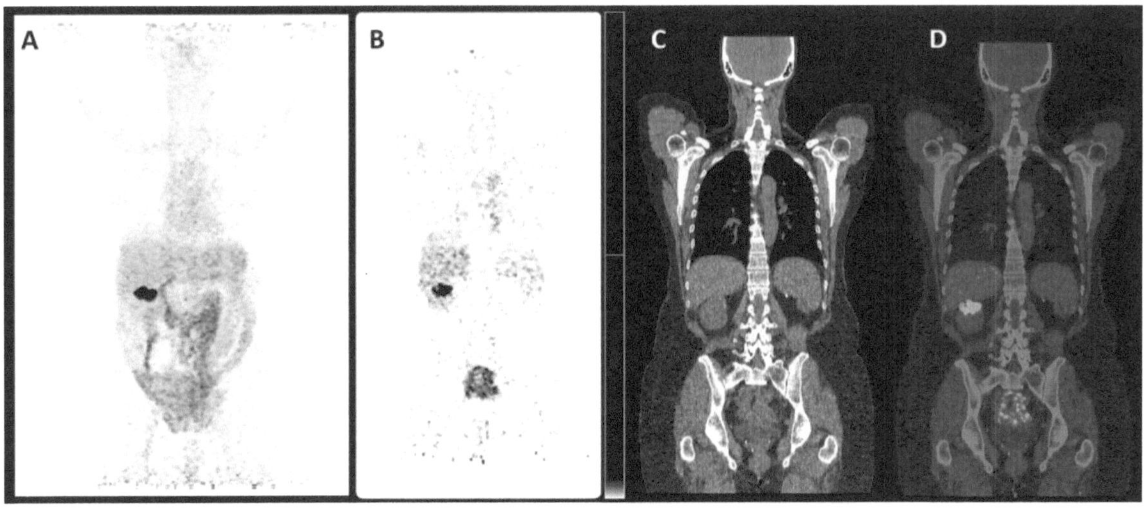

FIGURE 18.2 58-year-old female, who previously underwent left-side nephrectomy for benign reasons. One year before this scan she underwent partial right-sided nephrectomy for proven clear-cell renal cell carcinoma (Fuhrman grade II). Current [^{89}Zr]Zr-DFO-girentuximab scan is performed under suspicion of local recurrence; showing intense focal accumulation of [^{89}Zr]Zr-DFO-girentuximab in the upper pole of the right kidney (A, maximum intensity projection (MIP) image, B-D coronal images of PET, low-dose CT and fused, respectively). Physiological uptake of the ^{89}Zr-labelled antibody in spleen and liver, retained circulation of ^{89}Zr-labelled antibody in the circulation at 5 days post-injection. Colour image available at www.routledge.com/9781138593312.

predict biodistribution and dose-limiting organs during [^{90}Y]Y-DTPA-ibritumomab treatment [19]. Although not yet demonstrated, this application could be very useful for other radioimmunotherapy approaches as well, such as the combination of [^{89}Zr]Zr-DFO-girentuximab and [^{177}Lu]Lu-DOTA-girentuximab or [^{89}Zr]Zr-DFO-J519 and [^{177}Lu]Lu-DOTA-J519.[3, 8, 16, 17].

REFERENCES

[1] P. K. Borjesson *et al.*, "Performance of immuno-positron emission tomography with zirconium-89-labeled chimeric monoclonal antibody U36 in the detection of lymph node metastases in head and neck cancer patients," *Clin Cancer Res,* vol. 12, no. 7 Pt 1, pp. 2133–40, 2006, doi: 10.1158/1078-0432.CCR-05-2137.

[2] E. C. Dijkers *et al.*, "Biodistribution of 89Zr-trastuzumab and PET imaging of HER2-positive lesions in patients with metastatic breast cancer," *Clin Pharmacol Ther,* vol. 87, no. 5, pp. 586–92, 2010, doi: 10.1038/clpt.2010.12.

[3] N. Pandit-Taskar *et al.*, "A Phase I/II Study for Analytic Validation of 89Zr-J591 ImmunoPET as a Molecular Imaging Agent for Metastatic Prostate Cancer," *Clin Cancer Res,* vol. 21, no. 23, pp. 5277–85, 2015, doi: 10.1158/1078-0432.CCR-15-0552.

[4] B. N. McKnight and N. T. Viola-Villegas, "(89) Zr-ImmunoPET companion diagnostics and their impact in clinical drug development," *Journal of Labelled Compounds & Radiopharmaceuticals,* vol. 61, no. 9, pp. 727–738, 2018, doi: 10.1002/jlcr.3605.

[5] L. E. Lamberts *et al.*, "Antibody positron emission tomography imaging in anticancer drug development," *J Clin Oncol,* vol. 33, no. 13, pp. 1491–504, 2015, doi: 10.1200/JCO.2014.57.8278.

[6] W. Wei, D. Ni, E. B. Ehlerding, Q. Y. Luo, and W. Cai, "PET imaging of receptor Tyrosine Kinases in cancer," *Mol Cancer Ther,* vol. 17, no. 8, pp. 1625–36, 2018, doi: 10.1158/1535-7163.MCT-18-0087.

[7] N. Pandit-Taskar *et al.*, "(8)(9)Zr-huJ591 immuno-PET imaging in patients with advanced metastatic prostate cancer," *Eur J Nucl Med Mol Imaging,* vol. 41, no. 11, pp. 2093–105, 2014, doi: 10.1007/s00259-014-2830-7.

[8] M. C. H. Hekman *et al.*, "Positron Emission Tomography/Computed Tomography with (89)Zr-girentuximab Can Aid in Diagnostic Dilemmas of Clear Cell Renal Cell Carcinoma Suspicion," *European Urology,* vol. 74, no. 3, pp. 257–60, 2018, doi: 10.1016/j.eururo.2018.04.026.

[9] S. B. Gaykema, A. H. Brouwers, S. Hovenga, M. N. Lub-de Hooge, E. G. de Vries, and C. P. Schroder, "Zirconium-89-trastuzumab positron emission tomography as a tool to solve a clinical dilemma in a patient with breast cancer," *J Clin Oncol,* vol. 30, no. 6, pp. e74–75, 2012, doi: 10.1200/JCO.2011.38.0204.

[10] I. M. Desar *et al.*, "111In-bevacizumab imaging of renal cell cancer and evaluation of neoadjuvant treatment with the vascular endothelial growth factor receptor inhibitor sorafenib," *J Nucl Med,* vol. 51, no. 11, pp. 1707–15, 2010, doi: 10.2967/jnumed.110.078030.

[11] S. Heskamp, O. C. Boerman, J. D. Molkenboer-Kuenen, W. J. Oyen, W. T. van der Graaf, and H. W. van Laarhoven, "Bevacizumab reduces tumor targeting of anti-epidermal growth factor and anti-insulin-like growth factor 1 receptor antibodies," *Int J Cancer,* 2013, doi: 10.1002/ijc.28046.

[12] C. V. Pastuskovas *et al.*, "Effects of anti-VEGF on pharmacokinetics, biodistribution and tumor penetration of trastuzumab in a preclinical breast cancer model," *Mol Cancer Ther,* 2012, doi: 1535-7163.MCT-11-0742-T.

[13] S. B. Gaykema *et al.*, "89Zr-trastuzumab and 89Zr-bevacizumab PET to evaluate the effect of the HSP90 inhibitor NVP-AUY922 in metastatic breast cancer patients," *Clin Cancer Res,* vol. 20, no. 15, pp. 3945–54, 2014, doi: 10.1158/1078-0432.CCR-14-0491.

[14] G. Gebhart *et al.*, "Molecular imaging as a tool to investigate heterogeneity of advanced HER2-positive breast cancer and to predict patient outcome under trastuzumab emtansine (T-DM1): the ZEPHIR trial," *Ann Oncol,* vol. 27, no. 4, pp. 619–24, 2016, doi: 10.1093/annonc/mdv577.

[15] S. C. van Es *et al.*, "89Zr-bevacizumab PET: Potential Early Read Out for Efficacy of Everolimus in metastatic renal cell carcinoma patients," *J Nucl Med,* 2017, doi: 10.2967/jnumed.116.183475.

[16] S. T. Tagawa *et al.*, "Phase 1/2 study of fractionated dose lutetium-177-labeled anti-prostate-specific membrane antigen monoclonal antibody J591 ((177) Lu-J591) for metastatic castration-resistant prostate cancer," *Cancer,* vol. 125, no. 15, pp. 2561–69, 2019, doi: 10.1002/cncr.32072.

[17] C. H. Muselaers *et al.*, "Phase 2 Study of Lutetium 177-Labeled Anti-Carbonic Anhydrase IX Monoclonal Antibody Girentuximab in Patients with Advanced Renal Cell Carcinoma," *European Urology,* vol. 69, no. 5, pp. 767–70, 2016, doi: 10.1016/j.eururo.2015.11.033.

[18] L. R. Perk *et al.*, "(89)Zr as a PET surrogate radioisotope for scouting biodistribution of the therapeutic radiometals (90) Y and (177)Lu in tumor-bearing nude mice after coupling to the internalizing antibody cetuximab," *J Nucl Med,* vol. 46, no. 11, pp. 1898–1906, 2005.

[19] S. N. Rizvi *et al.*, "Biodistribution, radiation dosimetry and scouting of 90Y-ibritumomab tiuxetan therapy in patients with relapsed B-cell non-Hodgkin's lymphoma using 89Zr-ibritumomab tiuxetan and PET," *Eur J Nucl Med Mol Imaging,* vol. 39, no. 3, pp. 512–20, 2012, doi: 10.1007/s00259-011-2008-5.

[20] G. Vidarsson, G. Dekkers, and T. Rispens, "IgG subclasses and allotypes: from structure to effector functions," *Frontiers in Immunology,* vol. 5, p. 520, 2014, doi: 10.3389/fimmu.2014.00520.

[21] D. Vivier, S. K. Sharma, and B. M. Zeglis, "Understanding the in vivo fate of radioimmunoconjugates for nuclear imaging," *Journal of Labelled Compounds & Radiopharmaceuticals,* vol. 61, no. 9, pp. 672–92, 2018, doi: 10.1002/jlcr.3628.

[22] W. Wang, E. Q. Wang, and J. P. Balthasar, "Monoclonal antibody pharmacokinetics and pharmacodynamics," *Clin Pharmacol Ther,* vol. 84, no. 5, pp. 548–58, 2008, doi: 10.1038/clpt.2008.170.

[23] D. C. Roopenian and S. Akilesh, "FcRn: the neonatal Fc receptor comes of age," *Nature Reviews. Immunology,* vol. 7, no. 9, pp. 715–25, 2007, doi: 10.1038/nri2155.

[24] M. Tabrizi, G. G. Bornstein, and H. Suria, "Biodistribution mechanisms of therapeutic monoclonal antibodies in health and disease," *The AAPS Journal,* vol. 12, no. 1, pp. 33–43, 2010, doi: 10.1208/s12248-009-9157-5.

[25] J. T. Ryman and B. Meibohm, "Pharmacokinetics of monoclonal antibodies," *CPT Pharmacometrics Syst Pharmacol,* vol. 6, no. 9, pp. 576–88, 2017, doi: 10.1002/psp4.12224.

[26] A. Wright, Y. Sato, T. Okada, K. Chang, T. Endo, and S. Morrison, "In vivo trafficking and catabolism of IgG1 antibodies with Fc associated carbohydrates of differing structure," *Glycobiology,* vol. 10, no. 12, pp. 1347–55, 2000, doi: 10.1093/glycob/10.12.1347.

[27] A. T. Lucas, R. Robinson, A. N. Schorzman, J. A. Piscitelli, J. F. Razo, and W. C. Zamboni, "Pharmacologic considerations in the disposition of antibodies and antibody-drug conjugates in preclinical models and in patients," *Antibodies (Basel),* vol. 8, no. 1, 2019, doi: 10.3390/antib8010003.

[28] S.-P. Williams, "Tissue distribution studies of protein therapeutics using molecular probes: Molecular imaging," *The AAPS Journal,* vol. 14, no. 3, pp. 389–99, 2012.

[29] S. Heskamp, R. Raave, O. Boerman, M. Rijpkema, V. Goncalves, and F. Denat, "(89)Zr-Immuno-positron emission tomography in oncology: State-of-the-art (89)Zr radiochemistry," *Bioconjug Chem,* vol. 28, no. 9, pp. 2211–23, 2017, doi: 10.1021/acs.bioconjchem.7b00325.

[30] M. Conti and L. Eriksson, "Physics of pure and non-pure positron emitters for PET: A review and a discussion," *EJNMMI physics,* vol. 3, no. 1, p. 8, 2016.

[31] S. Todde *et al.*, "Guidance on validation and qualification of processes and operations involving radiopharmaceuticals," *EJNMMI Radiopharm Chem,* vol. 2, no. 1, p. 8, 2017, doi: 10.1186/s41181-017-0025-9.

[32] S. Todde *et al.*, "Guidance on validation and qualification of processes and operations involving radiopharmaceuticals," *EJNMMI Radiopharmacy and Chemistry,* vol. 2, no. 1, p. 8, 2017, doi: 10.1186/s41181-017-0025-9.

[33] S. Heskamp, H. W. van Laarhoven, W. T. van der Graaf, W. J. Oyen, and O. C. Boerman, "Radionuclide imaging of drug delivery for patient selection in targeted therapy," *Expert Opinion on Drug Delivery,* vol. 11, no. 2, pp. 175–85, 2014, doi: 10.1517/17425247.2014.870552.

[34] J. Fang, H. Nakamura, and H. Maeda, "The EPR effect: Unique features of tumor blood vessels for drug delivery, factors involved, and limitations and augmentation of the effect," *Advanced Drug Delivery Reviews,* vol. 63, no. 3, pp. 136–51, 2011, doi: 10.1016/j.addr.2010.04.009.

[35] S. I. Rudnick and G. P. Adams, "Affinity and avidity in antibody-based tumor targeting," *Cancer Biother Radiopharm,* vol. 24, no. 2, pp. 155–61, 2009, doi: 10.1089/cbr.2009.0627.

[36] F. C. van de Watering, M. Rijpkema, L. Perk, U. Brinkmann, W. J. Oyen, and O. C. Boerman, "Zirconium-89 labeled antibodies: A new tool for molecular imaging in cancer patients," *BioMed research international,* vol. 2014, p. 203601, 2014, doi: 10.1155/2014/203601.

[37] Y. W. Jauw *et al.*, "Immuno-positron emission tomography with Zirconium-89-Labeled monoclonal antibodies in oncology: What can we learn from initial clinical trials?," *Frontiers in Pharmacology,* vol. 7, p. 131, 2016, doi: 10.3389/fphar.2016.00131.

[38] M. A. Deri, B. M. Zeglis, L. C. Francesconi, and J. S. Lewis, "PET imaging with 89Zr: From radiochemistry to the clinic," *Nucl Med Biol,* vol. 40, no. 1, pp. 3–14, 2013, doi: http://dx.doi.org/10.1016/j.nucmedbio.2012.08.004.

[39] F. Bensch *et al.*, "(89)Zr-trastuzumab PET supports clinical decision making in breast cancer patients, when HER2 status cannot be determined by standard work up," *Eur J Nucl Med Mol Imaging,* vol. 45, no. 13, pp. 2300–06, 2018, doi: 10.1007/s00259-018-4099-8.

[40] S. R. Verhoeff *et al.*, "Lesion detection by [(89)Zr]Zr-DFO-girentuximab and [(18)F]FDG-PET/CT in patients with newly diagnosed metastatic renal cell carcinoma," *Eur J Nucl Med Mol Imaging,* vol. 46, no. 9, pp. 1931–39, 2019, doi: 10.1007/s00259-019-04358-9.

[41] F. Bensch *et al.*, "(89)Zr-atezolizumab imaging as a non-invasive approach to assess clinical response to PD-L1 blockade in cancer," *Nature Medicine,* vol. 24, no. 12, pp. 1852–58, 2018, doi: 10.1038/s41591-018-0255-8.

[42] A. N. Niemeijer *et al.*, "Whole body PD-1 and PD-L1 positron emission tomography in patients with non-small-cell lung cancer," *Nature Communications,* vol. 9, no. 1, p. 4664, 2018, doi: 10.1038/s41467-018-07131-y.

[43] L. R. Perk *et al.*, "Preparation and evaluation of (89)Zr-Zevalin for monitoring of (90)Y-Zevalin biodistribution with positron emission tomography," *Eur J Nucl Med Mol Imaging,* vol. 33, no. 11, pp. 1337–45, 2006, doi: 10.1007/s00259-006-0160-0.

[44] I. Bahce *et al.*, "Pilot study of (89)Zr-bevacizumab positron emission tomography in patients with advanced non-small cell lung cancer," *EJNMMI Res,* vol. 4, no. 1, p. 35, 2014, doi: 10.1186/s13550-014-0035-5.

[45] S. B. Gaykema *et al.*, "89Zr-bevacizumab PET imaging in primary breast cancer," *J Nucl Med,* vol. 54, no. 7, pp. 1014–18, 2013, doi: 10.2967/jnumed.112.117218.
[46] S. C. van Es *et al.*, "(89)Zr-Bevacizumab PET: Potential early indicator of Everolimus Efficacy in patients with metastatic renal cell carcinoma," *J Nucl Med,* vol. 58, no. 6, pp. 905–10, 2017, doi: 10.2967/jnumed.116.183475.
[47] S. J. van Asselt *et al.*, "Everolimus reduces (89)Zr-Bevacizumab tumor uptake in patients with neuroendocrine tumors," *J Nucl Med,* vol. 55, no. 7, pp. 1087–92, 2014, doi: 10.2967/jnumed.113.129056.
[48] G. A. Ulaner *et al.*, "First-in-human human epidermal growth factor receptor 2-targeted imaging using (89)Zr-Pertuzumab PET/CT: Dosimetry and clinical application in patients with breast cancer," *J Nucl Med,* vol. 59, no. 6, pp. 900–6, 2018, doi: 10.2967/jnumed.117.202010.
[49] K. Muylle *et al.*, "Tumour targeting and radiation dose of radioimmunotherapy with (90)Y-rituximab in CD20+ B-cell lymphoma as predicted by (89)Zr-rituximab immuno-PET: Impact of preloading with unlabelled rituximab," *Eur J Nucl Med Mol Imaging,* vol. 42, no. 8, pp. 1304–14, 2015, doi: 10.1007/s00259-015-3025-6.
[50] Y. W. Jauw *et al.*, "Performance of 89Zr-Labeled-Rituximab-PET as an imaging biomarker to assess CD20 targeting: A pilot study in patients with relapsed/refractory diffuse large B Cell lymphoma," *PLoS One,* vol. 12, no. 1, p. e0169828, 2017, doi: 10.1371/journal.pone.0169828.
[51] S. Bruijnen *et al.*, "B-cell imaging with zirconium-89 labelled rituximab PET-CT at baseline is associated with therapeutic response 24 weeks after initiation of rituximab treatment in rheumatoid arthritis patients," *Arthritis Res Ther,* vol. 18, no. 1, p. 266, 2016, doi: 10.1186/s13075-016-1166-z.
[52] B. de Keizer, K. G. Laban, and R. Kalmann, "Zirconium-89 labelled rituximab PET-CT imaging of Graves' orbitopathy," *Eur J Nucl Med Mol Imaging,* 2019, doi: 10.1007/s00259-019-04599-8.
[53] K. G. Laban, R. Kalmann, R. J. Leguit, and B. de Keizer, "Zirconium-89-labelled rituximab PET-CT in orbital inflammatory disease," *EJNMMI Res,* vol. 9, no. 1, p. 69, 2019, doi: 10.1186/s13550-019-0530-9.
[54] A. de Jong, R. Mous, G. A. van Dongen, O. S. Hoekstra, R. A. Nievelstein, and B. de Keizer, "(89) Zr-rituximab PET/CT to detect neurolymphomatosis," *Am J Hematol,* vol. 91, no. 6, pp. 649–50, 2016, doi: 10.1002/ajh.24328.

19 Radionuclide-based Diagnosis and Therapy of Prostate Cancer

Sven-Erik Strand, Mohamed Altai, Joanna Strand and David Ulmert

CONTENTS

DOI: 10.1201/9780429489501-19

19.1 INTRODUCTION

Prostate cancer (PCa) is the most frequent cancer in men in the Western world. One of every 8–10 patients diagnosed with PCa will die from this disease. Localized PCa is commonly treated by removing the prostate gland by surgery or ablating it with external beam radiation therapy (EBRT). Radiotherapy alone, or in combination with androgen deprivation, is commonly added if the tumour has extended the prostate gland (locally advanced disease). Metastatic disease is treated by inhibiting the androgen receptor (AR) pathway, which governs both healthy cells and most cancerous prostate cells. Prostate-specific antigen (PSA) concentration in blood is clinically applied for risk assessment and monitoring of PCa.

In the initial diagnostic process, PSA levels are utilized in combination with other factors (such as age and family history) for evaluating the need for prostate biopsies, which are inevitable for determining the presence of the disease. The use of PSA has increased the chances for curative treatment of aggressive disease by increasing the ability to find the aggressive disease at an earlier stage. Measurement of PSA levels in blood is also a simple and inexpensive method for detecting recurrence after curatively intended treatments and monitoring treatment of advanced adenocarcinoma. However, PSA is an imprecise test when distinguishing between early stages of aggressive diseases from indolent cases, which have resulted in high rates of unnecessary biopsies and overtreatment. Unfortunately, commonly available imaging modalities have not been shown to significantly improve diagnostic precision, either alone or in combination with PSA. Moreover, computed tomography (CT) is not applicable for evaluation of the local extent of PCa or lymph node involvement, mainly because of low sensitivity resulting from poor spatial resolution in soft tissues.

Magnetic resonance imaging (MRI) provides significantly better spatial resolution than CT, but diagnostic accuracy for primary tumour is confounded by benign processes (inflammation, hyperplasia, and fibrosis) and further hampered by high inter-observer and intra-observer variability. Although MRI is generally not recommended for clinical assessment of primary tumours, it may be used in high-risk patients for the investigation of local extension and seminal vesicle invasion.

Due to suboptimal soft tissue imaging methods, assessment of metastases in non-skeletal structures is based on the analysis of surgically dissected pelvic lymph nodes, which is the primary site for tumour dissemination. The accuracy of this invasive method depends on the anatomical extent of the dissection; expanding the dissected area and increasing the number of collected lymph nodes enhance diagnostic sensitivity.

The current standard methodology for detecting skeletal metastasis is most commonly performed by identifying sites of active bone remodelling. Dissemination of PCa to the bone is causing a chemical and physical stimulus released by the metastatic foci that most commonly lead to bone deposition. This aberrant morphology can be visualized using several clinical imaging modalities. The clinically approved methods either visualize aberrant tissue morphology (as in CT and MRI) or incorporation of radiotracers in the remodelling of bone matrix. The dominant approach to detecting metastatic involvement of the skeleton is by a radionuclide bone scan (RNB) using ^{99m}Technetium-radiolabeled bisphosphonates. The common issue associated with CT/MRI and RNB is lack of specificity; benign pathologies, such as benign degeneration and microfractures, mimic the signal of metastatic disease in images generated by both technologies. This is a significant problem, given that PCa patients suspected of having metastatic bone lesions are often elderly and present with these age-associated benign skeletal findings.

There is an unmet need for novel radiopharmaceuticals that are applicable for specific imaging and therapy of PCa. Several promising molecularly specific radiolabelled ligands and antibodies are currently under clinical evaluation. In this chapter, we summarize contemporary radionuclide-based clinical methods applicable for PCa diagnosis and therapy, but also some of the most promising compounds that are under preclinical and early clinical development.

19.2 PROSTATE ANATOMY AND PHYSIOLOGY

The human prostate is a secretory chestnut-shaped male accessory gland, located at the base of the urinary bladder, and encloses the upper part of the urethra, Figure 19.1. The prostate consists of muscle fibres (mostly unstriped), fibrous-, elastic-, vascular-, nerve-, connective, and glandular tissue.

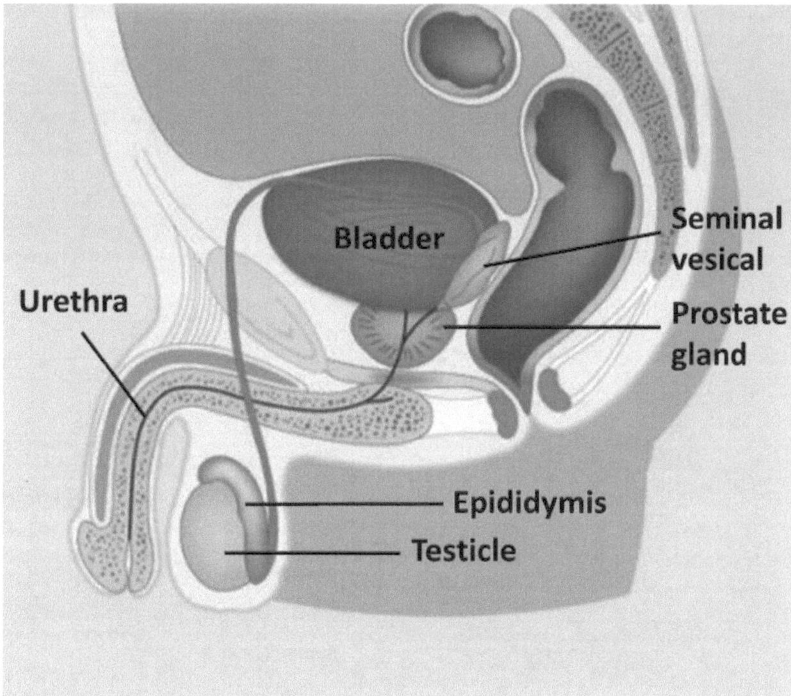

FIGURE 19.1 The anatomy of the male urinary tract with the prostate gland.

Although all major sex steroid hormones have been found to play a part in healthy prostate development and development of prostate pathologies, testosterone has the most impact. During ejaculation, sperm is transported from the testes along vas deferens (two bilateral tubes). The seminal vesicles, which are connected to the vas deferens at the entrance to the prostatic urethra, add a mildly alkaline fluid that contains gel-forming seminogelins that form an immobilizing matrix around the sperm. Next, prostate enzymes (PSA and hK2) that effectively liquefy the gel by cleaving the seminogelins are mixed in at the prostatic urethra. The prostatic fluid also contains high concentrations of zinc, which effectively inhibits the enzymatic activity from prematurely degrading the matrix. Normal PSA values in blood are age-related, with 40–49 years with approximate values such as < 2.5, 50–59 years, < 3.5, 60–69 years, < 4.5, 70 –79 years, < 5.5, above 80, > 7 ng/mL. However, enlarged prostate (benign prostate hyperplasia, BPH) can show PSA values of up to 20 ng/mL. The PSA production is schematically shown in Figure 19.2.

One of the earliest attempts to apply Positron Emission Tomography (PET) in PCa diagnostics was reported by Inaba [1]. He measured blood flow using ^{15}O-water PET and reported a significant difference in prostatic blood flow between BPH and cancer tissue (average blood flow was 15.7±7.4 mL/min/100 g in prostate tissue compared with 29.4±7.8 mL/min/100 g in PCa tissue). A subsequent study by Grkovski and colleagues, using ^{11}C-Choline-PET, also confirmed that the average blood flow rate in PCa is approximately 2–3 times higher than in normal prostate [2]. Similar results were shown in a study by Jochumsen and colleagues [3] using ^{82}Rb, a potassium analogue with intracellular trapping in metabolically active tissues at a rate proportional to tissue blood flow. ^{82}Rb is retained in the tissue, allowing both quantitative and semi-quantitative measurements. In their study they showed that ^{82}Rb uptake is higher in PCa than in normal prostate tissue and has potential as a non-invasive tool for evaluation of tumour aggressiveness.

19.3 PROSTATE CANCER

PCa (PCa) is the second most commonly occurring cancer in men and the fourth most commonly occurring cancer overall. There were 1.3 million new cases in 2018. The highest PCa rates are found in the Caribbean and Europe, while Central America, Australia/New Zealand, and Europe have the highest mortality rates (https://gco.iarc.fr/today/home), see Figure 19.3.

PCa incidence and mortality rates are strongly related to degree of PSA testing and age (> 65 years of age). It has also been shown that family history of PCa accounts for approximately 20 per cent of patients with PCa, which is not only an

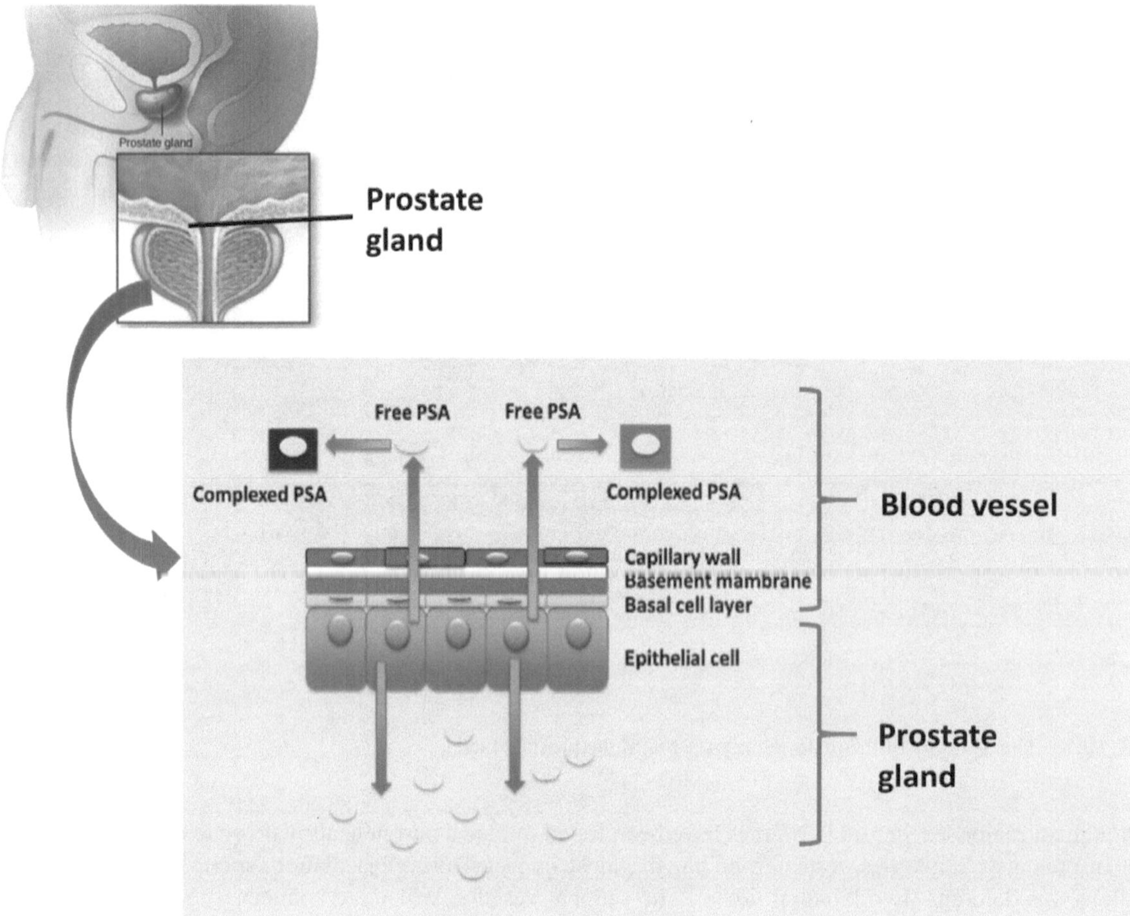

FIGURE 19.2 PSA production in the prostate gland and its leakage into blood. The blood levels of PSA are increased with PCa.

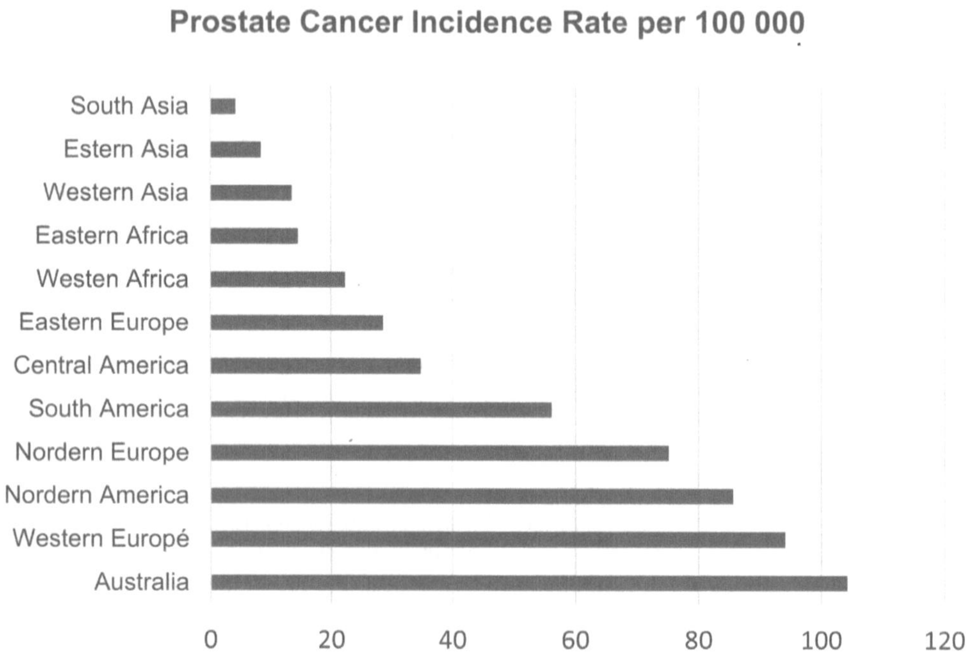

FIGURE 19.3 The global prostate cancer incidence for some regions in the world. Notable is the big difference between different regions.

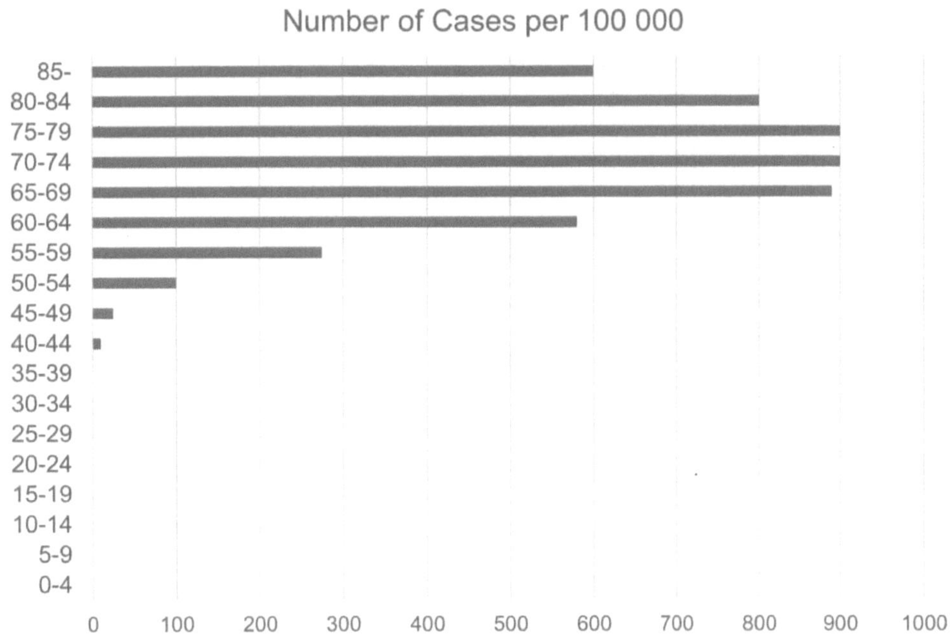

FIGURE 19.4 The incidence and deaths in prostate cancer in Sweden at different age intervals. The PCa incidence is highly age-related with high incidence at higher age. (Source: Redrawn from Swedish Cancer Society [5]).

effect of shared genes but also of a similar pattern of exposure to certain environmental carcinogens and common lifestyle habits [4]. In the United States, African-American men have the highest incidence rates and a more aggressive type of PCa compared to white men (https://gco.iarc.fr/today/home). According to the International Agency for Research on Cancer, approximately 345,000 men were diagnosed with PCa in Europe during 2012, and 72,000 men died of the disease. PCa is very rare before the age of 50 but is most common in men at age 70 and above, as can be seen in Figure 19.4, with data from Swedish Cancer Society (www.cancerfonden.se/om-cancer).

When there is clinical suspicion of PCa the patients first undergo rectal palpation of the prostate gland. If the gland is enlarged, asymmetric, or resistances can be palpable, there is definite suspicion of PCa. Furthermore, all patients with a clinical suspicion of PCa take a blood sample to measure PSA levels. If the PSA value is increased and/or palpation of the prostate indicates cancer suspicion the patient undergoes transrectal ultrasound, and biopsies are taken. If the PSA values are slightly increased. An alternative method for diagnosis has recently been introduced to the clinic, MR-imaging, followed by image-guided biopsies of the prostate gland. This method has as higher probability of detecting potentially serious cancer but avoids at least a quarter of unnecessary biopsies [6].

The results from tissue-sample analysis (biopsy) and imaging are then compiled into a tumour staging system, the TNM system. The system describes tumour invasiveness to surrounding tissue (T), spread to lymph nodes (N), and distant metastases (M). The T staging includes the following categories: T1 (incidental – PSA detected, and PCa not palpable by digital rectal examination); T2 (confined within the prostate gland, palpable); T3 (extending through the prostate capsule); and T4 (invading neighbouring organs) [5]. Further, the biopsy samples are scored according to the Gleason score grading system – the tumour is graded 1–5 according to the aggressiveness of the disease, with 1 being high differentiated cells and 5 being low differentiated cells. The overall Gleason score is calculated by adding the score of the most common grade and the highest grade seen in the biopsy samples. PSA levels, Gleason score, and tumour stage are the principal clinical parameters used to determine the likelihood of local or distant recurrent disease. PCa spreads mostly through lymphatic vessels, with pelvic nodes being the primary and most frequent site of metastases.

Localized PCa is treated with surgery and/or EBRT. For those patients having advanced androgen sensitive disease, the standard treatment is usually different forms of androgen deprivation therapy. If the disease progresses to castration-resistant PCa (CRPC), the prognosis becomes poor, with an expected survival of less than 19 months for patients with metastases. Commonly, CRPC is treated with the continuation of androgen deprivation, chemotherapy, and EBRT. When the tumour progresses to a lethal form, there are no effective long-term treatments.

Clinical trials have established the benefit of androgen-deprivation therapy (ADT) combined with radiotherapy in PCa. ADT sensitizes PCa to radiotherapy, inducing death at least in part through inhibition of DNA repair machinery [7].

Hypoxia initiated with androgen deprivation therapy (ADT) drives the selection of more malignant PCa tumours, that is, clonal selection, which perhaps can explain why patients usually relapse within 1–2 years with more malignant castrate-resistant PCa [8].

19.3.1 Principal for Targeting Diagnostics/Therapy

The main components of a radiopharmaceutical targeting PCa (and also other malignancies) are: (1) Targeting moiety/vehicle, a moiety recognizing a certain aberration or antigen on the tumour cells and that can either be retained on the cell surface or internalized; and (2) reporting/therapeutic unit (radioisotopes), see Figure 19.5. Exceptions for this model include some radioactive elements or ions that are taken up directly by the targeted organ/tissue due to the high natural affinity of that organ to the non-radioactive counterparts of these elements. Examples include ions such as ^{131}I (thyroid uptake)), $^{18}F^{-}$ (bone and skeletal tissue uptake), ^{89}Zr (bone uptake), and copper radioisotopes (liver uptake).

Most cancer-associated targets are antigens or enzymes, where antibodies and peptides are the preferred targeting moiety. To be able to image or give therapy, radionuclides with different decay modes have to be coupled to the targeting agent. Below, we will address commonly used radionuclides suitable for diagnosis and therapy. Thereafter, we will cover aspects of radiolabelling and review some of the most promising radiopharmaceuticals that are under preclinical and clinical development against PCa.

19.4 RADIONUCLIDES SUITABLE FOR DIAGNOSIS AND THERAPY OF PCA

The use of different radionuclides in nuclear medicine is governed by their mode of decay, that is, emissions. Radionuclides with predominant gamma emission following beta (β^-, β^+ and EC) decay are suitable for tumour imaging, whereas radionuclides with predominant beta (β^-) and alpha (α) decay are used for therapy. In the following, we will discuss some of the radionuclides used for diagnosis and therapy of PCa, their main production methods, and properties. These radionuclides can either be used directly (e.g., ^{223}Ra), incorporated into physiologically existing molecules (e.g., ^{18}FDG), or molecules targeting a specific aberration on PCa cells (e.g. ^{68}Ga-PSMA).

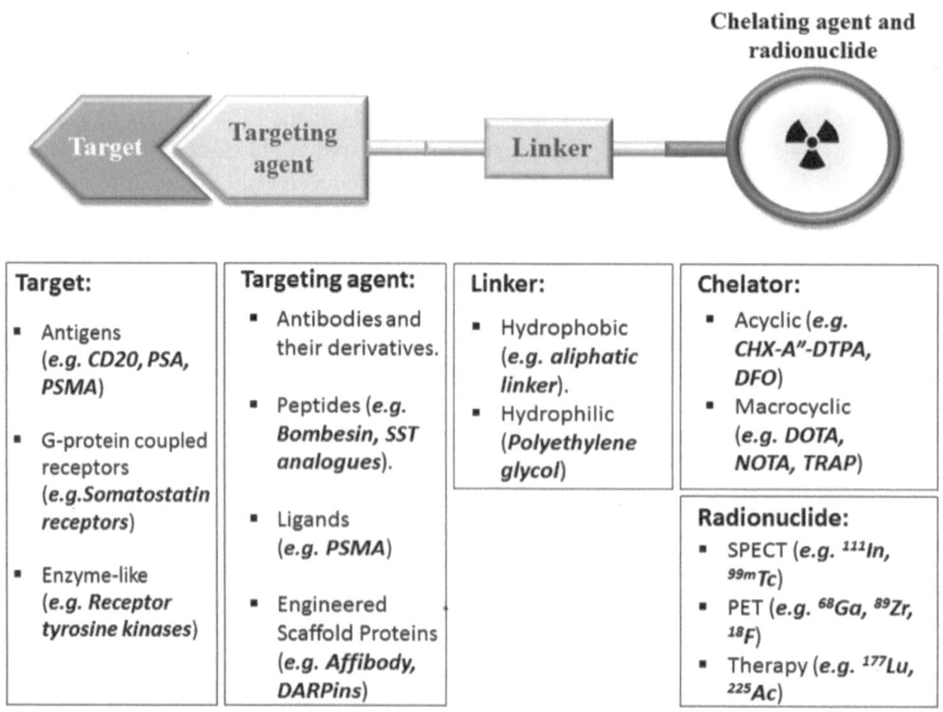

FIGURE 19.5 Illustration of the main components of a typical radiopharmaceutical used in nuclear medicine. Examples on targets, targeting agents, linkers, chelators, and radionuclides are given.

19.4.1 Radionuclides for Imaging in PCa

Before starting this section, we refer the reader to review Ghafoor and colleagues [9], for detailed information about various imaging modalities available for use in primary or recurrent PCa.

19.4.1.1 SPECT-radionuclides

Technetium-99m is the most readily available and commonly used radionuclide in nuclear medicine. This stems from the fact that ^{99m}Tc can be produced from decaying ^{99}Mo using a ^{99}Mo/^{99m}Tc generator, which is readily available at affordable prices. ^{99m}Tc decays by internal conversion to its ground state ^{99}Tc with a half-life of 6.04 h. The 140 keV emitted γ-photon is well suited for imaging using gamma cameras and SPECT. Indium-111 is also a γ-emitting radionuclide with two 172 keV and 245 keV gamma emissions and 2.8 days half-life. This relatively long half-life renders ^{111}In suitable for radiolabelling of different classes of targeting agents, with varying in vivo kinetics for imaging. One major advantage with indium isotopes is that they could easily be incorporated into different biomolecules using different chelating agents (see below). Other less available SPECT radionuclides include ^{67}Ga (Eγ= 93.3 keV (35.7%) and 184.6 keV (19.7%) with 3.3 days half-life and the radiohalogen ^{123}I (E_γ= 159 keV, $t_{1/2}$= 13.2 h) a decay product of the daughter ^{123}Xe produced after proton bombardment of ^{124}Xe in a (p, 2n) reaction.

In PCa ^{99m}Tc is mostly used to label phosphate compounds for imaging bone matrix. SPECT ^{99m}Tc methylenediphosphonate (^{99m}Tc-MDP) bone scintigraphy is the gold standard imaging modality used to detect bone metastases in PCa. ^{99m}Tc-MDP is absorbed in the site of sclerotic lesions (osteoblastic activity) following PCa metastasis. A major drawback is the lack of specificity to differentiate between different levels of osteoblastic activity.

19.4.1.2 PET-radionuclides

Positron Emission Tomography (PET) has become an important imaging modality to study physiological as well as pathological conditions in humans. Clinical PET cameras provide higher sensitivity and spatial resolution than SPECT as well as better quantification accuracy and dynamic scanning. With the development of whole-body PET cameras, that is, covering most of the body, detailed pharmacokinetics even at late time points (several half-lives of the radioisotope) are now feasible [10]. ^{18}F is the most used radionuclide for PET. Production at clinical cyclotrons, high abundance of positron decay (100%, E_{max}=0.64 MeV), and high spatial resolution (0.7 mm) favour the use of this radioisotope. Florine is a halogen – that is, it belongs to the elements of Group 7 in the Periodic Table, which have non-metallic properties. Other PET radionuclides from the same group include ^{124}I and ^{76}Br. Bromine-76 decays with a half-life of 16.2 h, emitting positrons in 54 per cent abundance with highest energy of 3.4 MeV. Unlike radioiodine, which often accumulates in the thyroid after in vivo metabolism of iodinated radiopharmaceuticals and thus can cause large, absorbed doses to this organ, radiobromine-labelled compounds pose no such concerns. Iodine-124 is also a long-lived PET radionuclide ($t_{1/2}$= 4.2 days), however, with a lower abundance of positron decay (23%). The higher positron energy of ^{76}Br and ^{124}I compared to that of ^{18}F provides inferior spatial resolution. Moreover, co-emission of high energy γ-photons also increases whole-body absorbed doses and decreases the image contrast.

The radiometals ^{68}Ga, ^{64}Cu, ^{44}Sc, ^{89}Zr, and ^{86}Y are considered as promising alternatives to ^{18}F. They are metals and can be coupled to peptides and antibodies with different chelators. The utilization of ^{68}Ga in oncological imaging has grown drastically due to the excellent properties of this radionuclide. Gallium-68 is a generator-eluted (^{68}Ge→^{68}Ga generator) short-lived ($t_{1/2}$=68 min) radionuclide with high abundance of positron decay (89%) and maximum energy of 1.92 MeV. The generator system ensures direct access to ^{68}Ga in clinical facilities for as long as one year (^{68}Ge half-life 270.8 d). This is an advantage over ^{18}F, which requires the availability of on-site or short-distance cyclotrons. As has been elaborated on elsewhere in this book there is a trade-off concerning particle range and spatial resolution. A good example of the physical properties concerning range and spatial resolution is given in Figure 19.6 comparing ^{18}F and ^{68}Ga [11]. In agreement with the positron range, the increasing relative spatial resolution is in the sequence (left to right) of ^{68}Ga>^{44}Sc>>^{64}Cu>^{18}F [12].

Copper isotopes are also gaining attention for both radionuclide-based imaging and therapy applications. Four copper radioisotopes are particularly promising for molecular imaging (^{60}Cu, ^{61}Cu, ^{62}Cu and ^{64}Cu) and two for targeted radionuclide therapy (^{64}Cu and ^{67}Cu). ^{64}Cu decays via positron, electron capture, and beta decays [$t_{1/2}$=12.7 h; E_{mean} β^+ = 0.28 MeV (17.86%); E_{mean} β^- =0.19 MeV (39.0%); EC (43.075%)]; this makes it suitable for both PET imaging and targeted radionuclide therapy, that is, theranostic radioisotope [13]. The beta particle emission implies a high absorbed dose, the electron capture decay is followed by the emission of Auger electrons with high linear energy transfer (LET); enhancing the therapy effect if the radionuclide is targeted near or within the cell nucleus. ^{64}Cu emits positrons of relatively

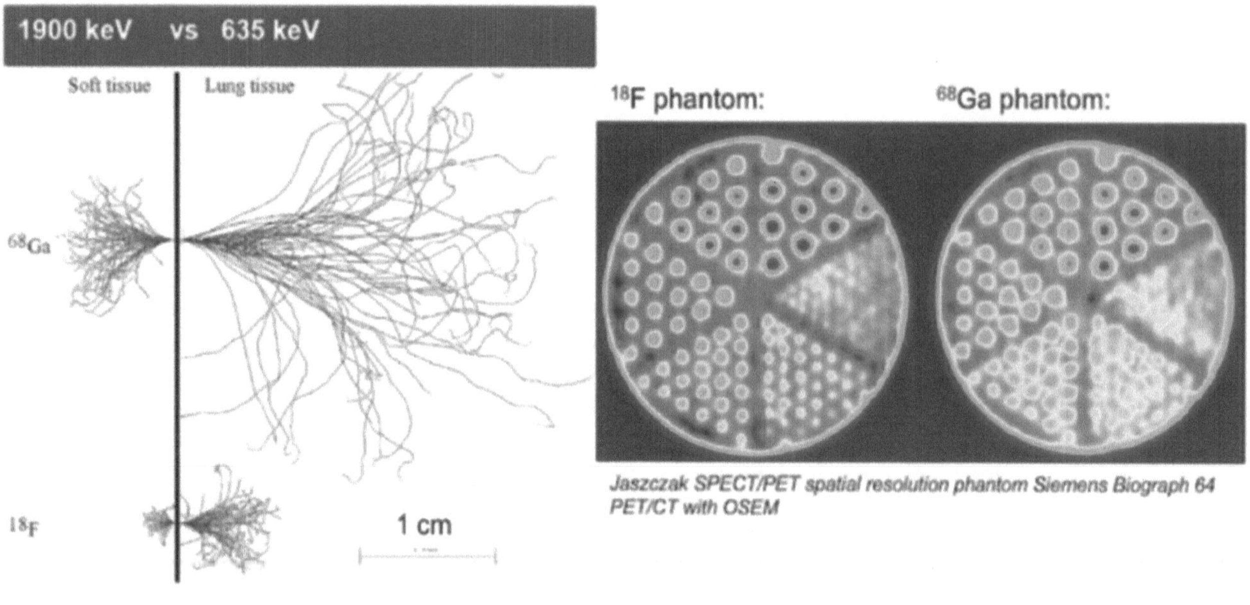

FIGURE 19.6 The range of positrons from the decay of ^{18}F and ^{68}Ga with impact on spatial resolution in PET imaging from reference, Sanchez-Crespo *et al.* [11]. Colour image available at www.routledge.com/9781138593312.

(Source: By permission from Sanches-Crespo [11].)

low energy (E_{av} = 278 keV), such as ^{18}F (E_{av} = 250 keV), therefore, the image spatial resolution of ^{64}Cu is expected to be better than that of ^{68}Ga, which emits positrons of a higher energy (E_{av} = 830 keV).

Scandium-44 is cyclotron-produced via the nuclear reaction $^{44}Ca(p,n)^{44}Sc$, or generator $^{45}Sc(p,2n)^{44}Ti \rightarrow {}^{44}Sc$. ^{44}Sc is a positron emitter with high abundance of decay (95 %), maximum positron energy of 1.47 MeV, and an accordingly shorter range in tissue compared to ^{68}Ga (7.6 vs. 9.8 mm). The longer half-life of ^{44}Sc (4 h) offers more flexibility for both radiolabelling and imaging time in comparison to ^{68}Ga and ^{18}F.

Recently, much interest has been directed towards the positron-emitter Zirconium-89 (E_{max}= 0.9 MeV, range 4.6 mm). ^{89}Zr has a half-life of 78.4 h and is produced by proton irradiation of natural yttrium-89. This relatively long half-life makes it suitable for the long kinetics exhibited by monoclonal antibodies (mAbs) in vivo. The major disadvantages of ^{89}Zr include limited availability and the high energy gamma emission (908.97 keV), which may limit the activity that can be administered into patients as well as compromise the image quality.

Yttrium-86 ($t_{1/2}$= 14.7 h, E_{max}= 1.25 MeV) is also of interest for radiolabelling proteins and peptides for PET imaging. This radionuclide presents the best diagnostic counterpart for the clinically used, therapeutic, pure β-emitter radioisotope, Yttrium-90. As these two radionuclides are chemically identical, differences in labelling chemistry and its influence on the properties of targeting agents can be omitted, thus allowing for the generation of high-fidelity companion diagnostic agents to ^{90}Y-radiolabled therapeutic agents.

The ultra-short half-life of ^{11}C ($t_{1/2}$= 20 min, E_{max}= 1.19 MeV) makes it only useful to label very small molecules with rapid kinetics. A comprehensive summary of radionuclides suitable for PET imaging is given by Nedrow and Anderson [14].

In Table 19.1 we provide a summary of radionuclides suitable for PET imaging.

19.4.2 Radionuclides for Therapy in PCa

Radionuclides suitable for application in cancer therapy are either β^-, α-, or Auger and conversion electron emitters.

19.4.2.1 Beta-emitters and Auger Electron Emitters

Beta-emitting radionuclides usually have a long path length (<12 mm) and low linear energy transfer (LET) (≈0.2 keV/μm), supporting their therapeutic effectiveness in medium to large tumours. The long β-particle range

TABLE 19.1
Common Radionuclides Suitable for PET Imaging of PCa

Radionuclide	Decay mode	Fraction ß+ (%)	Fraction Photons (%)/Energy(keV)	Half-life	E_{max} (ß+) (MeV)
^{18}F	β^+	100	0	110 m	0.64
^{68}Ga	β^+	88	3/1077	68 m	1.9
^{11}C	β^+	100	0	20 m	1.19
^{44}Sc	β^+	95	99/1157	3.9 h	1.47
^{61}Cu	β^+	61	12/284	3.3 h	1.215
^{64}Cu	β^+	18	0.6/1350	12.7 h	0.580
^{89}Zr	β^+	22	99/909	3.3 d	0.9
^{86}Y	β^+	33	83/1077	14.7 h	1.25

TABLE 19.2
Comparing the Decay Properties of ^{177}Lu and ^{131}I

	^{131}I	^{177}Lu
Half-life	8.02 d	6.73 d
Beta (energy/abundance)	608 keV (89.9%) 330 keV (7.3%) 250 keV (2.1%)	498 keV (78.6%) 385 keV (9.1%) 176 keV (12.2%)
Photons (energy/abundance)	636 keV (7.2%) 364 keV (81.7) 284 keV (6.14%)	113 keV (6.4%) 208 keV (11%)
Chemistry	Halogen	Metal chemistry

is advantageous in heterogeneous tumours as the effect expands to distant located cells, even if not targeted directly (crossfire effect). However, the crossfire effect can also result in the irradiation of healthy tissue surrounding the tumour site (i.e. bone metastases and bone marrow). For example, in targeted radioimmunotherapy using antibodies as targeting vectors and β-emitters, bone-marrow toxicity represents one of the most common side effects due to non-specific irradiation of hematopoietic stem cells as a result of the prolonged circulation of the radionuclide-carrying antibodies in the blood.

Iodine-131 is mostly produced by neutron-irradiation of a natural tellurium target in nuclear reactors and decays emitting both β^- particles and γ rays with a half-life of 8.02 days. The co-emitted high energy gamma (364 keV; 81.7%) also permits imaging. NaI (sodium/iodide) symporter-expressing tissues – for example, thyroid, stomach and salivary glands – are known to sequester ^{131}I (and other radioiodine isotopes), thus do not require the coupling of the radioisotope to a targeting vector. For thyroid studies, iodine isotopes are prepared as sodium salt, Na^{131}I, and administered orally or injected. However, lack of targeting may result in the accumulation of the radioactive isotope in non-diseased tissues such as the stomach and salivary gland.

Lutetium-177 is a medium-energy beta-emitter with a half-life of 6.7 days and a maximum energy of 0.5 MeV (maximum range in tissue of 2 mm). Several accompanying γ-photons (208 keV; 11% and 113 keV; 6.4%) permit diagnostic evaluation and image-based dosimetry during the treatment. ^{177}Lu can be considered as a metallic analogue of ^{131}I with many similarities in terms of mode of decay between the two isotopes, as shown in Table 19.2. ^{177}Lu is produced by irradiation of isotopically enriched ^{176}Lu or ^{176}Yb with reactor neutrons and is available at high specific activity for radiolabelling. High specific activities of ^{177}Lu can be produced via enriched ^{176}Lu as ^{176}Lu(n,γ)^{177}Lu. Together this unique combination of nuclear-physical and chemical properties of ^{177}Lu, ensured its place as one of the most promising and clinically relevant radionuclides in Onco-Nuclear Medicine.

The pure β^- emitter yttrium-90 ($t_{1/2}$=2.7 d; E_{max}=2.3 MeV) is produced from a $^{90}Sr/^{90}Y$ generator. Contrary to ^{131}I, the radionuclide yttrium-90 is a metal, and therefore it is only possible to utilize it clinically through coupling it to a targeting vector (protein, peptide). The high crossfire effect (maximum range 12 mm) and absence of emitted gammas renders ^{90}Y an attractive radionuclide for targeted therapy. However, the lack of gamma emissions makes it difficult to follow the kinetics of ^{90}Y -labelled agents in vivo using imaging (only poor spatial resolution Bremsstrahlung images can be acquired). This also limits the possibility of conducting patient-specific dosimetry [15, 16].

Rhenium-188 is a high-energy β^-emitter (E_{max}= 2.1 MeV; 72%), which decays with a 17 h half-life. ^{188}Re is produced with high specific activity using a $^{188}W/^{188}Re$ generator. It also has low-abundant gamma emission of 155 keV (15%), which could be used for SPECT imaging. The maximum tissue penetration of the emitted β^- particles is about 11 mm, making it more useful for treating larger tumours (i.e. those exceeding 1 cm in diameter). ^{186}Re is another isotope of rhenium, which decays with a half-life of 3.7 days emitting a low-energy β^- particle (E_{max}= 1.07MeV) as well as a low abundance (9%) γ-photon of 137 keV. It offers a good alternative for treating moderately sized tumours; however, the availability and specific activities are inferior to that of ^{188}Re.

Auger electrons have high LET (4–26 keV/μm) but a limited path length of 2–500 nm that restricts their efficacy to single cells, requiring the radionuclide to cross the cell membrane and reach the nucleus. This is not always possible considering that the radionuclide is mostly bound to a targeting vector that does not diffuse through the cell membrane. Therefore, the use of an internalizing targeting vector is a perquisite for efficient Auger electron therapy. A major advantage of Auger-emitting radionuclides over β-emitters is that major nonspecific radiotoxicity to the bone marrow, because of the circulating radioactivity, is not expected due to the short range of emitted electrons. On the other hand, this lack of "crossfire" effect from Auger electron-emitters does not permit killing of nontargeted cancer cells, although some reports claim a bystander effect on adjacent non-irradiated cells [17].

^{111}In has mostly been used for imaging but, because of its large emission of Auger electrons, therapeutic effects of the radionuclide when coupled to internalizing targeting agents have been observed. Indium-111 ($t_{1/2}$= 2.81 d) decays to ^{111}Cd by emitting Auger electrons K (119 keV/16%), L (2.7 keV/98%) and conversion electrons (124–245 keV/ 15 %). Additionally, two gammas, 171 keV (90%) and 245 keV (94%), and K-X-rays 23–26 keV (82%) can be used for imaging. Another high Auger electron emitting radionuclide is ^{114m}In ($t_{1/2}$ 49.5 d). It decays to ^{114}In that has beta emissions thus potentiating the therapeutic effect. The ^{114m}In emitted electrons include Auger electrons K (19–20 keV/ 6 %) L (2.7–2.8 keV/68%) and conversion electrons (162–190 keV/80%). Additionally, gammas of 190 keV/16%, 558 keV/3.2% and 725 keV/3.2% are also emitted by this radionuclide [18].

One interesting radionuclide is the radiolanthanide terbium-161 (^{161}Tb) because of its large Auger electron emission [19]. It can be produced by irradiating enriched ^{160}Gd targets with neutrons to generate the short-lived ^{161}Gd, which then decays to ^{161}Tb. It has similar chemical and physical properties to those of ^{177}Lu ($t_{1/2}$ = 6.7 d, $E_{\beta-\,av}$ = 134 keV, linear energy transfer (LET) of about 0.34 keV/μm). ^{161}Tb decays, with a half-life of 6.9 days, by the emission of low-energy β^- particles ($E_{\beta-\,av}$ = 154 keV) with a maximal tissue range of 0.29 mm and a LET of around 0.32 keV/μm. ^{161}Tb also emits photons, enabling planar or SPECT imaging. The decay process of ^{161}Tb also releases a significant number of Auger/ conversion electrons of an energy ≤50 keV (~12.4 e−, 46.5 keV per decay). This may be an advantage over ^{177}Lu, which in comparison emits a negligible number of Auger/conversion electrons (~1.11 e−, 1.0 keV per decay). Beta particles emitted during β^--decay have a LET of about 0.2 keV/μm and a tissue range of 0.5–12 mm. The high LET (~4–26 keV/ μm) and short tissue range (~2–500 nm) of Auger/conversion electrons may thus be of particular value for the treatment of single tumour cells [20]. Besides ^{161}Tb, other terbium isotopes that may be of clinical interest include ^{149}Tb, ^{152}Tb and ^{155}Tb. Sharing an identical chemical nature, these radioisotopes would permit using the same labelling protocol for the preparation of radiopharmaceuticals useful for PET (^{152}Tb) and SPECT (^{155}Tb) and for α-therapy (^{149}Tb) [21]. Examples of ß-emitting radionuclides suitable for targeted therapy of PCa are listed in Table 19.3.

19.4.2.2 Alpha Emitters

Alpha particle emitting radionuclides are also used for therapy as we will describe in the following text. However, for a comprehensive overview on alpha-emitters for radiotherapy, we refer the reader to a couple of reviews written by Poty and colleagues [22, 23].

α-particles have a moderate path length (50–100 μm) and high LET (80 keV/μm) with high cytotoxic effect that renders them especially suitable for therapy of small neoplasms or micrometastases. The ratio of energy deposited by α-particles' traversal of a cell nucleus, to that deposited by β-particles is typically about 400. The biological effect per absorbed dose unit is further enhanced by a factor of at least 3 (mostly a factor of 5 is used) due to the higher relative biological effectiveness (RBE) of α-radiation versus β-radiation. The α-particle absorbed dose reducing surviving

TABLE 19.3
Common Radionuclides for Therapy

Radionuclide	Decay mode	Half-life	Emax (MeV)	Mean range (mm)
^{90}Y	β	2.7 d	2.3	2.76
^{131}I	β,γ	8.0 d	0.81	0.40
^{177}Lu	β,γ	6.7 d	0.5	0.28
^{153}Sm	β,γ	2.0 d	0.8	0.53
^{190}Re	β,γ	3.8 d	1.1	0.92
^{188}Re	β,γ	17.0 h	2.1	2.43
^{186}Re	β,γ	3.7 d	1.07	0.9
^{67}Cu	β,γ	2.6 d	0.57	0.6
^{223}Ra	α	11.4 d	5.78	0.1
^{225}Ac	α	10 d	5.83	0.1
^{213}Bi	α	45.7 m	5.87	0.2
^{212}Bi	α	1.0 h	6.09	0.3
^{211}At	α	7.2 h	5.87	0.2
^{212}Pb	α	10.6 h	0.57	0.6
^{125}I	Auger	60.1 d	0.35	0.001-0.02
^{123}I	Auger	13.2 h	0.16	0.001-0.03
^{67}Ga	Auger,β, γ	3.3 d	0.18	0.001-0.04
^{195m}Pt	Auger	4.0 d	0.13	0.001-0.05
^{111}In	Auger	2.8 d	0.25	0.001-0.03
^{114m}In	Auger	49.5 d	0.36	0.001-0.02

fraction by a factor e^{-1} is reached with only 2 or 3 hits per cell, compared to more than 1000 hits for low LET radiation [24]. Moreover, like Auger electrons, due to their shorter range, α- particles will result in less crossfire effect to normal tissues. Figure 19.7 depicts how the range of α-and beta particles in comparison with size of micrometastases is important for the choice of radionuclide for therapy.

Astatine-211 is an α-particle emitter with a half-life of 7.2 h [25]. It can be cyclotron-produced by bombarding natural bismuth with a medium-energy α-particle beam (28–29.5 MeV) using the $^{209}Bi(\alpha,2n)^{211}At$ reaction. ^{211}At decays to stable ^{207}Pb, emitting 2 α-particles via two pathways. The first is through α-decay to ^{207}Bi which later decays to ^{207}Pb. In the second ^{211}At decays with electron capture to ^{211}Po thus emitting K x-rays before decaying to ^{207}Pb through another α-decay. Astatine is a non-metal and belongs to the halogen family; however, it differs from other halogens like iodine in forming a less-stable bond with carbon atoms present in targeting agents and thus requires different labelling protocols. The serial decay for ^{211}At is shown in Figure 19.8.

The main source of Actinium-225 is currently ^{229}Th generators ($t_{½}$=7340 y), which can be milked over a 3-wk period and allow the separation of ^{225}Ra and ^{225}Ac. ^{225}Ac ($t_{½}$ = 10.0 d; 5.8-MeV α-particle) decays sequentially through 6 dominant daughters to stable ^{209}Bi. Decay of a single ^{225}Ac atom yields 4 net α-disintegrations and 3 β-disintegrations together with the emission of 2 useful γ-emissions. The ^{225}Ac daughter ^{213}Bi ($t_{½}$ = 45.6 min; 97.8% β, 2.2% 6-MeV α-particle) is also a widely studied radionuclide for targeted α-therapy. Dissociated free ^{225}Ac accumulates primarily in the liver and bone. The ^{225}Ac daughters ^{221}Fr and ^{213}Bi will preferentially accumulate in the kidneys. Together, this makes these organs highly prone to toxic effects caused by ^{225}Ac-based radiopharmaceuticals. The decay scheme of ^{225}Ac can be seen in Figure 19.9 from Schwartz and colleagues [27].

Because of the recoil of the α-decay and the difference in chemical properties, the daughter nuclides, also α-emitters, can leave or dissociate from the chelator on the radioconjugate. For example, the first decay product of ^{225}Ac, is ^{221}Fr with a recoil kinetic energy of 120 keV which is significantly greater than the chemical bond energies of the nuclide with the chelating agent. ^{221}Fr by its way decays also to the radioactive ^{213}Bi. If this happens on the cell surface, the bloodstream can translocate the radioactive dissociated daughters into normal radiosensitive tissues, for example, the kidneys [27]. Therefore, to ensure that most of the resulting α-emissions are localized on the target, fast internalization into the tumour cells is desired for α-emitting radiopharmaceuticals. ^{227}Th and ^{223}Ra are both available on separation

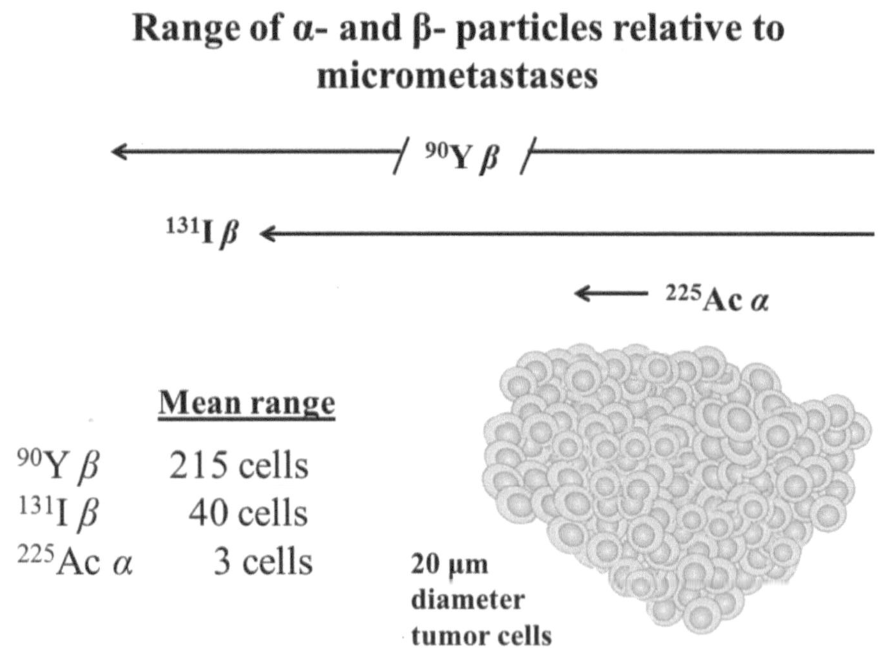

FIGURE 19.7 Range of α- and β- particles in comparison with size of micrometastases emphasizing importance of the choice of radionuclide for therapy.

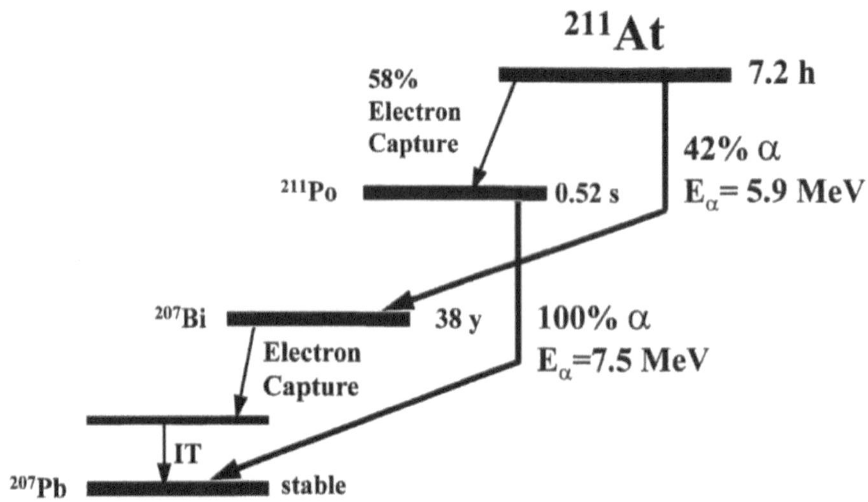

FIGURE 19.8 ^{211}At decay.

(Source: Reproduced with permission from from Vaidyanalutsky and Zalustsky [26].)

from their parent, ^{227}Ac ($t_{½}$ 21.7 d). Production of ^{223}Ra uses ^{227}Ac/^{227}Th-based generators. ^{227}Th ($t_{½}$ = 18.7 d; 6.0-MeV α-particle) and its daughter, ^{223}Ra ($t_{½}$ = 11.4 d; 5.7-MeV α-particle), releasing up to 4 high-energy α-particles before reaching stable ^{207}Pb. Biodistribution of ^{227}Th-citrate indicates high uptake in the femur and parietal bone. ^{223}Ra is an alkaline earth metal similar to calcium that, much as ^{227}Th preferentially accumulates in sites of bone mineralization, through binding to hydroxyapatite. If released, ^{223}Ra redistributes to the bone because of the α-recoil energy, resulting in an increased absorbed dose to the bone surface, which may be utilized in therapeutic protocols for treating bone metastasis. ^{223}Ra (as chloride salt, Xofigo) received marketing approval from the U.S. Food and Drug Administration (FDA) and the European Medicines Agency (EMA) for the treatment of castration-resistant PCa (CRPC) in patients

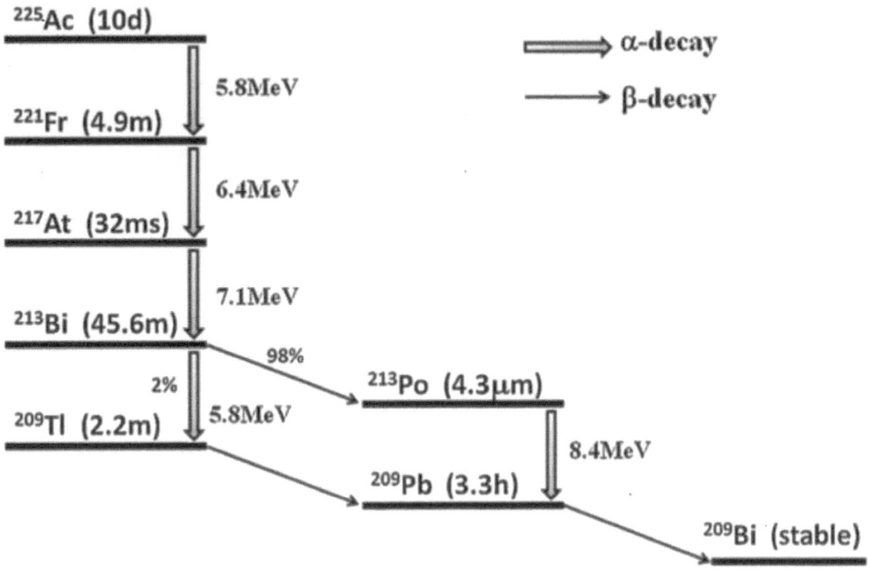

FIGURE 19.9 ^{225}Ac decay scheme.

(Source: Reproduced with permission from Schwartz et al. [27].)

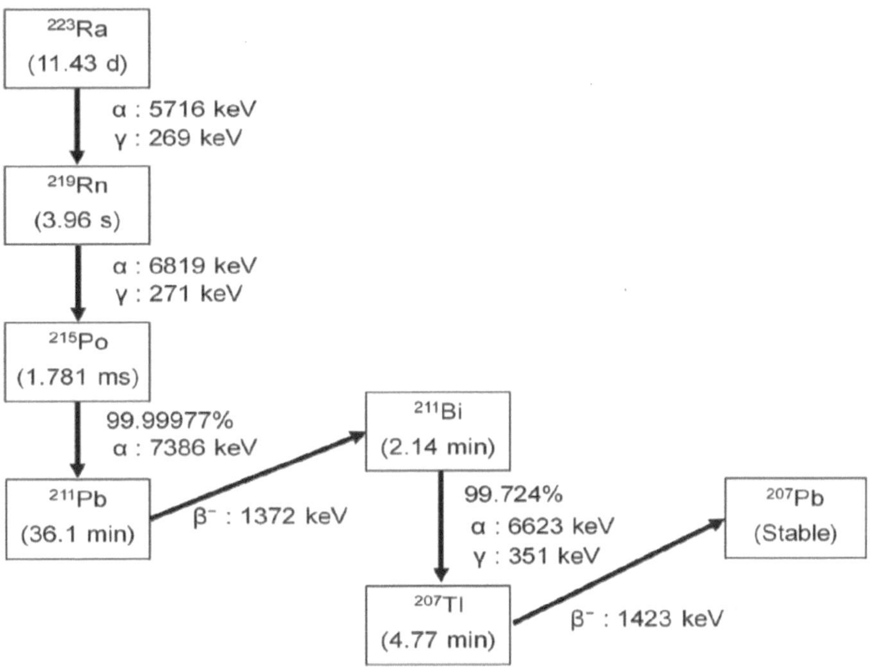

FIGURE 19.10 ^{223}Ra decay reproduced with permission from Humm et al. [29]

with symptomatic bone metastases without known visceral disease. Figure 19.10 gives the decay scheme of ^{223}Ra from Humm and colleagues [28].

^{224}Ra, ^{212}Pb, and ^{212}Bi are generator-produced α-emitters along with their long-lived parent, ^{228}Th. Severe radiolytic damage of the ^{228}Th-based generators had them replaced by ^{224}Ra-based generators, from which ^{212}Bi and ^{212}Pb can be eluted. ^{224}Ra ($t_{½}$ = 3.6 d; 5.7-MeV α-particle; 241-keV β-particle) decays into stable ^{208}Pb, producing 4 net α-particles and 2 β-particles, with the main recoil daughters being ^{212}Pb ($t_{½}$ = 10.6 h; 93.5-keV β-particle) and

^{212}Bi ($t_{½}$ = 60.6 min; 36% 6.1-MeV α-particle). During ^{212}Pb decay, γ emissions compete with internal conversion. Although free ^{212}Pb accumulates in the liver, bone, and kidneys, ^{212}Bi, like other bismuth isotopes, accumulates mainly in the kidneys.

19.5 RADIOCHEMISTRY – LABELLING METHODS FOR PCA RADIOPHARMACEUTICALS

A good overview of radiochemistry of radiometals and radiohalogens is given in Brandt and colleagues [30] and S. D. Wilburg and colleagues [31].

Radiochemistry plays a pivotal role in the development of radiolabelled agents for diagnostic and therapy applications. There are many factors to be considered when choosing the radionuclide and the optimal radiolabelling chemistry in preparing diagnostic or therapeutic agents. Long-term experience has revealed that a careful selection and optimization of radiolabelling chemistry can clearly make the difference between success and failure for the intended application. We have seen in the previous section that selection of the radionuclide is mainly governed by the intended application (imaging or therapy) as well as relatively easy, cost-efficient production and the radionuclide physical properties (half-life, high abundant co-emissions). In this section we will touch on the aspects of radiolabelling chemistry when developing radionuclide-based targeting agents. This largely includes description of the main radiolabelling strategies and how they would influence the choice of the radionuclide.

Radiochemistry in its simple definition means the stable attachment of radionuclides to targeting agents. This is usually done using several methods depending on the chemical nature of the radionuclide. Radionuclides usually belong to two main categories: those of non-metallic nature and radiometals. Radiohalogens represent the biggest class of the non-metallic radionuclides. Before investigating different radiotracers used in PCa, it would be useful for the reader to get an overview about the labelling approaches of targeting agents with these two main classes of radionuclides – that is, radiohalogens and radiometals.

19.5.1 RADIOHALOGENS

Halogens include elements of the group 7 of the periodic table (fluorine, chlorine, bromine, iodine, and astatine) as well as their isotopes. Radioiodine isotopes can be directly incorporated in the targeting agents through existing tyrosine or histidine residues (see Figure 19.11). Radioiodine isotopes (^{124}I, ^{123}I, ^{125}I and ^{131}I) are usually provided in a salt form (I^-) with a -1 valency. Oxidation of iodine, using oxidizing agents (Chloramine-T [32], Iodogen [33]), to the more reactive, electrophilic I+ (+1 valency) initiates its reaction with activated electron-rich aromatic rings such as that present in the

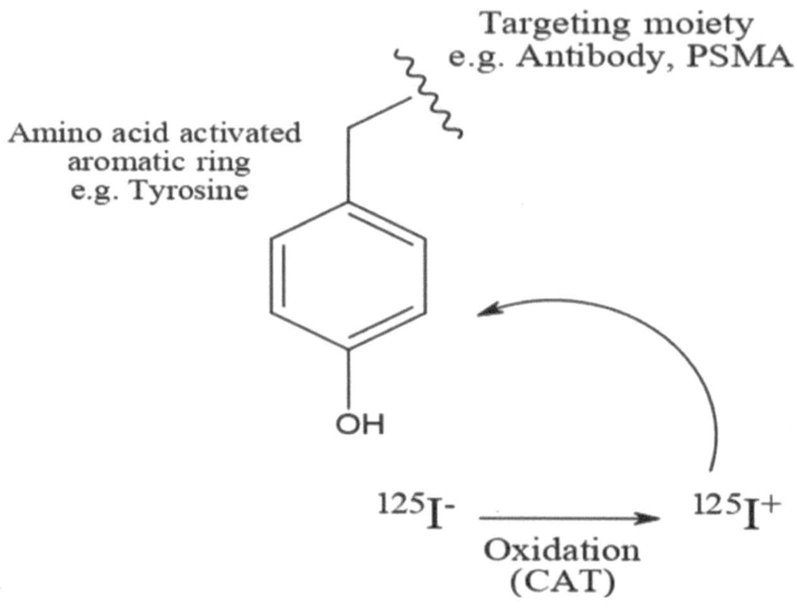

FIGURE 19.11 Iodination of an activated aromatic ring in tyrosine residues. It is chemically easier since the hydroxyl (–OH) group, already present on the ring, activates the electrophilic attack of iodine (I^+).

amino acid tyrosine. Reaction with other amino acids bearing electron-rich aromatic rings (phenylalanine, tryptophan, and histidine) is also possible, but less likely.

Direct radiohalogenation of targeting agents with radioiodine isotopes is usually a robust, fast labelling strategy with very high radiolabelling yields [31]. However, bromine and astatine are more resistant to oxidation, and attempts to directly radiobrominate and radioastatinate targeting agents have produced fairly low labelling yields and inconsistent results [34]. Moreover, the tyrosine residues, used for isotope binding in direct halogenation, may in some targeting agents be located in crucial target recognizing/binding sites on the targeting vector, thus comprising its immunoreactivity. Similarly, some disulphide bonds crucial for maintaining the targeting protein tertiary structure may be vulnerable to harsh reduction-oxidation conditions of direct radiohalogenation.

Indirect radiohalogenation has been proposed to overcome the limitations associated with the direct approach. In this method a bifunctional activated precursor is radiolabelled first. The precursor resembles tyrosine by having an activated aromatic ring, but, with an even better leaving group to ease substitution with radiohalogens. The first synthesized precursor was the N-succinimidyl ester of 3-(tri-n-butylstannyl) benzoate or ATE [35]. The radiolabelled precursor is then conjugated to N-terminus or Ɛ-amino groups of lysine or thiol groups of cysteine residues on targeting agents through its reactive N-succinimidyl or maleimide ester side chain. Due to the multistep radiolabelling, indirect radiohalogenation have lower labelling yields and specific activities, but a great advantage of this method is that it permits the manipulation of the properties of the final product by implementing different precursors having different properties – for example, hydrophilicity, lipophilicity. One example is increasing retention of radioactivity in tumour cells by using precursors that do not diffuse out of the cell, which is a prerequisite for efficient imaging and therapy [36, 37].

After binding to a carrier molecule (receptor, antigen), many targeting agents internalize inside cells by endocytosis and are transported to the lysosomes [38–40]. In the lysosomes, radiolabelled proteins undergo lysosomal degradation that leads to formation of small radiolabelled fragments (catabolites), while the carrier molecule is recycled back to the membrane [41]. Depending on the nature of a radionuclide and properties of a label, the radiocatabolites are either retained inside the cells, that is, residualizing labels, or excreted outside, that is, non-residualizing labels. Directly radiohalogenated targeting agents form non-residualizing (leaky) radiocatabolites that leak/diffuse freely outside the cell, through the cell membrane [42]. This washout of radioactivity would consequently mean lower absorbed dose to the tumour in therapeutic applications. Moreover, the catabolites leaking back into the circulation reduce imaging contrast as well as increasing unnecessary exposure to radiation of non-targeted organs. Indirect radioiodination using activated aromatic ring precursors resulted in improved tumour cell retention of radioactivity providing enhanced tumour-to-non-tumour ratios in comparison to directly radioiodinated agents [43]. Radioactivity in iodine-sequestering, non-targeted tissues is also reduced dramatically. The premise is that these precursors form positively charged catabolites post-lysosomal degradation that does not diffuse the cell membrane and is therefore highly retained "residualizing" [26, 44]. Indirect radiohalogenation of activated aromatic precursors permitted radiobromination [45, 46] and radioastatination [47, 48] of targeting agents without exposing the carrier protein to harsh labelling conditions. Alternatives to activated aromatic ring precursors have been proposed [49]. One approach involves loading bromine or astatine in precursors composed of boron clusters (boron cages). The boron-astatine and boron-bromine bonds are stronger than the carbon-astatine and carbon-bromine bonds in aromatic precursors, therefore this approach permits efficient and high stable labelling of targeting agents.

19.5.2 Radioactive Metals

The majority of radionuclides have metallic origins. Unlike halogens, metals are incapable of forming stable covalent bonds with carbon or other elements present in proteins and peptides. Chelators are used to bind metals to proteins and peptides Figure 19.12. A chelator in its simplest definition is a compound (electron donating multi-ligand) capable of binding (coordinating) metal ions forming a complex ring-like structure called "chelates." The term *chelate* is derived from the Greek word "chele," which means claw of a lobster and was originally developed to treat heavy-metal poisoning. Chelators in general form stable complexes with metals, a property defined by their high thermodynamic stability. Thermodynamic stability is a term used in chemistry to describe a chemical system that is neither consuming nor releasing heat energy. In the absence of a change in thermal energy, the substance is not undergoing a chemical reaction and is, therefore, stable – that is, does not release back the complexed metal in a reverse reaction. In addition to complexing metal ions, bifunctional chelators possess reactive functional groups that allow their conjugation to proteins or peptides. Common functional groups for bioconjugation involve activated esters and isothiocyanites for coupling to lysines and other side chain amines on the targeting protein and maleimides for coupling to cysteines [50]. Other

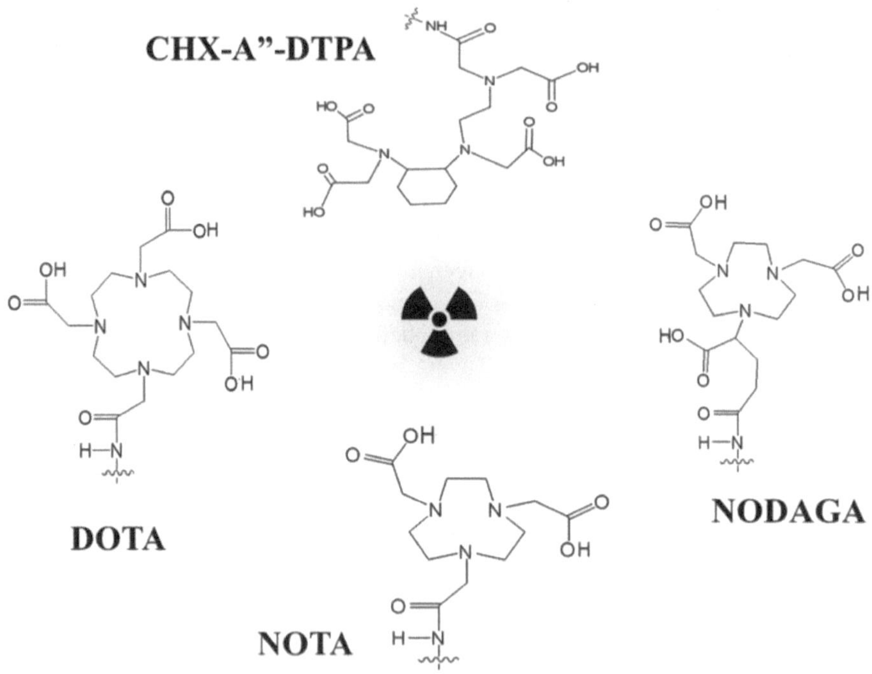

FIGURE 19.12 Structural overview of the commonly used bifunctional chelators used in nuclear medicine both for radiolabelling with diagnostic and therapy radionuclides.

promising and more specific methods for coupling have also been developed, implementing what is known as click chemistry [51, 52]. In some cases, a linker or spacer is used in order to increase the distance between the chelator-metal complex and the targeting vector binding site [53].

One of the main challenges in radiometal labelling is the presence of trace amounts of other competitive non-radioactive metals like iron, cadmium, zinc, and aluminium. These non-radioactive metal traces compete with radiometals for the complexation site on the chelator and, hence, reduce specific activity dramatically. Chelators can be classified into three main categories – acyclic, semirigid acyclic, and macrocyclic chelators [54]. Each of these possesses different properties when regarding the rate of complex formation, stability of formed complex and reaction conditions (pH, temperature). Acyclic chelators were the first to be developed and used for labelling targeting agents with radioactive metals – for example, DTPA. Acyclic chelators excel in their ability to rapidly coordinate metals with fast kinetics [54, 55]. This permits labelling of biomolecules in very short times at room temperature. When the metal-chelator complex is formed, the reaction is at equilibrium – that is, the forward (complex formation) and reverse reactions (complex dissociation) occur at equal rates at specific conditions. However, changing the conditions of the equilibrium, for example after administering the radiopharmaceutical in vivo, may bring stability problems. Dilution as well as the presence of abundant amounts of competing naturally existing chelating agents like plasma proteins (e.g. transferrin, ceruloplasmin) experienced in vivo may favour complex dissociation (reverse reaction). Fast kinetics of acyclic chelators of radiometal incorporation (on-rate) and consequently the low energetic barriers to the formation of the radiometal–chelate complexation can also mean fast radiometal decorporation (off-rates) and low energy barriers to radiometal release. In Figure 19.12 a structural overview of bifunctional chelates is given.

Semi-rigid acyclic chelators (e.g. CHX-A"-DTPA) show better stability in vivo compared to acyclic chelators. The cyclohexyl (CHX) backbone of CHX-A"-DTPA makes the chelator more pre-organized compared to DTPA on the metal ion binding site, thus enhancing kinetic inertness and capsizing the rapid metal decorporation in vivo experienced with acyclic DTPA. However, this also slows the radiolabelling reaction kinetics compared to those of DTPA and, hence, more warming (~37-40°C) and longer reaction times would be required to obtain high labelling yields. Figure 19.13 shows an example of labelling of an antibody via the CHX-A"-DTPA chelator with either ^{111}In for diagnosis or ^{177}Lu for therapy.

Macrocyclic chelators – for example, DOTA, NOTA, DOTAGA, NODAGA, and CB-TE2A – impose the highest degree of preorganization. They exhibit high thermodynamic stability like acyclic chelators but excel in being

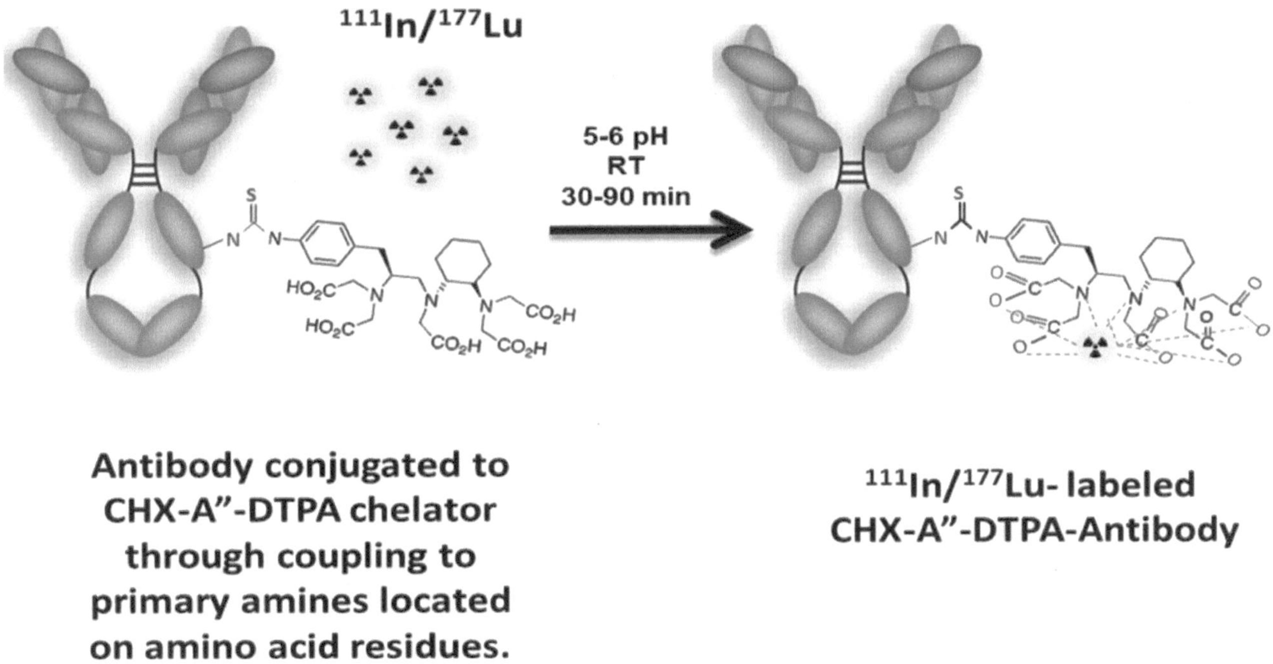

FIGURE 19.13 Example of labelling of an antibody via the CHX-A"-DTPA chelator with either ^{111}In for diagnosis or ^{177}Lu for therapy. Colour image available at www.routledge.com/9781138593312.

kinetically inert. Due to their high kinetic inertness, less radiometal decorporation occurs in vivo and, consequently, less accumulation of the radionuclide in normal non-targeted tissues is observed [54]. In order to increase the reaction kinetics, higher temperatures (higher than that required by CHX-A"-DTPA) and longer reaction times are required to obtain efficient labelling yields. These conditions might not be compatible with heat-sensitive targeting agents like antibodies [56].

As we described in the previous sections, the radiometal labelled targeting agents could be fragmented into 4 main components; the targeting vector, the linker or spacer, the chelating agent, and the radionuclide (see Figure 19.5). Each of these confers on the radiopharmaceutical different properties that could be used for optimizing its properties for the intended purpose. For example, it was shown that the radiometal-chelate complex may have a large impact on the targeting properties of radiolabelled protein and peptides [57–60]. The overall charge, hydrophilicity/hydrophobicity and geometry of the formed complex influence both target (receptor interaction, internalization rate) and also off-target interaction (blood kinetics, excretion pathway). This impact of the chelate-metal complex is more profound in smaller size, targeting agents and, as mentioned, can be used to modulate or optimize their in vivo properties. Such effect is less prominent in bulky agents like antibodies.

Technetium and rhenium are metals; however, they differ drastically from other metals such as lutetium or gallium in their coordination chemistry as well as on the types of ligands that can form stable complexes with them. Technetium and rhenium radioisotopes require reduction to a more reactive form (core) prior to forming complexes with ligands. For more detailed information about technetium and rhenium chemistry, we refer the reader to a good review written by Bartholomä and colleagues [61].

The difference between non-residualizing halogen radiolabel and residualizing metal radiolabel is shown in Figure 19.14. When the anti-hk2 antibody 11B6 was labelled with the non-residualizing halogen ^{125}I very low retention of radioactivity in PCa tumours was observed as shown using SPECT imaging. In the contrary, labelling with the residualizing ^{111}In radiometal revealed an enhanced tumour retention of radioactivity. As mentioned earlier, lysosomal degradation, following the antibody internalization by the cell will generate radioactive catabolites. Depending on the nature of these formed catabolites (charge, polarity) they could either leak/diffuse out of the cell, that is. non-residualizing ^{125}I halogen, or be retained in the cell compartment, that is, residualizing ^{111}In. This highly exemplifies the importance of choosing the proper labelling method reflecting the true biokinetics of the investigated biomolecule.

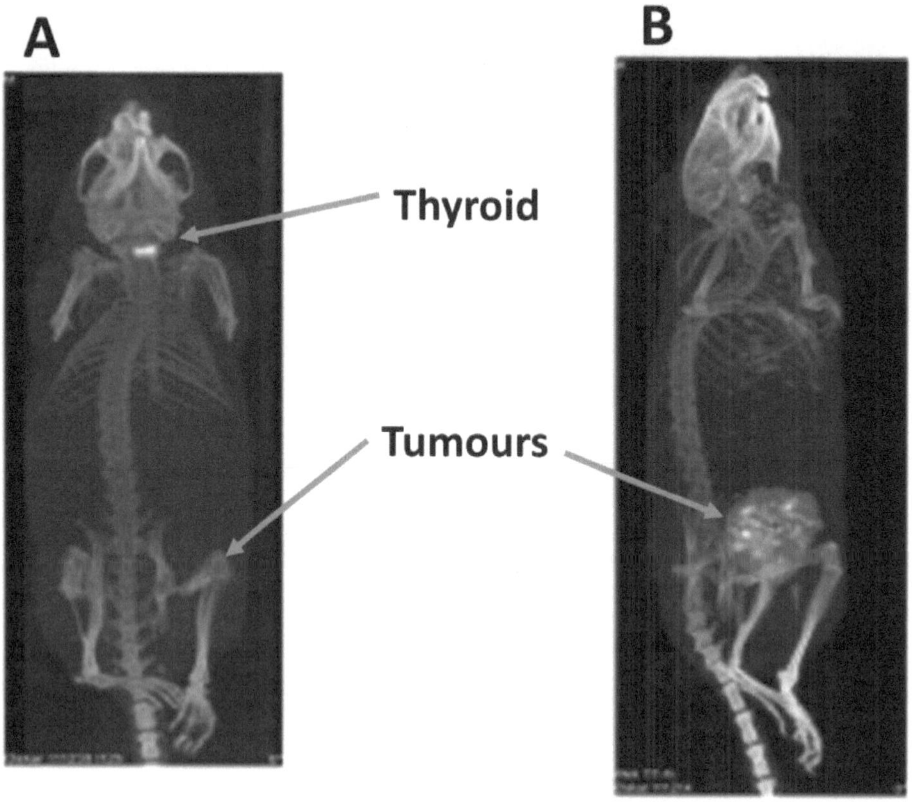

FIGURE 19.14 SPECT images of two PCa (LnCaP) tumour bearing mice injected with an internalizing antibody labelled with either (a) the halogen ^{125}I, or (b) the metal ^{111}In. After internalization into the tumour cells and degradation of the molecular complex in the lysosomes, the generated catabolites would have different fate depending on the nature of the label, that is, non-residualizing, ^{125}I or residualizing ^{111}In. The non-residualizing ^{125}I-catabolites diffuses out of the tumour cell back into the circulation and is re-uptaken by the thyroid through the Na/I symporters. On the contrary, ^{111}In catabolites are highly retained (residualizing) in the tumour cells after breakdown. Colour image available at www.routledge.com/9781138593312.

(Source: Image reproduced by courtesy of Thuy Tran.)

19.6 CLINICAL DIAGNOSTIC RADIOPHARMACEUTICALS

Many radiopharmaceuticals have been suggested for diagnosis of PCa by methods that complement clinical and PSA diagnostics. A summary of some of those commonly used radiopharmaceuticals is given below. These radiopharmaceuticals can be divided into three main groups – agents targeting the bone matrix, agents for imaging of metabolism and proliferation, and agents targeting receptors and membrane proteins.

19.6.1 Agents for Imaging the Bone Matrix

The physiology of $^{18}F^-$ has been elaborated on in Murray and colleagues [62]. ^{18}F-fluoride localizes to the inorganic part of the bone in areas of osteoblastic activity. Fluoride ions are incorporated into the hydroxyapatite crystal structure of the bone by substitution for hydroxyl (OH^-) ions. The relationship between osteoblastic and osteoclastic activity is believed to determine the amount of incorporation into the bone, given by

$$Ca_{10}\,(PO_4)_6(OH)_2 + F^-_2 \rightarrow Ca_{10}\,(PO_4)_6(F)_2 + 2OH^-$$

As an analogue of the hydroxyl group found in hydroxyapatite bone crystals, ^{18}F-NaF is an avid bone seeker. Ion exchange is its mechanism of uptake, and blood flow is the rate-limiting step in the transfer of fluoride ions from blood to bone tissue and is directly incorporated into the bone matrix, converting hydroxyapatite to fluoroapatite.

^{18}F-sodium fluoride PET offers many advantageous technical features with superior pharmacokinetics and image quality, good tumour-to-background ratio due to fast blood clearance, and high PET spatial resolution. It is rapidly cleared from plasma, with only 10 per cent of ^{18}F-NaF remaining in plasma at 1 h, which leads to high-quality (high target-to-background uptake ratio) skeletal images in less than 1 h p.i. Previous studies have shown that [^{18}F]Fluoride PET/CT imaging has better accuracy and detects more bone metastases and at an earlier time compared to ^{99m}Tc-phosphates in PCa patients. Fluoride PET is an indirect marker of bone metastatic growth, and signal is directly proportional to the formation of new bone. It does not, however, image soft-tissue or lymph node disease.

The clinical gold standard for detecting skeletal metastasis has been conventional ^{99m}Tc-methylene diphosphonate planar radionuclide bone scan (RNB) or SPECT imaging [63]. RNB has moderate sensitivity for osteoblastic bone lesions such as in PCa. The method shows low specificity due to skeletal uptake in degenerative changes and other benign conditions such as trauma and infection. A comprehensive and thorough review of phosphonate-based radiopharmaceuticals is given in Lange and colleagues [64].

Planar imaging, SPECT and PET with different sensitivities and spatial resolution together with different tumour-to-background rations has shown that ^{99m}Tc-MDP and ^{18}F- differ in sensitivity and specificity. The sensitivities of ^{99m}Tc-MDP planar scintigraphy, SPECT, and [^{18}F]NaF PET/CT are 70 per cent, 90 per cent, and 100 per cent, respectively. The specificities of ^{99m}Tc-MDP planar scintigraphy, SPECT, and [^{18}F]NaF PET/CT are 60 per cent, 80 per cent, and 100 per cent, respectively. A good summary is given in Koerber and colleagues [65].

19.6.2 Quantitative Imaging Biomarkers – Bone Scan Index

For bone scintigraphy, the bone scan index (BSI) has been developed, its automatic computation method evaluated by Ulmert and colleagues [66], and it can be used for quantitative analysis. BSI is defined as the percentage of skeletal mass affected by tumour calculated on a bone scintigram. Because bone scintigraphy is 2-D and lacks anatomic information, a standard weighting of bones is incorporated in the calculation. The commercial software EXINI Bone BSI (EXINI Diagnostics AB) allows for automated calculation of the BSI. Two other indices have been evaluated as BPI_{VOL}, which is the percentage of bone volume (including bone marrow) affected by tumour, and BPI_{SUV}, which additionally considers the target expression measured by the average SUV. In both indices, the anatomic information is extracted from the CT image, whereas the functional information is extracted from the PET image [67].

19.6.3 Agents for Imaging Metabolism and Proliferation

The nutrient demand of cancerous cells is much higher than that of normal tissues. Cancer cells show significant differences compared to normal cells in terms of uptake and use of body nutrients. This difference is a key for using imaging to differentiate malignant cells from normal ones.

19.6.3.1 ^{11}C-Choline/^{18}F-Choline

Choline is a precursor of phospholipids, major structural components of the plasma membrane, and as such choline is necessary for membrane synthesis and cell division. It is transported into cells via high-affinity sodium-dependent choline transporters, phosphorylated by choline kinase (which is overexpressed in PCa), and incorporated into phospholipids (phosphatidylcholine).

Several large studies have shown a sensitivity of 85–90 per cent, and specificity of about 90 per cent of choline scans. ^{11}C-Choline is relatively insensitive in patients who have biochemically recurrent PCa after surgery with PSA < 2 ng/mL. A pharmacokinetic evaluation in PCa patients with biochemical recurrence is given by Grkovski and colleagues [2].

Although choline-based PET/CT imaging has been widely used, numerous clinical studies have reported low sensitivity and specificity of this technique, especially at low PSA levels and high Gleason scores. In addition, ^{11}C- or ^{18}F-Choline is flow-limited and higher uptake may be present in benign prostate hyperplasia. ^{11}C- or ^{18}F-Choline PET/CT has good specificity for detecting lymph node metastases, but has low sensitivity, ranging from 10 to 73 per cent. In a prospective study, Poulsen and colleagues [68] demonstrated in 210 patients that ^{18}F-Choline PET/CT imaging is not sensitive enough when compared with lymph node dissection, and therefore should not replace this invasive procedure. Further studies demonstrated that disease detection rate using ^{18}F-Choline was related to serum PSA levels, showing values of 20 per cent, 44 per cent, and 81.8 per cent for PSA levels of PSA<1.1, PSA <5, and PSA >5 ng/mL[69].

19.6.3.2 ^{18}F-FDG

^{18}F-FDG has been extensively studied, but its role in the detection of metastatic PCa is limited because of the low glycemic activity of hormone-naive PCa cells [70]. For diagnosis of primary prostate tumours an incidental high ^{18}F-FDG uptake has been observed in the prostate gland, which is not related to known prostate pathology. It is suggested that aggressive primary tumours with Gleason score > 7 tend to display high ^{18}F-FDG uptake. PET imaging with ^{18}F-FDG may be useful in staging of those patients having aggressive primary tumours.

19.6.3.3 ^{18}F-Fluciclovine / ^{18}F-FACBC (Axium)

Amino acids are essential to cell metabolism and growth. Several amino acid transporter systems are overexpressed in PCa. Anti-1-amino-3-[^{18}F]Flurocyclobutane-1-carboxylic acid (^{18}F-fluciclovine-FACBC) is a non-naturally occurring amino acid, and its transport is primarily mediated by sodium-dependent amino acid transporters. Because the amino acid transporters that are most involved in ^{18}F-fluciclovine transport mediate influx and efflux of amino acids, washout of the radiotracer occurs over time. The specificity of ^{18}F-fluciclovine for PCa relies on altered metabolic pathways overexpressed in PCa. Studies have shown a detection rate of 40 per cent for patients with biochemical recurrence and a PSA level of 0.79 ng/mL or less. Direct comparison between ^{18}F-fluciclovine and ^{11}C-choline PET/CT has demonstrated overall superior imaging performance for ^{18}F-fluciclovine in biochemically recurrent PCa. A good overview of ^{18}F-Fluciclovine is found in Parent and Schuster and colleagues [71].

19.6.3.4 $^{64}CuCl_2$

Copper is an essential trace element and elevated levels of copper have been found in a wide range of tumour tissues, that is, PCa. Various transporters and binding proteins are found to be responsible for copper uptake in the cells. The human copper transporter 1 (hCTR1) is the main copper transporter into cells. ^{64}Cu is described as a "theranostic" radioisotope, as it is potentially useful in PET/CT imaging and in radionuclide therapy. Copper in the form of copper chloride salt ($^{64}CuCl_2$) was proposed as a diagnostic agent to detect PCa using PET. Righi and colleagues and Piccardo and colleagues [13, 72] both used $^{64}CuCl_2$ to detect PCa recurrence in 50 patients with biochemical relapse, after prostatectomy or external beam radiation therapy. This represented a potential advantage over ^{18}F-choline. One limitation of using $^{64}CuCl_2$ stems from the low abundance of positron decay of ^{64}Cu (18%). This accordingly means that the imaging time should be extended in order to achieve good image contrast.

19.6.4 Agents Targeting Receptors and Membrane Proteins

19.6.4.1 Targeting Gastrin-Releasing Peptide Receptors (GRPRs)

The high-density expression of gastrin-releasing peptide receptors (GRPRs) in several human cancer forms, as PCa, compared to their low expression in surrounding healthy tissues, provided an opportunity to develop several theranostic agents targeting this tumour marker. Bombesin, a 14 amino acids peptide, was initially isolated from the skin of the European frog *Bombina bombina* in the early 1970s. Among several identified bombesin agonists, [Leu13]BN was found to have the highest affinity towards GRPR, and high homology to its endogenous ligand gastrin-releasing peptide (GRP). Agonists trigger internalization into the cell upon binding to GRPR. When labelled with residualizing labels efficient internalization, enables prominent accumulation of radioactivity in the tumour. High-affinity antagonists on the other hand have low internalization, but were still found to have superior tumour-targeting and pharmacokinetic properties compared to agonists [73]. The GRPR antagonist [^{111}In]-RM1, showed superior in vivo targeting properties, that is, 3-fold higher tumour uptake, compared to the potent GRPR agonist [^{111}In]- AMBA81 [74]. C-terminus modification of RM1, has led to the development of the more potent GRPR antagonist (D-Phe-Gln-Trp-Ala-Val-Gly-His-Leu-NHEt). Efforts to optimize targeting properties of this antagonist by conjugating it to different macrocyclic chelators as well as radiolabelling with several clinically relevant radionuclides resulted in developing ^{68}Ga-DOTA-Sarabesin3 [75], which exhibited encouraging results in men [76]. By replacement of the C-terminal Leu13-Met14-NH2dipeptide of SB3 bySta13-Leu14-NH2, the novel GRPR antagonist NeoBOMB1 was generated [77]. NeoBOMB1, was labelled with ^{68}Ga (for PET), ^{111}In (for SPECT), and ^{177}Lu (for therapy). ^{68}Ga-Neo-BOMB1, ^{111}In-NeoBOMB1, and ^{177}Lu-NeoBOMB1 efficiently localized in PCa cell line PC-3. Successful visualization of PCa lesions in men was possible using ^{68}Ga-NeoBOMB1 and PET/CT imaging [77]. The statine-based high affinity GRPR antagonist RM26 (also known as JMV594) formed the basis for various tracers that are under development for imaging and therapy in PCa [78]. RM26 was radiolabelled with cobalt isotopes and evaluated in PCa tumour-bearing mice. The cobalt-radiolabelled NOTA-PEG2-RM26 demonstrated favourable biodistribution, which translated into high-contrast preclinical PET/CT (using ^{55}Co) and SPECT/CT (using ^{57}Co) images of PC-3 xenografts [79]. Recently a study was designed to analyse the safety, biodistribution, and radiation dosimetry of ^{68}Ga-RM26 in healthy human volunteers [80]. Moreover, a direct comparison between GRPR antagonist ^{68}Ga-RM26 and agonist ^{68}Ga-BBN was evaluated. This study demonstrated the safety and significant efficiency of ^{68}Ga-RM26. The antagonist ^{68}Ga-RM26 was shown to be better than the GRPR agonist as an imaging marker to evaluate GRPR expression in PCa.

19.6.4.2 Targeting Prostate-specific Membrane Antigen (PSMA)

Prostate-specific membrane antigen (PSMA) is a 100-kDa type II transmembrane glycoprotein that is overexpressed in nearly all prostate cancers, particularly poorly differentiated and metastatic lesions, with only 5–10 per cent of primary PCa lesions shown to be PSMA-negative. Its expression increases in higher-grade, metastatic, and androgen-insensitive tumours, whereas expression is largely absent in benign or hyperplastic prostate tissue. Expression increases with Gleason score and remains high during castration resistant PCa -. The name PSMA is a misnomer, as low PSMA expression also occurs in proximal small bowel, kidneys, salivary and lacrimal glands. In the salivary glands, staining of PSMA protein is found in focal hot spots of acinar cells, preferably along their luminal border but not homogeneously dispersed within the parenchyma. In the kidneys, the PSMA protein is found in a selected subset of tubular cells, again with a predominant luminal staining pattern [81]. Expression in malignancies is not limited to PCa; PSMA expression was also found in other types of malignancies, for example, for breast and colon carcinoma. A good overview of PSMA ligands for theranostic applications can be found in the following review articles [82, 83]. The introduction of ^{68}Ga-PSMA-11 represented the first clinical breakthrough in PET imaging of PSMA expression [84]. ^{68}Ga-PSMA-11 demonstrated great utility both for initial diagnosis and restaging of PCa. It is currently the most widely used imaging PSMA ligand. In a comparative study, Calais and colleagues showed in a head-to-head comparison between ^{68}Ga-PSMA-11 and ^{18}F-fluciclovine, in a series of 10 patients with PCa recurrence, that there were improved detection rates for ^{68}Ga-PSMA-11 compared to ^{18}F-fluciclovine [85]. The schematic Figure 19.15 is showing the uptake mechanism of some antibodies and small molecules used for diagnosis of PCa.

Several other radiolabelled small-molecule inhibitors of PSMA have been proposed. These include, but are not limited to, the urea-based 2-(3-(1carboxy-5-[(6-^{18}F-fluoro-pyridine-3-carbonyl)-amino]-pentyl)-ureido)pentanedioic acid (^{18}F-DCFPyL) [86] that binds to the enzymatic active site, the clinically SPECT/CT tracer ^{99m}Tc-MIP-1404 and the

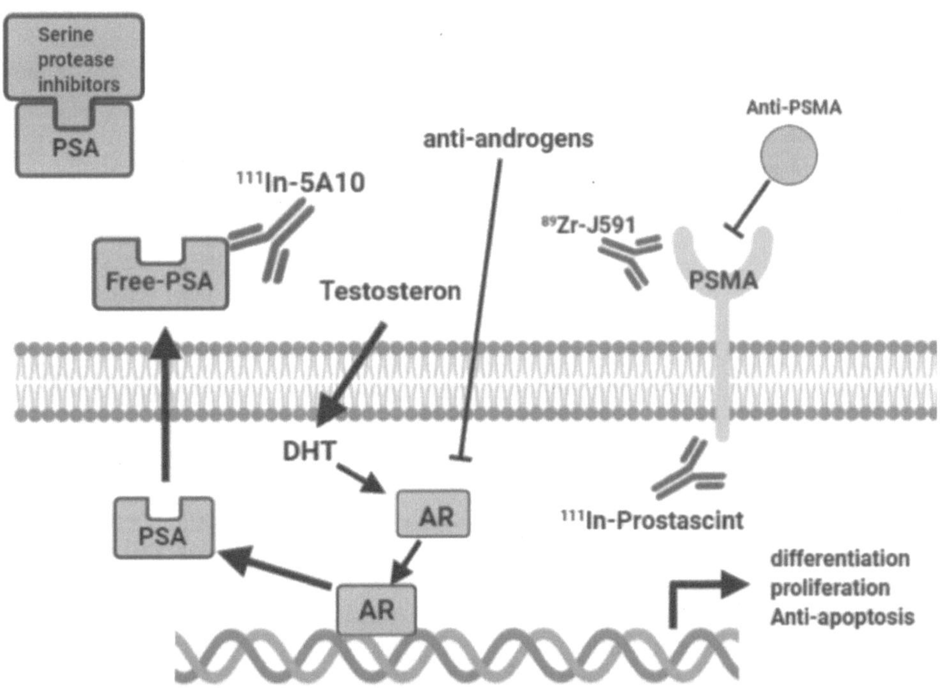

FIGURE 19.15 Schematic figure showing the uptake mechanism of some antibodies and small molecules used for diagnosis and therapy of PCa.

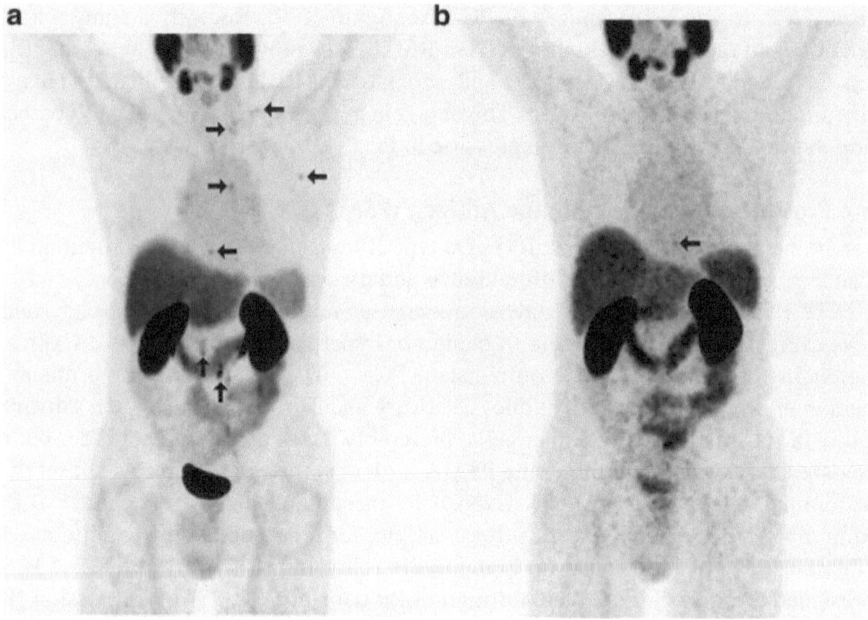

FIGURE 19.16 Here are shown PET images with a comparison between (a) MIP (maximum intensity projection) with [^{18}F]DCFPyL and (b) MIP with [^{68}Ga]Ga-PSMA-HBED-CC. [^{18}F]DCFPyL clearly demonstrates several supradiaphragmatic PSMA-positive lesions. [^{68}Ga]Ga-PSMA-HBED-CC PET/CT showed a supradiaphragmal lesion in the sternum.

(Source: Reproduced with permission from Dietlein et al. [92].)

PET tracers ^{68}Ga-PSMA-617, ^{68}Ga-PSMA I&T, and ^{124}I-MIP-1095. In a study involving 248 patients with biochemical recurrence after radical prostatectomy, ^{68}Ga-PSMA PET/CT showed a sensitivity of 66 per cent and specificity of 98.9 per cent for lymph node metastases. ^{68}Ga-PSMA PET/CT also outperformed planar bone scans for detection of osseous metastases. Large retrospective series reveal detection rates from 50 per cent to 58 per cent for patients with a PSA level of less than 0.5 ng/mL, and more than 95 per cent when the PSA level is more than 2 ng/mL [87] . The detection rate of ^{68}Ga-PSMA PET/CT imaging for recurrent PCa exceeds that of ^{18}F- and ^{11}C-choline. Guidelines for ^{68}Ga-PSMA imaging are given in a joint EANM and SNMMI publication [88].

PSMA-617, a ligand with optimized tumour cell internalization and low kidney retention with DOTA chelator, was developed for PSMA-targeted radioligand therapy. ^{177}Lu-PSMA-617 has been increasingly used for therapy of metastatic PCa patients (see next section). These prompted research to clinically investigate the potential of using a ^{68}Ga-labeled PSMA-617 as a companion diagnostic agent for PET imaging [89]. This study clearly demonstrated that ^{68}Ga-PSMA-617 is capable of detecting lesions of PCa with high contrast, especially in late time images.

Recently a bispecific heterodimer targeting both PSMA and GRPR for SPECT and PET diagnostic imaging of PCa was developed [90]. Simultaneous targeting of the PSMA and GRPR could improve the diagnostic accuracy in PCa.

In Rousseau and colleagues [91] is given examples of ^{18}F-DCPYL PET images of normal patient biodistribution (uptake by lacrimal glands, salivary glands, kidneys, liver, spleen, bowel, and bladder) and patient with metastatic PCa. In Figure 19.16 are shown results from a patient with a rising PSA level of 3.87 ng/mL. On PET images, comparison between a) MIP (maximum intensity projection) with [^{18}F]DCFPyL and b) MIP with [^{68}Ga]Ga-PSMA-HBED-CC. [^{18}F]DCFPyL PET/CT clearly demonstrates several additional supradiaphragmatic PSMA-positive lesions [92].

In the study by Rosseau and colleagues [91] using ^{18}F-DCFPyL PET/CT imaging, they also found that the proportion of positive scans increased with PSA level, with a PSA value for positive scans of 5.80 ± 6.87 ng/mL compared to negative scans with 1.86 ± 1.62 ng/mL. ^{18}F-DCFPyL PET/CT imaging improved decision making and changed management plans for most patients. In Dietlein and colleagues [92] is compared uptake in tumours compared to background in kidney or liver as SUV rations and results are shown in Figure 19.17.

In a metanalysis Hope and colleagues [93] searched the literature for how ^{68}Ga-PSMA-11 performed for the localization of metastatic PCa. With pathology as a gold standard, ^{68}Ga-PSMA-11 had a sensitivity and specificity of 74 per cent and 96 per cent. The detection rate was 63 per cent with a PSA less than 2.0 and 94 per cent with a PSA greater than 2.0 ng/mL.

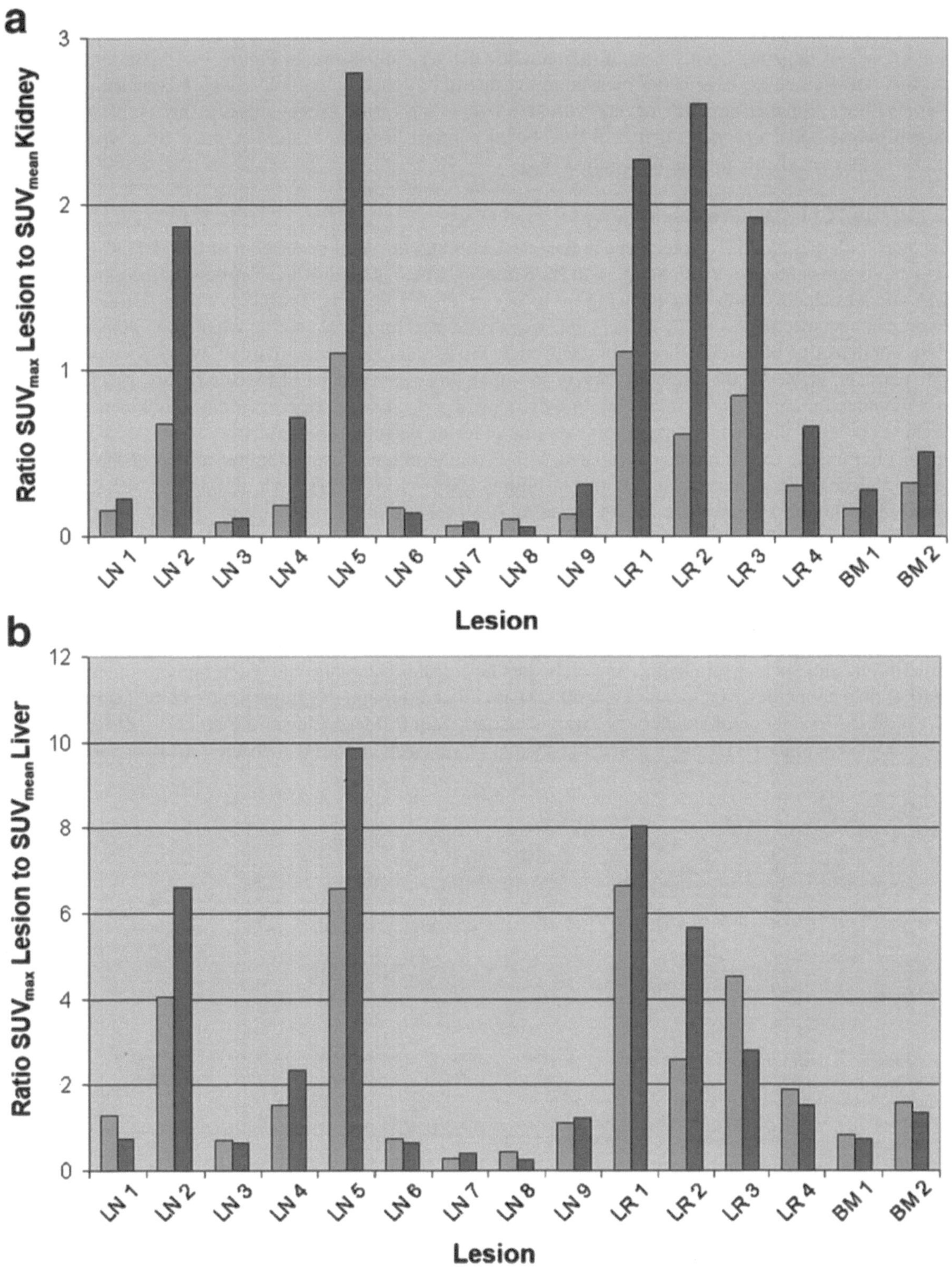

FIGURE 19.17 Tumour to background ratios where blue columns represent [68Ga]Ga-PSMA-HBED-CC, and the red columns represent [18F]DCFPyL. Ratios of SUVmax in lesions to SUVmean in (a) the kidney and (b) in the liver. LN lymph node metastasis, LR local relapse, BM bone metastasis.

(Source: Reproduced with permission from Dietlin et al. [92].)

19.6.5 Antibodies

Different classes of targeting agents used in radionuclide delivery are shown in Figure 19.18. Bulky targeting agents (such as full-size antibodies) have better bioavailability due to prolonged circulation, which is an advantage for therapeutic applications. Smaller targeting agents with size below that of the kidney cut-off, that is, 60 kDa, have better clearance from the body, a property that facilitates better contrast imaging at shorter times. Smaller agents also have rapid extravasation and better tumour penetration [94].

19.6.5.1 ^{111}In-7E11-C35 (Prostascint)

The first steps in targeting PSMA took place in the late 1980s and involved generation of the 7E11-C35 antibody (also known as capromab pendetide, ProstaScint; Aytu BioScience), which is specific to an epitope at the intracellular domain of PSMA and was labelled with ^{111}In, allowing for its use in SPECT imaging. ProstaScint demonstrated sensitivities of 52–62 per cent and specificities of 72–96 per cent, when evaluated for initial staging which then outperformed CT and MRI [95]. For detection of residual disease or recurrence, ProstaScint had a sensitivity of 49–77 per cent and specificity of 35–71 per cent. However, this antibody targets an intracellular epitope of PSMA (see Figure 19.15) that cannot be accessed in viable tumour cells, therefore limiting diagnostic performance. Moreover, the slow kinetics of ProstaScint in vivo causes delay in the time required to obtain optimal lesion detection (6–8 d).

Attempts to mitigate the limitations associated with ProstaScint have resulted in generation of the mAb J591 [96]. This antibody recognizes extracellular epitopes of human PSMA with a very high affinity (K_D = 1.83±1.21 nM) and is recognized as the most appropriate PSMA-targeting mAb for use in diagnostics and therapy. After PSMA binding the antibody-target complex is internalized thus enhancing the tumour retention of radioactivity when labelled with residualizing labels. The first J591 mAb developed was of murine origin. To overcome the issues of immunogenicity following repetitive administration a humanized version of the antibody was produced. J591 was mainly used for therapy applications using ^{177}Lu [97-99]. ^{177}Lu is also suitable for imaging. This offers a great advantage for the mAb J591 as both image-guided therapy and patient-specific dosimetry will be possible.

Smaller molecules such as minibodies and diabodies have faster clearance and reach higher tumour-to background ratios earlier than antibodies. Fragments below 60 kDa are filtered through the glomerular system, leading to significant urinary excretion. Evaluation of smaller J591 fragments targeting PSMA included the 80 kDa ^{89}Zr-labelled minibody IAB2M [100] for human PET imaging, and the 55kDa ^{99m}Tc-labelled J591C diabody [101] for preclinical SPECT

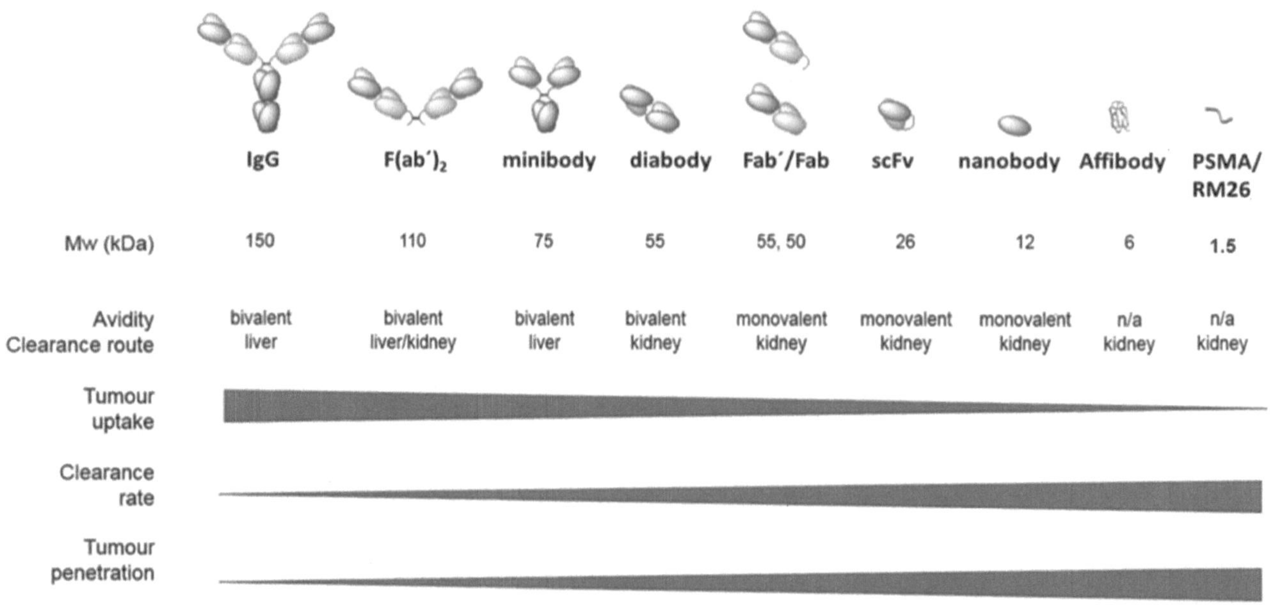

FIGURE 19.18 A schematic representation of different classes of targeting agents. Bulky targeting agents have better bioavailability due to prolonged circulation, which is an advantage for therapeutic applications. Smaller targeting agents with size below that of the kidney cutoff, that is, 60 kDa, have better clearance from the body a property that facilitates better contrast imaging at shorter times. Smaller agents also have rapid extravasation and better tumour penetration. Colour image available at www.routledge.com/9781138593312.

imaging, clearly showed that both agents have favourable pharmacokinetics compared to the parental antibody and early lesion detection was possible. However, small targeting agents (<60kDa) with significant renal excretion may be prone to high renal reabsorption from the primary urine by nephrotic tubular cells [102, 103]. For radiometal labelled targeting agents this is usually accompanied by high retention of radioactivity in the radiosensitive kidneys, due to the residualizing properties of the generated radiocatabolites inside the kidney cells. This makes the kidneys a dose-limiting organ for small- to medium-size protein-based targeting agents. Methods to reduce this renal reabsorption involve coinjection of the plasma expander gelofusine or positively charged amino acids such as arginine or lysine, although it may not be effective in some cases.

Combining different tracer will enable monitoring changes in PCa development. Harmon and colleagues [104] has elaborated on using imaging with different radiopharmaceuticals to reveal the disease state of PCa. They indicated how PSMA and fluoride can be used monitoring the decrease in PCa development to neuroendocrine stage. It has been reported that prostate-specific membrane antigen (PSMA) expression is decreased with neuroendocrine differentiation of PCa [105].

19.6.5.2 5A10 and 11B6 Mab´s Targeting PSA (KLK3) and hK2 (KLK2)

The two AR governed, prostate-tissue specific and phylogenetically related enzymes (80% amino acid homology) PSA (*KLK3*) and hK2 (*KLK2*) have been applied for PCa screening and disease monitoring. Both *KLK2* and *KLK3* are positively regulated by AR and specifically expressed at abundant levels in both healthy and malignant epithelial prostate cells. Because of their correlation with downstream AR-pathway activity, PSA and hK2 have come into focus as both diagnostic biomarkers and therapeutic targets of PCa in vivo. Since these proteins are secreted, attempts to target them have previously been avoided. In recent years, an antibody-based method for targeting tissue-associated PSA and hK2 has been successfully developed and applied for in vivo radio-immunotheranostics [106]. These mAbs have a high affinity for the catalytic clefts of hK2 and PSA, circumventing the complexed form of these enzymes in the blood (Figure 19.15). The mAb technology exploits the neonatal fc-receptor's inherited mechanism to route antigen-bound mAbs from recycling to lysosomal pathway compartments, resulting in internalization into target cells. This technological platform is currently at exciting stages of evaluation; a clinical trial is currently evaluating a hK2 targeted mAb (hu11B6), while the mAb targeting PSA is presently being studied preclinically.

The scientific advantages of PSA and hK2 targeted radiotheranostics compared to PSMA are the tissue specificity (no uptake in salivary glands, kidneys or nerve endings) and positive AR-regulation. Radionuclides are likely to impact tumour microenvironment and gene expression similar to the cytotoxic effect of irradiation [7, 107]. A proximal consequence of radiation-centred treatment is induction of AR, the central driver of PCa, and AR associated DNA repair genes [108, 109]. Similarly, a single injection of [^{225}Ac]hu11B6 results in significant induction of AR and AR target genes (*KLK3* [PSA], *TMPRSS2*, and *KLK2* [hK2]), while negatively regulated AR genes, such as FOLH1, that is, a gene that encodes PSMA downregulation [110]. This phenomenon decreases PSMA-RIT (radioimmunotherapy) efficacy, while igniting a feed-forward acceleration of PSA and hK2 targeted RIT due to the radiation-mediated increase in expression of PSA, hK2 and AR, respectively.

19.7 SKELETAL METASTASES –TARGETED TREATMENT FOR PAIN PALLIATION

Bone metastases often lead to pain and fractures. Pain palliation with analgesic or external beam irradiation can have serious side effects. ^{32}P, ^{89}Sr, and ^{223}Ra are so-called calcimimetic agents following calcium kinetics. Radionuclides can be labelled to phosphonates, with high affinity to bone minerals, be absorbed to calcium atoms and reduce pain by inducing apoptosis in osteoclasts. In therapy the most used phosphonate-based agents are ^{188}Re-HEDP, ^{153}Sm-EDTPM, and ^{177}Lu-EDTMP. For bone palliation, ^{89}Sr-chloride [111], ^{153}Sm-lexidronam, and ^{186}Re-etidronate, with pain relief being achieved in 60–80 per cent of patients [112]. Radionuclides used for phosphonate coupling and skeletal pain therapy are given in Lange and colleagues [64]. Table 19.4 lists the main physical characteristics for some radionuclides used for pain palliation in metastatic disease involving the bones.

For initial staging, ^{99m}Tc bone scans have sensitivities and specificities for osseous metastases of 46–89 per cent and 32–57 per cent, respectively [113]. A review paper by Van Dodewaard-de Jong and colleagues [114] states that beta-emitting radiopharmaceuticals are best for pain palliation and recommend the use of ^{153}Sm, ^{186}Re, and ^{188}Re because of lower haematological side effects. Alpha-emitters are gaining more interest in pain palliation protocols. Besides its pain-relieving effect, the α-emitter, ^{223}Ra, has proven also to prolong overall survival (see below) in PCa patients.

$^{89}SrCl_2$ has a physical half-life of 50.5 d, a maximum β-energy of 1.463 MeV (100%), and a range of 8 mm in tissue. The typical activity administered is 148 MBq, and pain relief usually begins in 10–20 days and lasts up to 6 months.

TABLE 19.4
Main Physical Characteristics of Key Radionuclides for Bone-palliation

	Physical half-life	Maximal range in soft tissue
^{166}Ho	1.1 days	9 mm
^{177}Lu	6.7 days	2 mm
^{32}P	14.3 days	8 mm
^{186}Re	3.7 days	5 mm
^{188}Re	0.7 days	10 mm
^{105}Rh	1.5 days	2 mm
^{153}Sm	1.9 days	4 mm
^{90}Y	2.7 days	11 mm
^{117m}Sn	13.6 days	300 µm
^{225}Ac	10.0 days	100 µm
^{211}At	7.2 h	70 µm
^{212}Bi	1.1 h	90 µm
^{227}Th	18.7 days	100 µm

^{153}Sm-EDTMP emits both medium-energy β-particles and a photon (which allows for imaging), and has a physical half-life of 46.3 h. The recommended activity of ^{153}Sm-EDTMP is 37 MBq/kg. ^{186}Re is a beta emitter with a half-life of 3.72 days, a maximum beta energy of 1.07 MeV, and ranges of 1.1 mm in soft tissue and 0.5 mm in bone. The emission at 137 keV photons can be used for imaging. Chelated to hydroxyethylidene-diphosphonate (HEDP), ^{186}Re-HEDP binds to hydroxyapatite crystals in bone. Average pain response rates from 50 to 89.5 per cent, and durations of pain relief for 6–10 weeks have been observed [115].

19.8 THERAPY RADIOPHARMACEUTICALS

The main problem with castrate-resistant PCa with metastatic spread is that conventional treatments with surgery, EBRT, or androgen deprivation treatments are ineffective. For example, novel agents' abiraterone and enzalutamide provide limited survival benefits of only 3.9 and 4.8 months, respectively. Thus, the development of more effective therapies targeting metastases is needed.

19.8.1 ^{223}Radium

Radionuclide therapy with the α-emitting bone seeker ^{223}Ra, initially intended for pain palliation, has shown some therapy efficacy. ^{223}Ra-dichloride is accumulating in areas of increased bone turnover and due to the high LET causing cytotoxicity independent of absorbed dose rate, cell-cycle growth phase, and oxygen concentration. ^{223}Ra localizes to areas of osteoblastic activity, and the hydroxyapatite crystal structure is thought to be the target for the ^{223}Ra ions. These ions are taken up into bone by ionic exchange with the calcium ions. $^{223}RaCl_2$ results in reduction of bone pain and overall survival advantage of 3.6 months in castration-resistant PCa patients with bone metastases. The use of ^{223}Ra is limited to the treatment of bone metastases and not generalized metastasized PCa. The palliation/therapy activity is usually six cycles of 50 kBq/kg, given with 4 weeks intervals. The safety and efficacy of palliation with $^{223}RaCl_2$ is associated with low myelosuppression rates and few adverse events. ^{223}Ra (as chloride salt, Xofigo) has received marketing approval from the US Food and Drug Administration (FDA) and the European Medicines Agency (EMA) for the treatment of castration-resistant PCa (CRPC) with bone metastases in patients with symptomatic bone metastases and without known visceral disease. In Bieth and colleagues, a Figure 19 is shown, with ^{68}Ga-PSMA PET images of a patient before and after $^{223}RaCl_2$ radionuclide therapy. Bone lesions showing response to therapy, seen as decreasing PSMA expression, are clearly shown. However, also a few new lesions appear on late PET images [67]. In Figure 19.19 is shown whole body images of ^{223}Ra uptake up to 6 days after injection [116].

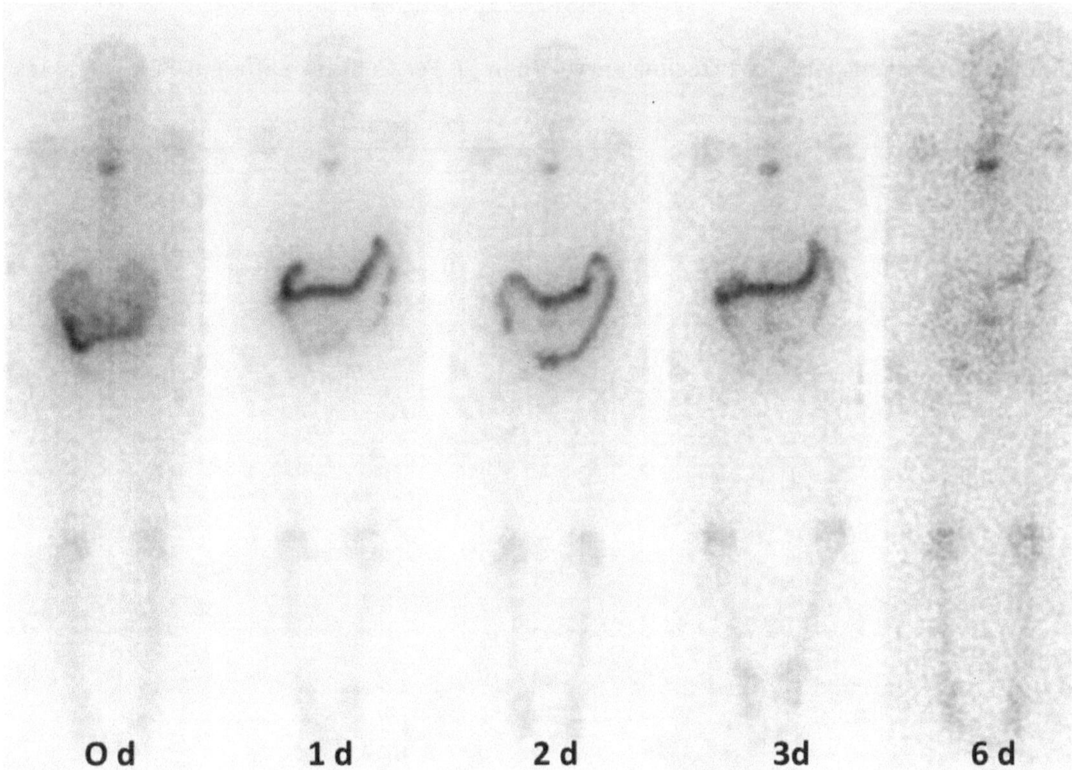

FIGURE 19.19 Whole body anterior images of a patient injected with ^{223}Ra at different time points after injection.

(Source: Reproduced from Flux [116].)

19.8.2 PSMA Therapy

Therapy with PSMA targeting radiopharmaceuticals has been performed implementing the β-emitter ^{177}Lu or the α-emitter ^{225}Ac. Promising results for response rates and safety profile after therapy with ^{177}Lu-PSMA-617 in patients with metastatic castration-resistant PCa (mCRPC) have been reported [117]. Here, ^{177}Lu-PSMA-617 therapy showed low haematoxicity and no severe nephrotoxicity with only mild to moderate xerostomia in 8 per cent of the patients with an administered activity of as high as 6 GBq.

Initial experience with targeted ^{225}Ac-PSMA-617 α-therapy was reported in a pilot clinical study involving one patient for whom treatment with β-emitters was contraindicated (patient A) and one patient resistant to ^{177}Lu-PSMA-617 (patient B) [81]. Results from this study showed tumour progression in both patients, with one patient having a drop in PSA level from more than 3,000 ng/mL to 0.26 ng/mL after 3 cycles of treatment. This study clearly demonstrated the high potential of ^{225}Ac-PSMA-617 to significantly benefit mCRPC patients with diffuse red marrow infiltration and resistance to other therapies. Moreover, targeted therapy with ^{225}Ac-PSMA was clearly superior to ^{177}Lu-PSMA, in terms of both therapeutic efficacy and adverse effects.

As we have seen in the previous example, changes in PSA may act as a good indicator for the efficacy of targeted therapy. In several studies, the number of patients having a drop in PSA by more than 50 per cent has been investigated, and in Table 19.5 some of those results are listed. As can be seen there are very few examples of therapies rendering results with numbers above 50 per cent of the patients.

In Kratochwil [118] is given an example on the effect of α-therapy with ^{225}Ac-anti PSMA-617. The reduction in number of metastases, together with a dramatic decrease in PSA levels are clearly shown. The PET images are performed with ^{68}Ga-PSMA-11. In Figure 19.20 [^{68}Ga]Ga-PSMA-11, PET images of a patient with castration-resistant prostate cancer are shown [119]. Impressive decrease in tumour burden after two cycles of α-emitting [^{225}Ac]Ac-PSMA-617 can be observed.

TABLE 19.5
Fraction of Patients with PSA Decline Larger Than 50 Per Cent for Different PCa Therapies

Drug	PSA decline > 50%
^{225}Ac-PSMA-617	63 %
^{177}Lu-PSMA-617	30-59 %
^{223}Ra	10 %
Abiraterone	29 %
Enzalutamide	54 %

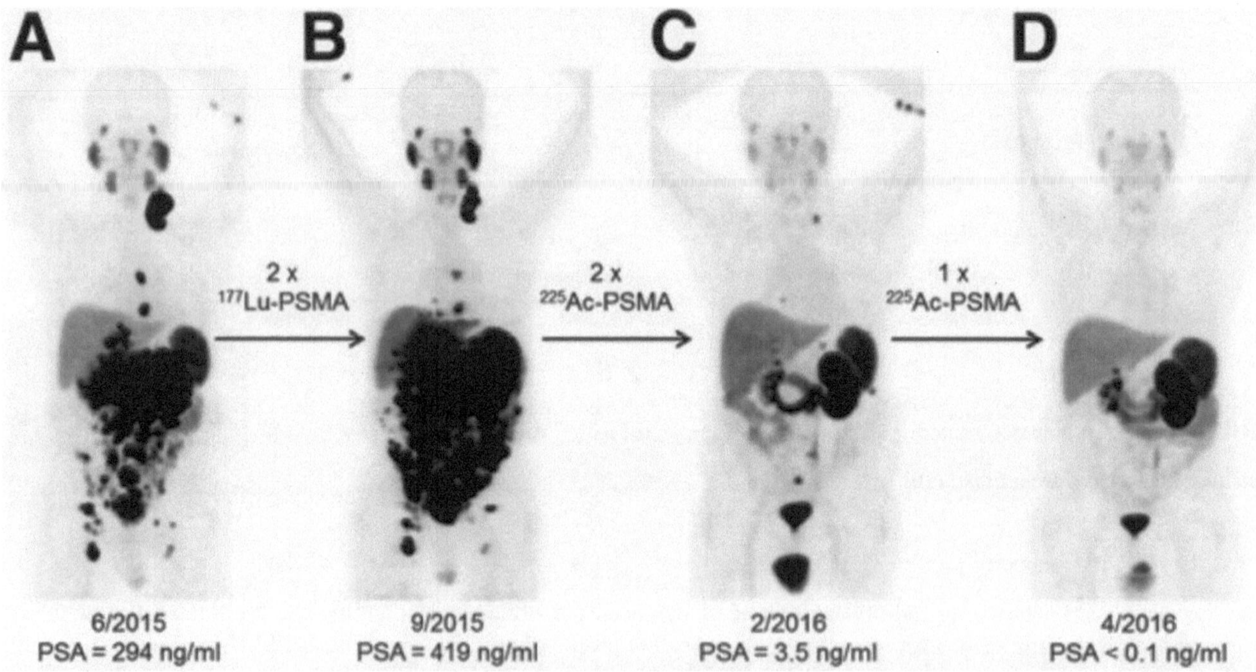

FIGURE 19.20 Effect of therapy with either beta or alpha emitter in PCa patient. [^{68}Ga]Ga-PSMA-11 PET images of a patient with CRPC. (A) Initial tumour burden (B) Progression after 2 cycles of [^{177}Lu]Lu-PSMA-617 (C,D) Decrease in tumour burden after two cycles of [^{225}Ac]Ac-PSMA-617.

(Source: Reproduced with permission [119].)

19.9 DOSIMETRY

19.9.1 Diagnostic Imaging Dosimetry

For diagnostic imaging, dosimetry reports for some of the clinically used PCa radiopharmaceuticals have been published, and below is a short summary with some dosimetry values given in Table 19.6. The minibody ^{89}Zr-IAB2M dosimetry has been evaluated in patients given 185 MBq, and biokinetics was evaluated with whole-body PET imaging and serum clearance. The absorbed doses were 1.67, 1.36, and 0.32 mGy/MBq to liver, kidney, and marrow, respectively, with an effective dose of 0.41mSv/MBq [100]. ^{64}Cu-PSMA-617 has been reported to have the absorbed doses as total body 0.014, liver 0.3, kidneys 0.07, and salivary glands 1.0 mGy/MBq. Usually about 250 MBq was administered [69].

^{68}Ga-PSMA for diagnosis has been reported with about 0.02 mSv/MBq [88] meaning 3 mSv after 150 MBq injected activity. Kidneys will receive about 0.2 mSv/MBq corresponding to 30 mSv. Sandgren and colleagues [120] investigated the dosimetry for ^{68}Ga-PSMA-11 in PCa patients with PET scanning up to 255 min p.i. and sampling both blood and urine. The effective dose was 0.022 mSv/MBq. The absorbed dose to the kidneys was 0.24 mGy/MBq and salivary

TABLE 19.6
Absorbed Doses for Some PCa Diagnostic Radiopharmaceuticals in Critical Organs, Reported in the Literature

	Absorbed dose per administered activity (mGy/MBq)				Effective Dose (mSv/MBq)	
	Salivary Glands	Liver	Kidneys	Bone marrow	Total Body	Reference
PSMA-ligands						
^{64}Cu-PSMA-617	1.0	0.3	0.07		0.03	[69]
^{68}Ga-PSMA-11	0.09		0.24		0.02	[120]
^{68}Ga-PSMA-11		0.03	0.2	0.009	0.02	[89]
^{18}F-JK-PSMA-7	0.05	0.08	0.2		0.01	[121]
^{18}F-PSMA-1007	0.09	0.06	0.2	0.01	0.02	[122]
^{18}FDCPyL	0.03	0.04	0.09	0.01	0.01	[123]
^{18}F-DCFB		0.02	0.03		0.02	[124]
^{18}F-CTT1057	0.02	0.02	0.07	0.007	0.02	[125]
^{18}F-PSMA-11	0.035	0.017	0.085	0.0079	0.013	[126]
ANTIBODIES						
^{89}Zr-DFO-MSTP2109A		1.2	0.73	0.43	0.44	[127]
^{111}In-DOTA-5D3	0.12			0.04		[128]
MINIBODY						
^{89}Zr-IAB2M		1.7	1.4	0.3	0.4	[100]

glands 0.09 mGy/MBq. A mean activity of 160 MBq was administered. In a study by Zlatopolskiy and colleagues [121] they developed a new anti PSMA tracer [^{18}F]F-JK-PSMA-7, for PET/CT imaging of PCa, and Hohberg and colleagues [129] performed dosimetry on patients. The effective dose from [^{18}F]F-JK-PSMA-7 for the whole body was calculated to be 0.01mGy/MBq. The highest absorbed dose was observed in the kidneys, 0.18mGy/MBq, followed by the liver, 0.08mGy/MBq, salivary glands, 0.05mGy/MBq, spleen, 0.02mGy/MBq, and lungs, 0.01mGy/MBq.

^{89}Zr-DFO-MSTP2109A an antibody conjugate to image six-transmembrane epithelial antigen of prostate (STEAP) comprises a cell surface marker that is highly expressed in PCa but also present in other cancers, and has little cross-reactivity to other normal tissues. The dosimetry of ^{89}Zr-DFOMSTP2109A in patients has been reported in Donoghue and colleagues [127]. Patients underwent blood sampling, whole body PET, and single detector whole-body counting for biokinetics data collection. The absorbed doses with highest values were liver, heart wall, lung, kidney, and spleen, with mean values of 1.18, 1.11, 0.78, 0.73, and 0.71 mGy/MBq, respectively. An imaging activity of 184 MBq (range 174–186 MBq) was administered. In Figure 19.21 a nice example is given of using different PCa targeting radiopharmaceuticals diagnosing disease spread in a patient with metastatic PCa to the bone. (A) Planar scintillation camera image of ^{99m}Tc[Tc]-MDP bone scan; (B) ^{18}F[F]DG PET image of uptake in metastatic bone disease similar to the bone scan; (C) Serial PET images obtained over 6 days after injection of [89Zr]Zr-DFO-MSTP2109A, showing prolonged blood pool activity. Increasing uptake in bone metastases is seen over time, with little uptake in the initial scan and marked uptake by 2 days, which persists with better contrast at 6 days.

^{18}F-CTT1057 was developed based on a phosphoramidate scaffold, unlike most other PSMA agents with a urea backbone. In a patient study, dosimetry was performed. The average effective dose was 0.023 mSv/MBq. The kidneys exhibited the highest absorbed dose, 0.067 mGy/MBq. The absorbed dose of the salivary glands was 0.015 mGy/MBq.

In their study of ^{18}F-PSMA-11, Piron and colleagues [126] [130] have made a comprehensive comparison of the mean effective absorbed doses for various PSMA diagnostic radiopharmaceuticals such as ^{18}F-DCFPyL, ^{18}F-DCFBC, ^{68}Ga- PSMA-617, ^{18}F-PSMA-1007, ^{68}Ga-PSMA-11, and ^{18}F-PSMA-11. A summary is given in Figure 19.22.

The same authors made a comparison of the absorbed doses per activity unit (mGy/MBq) in different organs for various PSMA PET radiopharmaceuticals: ^{18}F-DCFPyL, ^{18}F-DCFBC, ^{68}Ga-PSMA-617, ^{18}F-PSMA-1007, ^{68}Ga-PSMA-11, and ^{18}F-PSMA-11, and some results are displayed in Figure 19.23.

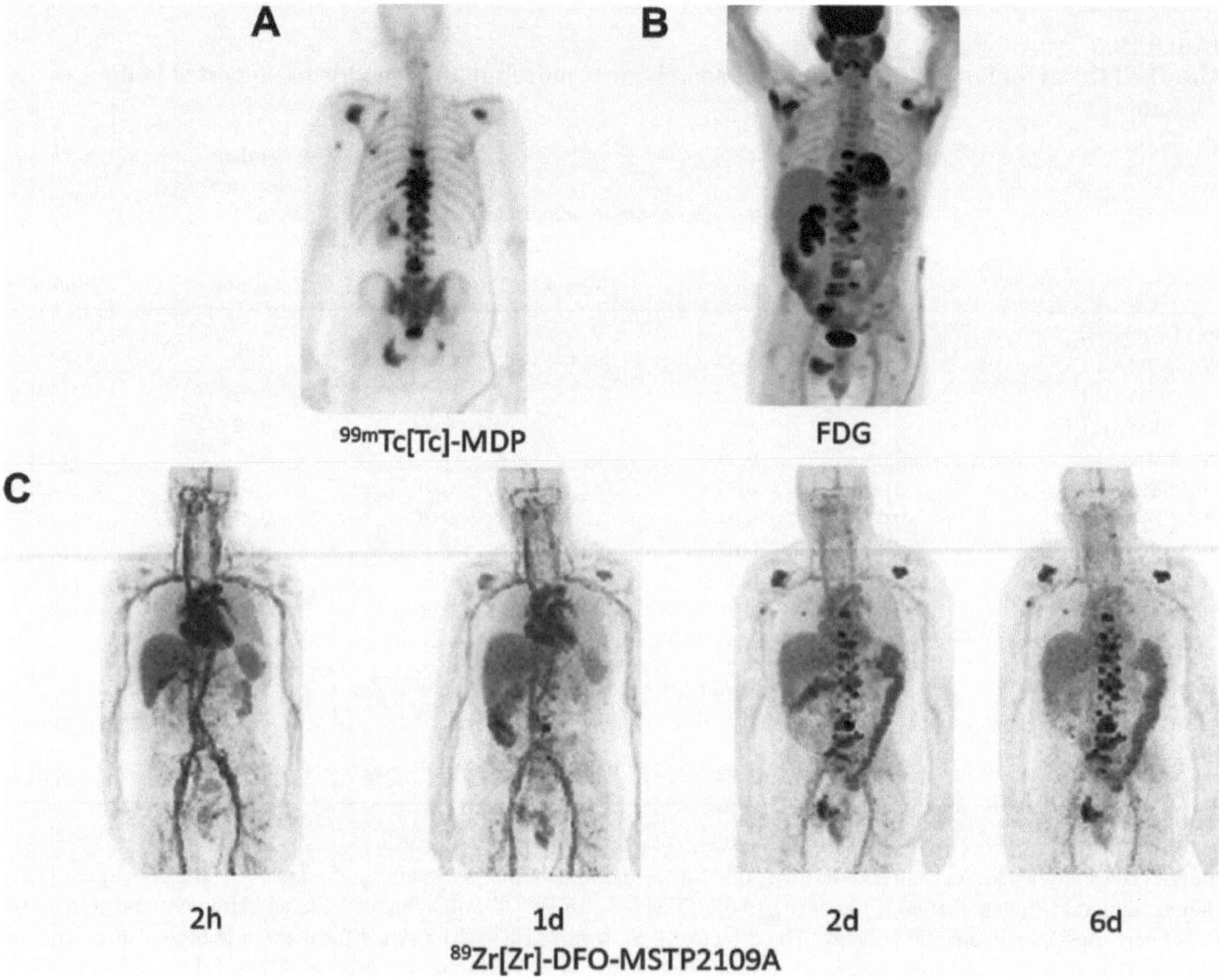

FIGURE 19.21 An example of using different PCa targeting radiopharmaceuticals diagnosing disease spread in a patient with metastatic prostate cancer to the bone. (A) Planar scintillation camera image of ^{99m}Tc[Tc]-MDP. (B) ^{18}F[F]DG PET image of uptake in metastatic bone disease similar to the bone scan. (C) Serial PET images obtained over 6 days after injection of [^{89}Zr]Zr-DFO-MSTP2109A Anti-STEAP antibody, showing prolonged blood pool activity. Increasing uptake in bone metastases is seen over time, with little uptake in the initial scan and marked uptake by 2 days, which persists with better contrast at 6 days.

(Source: Reproduced from O'Donoghue *et al.* with permission [127].)

^{89}Zr dosimetry results in relatively high normal organ-absorbed doses because of the high-energy photon emission. In the literature, the following estimates for different antibody conjugates are given. ^{89}Zr-DFOUC36 has an average absorbed dose to liver, kidney, lung, and spleen, 98 per cent of that for ^{89}Zr-DFOMSTP2109A. ^{89}Zr-DFO-huJ591 showed high liver uptake, resulting in a 76 per cent higher liver absorbed dose than that for [^{89}Zr]Zr-DFO-MSTP2109A. Otherwise, absorbed doses to kidney, lung, and spleen from ^{89}Zr-DFO-huJ591 were similar to that of ^{89}Zr-DFO-MSTP2109A [127].

19.9.2 Therapy Dosimetry

Few radionuclide treatments to date for PCa have involved the use of dosimetry either to plan treatment or to retrospectively ascertain the absorbed dose delivered during treatment. Thus, few reports exist for correlation between absorbed dose and biological effect. A summary is given in Table 19.7. In PSMA-targeted therapy, standard activity and peptide amount are usually administered, although the tumour burden varies considerably (from millilitres to several litres). Optimization of therapy with mathematical models has been reported by Begum and colleagues [131], and it was shown

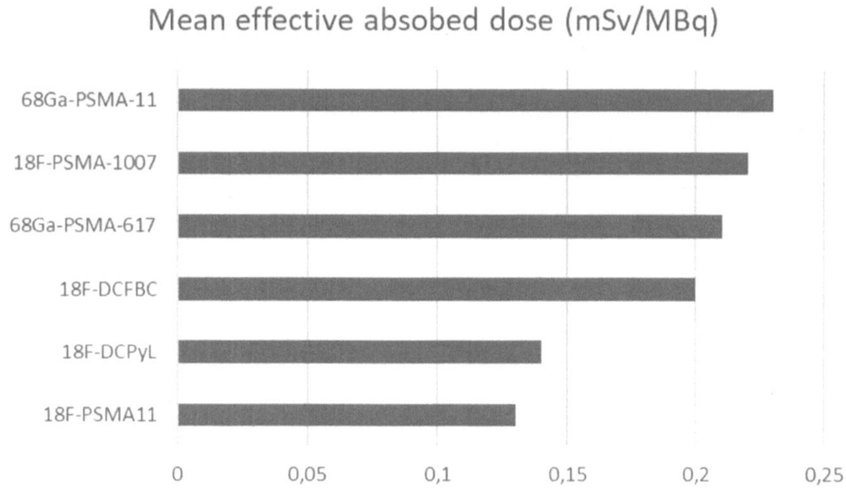

FIGURE 19.22 Comparison of mean effective absorbed doses of various developed PSMA radiopharmaceuticals: ^{18}F-DCFPyL, ^{18}F-DCFBC, ^{68}Ga- PSMA-617, ^{18}F-PSMA-1007, ^{68}Ga-PSMA-11, and ^{18}F-PSMA-11.

(Source: Redrawn from reference [130].)

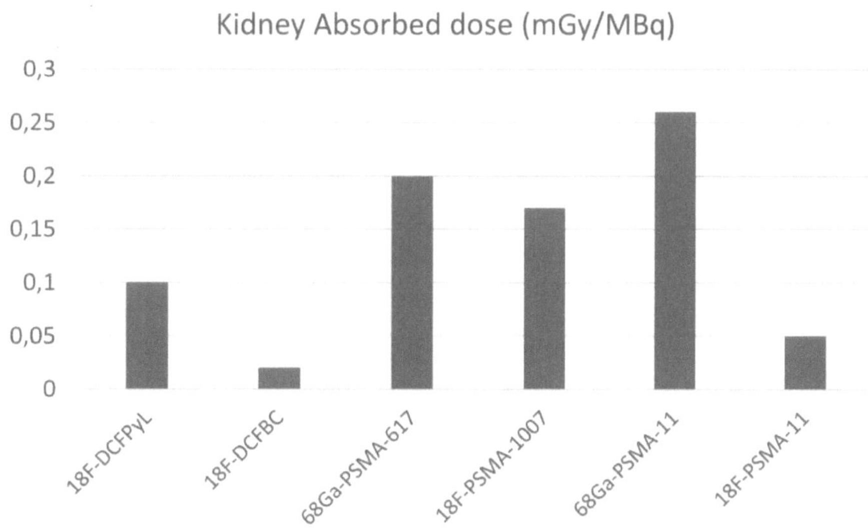

FIGURE 19.23 Comparison of absorbed dose in kidneys for various PSMA radiopharmaceuticals ^{18}F-DCFPyL, ^{18}F-DCFBC, ^{68}Ga-PSMA-617, ^{18}F-PSMA-1007, ^{68}Ga-PSMA-11, and ^{18}F-PSMA-11.

(Source: Redrawn from Piron *et al.* [130].)

that the biologically effective doses (BEDs) to normal tissues and tumour tissue could vary considerably in CRPC patients treated with ^{177}Lu-PSMA. For example, large tumour burdens most presumably lead to less effective treatment when a standard activity and amount of peptide was used. Their results suggest that in patients with large PSMA-positive tumour volumes, higher activities and peptide amounts can be safely administered to maximize tumour BEDs without exceeding the tolerable BED to risk organs.

In a therapy study of ^{177}Lu-PSMA, dosimetry was evaluated using whole-body imaging acquired at 5 time points from 0.5 to 118 h after injection. ^{177}Lu-PSMA demonstrated high tumour-absorbed doses (median, 3.3 mGy/MBq) compared with the levels in normal organs. Parotid glands received higher absorbed doses (1.3 mGy/MBq) than kidneys (0.8 mGy/MBq). The maximum absorbed dose delivered to a paraaortic lymph node metastasis was 468 Gy. In a comparison, the

TABLE 19.7
Absorbed Doses for PCa Therapeutic Radiopharmaceuticals in Some Organs Reported in the Literature

	Salivary Glands	Liver	Kidneys	Red marrow	Total Body	Tumour	Reference
PSMA-ligands Beta emitters (mGy/MBq)							
^{177}Lu-PSMA	1.3		0.8	0.014	0.02	3.3	[132]
^{177}Lu-PSMA-617	1.4		0.8	0.03		6.6	[133]
^{177}Lu-PSMA-617	1.90		0.82	0.03			[134]
^{177}Lu-PSMA-617	1.01-0.50		0.62	0.04			[135]
^{177}Lu-PSMA-617	0.44-0.58	0.1	0.39	0.11			[136]
^{177}Lu-PSMA-I&T	0.55-3.8	0.12	0.72			3.3	[137]
^{177}Lu-DOTA-5D3	0.24	0.97		0.04			[128]
^{177}Lu-DOTA^{ZOL}	0.019	0.018	0.490	0.461	0.092		[138]
Alpha-emitters (mSv/MBq)							
^{225}Ac-PSMA-617	2.3		0.7	0.05		5.7	[81]
^{213}Bi-PSMA-617	8.1	1.2	8.1	0.52		6.3	
^{223}Ra		180	16				[139]
^{223}Ra		180	17	370			[140]
^{223}Ra		10	30	10-2040			[141]
ANTIBODIES Beta emitters (mGy/MBq)							
^{177}Lu-DOTA-J591			1.40	0.32	0.19		[142]
^{177}Lu-DOTA-J591				0.3			[128]
^{131}I-MIP-1095	4.62	1.47	1.45	0.31	0.37		[143]
Gastrin							
^{177}Lu-RM2		0.05	0.35	0.02		6.2	[144]

mean absorbed doses of ^{177}Lu-PSMA delivered to the whole body, red marrow, and kidneys (0.02, 0.014, and 0.8 mGy/MBq, respectively) were in a comparison found to be significantly lower than those of ^{177}Lu-DOTA-J591 (0.19, 0.32, and 1.40 mGy/MBq, respectively) [132].

Dosimetry for ^{225}Ac-PSMA-617 has been published in Kratochwil and colleagues [81] with biokinetics extrapolated from earlier studies of ^{177}Lu-PSMA-617 [133]. Biokinetics from whole body and organs were extrapolated from the 6.7 d half-life for ^{177}Lu to the half-life for 9.9 d of ^{225}Ac. The residency times of ^{225}Ac were then forwarded as residency times of ^{221}Fr, ^{217}At, ^{213}Bi, ^{213}Po, ^{209}Tl, and ^{209}Pb, assuming instant decay of the daughter nuclides and no translocation. The relative biologic efficacy (RBE) was assumed to be 5. The dosimetry results were for salivary glands (2.3 Sv/MBq), kidneys (0.7 Sv/MBq), and red marrow (0.05 Sv/MBq).

In another study [145], Kratochwill and colleagues used ^{68}Ga-PSMA-617 extrapolating ^{68}Ga ($T_{1/2}$ 68 min) to the half-life of ^{213}Bi (45.6 m). The same residence time of ^{213}Bi was used to calculate equivalent absorbed doses for all daughter nuclides (^{209}Tl, ^{213}Po and ^{209}Pb) and with RBE 5 the equivalent absorbed doses of 8.1 Sv/GBq for salivary glands, 8.1 Sv/GBq for kidneys and 0.52 Sv/GBq for red marrow, liver (1.2 Sv/GBq), spleen (1.4 Sv/GBq), and bladder (0.28 Sv/GBq).

Banerjee and colleagues (Banerjee *et al.* 2019) have developed a new monoclonal antibody, ^{111}In-DOTA-5D3, specific for PSMA. They claim that PSMA small-molecule agents have side effects, such as causing damage to the salivary glands and renal toxicity that can be overcome with the use of antibodies. They calculated dosimetry based on their preclinical biokinetics and extrapolated to ^{177}Lu labelled 5D3. In addition, a comparison with the huJ591 antibody dosimetry was done. The absorbed doses of ^{177}Lu for the kidneys, lacrimal glands, parotid glands, and submandibular glands were 0.10–0.17 mGy/ MBq, whereas the liver-absorbed dose was 0.97 mGy/MBq. The estimated bone-marrow absorbed dose for both ^{111}In-DOTA-5D3 and ^{177}Lu-DOTA-5D3 was 0.04 mGy/MBq. The estimated red marrow absorbed dose of ^{177}Lu-DOTA-5D3 (0.04 mGy/MBq), was slightly lower than that of ^{177}Lu-DOTA-J591 (0.3 ± 0.1 mGy/MBq). The absorbed dose to the salivary glands, 0.24 mGy/MBq from ^{177}Lu-DOTA-5D3 is much lower than that from ^{177}LuPSMA-617, 1.4 mGy/MBq. Additionally, the absorbed dose within the lacrimal glands for ^{111}Lu-DOTA-5D3 (0.12 mGy/ MBq) was significantly lower than that of ^{177}Lu-PSMA-617 (2.82 mGy/MBq).

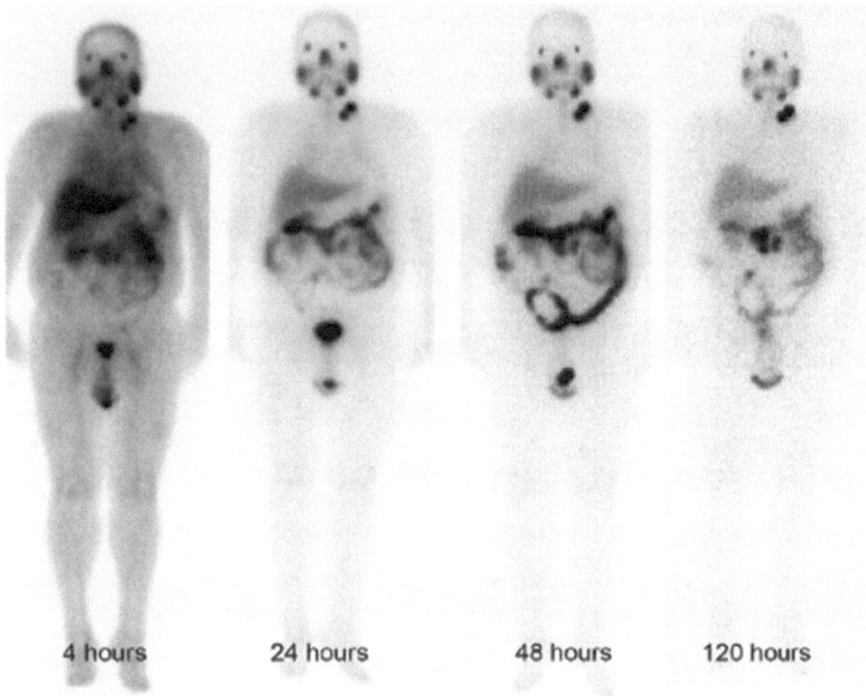

FIGURE 19.24 ^{177}Lu-PSMA-617 whole body anterior images of a patient at different time points post injection.

(Source: Reproduced from Kabasakal *et al.* [134] with permission.)

In a study by Kabasakal and colleagues [134] they used whole-body planar imaging at 4 to 120 h p.i. and a SPECT/CT study at 24h p.i. for ^{177}Lu-PSMA-617. The SPECT image was quantified, and the whole-body images using the conjugate view method was used accordingly to obtain the activity in different organs. Their estimates of the absorbed doses for the kidney, parotid gland, and bone marrow were 0.82, 1.90, and 0.03 Gy/GBq, respectively. Based on those data, the following activities could be administered without reaching the different organs tolerable absorbed dose, parotid gland 21.4 GBq/30Gy, Kidney 32.9GBq/23 Gy, bone marrow 73.8 GBq/2 Gy and liver 254.2 GBq/32Gy. In Figure 19.24 is displayed ^{177}Lu-PSMA-617 whole body planar images of a patient at different time points post injection – from Kabasakal L and colleagues [134].

^{177}Lu-PSMA-617 dosimetry was evaluated by Scarpa and colleagues [135] using planar anterior and posterior whole-body planar imaging. In addition, SPECT/CT imaging of the abdomen was performed at 24 hours to rule out overlays between different organs/tumours and to evaluate organ and tumour volumes. The geometric mean of the anterior and posterior projections was calculated for quantification of activity. The dosimetry values obtained were an average kidney absorbed dose of 0.60 ± 0.36 Gy/GBq and an average red bone marrow absorbed dose of 0.04 ± 0.03 Gy/GBq. The mean absorbed dose to the parotid glands was 0.56 ± 0.25 Gy/GBq, to the sub-mandibular glands, 0.50± 0.15 Gy/GBq, and to the lacrimal glands, 1.01 ± 0.69 Gy/GBq.

Okamotoo and colleagues [137] calculated dosimetry for ^{177}Lu-PSMA I&T for the whole body; kidneys; liver, parotid, submandibular, and lacrimal glands based on whole body imaging up to 7 d p.i. with activities of about 7.3 GBq/cycle given in 4 cycles. The mean organ-absorbed doses were 5.36± 1.6 Gy (0.72 Gy/GBq) for the kidneys; 0.89 ± 0.42 Gy (0.12 Gy/GBq) for the liver; and 4.0 ± 1.1 Gy (0.55 Gy/GBq) for the parotid, 4.8 ± 2.8 Gy (0.64 Gy/GBq) for the submandibular, and 27 ± 10 Gy (3.8 Gy/GBq) for the lacrimal glands. Also, tumour-absorbed doses were calculated where all tumours had a mean absorbed dose per cycle of 23 ± 20 Gy (3.3 Gy/GBq). Mean absorbed dose for bone, lymph node, liver, and lung metastases were 26 6±20 Gy (3.4 Gy/GBq), 24 ± 16 Gy (3.2 Gy/GBq), 8.5 ± 4.7 Gy (1.28 Gy/GBq), and 13 ± 7.4 Gy (1.7 Gy/GBq). They had an interesting observation, also reported by others for PSMA targeting therapies, that the absorbed dose to the tumour decreased with repeated cycles, whereas it was constant for organs, see Figure 19.25.

Voxel-based dosimetry has been performed by Hofman and colleagues [136] on men with ^{68}Ga-PSMA-11 PET imaging preceding therapy with ^{177}Lu-PSMA-617, and the patients underwent quantitative SPECT/CT at 4, 24, and 96

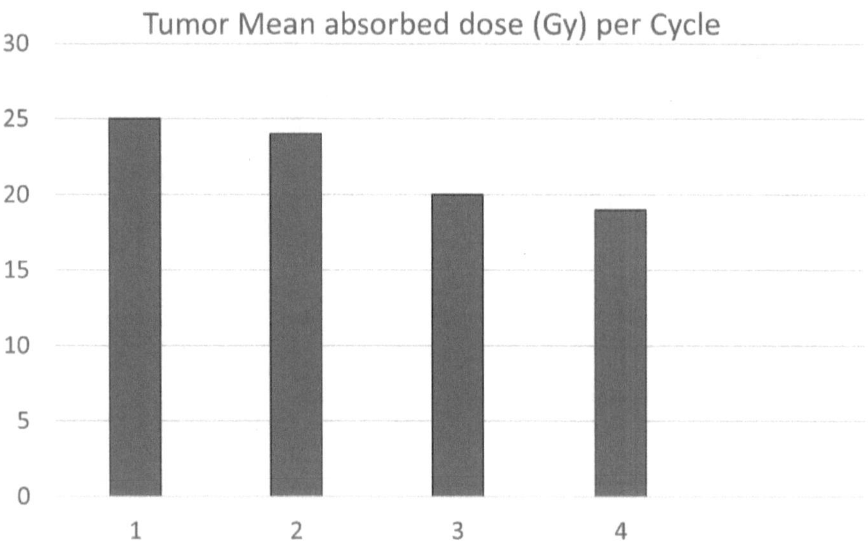

FIGURE 19.25 Mean absorbed doses for tumours after therapies with 7.3 GBq given in 4 fractions of ^{177}Lu-PSMA I&T. Interesting is the decrease with fraction of tumour absorbed dose with increasing cumulative absorbed dose.

(Source: Data from Okamoto *et al.* [137].)

h. Biokinetics was estimated at a voxel level and absorbed dose calculated using voxel S values. They obtained mean absorbed dose to kidneys, submandibular and parotid glands, liver, spleen, and bone marrow of 0.39, 0.44, 0.58, 0.1, 0.06, and 0.11 Gy/MBq, respectively. The patients received 5.7-8.7GBq (mean 7.8 GBq) in 4 cycles. They reported skeletal and lymph node metastases absorbed doses as high as 74 and 93 Gy with a mean of 41 and 29 Gy, respectively.

Dosimetry of kidney and tumours was investigated in patients with only one kidney [146]. The single functioning kidney presented high activity of ^{177}Lu-PSMA-617. The patients were given between 2 to 6 cycles of therapy with between 15–34 GBq. The mean injected activity per cycle was 6.39 ± 1.05 GBq (range, 3.9–8.1 GBq). The kidney absorbed dose per cycle was 5.3 ± 2.1 Gy (0.81 ± 0.32 Gy/GBq). Also, the absorbed dose to tumours was calculated with a mean absorbed doses to tumour lesions of 94.3 ± 23 Gy (13.75 6 31.59 Gy/GBq). Interestingly, in patients with a single functioning kidney, ^{177}Lu-PSMA-617 PRLT was well tolerated, without any signs of acute or subacute nephrotoxicity, during a mean follow-up of nearly 2 years.

In a preclinical study, the dosimetry estimates after administration of ^{225}Ac labelled whole antibodies taken into concern the combination of equilibrium ^{225}Ac and non-equilibrium ^{213}Bi to the entire mouse kidney has been investigated by Schwartz and colleagues [27]. The absorbed dose to the mouse kidney by the administration of 22.2 kBq of ^{225}Ac-huM195 was 17 Gy (0.8 Gy/ kBq). The contribution of non-equilibrium ^{213}Bi was 10 Gy. The absorbed dose estimates were 15 Gy to the cortex and 23 Gy to the medulla. It is noteworthy that a large component (18 Gy) of the estimated absorbed dose to the renal medulla was delivered by non-equilibrium excess ^{213}Bi generated from extra-renal decays of ^{225}Ac. The projected absorbed dose to kidney for patients would then be 0.28 Gy/MBq of which 0.11 Gy/MBq is due to activity arriving in the form of ^{225}Ac and 0.17 Gy/MBq to non-equilibrium excess ^{213}Bi [27].

The relatively low emission rate of photons for ^{223}Ra can be used for imaging. ^{223}Ra has a complicated decay scheme, with a series of six daughter products, before decaying to stable lead. The total emitted energy is 28.2 MeV, of which 95 per cent is from α-emissions, 3.2 per cent from beta particles and 2 per cent from gamma emissions. The absorbed dose to bone surface was 2.3-13.1 Gy/MBq from α-emission, whereas as it was as low as 9-51 mGy/MBq from beta emission [116]. In Table 19.8 dosimetry values are given for different bone seeking therapeutic radiopharmaceuticals.

Murray and colleagues [62] found that absorbed doses to metastatic bone lesions from $^{223}RaCl_2$ ranged from 0.6 Gy to 44.1 Gy. For individual patients, there was a difference of 5.3 (range 2.5–11.0) between the maximum and minimum tumour absorbed dose. Interestingly, they found that baseline uptake of ^{18}F-fluoride in bone metastases is significantly correlated with corresponding ^{223}Ra absorbed dose and response to treatment.

An interesting example of patient-specific dosimetry, with whole body single detector measurements and SPECT activity quantifications, as a predictive marker of survival and as a tool for treatment planning of bone metastases is given

TABLE 19.8
Dosimetry for Bone Palliation Radiopharmaceuticals From Ref [116]

Radionuclide	Administered activity (MBq)	Mean absorbed dose (Gy)	
		Bone surface	Red marrow
^{89}Sr	150	2.6	1.7
^{153}Sm	2590	17.6	3.9
^{186}Re	1295	1.8	1.7
^{223}Ra	3.9	54-303	4-23

in [115]. Using whole-body measurements with a calibrated scintillation detector, anterior-posterior measurements were done with the patient at 2 m distance. Patients were given 5 GBq of ^{186}Re-HEDP and the mean whole-body absorbed dose was 0.33 Gy and metastases absorbed doses ranged from 4 to 78 Gy.

Interesting findings were observed in a preclinical study using ^{125}I (Auger emitter range <10 µm), suitable for treating micrometastases. ^{125}I-DCIBzL a highly specific small molecule targeting the prostate-specific membrane antigen (PSMA) was used. Internalization was seen using confocal microscopy revealing ^{125}I activity in the perinuclear area and plasma membrane. For small-scale dosimetry it is interesting that PSMA is specifically localized to the mitotic spindle poles, which would bring the Auger emitter close to DNA during mitosis, especially in anaphase when the condensed chromosomes are pulled toward the spindle poles. After treatment, DNA damage increased and clonogenic survival decreased. The therapy with the Auger emitter ^{125}I-DCIBzL yielded highly specific antitumour efficacy in animals with small tumours [147].

19.9.3 High-Resolution Activity and Absorbed Dose Distributions

The distribution of radiopharmaceuticals is heterogeneous in all tissues. The spatial resolution in planar, SPECT and PET imaging is not better than about 10 mm, meaning that the exact localization of the activity distribution in tissues and tumours cannot be revealed accurately influencing the interpretation of physiology and, more important, making detailed dosimetry inaccurate. Besides the traditional film autoradiography with a spatial resolution in the order of a few µm beta-cameras [148] [149] and alpha-cameras [150] have been developed for imaging of activity distributions on microscopic scale. A nice example on imaging of ^{223}Ra uptake in different PCa tumours in mice skeletons is given in Figure 19.26, taken from Abou and colleagues [151].

Radiolabelled peptides and antibodies distribute unevenly in solid tumours. The heterogeneous activity distribution then will generate a likewise heterogeneous absorbed-dose distribution. One such example is given in Figure 19.27 with a simulation of ^{90}Y, ^{177}Lu and 7 MeV α-particles on a tumour section from a preclinical study of an antibody distribution [152]. The autoradiography shows images a heterogeneous activity distribution. The absorbed dose rate distribution for ^{177}Lu at the time of dissection 10 h post injection showed a maximum value of 2.9±0.4 Gy/h (mean ± SD), compared to 6.0±0.9 and 159±25 Gy/h for the hypothetical ^{90}Y and 7 MeV α-particle cases, which should be compared to the mean absorbed dose rate values of 0.13, 0.53, and 6.43 Gy/h for ^{177}Lu, ^{90}Y, and α-particles, respectively. In Figure 19.27 an example is given of a heterogeneous activity distribution generating a likewise heterogeneous absorbed dose distribution from ^{90}Y, ^{177}Lu and 7 MeV α-particles on a tumour section with antibody uptake [152].

The higher beta particle energy of ^{90}Y reveals less heterogeneous absorbed dose rate distribution and a higher mean absorbed dose rate compared to ^{177}Lu. The α-particle results indicate the possibility of a higher maximum absorbed dose rate, although with a more heterogeneous absorbed dose rate distribution.

^{211}At-PSMA has shown to yield significantly improved survival in mice bearing PCa micrometastases after systemic administration but showed uptake in renal proximal tubules resulting in late nephrotoxicity. Dosimetry calculations for the kidney gave the mean absorbed dose of 24.6 mGy/kBq. When the activity distribution derived from α-camera images were inserted in a nephron dosimetry model an absorbed dose up to 123 Gy (or 123 mGy/ kBq) in portions of the proximal tubules was determined. This is about 5-fold higher than the mean absorbed dose to the whole kidney and higher than glomerular absorbed doses (which were up to 42 Gy) [153].

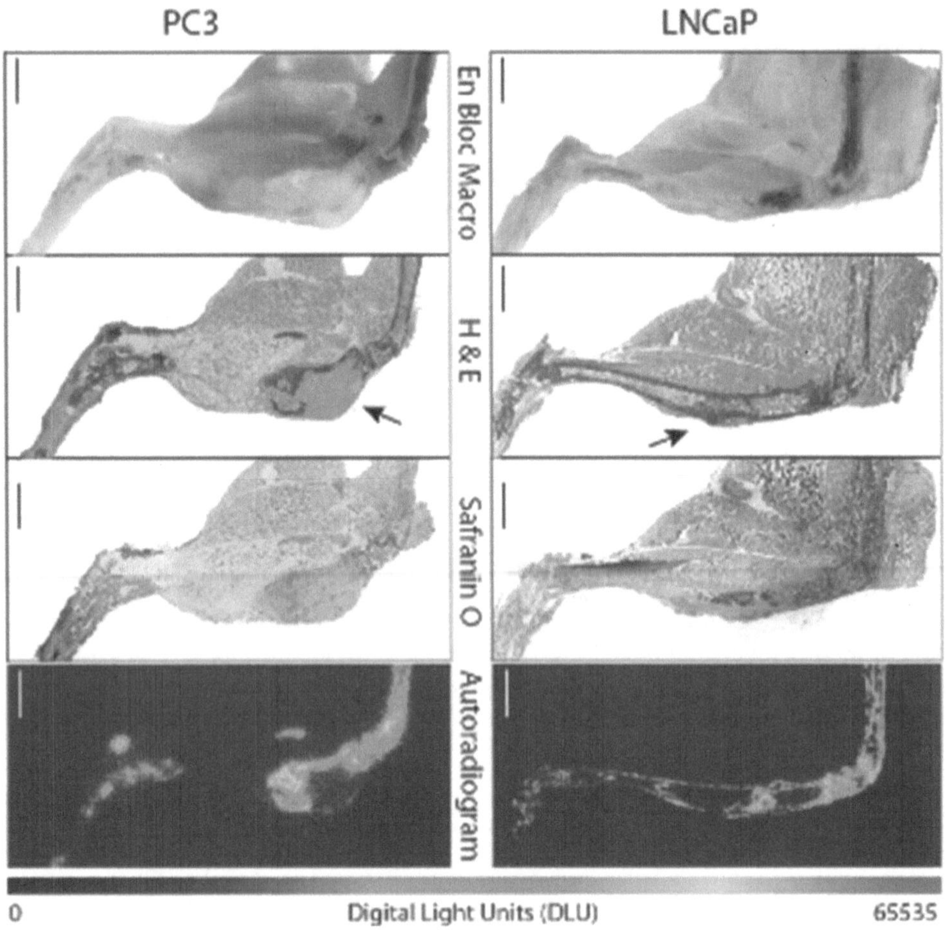

FIGURE 19.26 Imaging of ^{223}Ra uptake in different PCa tumours (osteoblastic (LNCaP) and osteolytic (PC3)) inoculated in mice skeleton. From top photograph of embedded sample, haematoxylin and eosin (H&E) stain – tumour indicated by arrow, Safranin-O staining for calcified (green) and uncalcified (orange-brown) areas, autoradiography. Areas of intense uptake colocalize with active bone modelling/remodelling sites (at the interface of the calcified and uncalcified compartments) in the distal femur, proximal tibia, and the bone surfaces adjacent to the bone metastasis. Colour image available at www.routledge.com/9781138593312.

(Source: From Abou *et al.* [151].)

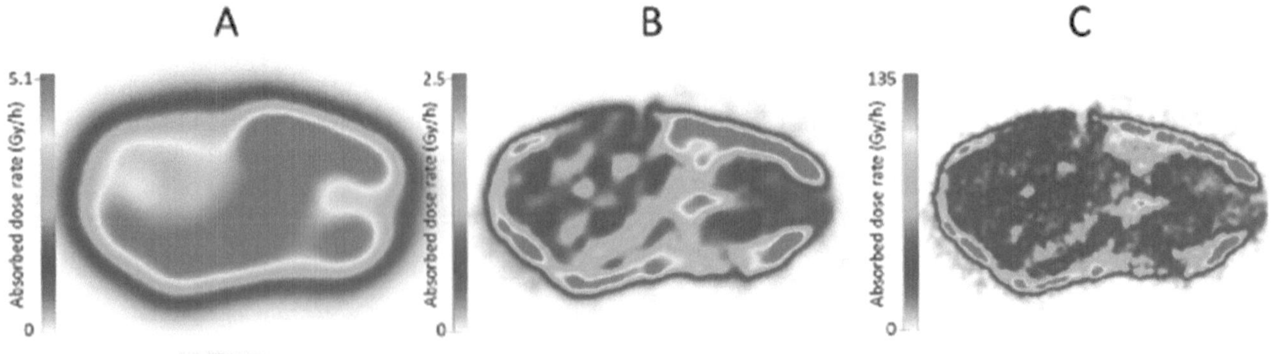

FIGURE 19.27 Example of a heterogeneous activity distribution generating a likewise heterogeneous absorbed dose distribution ^{90}Y, ^{177}Lu and 7 MeV α-particles on a tumour section from a preclinical study of an antibody. Colour image available at www.routledge.com/9781138593312.

(Reproduced from Timmermand et al. [152] with permission.)

19.10 ADVERSE EFFECTS: MYELOTOXICITY, NEPHROTOXICITY, XEROSTOMIA

Therapy with PSMA remains a noncurative treatment, thus the prolongation of life must be balanced against the direct toxicities of the treatment and their impact on quality of life. The uptake of radiolabelled small molecule–based PSMA radiopharmaceuticals is high in the salivary glands, suggesting both specific and nonspecific uptake. Interestingly radiolabelled anti-PSMA antibodies show only low uptake in the salivary glands. Xerostomia is among the most common side effects particularly for α-therapy with, for example, ^{225}Ac-PSMA, whereas ^{177}Lu labelled PSMA rarely leads to symptoms. In Table 19.7 dosimetry values for salivary glands are given. The range of absorbed dose to the salivary glands from ^{177}Lu-PSMA-617 varied from 0.8 to 2.5 Gy/GBq. As stated by Taïeb and colleagues [154] because ^{177}Lu- and ^{225}Ac-labeled PSMA therapies are mainly used after failure of other treatments, and therefore treated patients usually have a limited life expectancy, which offsets the duration of possible side effects.

Langbein and colleagues [155] have reported on xerostomia after PSMA radioligand therapy. ^{177}Lu-PSMA revealed only mild to moderate reversible Xerostomia, whereas for ^{225}Ac-PSMA, severe xerostomia occurred frequently and became the dose-limiting toxicity. They state that PSMA-targeting antibodies such as J591 instead of small molecules such as PSMA-617 might be able to lower the sialotoxicity (salivary gland toxicity) of ^{225}Ac-PSMA because there is no significant salivary gland uptake shown on ^{89}Zr-J591 or ^{177}Lu-J591 imaging. This effect comes, unfortunately, at the cost of increased myelotoxicity due to longer blood circulation for whole antibodies.

In a study by Soeda and colleagues [156], it was found that one can obtain a better therapeutic ratio while minimizing adverse effects on salivary glands by setting an appropriate peptide concentration in PSMA-targeted therapy. They showed that a decrease in the concentration level of the injected ^{18}F-PSMA-1007 solution resulted in decreased uptake in tumour and normal organs, especially in salivary gland uptake.

19.11 METHODS TO ENHANCE THERAPEUTIC RATIO (INCREASE TUMOUR TOXICITY AND REDUCE NORMAL ORGAN TOXICITY)

It has been shown that androgen receptor (AR) inhibition can enhance the PSMA uptake (^{68}Ga-PSMA-11) in preclinical and clinical studies. The increase of PSMA expression may consequently increase the number of lesions visualized [123]; PSMA uptake increased 1.5-fold to 2.0-fold in a xenograft mouse model. Patient imaging demonstrated a 7-fold increase in PSMA uptake post AR inhibition.

High uptake in kidneys has been reported for small molecules and peptides. Due to having a small size below the kidney cut-off (<60 kDa), these radiolabelled small molecules are readily filtered by the glomerulus into the nephrotic tubules. The tubular cells reabsorb these filtered molecules, where they are subjected to lysosomal degradation and processing. As we have discussed earlier, non-residualizing radio-catabolites are retained inside the cell and cannot leak back to the primary urine. Therefore, the absorbed doses must be limited to avoid kidney damage. The human radical scavenger and antioxidant, a1-microglobulin (A1M), has been shown to have pharmacokinetics similar to peptides. In a preclinical study of its nephroprotective effects ^{177}Lu-DOTATATE resulted in increased formation of DNA double-strand breaks in the renal cortex, upregulated expression of apoptosis and stress response-related genes, and proteinuria (albumin in urine) – all of which were significantly suppressed by coadministration of A1M (7 mg/kg). At six, 12, and 24 weeks post-^{177}Lu-DOTATATE injections, there was an increase in animal deaths, kidney lesions, glomerular loss, upregulation of stress genes, proteinuria, and plasma markers of reduced kidney function. All of the aforementioned adverse effects were suppressed by coadministration of A1M, thus clearly demonstrating that A1M effectively inhibits radiation-induced renal damage [157]. It has also been shown that A1M will not affect therapeutic efficacy [158].

One problem in radioimmunotherapy with antibodies is the binding-site-barrier, which limits the penetration into larger tumours. Binding-site barrier results from targeting agents of high affinity to a certain target being trapped by the target in the outer rim of the tumour, thus reducing intratumoural penetration. One approach to overcome this binding-site barrier is via the mass effect, where preload of excess of cold antibody is injected to saturate the target on the tumour outer surface, thereby facilitating penetration of the radiolabelled counterpart into the tumour core [159]. Another approach is postloading, as elaborated by Palm and colleagues [160]. The authors suggest that a timed manipulation of the α-radioimmunotherapy–specific activity in vivo may present a way to maximize the effectiveness against treatment-resistant microtumours with minimal additional toxicity.

The activity of ^{177}Lu/^{225}Ac-PSMA administered to patients is limited by toxicity to normal organs; high uptake is observed in the lacrimal glands, parotid glands, submandibular glands, and renal cortex with hematotoxicity, xerostomia, and renal dysfunction. In a preclinical study it was shown that by using monosodium glutamate (MSG) it was possible to reduce the radioactivity uptake by 50 per cent in the kidneys and salivary glands [161].

A method ECAT (extracorporeal affinity adsorption treatment) can be used to deplete the circulation from labelled antibody, thus reducing the absorbed dose to blood-rich organs such as the bone marrow [162]. A successful clinical implementation of this method has been published by Lindén and colleagues [163]. The reduction of blood-borne radioactivity was 96 per cent and that of the whole body was 49 per cent. The tumour uptake of radioactivity was mostly retained using this method.

Successful application of small peptides in a theranostic approach is often hampered by their fast in vivo degradation by proteolytic enzymes, such as neutral endopeptidase (NEP). Co-injection of NEP inhibitors may help optimize in vivo targeting properties of such peptides. K. L. Chatalic and colleagues have evaluated the effect of co-injecting the NEP inhibitor Phosphoramidon (PA) and the GRPR radiolabelled antagonist [164]. Co-injection of PA (300 μg) led to stabilization of ^{177}Lu-JMV4168 in murine peripheral blood. In PC-3 tumour-bearing mice, a PA co-injection led to a two-fold increase in tumour uptake of ^{68}Ga-/^{177}Lu-JMV4168, 1 h after injection. This data clearly demonstrated that stabilization of small molecule targeting agents in vivo may enhance the theranostic potential of such radioconjugates.

19.12 PCA IMAGING GUIDING DOSE PLANNING IN EXTERNAL BEAM RADIATION (ERBT) THERAPY

Radiotherapy and radical prostatectomy are the main treatment options for patients with localized PCa. A rising level of PSA after radical prostatectomy indicates PCa recurrence, and these patients may still be cured with EBRT. The target volume should completely encompass the extent of PCa and, in this way, accurate estimation of the location of the tumour is important for radiotherapy planning. A comprehensive review can be found in Calais and colleagues [165]. In this paper they summarized studies assessing the impact of ^{18}F-choline or ^{68}Ga-PSMA on dose planning and found a 13–44 per cent change in dose planning due to results from PCa imaging. In another study, the integration of ^{68}Ga-PSMA-PET-imaging into the EBRT treatment planning process is useful for detailed target volume planning [166]. The performance of a ^{68}Ga-PSMA-PET frequently leads to changes in the staging, altering the EBRT treatment regimen and the subsequent target volume (an example is shown in Figure 19.28).

The impact of PSMA imaging on EBRT treatment planning in PCa patients with biochemical recurrence has been reviewed by Ekmekciolu and colleagues [167]. Biochemical recurrence (BCR) is generally considered the first sign of treatment failure. Conventional imaging techniques are not accurate enough to determine the location of recurrence. The group concluded that PSMA imaging demonstrated a high clinical impact in patients with BCR, with modifications to the original treatment plan occurring among half the patients. Detecting recurrence in BCR can prevent unnecessary toxicity and lead to individualized therapy. A multicentre study of 270 PCa patients with biochemical recurrence (with low PSA values below 1.0 ng/mL) after radical prostatectomy, showed an almost 20 per cent impact on the delineation of the target volume for EBRT [168].

^{18}F-FACBC (fluciclovine) PET-CT imaging in PCa patients has been evaluated for ERBT treatment field delineation. Overall radiotherapy decision was changed in about 41 per cent of the patients. Specifically, about 5 per cent had the decision for radiotherapy withdrawn due to positive extra-pelvic findings. Radiotherapy field decisions were changed in 36 per cent of the cases. Also, 73 per cent of the patients had fields changed from prostate bed only to both prostate bed and pelvis, while 27 per cent had fields changed from both prostate bed and pelvis to prostate bed only [169].

In a larger study involving 121 patients, Koerber and colleagues [65] found a change in staging and radiotherapeutic management in 40.5 per cent and 51.2 per cent of the patients, respectively. In treatment-naive patients, a change in staging and radiotherapeutic plan occurred in 26.0 per cent and 44.0 per cent respectively. For patients with PSA persistence or recurrence, staging and radiotherapeutic management were changed in 50.7 per cent and 56.3 per cent of patients, respectively. A study was done to assess the ability of PET/CT imaging, using ^{18}F-fluoromethylcholine (^{18}F-FCH) and ^{68}Ga-HBED-CC (PSMA-11) radiotracers, and pelvic multiparametric MRI to identify men who will best benefit from EBRT. The actual treatment differed from that planned before imaging in 47 per cent of men enrolled. Schmidt-Hegemann and colleagues reported that compared with CT, the ^{68}Ga-PSMA PET/CT resulted in a change in treatment in 62 per cent of 172 PCa patients who were referred for primary definitive radiotherapy (22 patients), for prostate-specific antigen (PSA) persistence (88 patients), and PSA recurrence after radical prostatectomy (62 patients) [170].

Focal treatment of the prostate with high-intensity focused ultrasound (HIFU) is a promising new modality for the treatment of localized PCa and has shown limited side effects, such as urinary incontinence and erectile dysfunction. MRI is often difficult because of signal alterations of the treated prostate. In a study with ^{68}Ga-PSMA-11 this method showed promise to monitor treatment success [171].

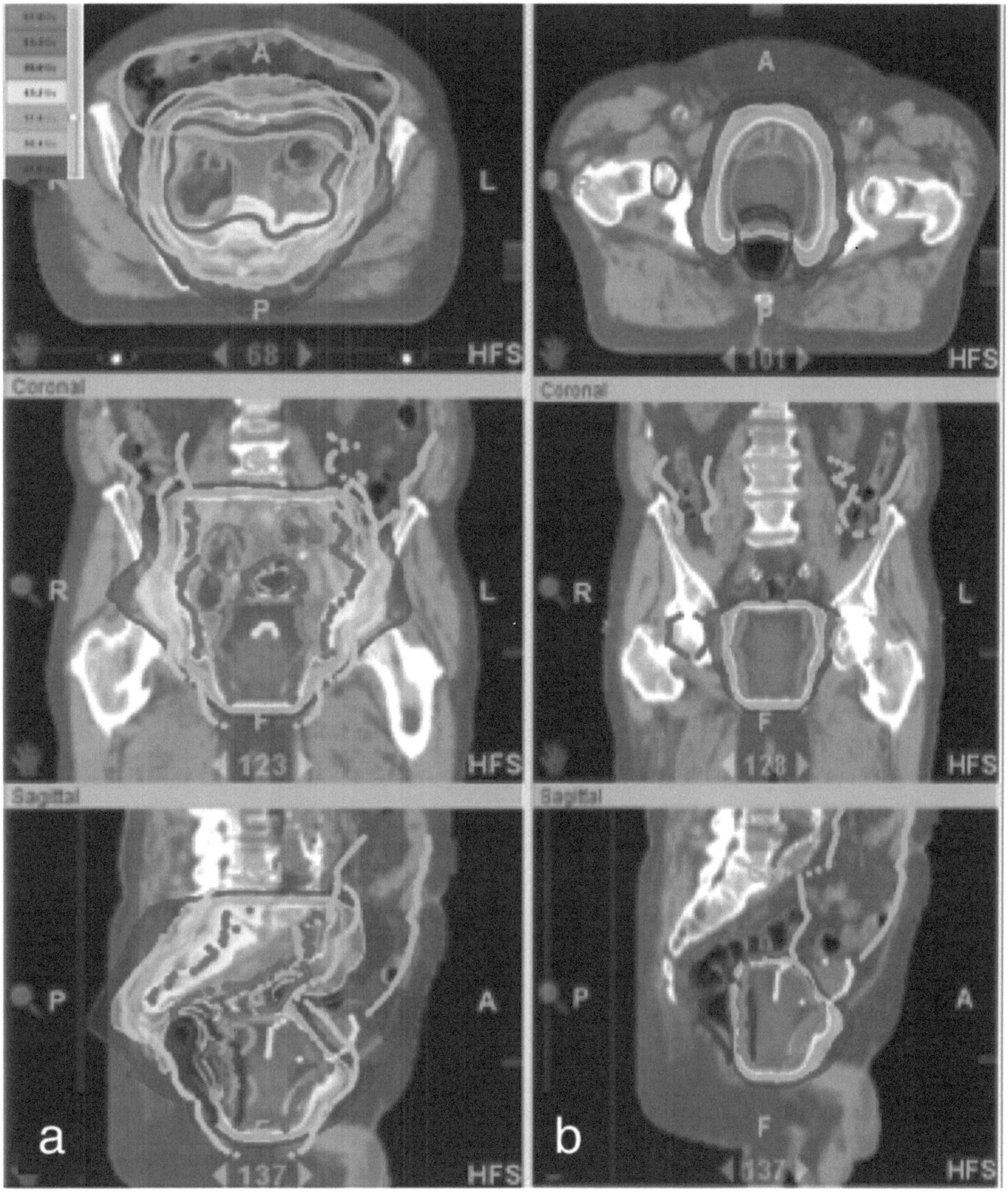

FIGURE 19.28 The impact on a treatment plan due to PCa PET imaging. Treatment plan (a), including lymph node irradiation up to 45 Gy à 1.8 Gy. Based on the information obtained by ^{68}Ga-PSMA-PET imaging, PET-positive nodes receive a simultaneous integrated boost to 54 Gy à 2.17 Gy. In comparison with an example of the absorbed dose distribution (b) without the knowledge through the ^{68}Ga-PSMA-PET; the lymph node involvement would not have been suspected and, therefore, only the prostate itself would have been treated. Colour image available at www.routledge.com/9781138593312.

(Source: Reproduced from Dewes *et al.* [166] with permission.)

19.13 FUTURE ASPECTS

19.13.1 Combining High and Low LET Emitters

An interesting discussion is whether to favour α- or β- emitters for therapy, which has been elaborated in Haberkorn and colleagues [172], who suggest that both emitters may be used for therapy. The idea is that an α-emitter is applied for patients in whom ^{177}Lu therapy has failed or who have disseminated disease of the bone marrow. Another strategy could involve one carrier molecule labelled with an α-emitter and a β-emitter. Another approach would be the administration of two different radiopharmaceuticals for two different targets, that is, co-targeting. Then the β-emitter may be used for debulking of large tumour masses, and the α-emitter may target critical subpopulations in a tumour such as cells with stem cell properties. Shown in Haberkorn and colleagues [172] is an example of possible stratification of therapy according to bone marrow involvement. In the figure is shown PET images with PSMA in 2 patients: One patient lacking diffuse bone marrow infiltration subsequently planned for treatment with ^{177}Lu-PSMA, whereas another patient with infiltration of bone marrow was instead to be treated with ^{225}Ac-PSMA.

19.13.2 Guided Biopsies and/or Surgery

The goals of surgery are to completely resect the primary tumour and metastatic lymph nodes. PCa is challenging for surgeons to identify the malignant lesions intraoperatively. Thus, surgeons have to risk injury to nerves, sphincter, or bladder to increase the chances of achieving a complete resection. Conversely, small islands of cancer may be missed, leading to recurrence in these patients Less invasive surgical approaches would be more feasible if PCa could be localized more accurately. An attractive strategy to achieve localization of PCa comprises dual-modality imaging probes that are labelled with a fluorescent dye and a positron emission radioisotope. The radioisotope is used for preoperative PET imaging and surgical planning and, during surgery, the fluorescent dye is used for intraoperative guidance using real-time fluorescence imaging [106, 173]. Guided biopsies using ^{89}Zr-labeled agents will be feasible due to its longer physical half-life of 78.4 h. After a ^{89}Zr diagnostic PET scan is interpreted, lesions retain activity for several days, enabling PET-guided biopsy later, and this has been shown to be safe and effective [174].

PCa is spread through the lymphatic system, and the first draining (sentinel) lymph nodes represent the first sites that may contain metastatic PCa cells. Localization of those nodes at surgery helps the surgeon in finding and resectioning suspicious nodes. Colloids have been used as ^{99m}Tc-nanocolloid. After local injection within the prostate, ^{99m}Tc-nanocolloid follows the lymphatic drainage, and an intraoperative gamma-probe helps to identify sentinel lymph nodes. The method was extended with intraoperative fluorescence imaging after the development of so-called hybrid tracers that contain not only a radiolabel but also a fluorescent label (e.g., indocyanine green ^{99m}Tc-nanocolloid). With the development of, for example, ^{99m}Tc-PSMA the radioguided surgery can rely on more precise guidance because of the active uptake in PCa cells instead of passive entrapment in lymph nodes with radiolabeled colloids. This technique has been reported by Maurer and colleagues [175]. In another publication by Maurer is an example of radioguided surgery after injection of ^{99m}Tc-labeled PSMA in patients with recurrent PCa. The intraoperative measurements with a γ-probe detected ^{99m}Tc-PSMA accumulation post-surgery, later confirmed by the ex vivo measurement [176]. A nice overview is given in Maurer and colleagues [177] and, in Figure 19.29, the steps in imaging and radioguided surgery are depicted.

19.14 CONCLUSION

The recent development of targeting radiopharmaceuticals for PCa has shown promise for the development of theranostic methods, enhancing diagnostic accuracy and making treatment of spread PCa feasible. Not only for more patient-specific handling of treatment based on imaging, but also for improved guiding of surgeons and radiation therapists, the imaging methods already have had a great impact on choice of therapy and dose planning in EBRT. Although the radionuclide therapies have not yet shown high numbers of long-lasting complete remissions, the data already obtained are very encouraging. To further enhance the development of new theranostic methods for PCa, development of new targeting agents, precise small-scale dosimetry, and careful study of radiobiological effects are most needed.

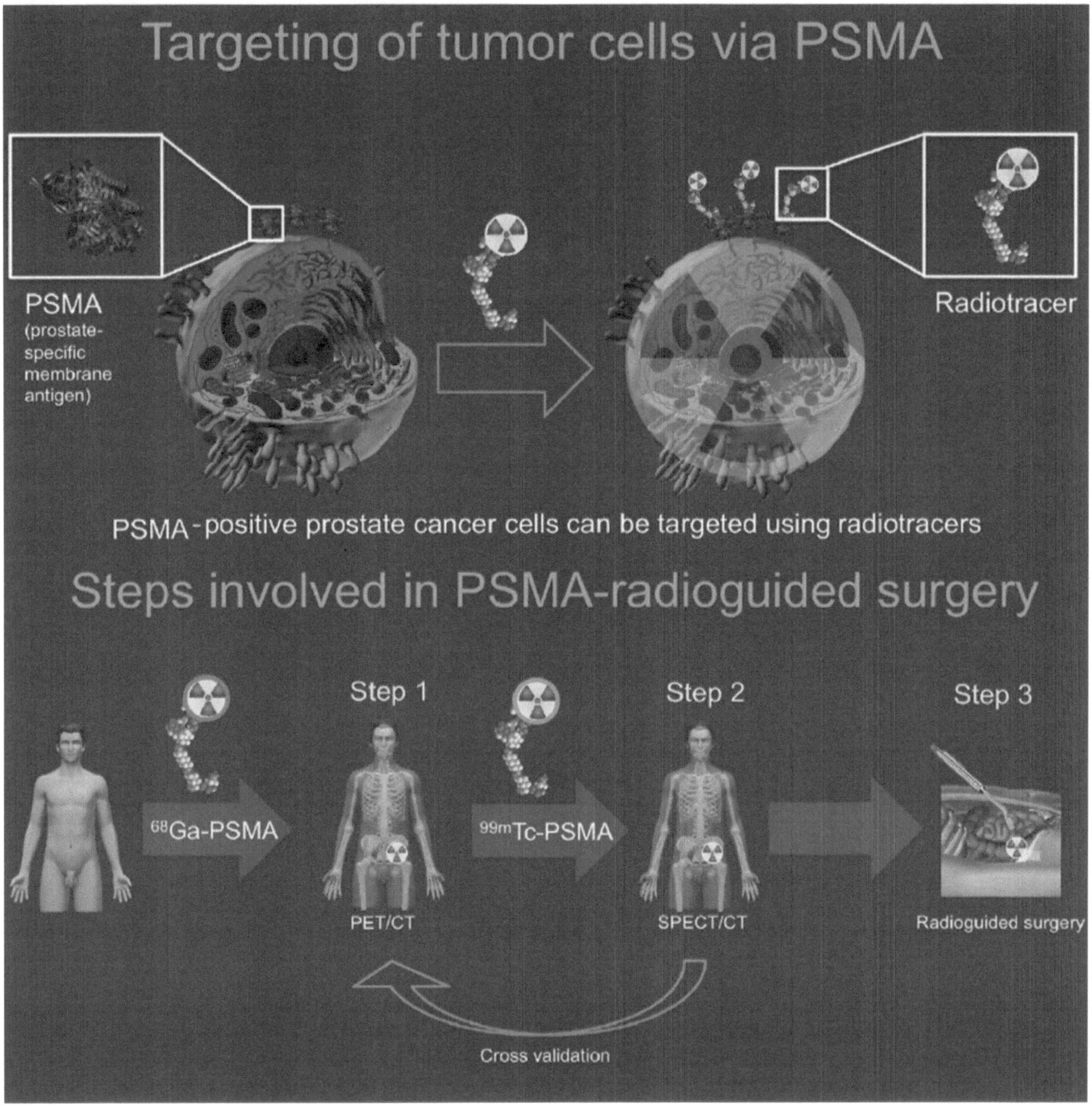

FIGURE 19.29 Overview of different steps for radioguided surgery: (1) selection of patients based on PET imaging and clinical history; (2) injection of ^{99m}Tc-PSMA-I&S and subsequent SPECT imaging to confirm tumour uptake of preoperative ^{68}Ga-PSMA-11 findings; (3) PSMA-radioguided surgery with gamma probe measurements to outline metastatic PCa tumours. Colour image available at www.routledge.com/9781138593312.

(Source: Reproduced with permission from Maurer *et al.*)

REFERENCES

[1] T. Inaba, "Quantitative measurements of prostatic blood flow and blood volume by positron emission tomography," *J Urol,* vol. 148, no. 5, pp. 1457–60, 1992, doi: 10.1016/s0022-5347(17)36939-2.

[2] M. Grkovski *et al.*, "(11)C-Choline Pharmacokinetics in Recurrent Prostate Cancer," *J Nucl Med,* vol. 59, no. 11, pp. 1672–78, 2018, doi: 10.2967/jnumed.118.210088.

[3] M. R. Jochumsen *et al.*, "Quantitative Tumor Perfusion Imaging with (82)Rb PET/CT in Prostate Cancer: Analytic and Clinical Validation," *J Nucl Med,* vol. 60, no. 8, pp. 1059–1065, 2019, doi: 10.2967/jnumed.118.219188.

[4] R. Eeles and H. Ni Raghallaigh, "Men with a susceptibility to prostate cancer and the role of genetic based screening," *Transl Androl Urol,* vol. 7, no. 1, pp. 61–69, 2018, doi: 10.21037/tau.2017.12.30.
[5] *Cancer i siffror 2018 (in Swedish),* vol. 2018-6-10, 2018.
[6] V. Kasivisvanathan *et al.*, "MRI-targeted or Standard Biopsy for Prostate-Cancer Diagnosis," *N Engl J Med,* vol. 378, no. 19, pp. 1767–77, 2018, doi: 10.1056/NEJMoa1801993.
[7] D. E. Spratt *et al.*, "Androgen Receptor Upregulation Mediates Radioresistance after Ionizing Radiation," *Cancer Res,* vol. 75, no. 22, pp. 4688–96, 2015, doi: 10.1158/0008-5472.CAN-15-0892.
[8] N. M. Byrne, H. Nesbitt, L. Ming, S. R. McKeown, J. Worthington, and D. J. McKenna, "Androgen deprivation in LNCaP prostate tumour xenografts induces vascular changes and hypoxic stress, resulting in promotion of epithelial-to-mesenchymal transition," *Br J Cancer,* vol. 114, no. 6, pp. 659–68, 2016, doi: 10.1038/bjc.2016.29.
[9] S. Ghafoor, I. A. Burger, and A. H. Vargas, "Multimodality Imaging of Prostate Cancer," *J Nucl Med,* vol. 60, no. 10, pp. 1350–58, 2019, doi: 10.2967/jnumed.119.228320.
[10] E. Berg *et al.*, "Total-Body PET and Highly Stable Chelators Together Enable Meaningful (89)Zr-Antibody PET Studies Up to 30 Days after Injection," *J Nucl Med,* vol. 61, no. 3, pp. 453–460, 2020, doi: 10.2967/jnumed.119.230961.
[11] A. Sanchez-Crespo, "Comparison of Gallium-68 and Fluorine-18 imaging characteristics in positron emission tomography," *Appl Radiat Isot,* vol. 76, pp. 55–62, 2013, doi: 10.1016/j.apradiso.2012.06.034.
[12] M. Bunka *et al.*, "Imaging quality of (44)Sc in comparison with five other PET radionuclides using Derenzo phantoms and preclinical PET," *Appl Radiat Isot,* vol. 110, pp. 129–133, 2016, doi: 10.1016/j.apradiso.2016.01.006.
[13] S. Righi *et al.*, "Biokinetic and dosimetric aspects of (64)CuCl2 in human prostate cancer: Possible theranostic implications," *EJNMMI Res,* vol. 8, no. 1, p. 18, 2018, doi: 10.1186/s13550-018-0373-9.
[14] J. R. A. Nedrow, C.J., "Emergeing Radiometals for PET imaging," *Encyclopedia of Inorganic and Bioinorganic Chemistry,* pp. 1–11, 2016.
[15] D. Minarik *et al.*, "90Y Bremsstrahlung imaging for absorbed-dose assessment in high-dose radioimmunotherapy," *J Nucl Med,* vol. 51, no. 12, pp. 1974–78, 2010, doi: 10.2967/jnumed.110.079897.
[16] D. Minarik, M. Ljungberg, P. Segars, and K. S. Gleisner, "Evaluation of quantitative planar 90Y bremsstrahlung whole-body imaging," *Phys Med Biol,* vol. 54, no. 19, pp. 5873–83, 2009, doi: 10.1088/0031-9155/54/19/014.
[17] D. Brady, J. M. O'Sullivan, and K. M. Prise, "What is the Role of the Bystander Response in Radionuclide Therapies?" *Front Oncol,* vol. 3, p. 215, 2013, doi: 10.3389/fonc.2013.00215.
[18] V. Tolmachev, P. Bernhardt, E. Forssell-Aronsson, and H. Lundqvist, "114mIn, a candidate for radionuclide therapy: Low-energy cyclotron production and labeling of DTPA-D-phe-octreotide," *Nucl Med Biol,* vol. 27, no. 2, pp. 183–88, 2000.
[19] C. Champion, M. A. Quinto, C. Morgat, P. Zanotti-Fregonara, and E. Hindie, "Comparison between Three Promising ss-emitting Radionuclides, (67)Cu, (47)Sc and (161)Tb, with Emphasis on Doses Delivered to Minimal Residual Disease," *Theranostics,* vol. 6, no. 10, pp. 1611–18, 2016, doi: 10.7150/thno.15132.
[20] S. Haller *et al.*, "Contribution of Auger/conversion electrons to renal side effects after radionuclide therapy: Preclinical comparison of (161)Tb-folate and (177)Lu-folate," *EJNMMI Res,* vol. 6, no. 1, p. 13, 2016, doi: 10.1186/s13550-016-0171-1.
[21] C. Muller *et al.*, "A unique matched quadruplet of terbium radioisotopes for PET and SPECT and for alpha- and beta- radionuclide therapy: An in vivo proof-of-concept study with a new receptor-targeted folate derivative," *J Nucl Med,* vol. 53, no. 12, pp. 1951–59, 2012, doi: 10.2967/jnumed.112.107540.
[22] S. Poty, L. C. Francesconi, M. R. McDevitt, M. J. Morris, and J. S. Lewis, "alpha-Emitters for Radiotherapy: From Basic Radiochemistry to Clinical Studies-Part 1," *J Nucl Med,* vol. 59, no. 6, pp. 878–84, 2018, doi: 10.2967/jnumed.116.186338.
[23] S. Poty, L. C. Francesconi, M. R. McDevitt, M. J. Morris, and J. S. Lewis, "Alpha-Emitters for Radiotherapy: From Basic Radiochemistry to Clinical Studies, Part 2," *J Nucl Med,* vol. 59, no. 7, pp. 1020–27, 2018, doi: 10.2967/jnumed.117.204651.
[24] J. L. Humm, "Dosimetric aspects of radiolabeled antibodies for tumor therapy," *J Nucl Med,* vol. 27, no. 9, pp. 1490–97, 1986.
[25] G. Vaidyanathan and M. R. Zalutsky, "Targeted therapy using alpha emitters," *Phys Med Biol,* vol. 41, no. 10, pp. 1915–31, 1996, doi: 10.1088/0031-9155/41/10/005.
[26] G. Vaidyanathan, D. J. Affleck, J. Li, P. Welsh, and M. R. Zalutsky, "A polar substituent-containing acylation agent for the radioiodination of internalizing monoclonal antibodies: N-succinimidyl 4-guanidinomethyl-3-[131I]iodobenzoate ([131I] SGMIB)," *Bioconjug Chem,* vol. 12, no. 3, pp. 428–38, 2001, doi: 10.1021/bc0001490.
[27] J. Schwartz *et al.*, "Renal uptake of bismuth-213 and its contribution to kidney radiation dose following administration of actinium-225-labeled antibody," *Phys Med Biol,* vol. 56, no. 3, pp. 721–33, 2011, doi: 10.1088/0031-9155/56/3/012.
[28] J. L. Humm, O. Sartor, C. Parker, O. S. Bruland, and R. Macklis, "Radium-223 in the treatment of osteoblastic metastases: A critical clinical review," *Int J Radiat Oncol Biol Phys,* vol. 91, no. 5, pp. 898–906, 2015, doi: 10.1016/j.ijrobp.2014.12.061.
[29] Y. Ooe, T. Tsukada, Y. Yamasaki, M. Kaji, and K. Shimizu, "[A Case of Synchronous and Solitary Gallbladder Metastasis from Gastric Cancer]," *Gan To Kagaku Ryoho,* vol. 46, no. 11, pp. 1765–69, 2019.
[30] M. Brandt, J. Cardinale, M. L. Aulsebrook, G. Gasser, and T. L. Mindt, "An Overview of PET Radiochemistry, Part 2: Radiometals," *J Nucl Med,* vol. 59, no. 10, pp. 1500–06, 2018, doi: 10.2967/jnumed.117.190801.
[31] D. S. Wilbur, "Radiohalogenation of proteins: An overview of radionuclides, labeling methods, and reagents for conjugate labeling," *Bioconjug Chem,* vol. 3, no. 6, pp. 433–70, 1992, doi: 10.1021/bc00018a001.

[32] W. M. Hunter and F. C. Greenwood, "Preparation of iodine-131 labelled human growth hormone of high specific activity," *Nature,* vol. 194, pp. 495–96, 1962, doi: 10.1038/194495a0.

[33] P. J. Fraker and J. C. Speck, Jr., "Protein and cell membrane iodinations with a sparingly soluble chloroamide, 1,3,4,6-tetrachloro-3a,6a-diphrenylglycoluril," *Biochem Biophys Res Commun,* vol. 80, no. 4, pp. 849–57, 1978, doi: 10.1016/0006-291x(78)91322-0.

[34] A. Lovqvist, A. Sundin, H. Ahlstrom, J. Carlsson, and H. Lundqvist, "76Br-labeled monoclonal anti-CEA antibodies for radioimmuno positron emission tomography," *Nucl Med Biol,* vol. 22, no. 1, pp. 125–31, 1995, doi: 10.1016/0969-8051(94)e0065-q.

[35] M. R. Zalutsky and A. S. Narula, "Radiohalogenation of a monoclonal antibody using an N-succinimidyl 3-(tri-n-butylstannyl)benzoate intermediate," *Cancer Res,* vol. 48, no. 6, pp. 1446–50, 1988.

[36] P. K. Garg, S. Garg, and M. R. Zalutsky, "Fluorine-18 labeling of monoclonal antibodies and fragments with preservation of immunoreactivity," *Bioconjug Chem,* vol. 2, no. 1, pp. 44–49, 1991, doi: 10.1021/bc00007a008.

[37] C. J. Reist, P. K. Garg, K. L. Alston, D. D. Bigner, and M. R. Zalutsky, "Radioiodination of internalizing monoclonal antibodies using N-succinimidyl 5-iodo-3-pyridinecarboxylate," *Cancer Res,* vol. 56, no. 21, pp. 4970–77, 1996.

[38] R. J. Kyriakos, L. B. Shih, G. L. Ong, K. Patel, D. M. Goldenberg, and M. J. Mattes, "The fate of antibodies bound to the surface of tumor cells in vitro," *Cancer Res,* vol. 52, no. 4, pp. 835–42, 1992.

[39] M. J. Mattes, G. L. Griffiths, H. Diril, D. M. Goldenberg, G. L. Ong, and L. B. Shih, "Processing of antibody-radioisotope conjugates after binding to the surface of tumor cells," *Cancer,* vol. 73, no. 3 Suppl, pp. 787–93, 1994, doi: 10.1002/1097-0142(19940201)73:3+<787::aid-cncr2820731307>3.0.co;2-5.

[40] L. B. Shih *et al.*, "The processing and fate of antibodies and their radiolabels bound to the surface of tumor cells in vitro: A comparison of nine radiolabels," *J Nucl Med,* vol. 35, no. 5, pp. 899–908, 1994.

[41] D. Vivier, S. K. Sharma, and B. M. Zeglis, "Understanding the in vivo fate of radioimmunoconjugates for nuclear imaging," *J Labelled Comp Radiopharm,* vol. 61, no. 9, pp. 672–92, 2018, doi: 10.1002/jlcr.3628.

[42] O. W. Press *et al.*, "Comparative metabolism and retention of iodine-125, yttrium-90, and indium-111 radioimmunoconjugates by cancer cells," *Cancer Res,* vol. 56, no. 9, pp. 2123–29, 1996.

[43] V. Tolmachev, A. Orlova, and H. Lundqvist, "Approaches to improve cellular retention of radiohalogen labels delivered by internalising tumour-targeting proteins and peptides," *Curr Med Chem,* vol. 10, no. 22, pp. 2447–60, 2003, doi: 10.2174/0929867033456666.

[44] S. Shankar, G. Vaidyanathan, D. Affleck, P. C. Welsh, and M. R. Zalutsky, "N-succinimidyl 3-[(131)I]iodo-4-phosphonomethylbenzoate ([(131)I]SIPMB), a negatively charged substituent-bearing acylation agent for the radioiodination of peptides and mAbs," *Bioconjug Chem,* vol. 14, no. 2, pp. 331–41, 2003, doi: 10.1021/bc025636p.

[45] J. Hoglund, V. Tolmachev, A. Orlova, H. Lundqvist, and A. Sundin, "Optimized indirect (76)Br-bromination of antibodies using N-succinimidyl para-[76Br]bromobenzoate for radioimmuno PET," *Nucl Med Biol,* vol. 27, no. 8, pp. 837–43, 2000, doi: 10.1016/s0969-8051(00)00153-0.

[46] E. Mume, A. Orlova, P. U. Malmstrom, H. Lundqvist, S. Sjoberg, and V. Tolmachev, "Radiobromination of humanized anti-HER2 monoclonal antibody trastuzumab using N-succinimidyl 5-bromo-3-pyridinecarboxylate, a potential label for immunoPET," *Nucl Med Biol,* vol. 32, no. 6, pp. 613–22, 2005, doi: 10.1016/j.nucmedbio.2005.04.010.

[47] M. R. Zalutsky, P. K. Garg, H. S. Friedman, and D. D. Bigner, "Labeling monoclonal antibodies and F(ab')2 fragments with the alpha-particle-emitting nuclide astatine-211: Preservation of immunoreactivity and in vivo localizing capacity," *Proc Natl Acad Sci U S A,* vol. 86, no. 18, pp. 7149–53, 1989, doi: 10.1073/pnas.86.18.7149.

[48] D. S. Wilbur, R. L. Vessella, J. E. Stray, D. K. Goffe, K. A. Blouke, and R. W. Atcher, "Preparation and evaluation of para-[211At]astatobenzoyl labeled anti-renal cell carcinoma antibody A6H F(ab')2. In vivo distribution comparison with para-[125I]iodobenzoyl labeled A6H F(ab')2," *Nucl Med Biol,* vol. 20, no. 8, pp. 917–27, 1993, doi: 10.1016/0969-8051(93)90092-9.

[49] D. S. Wilbur *et al.*, "Reagents for astatination of biomolecules. 4. Comparison of maleimido-closo-decaborate(2-) and meta-[(211)At]astatobenzoate conjugates for labeling anti-CD45 antibodies with [(211)At]astatine," *Bioconjug Chem,* vol. 20, no. 10, pp. 1983–91, 2009, doi: 10.1021/bc9000799.

[50] M. S. Cooper, E. Sabbah, and S. J. Mather, "Conjugation of chelating agents to proteins and radiolabeling with trivalent metallic isotopes," *Nat Protoc,* vol. 1, no. 1, pp. 314–17, 2006, doi: 10.1038/nprot.2006.49.

[51] P. Adumeau, S. K. Sharma, C. Brent, and B. M. Zeglis, "Site-specifically Labeled Immunoconjugates for Molecular Imaging – Part 2: Peptide Tags and Unnatural Amino Acids," *Mol Imaging Biol,* vol. 18, no. 2, pp. 153–65, 2016, doi: 10.1007/s11307-015-0920-y.

[52] P. Adumeau, S. K. Sharma, C. Brent, and B. M. Zeglis, "Site-specifically Labeled Immunoconjugates for Molecular Imaging – Part 1: Cysteine Residues and Glycans," *Mol Imaging Biol,* vol. 18, no. 1, pp. 1–17, 2016, doi: 10.1007/s11307-015-0919-4.

[53] P. Antunes, M. Ginj, M. A. Walter, J. Chen, J. C. Reubi, and H. R. Maecke, "Influence of different spacers on the biological profile of a DOTA-somatostatin analogue," *Bioconjug Chem,* vol. 18, no. 1, pp. 84–92, 2007, doi: 10.1021/bc0601673.

[54] E. W. Price and C. Orvig, "Matching chelators to radiometals for radiopharmaceuticals," *Chem Soc Rev,* vol. 43, no. 1, pp. 260–90, 2014, doi: 10.1039/c3cs60304k.

[55] T. W. Price, J. Greenman, and G. J. Stasiuk, "Current advances in ligand design for inorganic positron emission tomography tracers (68)Ga, (64)Cu, (89)Zr and (44)Sc," *Dalton Trans,* vol. 45, no. 40, pp. 15702–24, 2016, doi: 10.1039/c5dt04706d.

[56] T. A. Kaden, "Labelling monoclonal antibodies with macrocyclic radiometal complexes. A challenge for coordination chemists," *Dalton Trans,* no. 30, pp. 3617–23, 2006, doi: 10.1039/b606410h.

[57] Z. Varasteh *et al.*, "The effect of macrocyclic chelators on the targeting properties of the 68Ga-labeled gastrin releasing peptide receptor antagonist PEG2-RM26," *Nucl Med Biol,* vol. 42, no. 5, pp. 446–54, 2015, doi: 10.1016/j.nucmedbio.2014.12.009.

[58] K. Westerlund *et al.*, "Increasing the Net Negative Charge by Replacement of DOTA Chelator with DOTAGA Improves the Biodistribution of Radiolabeled Second-Generation Synthetic Affibody Molecules," *Mol Pharm,* vol. 13, no. 5, pp. 1668–78, 2016, doi: 10.1021/acs.molpharmaceut.6b00089.

[59] M. Lin, M. J. Welch, and S. E. Lapi, "Effects of chelator modifications on (68)Ga-labeled [Tyr (3)]octreotide conjugates," *Mol Imaging Biol,* vol. 15, no. 5, pp. 606–13, 2013, doi: 10.1007/s11307-013-0627-x.

[60] M. Asti *et al.*, "Influence of different chelators on the radiochemical properties of a 68-Gallium labelled bombesin analogue," *Nucl Med Biol,* vol. 41, no. 1, pp. 24–35, 2014, doi: 10.1016/j.nucmedbio.2013.08.010.

[61] M. D. Bartholoma, A. S. Louie, J. F. Valliant, and J. Zubieta, "Technetium and gallium derived radiopharmaceuticals: Comparing and contrasting the chemistry of two important radiometals for the molecular imaging era," *Chem Rev,* vol. 110, no. 5, pp. 2903–20, 2010, doi: 10.1021/cr1000755.

[62] I. Murray *et al.*, "The potential of (223)Ra and (18)F-fluoride imaging to predict bone lesion response to treatment with (223)Ra-dichloride in castration-resistant prostate cancer," *Eur J Nucl Med Mol Imaging,* vol. 44, no. 11, pp. 1832–44, 2017, doi: 10.1007/s00259-017-3744-y.

[63] C. Wassberg, M. Lubberink, J. Sorensen, and S. Johansson, "Repeatability of quantitative parameters of 18F-fluoride PET/CT and biochemical tumour and specific bone remodelling markers in prostate cancer bone metastases," *EJNMMI Res,* vol. 7, no. 1, p. 42, 2017, doi: 10.1186/s13550-017-0289-9.

[64] R. Lange, R. Ter Heine, R. F. Knapp, J. M. de Klerk, H. J. Bloemendal, and N. H. Hendrikse, "Pharmaceutical and clinical development of phosphonate-based radiopharmaceuticals for the targeted treatment of bone metastases," *Bone,* vol. 91, pp. 159–79, 2016, doi: 10.1016/j.bone.2016.08.002.

[65] S. A. Koerber *et al.*, "(68)Ga-PSMA-11 PET/CT in Primary and Recurrent Prostate Carcinoma: Implications for Radiotherapeutic Management in 121 Patients," *J Nucl Med,* vol. 60, no. 2, pp. 234–240, 2018, doi: 10.2967/jnumed.118.211086.

[66] D. Ulmert *et al.*, "A novel automated platform for quantifying the extent of skeletal tumour involvement in prostate cancer patients using the Bone Scan Index," *Eur Urol,* vol. 62, no. 1, pp. 78–84, 2012, doi: 10.1016/j.eururo.2012.01.037.

[67] M. Bieth *et al.*, "Exploring New Multimodal Quantitative Imaging Indices for the Assessment of Osseous Tumor Burden in Prostate Cancer Using (68)Ga-PSMA PET/CT," *J Nucl Med,* vol. 58, no. 10, pp. 1632–37, 2017, doi: 10.2967/jnumed.116.189050.

[68] M. H. Poulsen *et al.*, "[18F]fluoromethylcholine (FCH) positron emission tomography/computed tomography (PET/CT) for lymph node staging of prostate cancer: A prospective study of 210 patients," *BJU Int,* vol. 110, no. 11, pp. 1666–71, 2012, doi: 10.1111/j.1464-410X.2012.11150.x.

[69] B. Grubmuller *et al.*, "(64)Cu-PSMA-617 PET/CT Imaging of Prostate Adenocarcinoma: First In-Human Studies," *Cancer Biother Radiopharm,* 2016, doi: 10.1089/cbr.2015.1964.

[70] H. Jadvar, "Is There Use for FDG-PET in Prostate Cancer?" *Semin Nucl Med,* vol. 46, no. 6, pp. 502–6, 2016, doi: 10.1053/j.semnuclmed.2016.07.004.

[71] E. E. Parent and D. M. Schuster, "Update on (18)F-Fluciclovine PET for Prostate Cancer Imaging," *J Nucl Med,* vol. 59, no. 5, pp. 733–39, 2018, doi: 10.2967/jnumed.117.204032.

[72] A. Piccardo *et al.*, "(64)CuCl2 PET/CT in Prostate Cancer Relapse," *J Nucl Med,* vol. 59, no. 3, pp. 444–51, 2018, doi: 10.2967/jnumed.117.195628.

[73] R. Cescato *et al.*, "Bombesin receptor antagonists may be preferable to agonists for tumor targeting," *J Nucl Med,* vol. 49, no. 2, pp. 318–26, 2008, doi: 10.2967/jnumed.107.045054.

[74] R. Mansi *et al.*, "Evaluation of a 1,4,7,10-tetraazacyclododecane-1,4,7,10-tetraacetic acid-conjugated bombesin-based radioantagonist for the labeling with single-photon emission computed tomography, positron emission tomography, and therapeutic radionuclides," *Clin Cancer Res,* vol. 15, no. 16, pp. 5240–49, 2009, doi: 10.1158/1078-0432.CCR-08-3145.

[75] D. H. Coy *et al.*, "Short-chain pseudopeptide bombesin receptor antagonists with enhanced binding affinities for pancreatic acinar and Swiss 3T3 cells display strong antimitotic activity," *J Biol Chem,* vol. 264, no. 25, pp. 14691–97, 1989.

[76] T. Maina *et al.*, "Preclinical and first clinical experience with the gastrin-releasing peptide receptor-antagonist [(6)(8)Ga]SB3 and PET/CT," *Eur J Nucl Med Mol Imaging,* vol. 43, no. 5, pp. 964–73, 2016, doi: 10.1007/s00259-015-3232-1.

[77] B. A. Nock *et al.*, "Theranostic Perspectives in Prostate Cancer with the Gastrin-Releasing Peptide Receptor Antagonist NeoBOMB1: Preclinical and First Clinical Results," *J Nucl Med,* vol. 58, no. 1, pp. 75–80, 2017, doi: 10.2967/jnumed.116.178889.

[78] Z. Varasteh *et al.*, "In vitro and in vivo evaluation of a (18)F-labeled high affinity NOTA conjugated bombesin antagonist as a PET ligand for GRPR-targeted tumor imaging," *PLoS One,* vol. 8, no. 12, p. e81932, 2013, doi: 10.1371/journal.pone.0081932.

[79] B. Mitran *et al.*, "Selection of an optimal macrocyclic chelator improves the imaging of prostate cancer using cobalt-labeled GRPR antagonist RM26," *Sci Rep,* vol. 9, no. 1, p. 17086, 2019, doi: 10.1038/s41598-019-52914-y.

[80] J. Zhang *et al.*, "PET Using a GRPR Antagonist (68)Ga-RM26 in Healthy Volunteers and Prostate Cancer Patients," *J Nucl Med,* vol. 59, no. 6, pp. 922–928, 2018, doi: 10.2967/jnumed.117.198929.

[81] C. Kratochwil *et al.*, "Targeted alpha-Therapy of Metastatic Castration-resistant Prostate Cancer with (225)Ac-PSMA-617: Dosimetry Estimate and Empiric Dose Finding," *J Nucl Med,* vol. 58, no. 10, pp. 1624–31, 2017, doi: 10.2967/jnumed.117.191395.

[82] T. Wustemann, U. Haberkorn, J. Babich, and W. Mier, "Targeting prostate cancer: Prostate-specific membrane antigen based diagnosis and therapy," *Med Res Rev,* vol. 39, no. 1, pp. 40–69, 2019, doi: 10.1002/med.21508.

[83] F. Arsenault, J. M. Beauregard, and F. Pouliot, "Prostate-specific membrane antigen for prostate cancer theranostics: From imaging to targeted therapy," *Curr Opin Support Palliat Care,* vol. 12, no. 3, pp. 359–65, 2018, doi: 10.1097/SPC.0000000000000357.

[84] A. Afshar-Oromieh, U. Haberkorn, M. Eder, M. Eisenhut, and C. M. Zechmann, "[68Ga]Gallium-labelled PSMA ligand as superior PET tracer for the diagnosis of prostate cancer: Comparison with 18F-FECH," *Eur J Nucl Med Mol Imaging,* vol. 39, no. 6, pp. 1085–86, 2012, doi: 10.1007/s00259-012-2069-0.

[85] J. Calais, W. P. Fendler, K. Herrmann, M. Eiber, and F. Ceci, "Comparison of (68)Ga-PSMA-11 and (18)F-Fluciclovine PET/CT in a Case Series of 10 Patients with Prostate Cancer Recurrence," *J Nucl Med,* vol. 59, no. 5, pp. 789–94, 2018, doi: 10.2967/jnumed.117.203257.

[86] S. P. Rowe *et al.*, "PSMA-Based [(18)F]DCFPyL PET/CT Is Superior to Conventional Imaging for Lesion Detection in Patients with Metastatic Prostate Cancer," *Mol Imaging Biol,* vol. 18, no. 3, pp. 411–19, 2016, doi: 10.1007/s11307-016-0957-6.

[87] M. Eiber *et al.*, "Evaluation of Hybrid (6)(8)Ga-PSMA Ligand PET/CT in 248 Patients with Biochemical Recurrence After Radical Prostatectomy," *J Nucl Med,* vol. 56, no. 5, pp. 668–74, 2015, doi: 10.2967/jnumed.115.154153.

[88] W. P. Fendler *et al.*, "(68)Ga-PSMA PET/CT: Joint EANM and SNMMI procedure guideline for prostate cancer imaging: Version 1.0," *Eur J Nucl Med Mol Imaging,* vol. 44, no. 6, pp. 1014–24, 2017, doi: 10.1007/s00259-017-3670-z.

[89] A. Afshar-Oromieh *et al.*, "The Theranostic PSMA Ligand PSMA-617 in the Diagnosis of Prostate Cancer by PET/CT: Biodistribution in Humans, Radiation Dosimetry, and First Evaluation of Tumor Lesions," *J Nucl Med,* vol. 56, no. 11, pp. 1697–705, 2015, doi: 10.2967/jnumed.115.161299.

[90] B. Mitran *et al.*, "Bispecific GRPR-Antagonistic Anti-PSMA/GRPR Heterodimer for PET and SPECT Diagnostic Imaging of Prostate Cancer," *Cancers (Basel),* vol. 11, no. 9, 2019, doi: 10.3390/cancers11091371.

[91] E. Rousseau *et al.*, "A Prospective Study on (18)F-DCFPyL PSMA PET/CT Imaging in Biochemical Recurrence of Prostate Cancer," *J Nucl Med,* vol. 60, no. 11, pp. 1587–93, 2019, doi: 10.2967/jnumed.119.226381.

[92] M. Dietlein *et al.*, "Comparison of [(18)F]DCFPyL and [(68)Ga]Ga-PSMA-HBED-CC for PSMA-PET Imaging in Patients with Relapsed Prostate Cancer," *Mol Imaging Biol,* vol. 17, no. 4, pp. 575–84, 2015, doi: 10.1007/s11307-015-0866-0.

[93] T. A. Hope, J. Z. Goodman, I. E. Allen, J. Calais, W. P. Fendler, and P. R. Carroll, "Metaanalysis of (68)Ga-PSMA-11 PET Accuracy for the Detection of Prostate Cancer Validated by Histopathology," *J Nucl Med,* vol. 60, no. 6, pp. 786–93, 2019, doi: 10.2967/jnumed.118.219501.

[94] M. M. Schmidt and K. D. Wittrup, "A modeling analysis of the effects of molecular size and binding affinity on tumor targeting," *Mol Cancer Ther,* vol. 8, no. 10, pp. 2861–71, 2009, doi: 10.1158/1535-7163.MCT-09-0195.

[95] H. M. Lamb and D. Faulds, "Capromab pendetide. A review of its use as an imaging agent in prostate cancer," *Drugs Aging,* vol. 12, no. 4, pp. 293–304, 1998, doi: 10.2165/00002512-199812040-00004.

[96] H. Liu *et al.*, "Monoclonal antibodies to the extracellular domain of prostate-specific membrane antigen also react with tumor vascular endothelium," *Cancer Res,* vol. 57, no. 17, pp. 3629–34, 1997.

[97] S. T. Tagawa *et al.*, "Phase II study of Lutetium-177-labeled anti-prostate-specific membrane antigen monoclonal antibody J591 for metastatic castration-resistant prostate cancer," *Clin Cancer Res,* vol. 19, no. 18, pp. 5182–91, 2013, doi: 10.1158/1078-0432.CCR-13-0231.

[98] S. T. Tagawa *et al.*, "Bone marrow recovery and subsequent chemotherapy following radiolabeled anti-prostate-specific membrane antigen monoclonal antibody j591 in men with metastatic castration-resistant prostate cancer," *Front Oncol,* vol. 3, p. 214, 2013, doi: 10.3389/fonc.2013.00214.

[99] S. Vallabhajosula *et al.*, "Radioimmunotherapy of Metastatic Prostate Cancer with (1)(7)(7)Lu-DOTAhuJ591 Anti Prostate Specific Membrane Antigen Specific Monoclonal Antibody," *Curr Radiopharm,* vol. 9, no. 1, pp. 44–53, 2016, doi: 10.2174/1874471008666150313114005.

[100] N. Pandit-Taskar *et al.*, "First-in-Human Imaging with 89Zr-Df-IAB2M Anti-PSMA Minibody in Patients with Metastatic Prostate Cancer: Pharmacokinetics, Biodistribution, Dosimetry, and Lesion Uptake," *J Nucl Med,* vol. 57, no. 12, pp. 1858–164, 2016, doi: 10.2967/jnumed.116.176206.

[101] F. Kampmeier, J. D. Williams, J. Maher, G. E. Mullen, and P. J. Blower, "Design and preclinical evaluation of a 99mTc-labelled diabody of mAb J591 for SPECT imaging of prostate-specific membrane antigen (PSMA)," *EJNMMI Res,* vol. 4, no. 1, p. 13, 2014, doi: 10.1186/2191-219X-4-13.

[102] J. Garousi, A. Vorobyeva, and M. Altai, "Influence of Several Compounds and Drugs on the Renal Uptake of Radiolabeled Affibody Molecules," *Molecules,* vol. 25, no. 11, p. 2673, 2020, doi: 10.3390/molecules25112673.

[103] M. Altai, J. Garousi, S. S. Rinne, A. Schulga, S. Deyev, and A. Vorobyeva, "On the prevention of kidney uptake of radiolabeled DARPins," *EJNMMI Res,* vol. 10, no. 1, p. 7, 2020, doi: 10.1186/s13550-020-0599-1.

[104] S. A. Harmon *et al.*, "A Prospective Comparison of (18)F-Sodium Fluoride PET/CT and PSMA-Targeted (18)F-DCFBC PET/CT in Metastatic Prostate Cancer," *J Nucl Med,* vol. 59, no. 11, pp. 1665–71, 2018, doi: 10.2967/jnumed.117.207373.

[105] S. P. Rowe, M. G. Pomper, and M. A. Gorin, "Molecular Imaging of Prostate Cancer: Choosing the Right Agent," *J Nucl Med,* vol. 59, no. 5, pp. 787–88, 2018, doi: 10.2967/jnumed.117.206318.

[106] D. L. Thorek *et al.*, "Internalization of secreted antigen-targeted antibodies by the neonatal Fc receptor for precision imaging of the androgen receptor axis," *Sci Transl Med,* vol. 8, no. 367, p. 367ra167, 2016, doi: 10.1126/scitranslmed.aaf2335.

[107] B. M. Olson *et al.*, "Prostate Cancer Cells Express More Androgen Receptor (AR) Following Androgen Deprivation, Improving Recognition by AR-Specific T Cells," *Cancer Immunol Res,* vol. 5, no. 12, pp. 1074–85, 2017, doi: 10.1158/2326-6066.CIR-16-0390.

[108] J. F. Goodwin *et al.*, "A hormone-DNA repair circuit governs the response to genotoxic insult," *Cancer Discov,* vol. 3, no. 11, pp. 1254–71, 2013, doi: 10.1158/2159-8290.CD-13-0108.

[109] W. R. Polkinghorn *et al.*, "Androgen receptor signaling regulates DNA repair in prostate cancers," *Cancer Discov,* vol. 3, no. 11, pp. 1245–53, 2013, doi: 10.1158/2159-8290.CD-13-0172.

[110] M. R. McDevitt *et al.*, "Feed-forward alpha particle radiotherapy ablates androgen receptor-addicted prostate cancer," *Nat Commun,* vol. 9, no. 1, p. 1629, 2018, doi: 10.1038/s41467-018-04107-w.

[111] J. Tennvall, L. Darte, R. Lundgren, and A. M. el Hassan, "Palliation of multiple bone metastases from prostatic carcinoma with strontium-89," *Acta Oncol,* vol. 27, no. 4, pp. 365–69, 1988, doi: 10.3109/02841868809093556.

[112] Y. Du and S. Dizdarevic, "Molecular radiotheragnostics in prostate cancer," *Clin Med (Lond),* vol. 17, no. 5, pp. 458–61, 2017, doi: 10.7861/clinmedicine.17-5-458.

[113] E. Even-Sapir, U. Metser, E. Mishani, G. Lievshitz, H. Lerman, and I. Leibovitch, "The detection of bone metastases in patients with high-risk prostate cancer: 99mTc-MDP Planar bone scintigraphy, single- and multi-field-of-view SPECT, 18F-fluoride PET, and 18F-fluoride PET/CT," *J Nucl Med,* vol. 47, no. 2, pp. 287–97, 2006.

[114] J. M. Jong *et al.*, "Radiopharmaceuticals for Palliation of Bone Pain in Patients with Castration-resistant Prostate Cancer Metastatic to Bone: A Systematic Review," *Eur Urol,* vol. 70, no. 3, pp. 416–26, 2016, doi: 10.1016/j.eururo.2015.09.005.

[115] A. M. Denis-Bacelar *et al.*, "Phase I/II trials of (186)Re-HEDP in metastatic castration-resistant prostate cancer: Post-hoc analysis of the impact of administered activity and dosimetry on survival," *Eur J Nucl Med Mol Imaging,* vol. 44, no. 4, pp. 620–29, 2017, doi: 10.1007/s00259-016-3543-x.

[116] G. D. Flux, "Imaging and dosimetry for radium-223: The potential for personalized treatment," *Br J Radiol,* vol. 90, no. 1077, p. 20160748, 2017, doi: 10.1259/bjr.20160748.

[117] K. Rahbar *et al.*, "German Multicenter Study Investigating 177Lu-PSMA-617 Radioligand Therapy in Advanced Prostate Cancer Patients," *J Nucl Med,* vol. 58, no. 1, pp. 85–90, 2017, doi: 10.2967/jnumed.116.183194.

[118] C. Kratochwil *et al.*, "225Ac-PSMA-617 for PSMA-targeted alpha-Radiation Therapy of Metastatic Castration-Resistant Prostate Cancer," *J Nucl Med,* vol. 57, no. 12, pp. 1941–44, 2016, doi: 10.2967/jnumed.116.178673.

[119] B. J. B. Nelson, J. D. Andersson, and F. Wuest, "Targeted Alpha Therapy: Progress in Radionuclide Production, Radiochemistry, and Applications," *Pharmaceutics,* vol. 13, no. 1, p. 49, 2020, doi: 10.3390/pharmaceutics13010049.

[120] K. Sandgren *et al.*, "Radiation dosimetry of [(68)Ga]PSMA-11 in low-risk prostate cancer patients," *EJNMMI Phys,* vol. 6, no. 1, p. 2, 2019, doi: 10.1186/s40658-018-0239-2.

[121] B. D. Zlatopolskiy *et al.*, "Discovery of (18)F-JK-PSMA-7, a PET Probe for the Detection of Small PSMA-Positive Lesions," *J Nucl Med,* vol. 60, no. 6, pp. 817–23, 2019, doi: 10.2967/jnumed.118.218495.

[122] F. L. Giesel *et al.*, "F-18 labelled PSMA-1007: Biodistribution, radiation dosimetry and histopathological validation of tumor lesions in prostate cancer patients," *Eur J Nucl Med Mol Imaging,* vol. 44, no. 4, pp. 678–88, 2017, doi: 10.1007/s00259-016-3573-4.

[123] Z. Szabo *et al.*, "Initial Evaluation of [(18)F]DCFPyL for Prostate-Specific Membrane Antigen (PSMA)-Targeted PET Imaging of Prostate Cancer," *Mol Imaging Biol,* vol. 17, no. 4, pp. 565–74, 2015, doi: 10.1007/s11307-015-0850-8.

[124] S. Y. Cho *et al.*, "Biodistribution, tumor detection, and radiation dosimetry of 18F-DCFBC, a low-molecular-weight inhibitor of prostate-specific membrane antigen, in patients with metastatic prostate cancer," *J Nucl Med,* vol. 53, no. 12, pp. 1883–91, 2012, doi: 10.2967/jnumed.112.104661.

[125] S. C. Behr *et al.*, "Phase I Study of CTT1057, an (18)F-Labeled Imaging Agent with Phosphoramidate Core Targeting Prostate-Specific Membrane Antigen in Prostate Cancer," *J Nucl Med,* vol. 60, no. 7, pp. 910–16, 2019, doi: 10.2967/jnumed.118.220715.

[126] S. Piron *et al.*, "Optimization of PET protocol and interrater reliability of (18)F-PSMA-11 imaging of prostate cancer," *EJNMMI Res,* vol. 10, no. 1, p. 14, 2020, doi: 10.1186/s13550-020-0593-7.

[127] J. A. O'Donoghue *et al.*, "Pharmacokinetics and Biodistribution of a [(89)Zr]Zr-DFO-MSTP2109A Anti-STEAP1 Antibody in Metastatic Castration-Resistant Prostate Cancer Patients," *Mol Pharm,* vol. 16, no. 7, pp. 3083–90, 2019, doi: 10.1021/acs.molpharmaceut.9b00326.

[128] S. R. Banerjee *et al.*, "Evaluation of (111)In-DOTA-5D3, a Surrogate SPECT Imaging Agent for Radioimmunotherapy of Prostate-Specific Membrane Antigen," *J Nucl Med,* vol. 60, no. 3, pp. 400–6, 2019, doi: 10.2967/jnumed.118.214403.

[129] M. Hohberg *et al.*, "Biodistribution and radiation dosimetry of [(18)F]-JK-PSMA-7 as a novel prostate-specific membrane antigen-specific ligand for PET/CT imaging of prostate cancer," *EJNMMI Res,* vol. 9, no. 1, p. 66, 2019, doi: 10.1186/s13550-019-0540-7.

[130] S. Piron *et al.*, "Radiation Dosimetry and Biodistribution of (18)F-PSMA-11 for PET Imaging of Prostate Cancer," *J Nucl Med,* vol. 60, no. 12, pp. 1736–42, 2019, doi: 10.2967/jnumed.118.225250.

[131] N. J. Begum *et al.*, "The Effect of Total Tumor Volume on the Biologically Effective Dose to Tumor and Kidneys for (177) Lu-Labeled PSMA Peptides," *J Nucl Med,* vol. 59, no. 6, pp. 929–33, 2018, doi: 10.2967/jnumed.117.203505.

[132] R. P. Baum *et al.*, "177Lu-Labeled Prostate-Specific Membrane Antigen Radioligand Therapy of Metastatic Castration-Resistant Prostate Cancer: Safety and Efficacy," *J Nucl Med,* vol. 57, no. 7, pp. 1006–13, 2016, doi: 10.2967/jnumed.115.168443.

[133] C. Kratochwil *et al.*, "PSMA-Targeted Radionuclide Therapy of Metastatic Castration-Resistant Prostate Cancer with 177Lu-Labeled PSMA-617," *J Nucl Med,* vol. 57, no. 8, pp. 1170–76, 2016, doi: 10.2967/jnumed.115.171397.

[134] L. Kabasakal *et al.*, "Lu-177-PSMA-617 Prostate-Specific Membrane Antigen Inhibitor Therapy in Patients with Castration-Resistant Prostate Cancer: Stability, Bio-distribution and Dosimetry," *Mol Imaging Radionucl Ther,* vol. 26, no. 2, pp. 62–68, 2017, doi: 10.4274/mirt.08760.

[135] L. Scarpa *et al.*, "The (68)Ga/(177)Lu theragnostic concept in PSMA targeting of castration-resistant prostate cancer: Correlation of SUVmax values and absorbed dose estimates," *Eur J Nucl Med Mol Imaging,* vol. 44, no. 5, pp. 788–800, 2017, doi: 10.1007/s00259-016-3609-9.

[136] J. Violet *et al.*, "Dosimetry of (177)Lu-PSMA-617 in Metastatic Castration-Resistant Prostate Cancer: Correlations Between Pretherapeutic Imaging and Whole-Body Tumor Dosimetry with Treatment Outcomes," *J Nucl Med,* vol. 60, no. 4, pp. 517–23, 2019, doi: 10.2967/jnumed.118.219352.

[137] S. Okamoto *et al.*, "Radiation Dosimetry for (177)Lu-PSMA I&T in Metastatic Castration-Resistant Prostate Cancer: Absorbed Dose in Normal Organs and Tumor Lesions," *J Nucl Med,* vol. 58, no. 3, pp. 445–50, 2017, doi: 10.2967/jnumed.116.178483.

[138] A. Khawar *et al.*, "Biodistribution and post-therapy dosimetric analysis of [(177)Lu]Lu-DOTA(ZOL) in patients with osteoblastic metastases: First results," *EJNMMI Res,* vol. 9, no. 1, p. 102, 2019, doi: 10.1186/s13550-019-0566-x.

[139] O. S. Bruland, S. Nilsson, D. R. Fisher, and R. H. Larsen, "High-linear energy transfer irradiation targeted to skeletal metastases by the alpha-emitter 223Ra: Adjuvant or alternative to conventional modalities?" *Clin Cancer Res,* vol. 12, no. 20, Pt 2, pp. 6250s-57s, 2006, doi: 10.1158/1078-0432.CCR-06-0841.

[140] M. Lassmann and D. Nosske, "Dosimetry of 223Ra-chloride: Dose to normal organs and tissues," *Eur J Nucl Med Mol Imaging,* vol. 40, no. 2, pp. 207–12, 2013, doi: 10.1007/s00259-012-2265-y.

[141] S. J. Chittenden *et al.*, "A Phase 1, Open-Label Study of the Biodistribution, Pharmacokinetics, and Dosimetry of 223Ra-Dichloride in Patients with Hormone-Refractory Prostate Cancer and Skeletal Metastases," *J Nucl Med,* vol. 56, no. 9, pp. 1304–9, 2015, doi: 10.2967/jnumed.115.157123.

[142] S. Vallabhajosula *et al.*, "Pharmacokinetics and biodistribution of 111In- and 177Lu-labeled J591 antibody specific for prostate-specific membrane antigen: prediction of 90Y-J591 radiation dosimetry based on 111In or 177Lu?" *J Nucl Med,* vol. 46, no. 4, pp. 634–41, 2005.

[143] C. M. Zechmann *et al.*, "Radiation dosimetry and first therapy results with a (124)I/ (131)I-labeled small molecule (MIP-1095) targeting PSMA for prostate cancer therapy," *Eur J Nucl Med Mol Imaging,* vol. 41, no. 7, pp. 1280–92, 2014, doi: 10.1007/s00259-014-2713-y.

[144] J. Kurth, B. J. Krause, S. M. Schwarzenbock, C. Bergner, O. W. Hakenberg, and M. Heuschkel, "First-in-human dosimetry of gastrin-releasing peptide receptor antagonist [(177)Lu]Lu-RM2: A radiopharmaceutical for the treatment of metastatic castration-resistant prostate cancer," *Eur J Nucl Med Mol Imaging,* vol. 47, no. 1, pp. 123–35, 2020, doi: 10.1007/s00259-019-04504-3.

[145] C. Kratochwil *et al.*, "Targeted alpha therapy of mCRPC: Dosimetry estimate of (213)Bismuth-PSMA-617," *Eur J Nucl Med Mol Imaging,* vol. 45, no. 1, pp. 31–37, 2018, doi: 10.1007/s00259-017-3817-y.

[146] J. Zhang *et al.*, "(177)Lu-PSMA-617 Radioligand Therapy in Metastatic Castration-Resistant Prostate Cancer Patients with a Single Functioning Kidney," *J Nucl Med,* vol. 60, no. 11, pp. 1579–86, 2019, doi: 10.2967/jnumed.118.223149.

[147] A. P. Kiess *et al.*, "Auger Radiopharmaceutical Therapy Targeting Prostate-Specific Membrane Antigen," *J Nucl Med,* vol. 56, no. 9, pp. 1401–7, 2015, doi: 10.2967/jnumed.115.155929.

[148] K. Ljunggren and S. E. Strand, "Beta camera for static and dynamic imaging of charged-particle emitting radionuclides in biologic samples," *J Nucl Med,* vol. 31, no. 12, pp. 2058–63, 1990.

[149] A. Orbom *et al.*, "Characterization of a double-sided silicon strip detector autoradiography system," *Med Phys,* vol. 42, no. 2, pp. 575–84, 2015, doi: 10.1118/1.4905049.

[150] T. Back and L. Jacobsson, "The alpha-camera: A quantitative digital autoradiography technique using a charge-coupled device for ex vivo high-resolution bioimaging of alpha-particles," *J Nucl Med,* vol. 51, no. 10, pp. 1616–23, 2010, doi: 10.2967/jnumed.110.077578.

[151] D. S. Abou, D. Ulmert, M. Doucet, R. F. Hobbs, R. C. Riddle, and D. L. Thorek, "Whole-body and Microenvironmental Localization of Radium-223 in Naive and Mouse Models of Prostate Cancer Metastasis," *J Natl Cancer Inst,* vol. 108, no. 5, p. 380, 2016, doi: 10.1093/jnci/djv380.

[152] O. V. Timmermand, J. Nilsson, S. E. Strand, and J. Elgqvist, "High resolution digital autoradiographic and dosimetric analysis of heterogeneous radioactivity distribution in xenografted prostate tumors," *Med Phys,* vol. 43, no. 12, p. 6632, 2016, doi: 10.1118/1.4967877.

[153] A. P. Kiess *et al.*, "(2S)-2-(3-(1-Carboxy-5-(4-211At-Astatobenzamido)Pentyl)Ureido)-Pentanedioic Acid for PSMA-Targeted alpha-Particle Radiopharmaceutical Therapy," *J Nucl Med,* vol. 57, no. 10, pp. 1569–75, 2016, doi: 10.2967/jnumed.116.174300.

[154] D. Taieb, J. M. Foletti, M. Bardies, P. Rocchi, R. J. Hicks, and U. Haberkorn, "PSMA-targeted Radionuclide Therapy and Salivary Gland Toxicity: Why Does It Matter?" *J Nucl Med,* vol. 59, no. 5, pp. 747–48, 2018, doi: 10.2967/jnumed.118.207993.

[155] T. Langbein, G. Chausse, and R. P. Baum, "Salivary Gland Toxicity of PSMA Radioligand Therapy: Relevance and Preventive Strategies," *J Nucl Med,* vol. 59, no. 8, pp. 1172–73, 2018, doi: 10.2967/jnumed.118.214379.

[156] F. Soeda *et al.*, "Impact of (18)F-PSMA-1007 Uptake in Prostate Cancer Using Different Peptide Concentrations: Preclinical PET/CT Study on Mice," *J Nucl Med,* vol. 60, no. 11, pp. 1594–99, 2019, doi: 10.2967/jnumed.118.223479.

[157] A. Kristiansson *et al.*, "Protection of Kidney Function with Human Antioxidation Protein alpha1 Microglobulin in a Mouse (177)Lu-DOTATATE Radiation Therapy Model," *Antioxid Redox Signal,* 2018, doi: 10.1089/ars.2018.7517.

[158] C. K. Andersson, E. Shubbar, E. Schuler, B. Akerstrom, M. Gram, and E. B. Forssell-Aronsson, "Recombinant alpha1-Microglobulin Is a Potential Kidney Protector in (177)Lu-Octreotate Treatment of Neuroendocrine Tumors," *J Nucl Med,* vol. 60, no. 11, pp. 1600–4, 2019, doi: 10.2967/jnumed.118.225243.

[159] M. Garkavij *et al.*, "Enhanced radioimmunotargeting of 125I-labeled L6-biotin monoclonal antibody (MAb) by combining preload of cold L6 MAb and subsequent immunoadsorption in rats," *Cancer Res,* vol. 55, no. 23 Suppl, pp. 5874s–80s, 1995.

[160] S. Palm, T. Back, S. Lindegren, R. Hultborn, L. Jacobsson, and P. Albertsson, "Model of Intraperitoneal Targeted alpha-Particle Therapy Shows That Posttherapy Cold-Antibody Boost Enhances Microtumor Radiation Dose and Treatable Tumor Sizes," *J Nucl Med,* vol. 59, no. 4, pp. 646–51, 2018, doi: 10.2967/jnumed.117.201285.

[161] E. Rousseau *et al.*, "Monosodium Glutamate Reduces (68)Ga-PSMA-11 Uptake in Salivary Glands and Kidneys in a Preclinical Prostate Cancer Model," *J Nucl Med,* vol. 59, no. 12, pp. 1865–68, 2018, doi: 10.2967/jnumed.118.215350.

[162] S. E. Strand *et al.*, "A general extracorporeal immunoadsorption method to increase tumor-to-tissue ratio," *Cancer,* vol. 73, no. 3 Suppl, pp. 1033–37, 1994, doi: 10.1002/1097-0142(19940201)73:3+<1033::aid-cncr2820731342>3.0.co;2-q.

[163] O. Linden *et al.*, "A novel platform for radioimmunotherapy: Extracorporeal depletion of biotinylated and 90Y-labeled rituximab in patients with refractory B-cell lymphoma," *Cancer Biother Radiopharm,* vol. 20, no. 4, pp. 457–66, 2005, doi: 10.1089/cbr.2005.20.457.

[164] K. L. Chatalic *et al.*, "In Vivo Stabilization of a Gastrin-Releasing Peptide Receptor Antagonist Enhances PET Imaging and Radionuclide Therapy of Prostate Cancer in Preclinical Studies," *Theranostics,* vol. 6, no. 1, pp. 104–17, 2016, doi: 10.7150/thno.13580.

[165] J. Calais, M. Cao, and N. G. Nickols, "The Utility of PET/CT in the Planning of External Radiation Therapy for Prostate Cancer," *J Nucl Med,* vol. 59, no. 4, pp. 557–67, 2018, doi: 10.2967/jnumed.117.196444.

[166] S. Dewes *et al.*, "Integration of (68)Ga-PSMA-PET imaging in planning of primary definitive radiotherapy in prostate cancer: a retrospective study," *Radiat Oncol,* vol. 11, p. 73, 2016, doi: 10.1186/s13014-016-0646-2.

[167] O. Ekmekcioglu, M. Busstra, N. D. Klass, and F. Verzijlbergen, "Bridging the Imaging Gap: PSMA PET/CT Has a High Impact on Treatment Planning in Prostate Cancer Patients with Biochemical Recurrence-A Narrative Review of the Literature," *J Nucl Med,* vol. 60, no. 10, pp. 1394–98, 2019, doi: 10.2967/jnumed.118.222885.

[168] J. Calais *et al.*, "(68)Ga-PSMA-11 PET/CT Mapping of Prostate Cancer Biochemical Recurrence After Radical Prostatectomy in 270 Patients with a PSA Level of Less Than 1.0 ng/mL: Impact on Salvage Radiotherapy Planning," *J Nucl Med,* vol. 59, no. 2, pp. 230–37, 2018, doi: 10.2967/jnumed.117.201749.

[169] O. O. Akin-Akintayo *et al.*, "Change in Salvage Radiotherapy Management Based on Guidance with FACBC (Fluciclovine) PET/CT in Postprostatectomy Recurrent Prostate Cancer," *Clin Nucl Med,* vol. 42, no. 1, pp. e22–e28, 2017, doi: 10.1097/RLU.0000000000001379.

[170] N. S. Schmidt-Hegemann *et al.*, "Impact of (68)Ga-PSMA PET/CT on the Radiotherapeutic Approach to Prostate Cancer in Comparison to CT: A Retrospective Analysis," *J Nucl Med,* vol. 60, no. 7, pp. 963–70, 2019, doi: 10.2967/jnumed.118.220855.

[171] I. A. Burger *et al.*, "(68)Ga-PSMA-11 PET/MR Detects Local Recurrence Occult on mpMRI in Prostate Cancer Patients after HIFU," *J Nucl Med,* vol. 60, no. 8, pp. 1118–23, 2019, doi: 10.2967/jnumed.118.221564.

[172] U. Haberkorn, F. Giesel, A. Morgenstern, and C. Kratochwil, "The Future of Radioligand Therapy: alpha, beta, or Both?" *J Nucl Med,* vol. 58, no. 7, pp. 1017–18, 2017, doi: 10.2967/jnumed.117.190124.

[173] H. Zhang *et al.*, "Dual-modality Imaging of Prostate Cancer with a Fluorescent and Radiogallium-labeled Gastrin-releasing Peptide Receptor Antagonist," *J Nucl Med,* vol. 58, no. 1, pp. 29–35, 2017, doi: 10.2967/jnumed.116.176099.

[174] F. H. Cornelis *et al.*, "Long-Half-Life (89)Zr-Labeled Radiotracers Can Guide Percutaneous Biopsy within the PET/CT Suite without Reinjection of Radiotracer," *J Nucl Med,* vol. 59, no. 3, pp. 399–402, 2018, doi: 10.2967/jnumed.117.194480.

[175] T. Maurer, F. W. B. van Leeuwen, M. Schottelius, H. J. Wester, and M. Eiber, "Entering the era of molecular-targeted precision surgery in recurrent prostate cancer," *J Nucl Med,* 2018, doi: 10.2967/jnumed.118.221861.

[176] T. Maurer *et al.*, "Prostate-specific Membrane Antigen-Guided Surgery," *J Nucl Med,* vol. 61, no. 1, pp. 6–12, 2020, doi: 10.2967/jnumed.119.232330.

[177] T. Maurer *et al.*, "(99m)Technetium-based Prostate-specific Membrane Antigen-radioguided Surgery in Recurrent Prostate Cancer," *Eur Urol,* vol. 75, no. 4, pp. 659–66, 2019, doi: 10.1016/j.eururo.2018.03.013.

20 Peptide Receptor Radionuclide Therapy for Neuroendocrine Tumours

Anna Sundlöv and Katarina Sjögreen Gleisner

CONTENTS

20.1 NEUROENDOCRINE TUMOURS

Neuroendocrine tumours (NETs) are malignant neoplasms originating from the diffuse neuroendocrine system, which is one of the human body's internal communication systems. This system is present in virtually all the epithelialized organs of the body, in the form of neuroendocrine cells interspersed among other cell types of the epithelium, and also includes endocrine organs such as the pituitary gland, thyroid, pancreas, and adrenal glands. NETs are thus tumours originating from these specialized neuroendocrine cells. Although NETs are in part defined by their organ of origin, they have more in common with other NETs than with other tumours of the same organ. They have a wide range of clinical presentations, depending on location of the primary tumours and metastases, hormonal production, and proliferation rate.

NETs that secrete one or more hormones may cause typical hormonal syndromes. The most well-known is the carcinoid syndrome, which is caused by tumoural secretion of vaso-active substances of which the most important is serotonin. The carcinoid syndrome is a triad of flushing (redness and heat in the skin), diarrhea and carcinoid heart disease (fibrosis of the right heart valves causing congestive heart failure). Another very typical, but also very rare, hormonal syndrome is caused by pancreatic insulinomas – NETs that continuously produce insulin – causing life-threatening episodes of hypoglycemia, which worsen as the tumour progresses. Other pancreatic NETs may secrete glucagon or gastrin, among other hormones, each with its specific clinical presentation. Bronchopulmonary NETs sometimes secrete ACTH, which causes the ectopic Cushing syndrome (hyperglycemia, muscle wasting, redistribution of body fat, osteoporosis, etc.) by stimulating the production of cortisol by the adrenal glands.

NETs of the gastrointestinal tract (including pancreas) are classified according to their proliferative activity into grade 1–3, based on the current WHO classification system [1]. The proliferation rate is analysed through immunohistochemical staining of tumour tissue for quantification of the protein Ki67, which is a marker of cell proliferation. If <3 per cent of the tumour cells are positive for Ki67, the tumour is classified as G1. If the Ki67-index is >20 per cent, it is a G3 tumour. The G2 tumours have a Ki67-index between 3 per cent and 20 per cent. Among the tumours there are two subcategories: NET (NE *tumour*) G3 for the well-differentiated tumours, and NEC (NE *carcinoma*) G3 for the poorly differentiated ones. NET G3 has a better prognosis than NEC. The correct overall term for both NET and NEC is NEN – neuroendocrine neoplasia.

DOI: 10.1201/9780429489501-20

Grade 1 (G1) tumours have a very low proliferation rate and are thus the most indolent; even patients with metastatic disease can live for 10 years or more. The grade 3 (G3) NECs, on the contrary, are one of the most aggressive cancers known to man, with a median overall survival of about one year for metastatic disease. In between these two extremes we find the grade 2 (G2) tumours, which are often slightly less aggressive than other common cancers, but still have a median overall survival of around five years.

In summary, neuroendocrine tumours are defined by their organ of origin, the proliferation rate, and their hormonal activity. All these factors, together with the stage of the disease (local tumour growth only, regional spread or distant metastases), need to be kept in mind when deciding which treatment the patient should receive.

20.2 THE PRRT-TARGET: THE SOMATOSTATIN RECEPTOR

The somatostatin receptor (SSTR) is a transmembrane, G-protein-coupled receptor that is expressed by neuroendocrine cells present in most tissues and organs throughout the body. NETs, being tumours made up of neuroendocrine cells, have an especially high receptor density. There are five subtypes of the SSTR, and different tissues express different combinations of receptor subtypes. The natural ligand, somatostatin, is a hormone secreted by the hypothalamus, whose function is to inhibit the hormonal secretion from neuroendocrine cells [2]. The same effect can be achieved through injection of a synthetic analogue of somatostatin (SSA), which has been used since the 1980s to treat acromegaly (hypersecretion of growth hormone) and NETs.

Approximately 80 per cent of well-differentiated NETs express the SSTR. The subreceptor type 2 is the most common, followed by SSTR1 and 5 [2]. All currently available SSAs, octreotide, lanreotide and the newer analogue pasireotide, have a high binding affinity to both SSTR2 and 5 with the consequent effect of reducing both the hormonal secretion and the tumour growth [3–5]. The octreotide molecule has been further modified to be able to bind radioactive isotopes to it. By choosing a gamma- or positron-emitting radioisotope it is possible to image the degree of receptor expression and the distribution of the tumour in the patient with the help of SPECT (single photon emission computed tomography) or PET (positron emission tomography) imaging. Analogously, the SSA can be labelled with beta-emitting radionuclides such as ^{177}Lu (Lutetium) or ^{90}Y (Yttrium) for therapeutic purposes. This is what is known as peptide receptor radionuclide therapy (PRRT).

The advent of the radiolabelled synthetic SSAs led to the development of the diagnostic nuclear medicine method of somatostatin receptor scintigraphy using ^{111}In (Indium). Since ^{111}In emits Auger electrons in its radioactive decay, it can potentially be used for therapeutic purposes. The first trials in humans with high injected activities of ^{111}In-pentreotide demonstrated a small but encouraging therapeutic effect [6]. Currently, SSTR scintigraphy is rapidly being replaced by SSTR PET/CT using ^{68}Ga (Gallium)-labelled DOTA-TOC or DOTA-TATE, which is used not only for staging of NET, but also for selection of patients for PRRT (Figure 1). For therapeutic purposes, the real breakthrough was the development of DOTA-chelated, beta-emitter labelled compounds such as ^{90}Y-DOTA-TOC and ^{177}Lu-DOTA-TATE, the latter being the compound that has reached the widest clinical use. The safety and efficacy of ^{177}Lu-DOTA-TATE vis a vis standard therapy (high-dose octreotide) was demonstrated in the phase III trial NETTER-1: The patients in the PRRT arm reached a median progression-free survival (mPFS) of 28 months, whereas those in the control arm had a mPFS of only 8 months [7]. Based on these results, PRRT with ^{177}Lu-DOTA-TATE was approved by the European Medicines Agency and the US Food and Drug Administration for the treatment of gastro-entero-pancreatic NETs.

20.3 INTERNAL DOSIMETRY IN PRRT

Both ^{177}Lu and ^{90}Y undergo β−decay, and thus emit electrons. An advantage of ^{177}Lu is that it also emits gamma photons with energies suitable for SPECT imaging (208 keV and 113 keV). Although ^{90}Y lacks the emission of gamma photons, the bremsstrahlung emitted from electrons slowing down can be measured with a gamma camera, and it was rather recently discovered that owing to the low, yet measurable branching for β+emission, ^{90}Y can also be measured by PET/CT [8]. The half-lives of ^{177}Lu and ^{90}Y are 6.6 days and 2.7 days, respectively, and are suitable for PRRT in that they match the biological half-life of the peptide molecules used for targeting of NETs.

The absorbed doses (ADs) delivered by internally distributed radioactive drugs are estimated by means of internal dosimetry. Parameters that govern the AD are the activity distribution in different organs and tissues over time, the radiation energy emitted in each radioactive decay, how this energy is transported between body regions, and the mass of the tissues in which the radiation energy is imparted. The expressions used for internal dosimetry calculations were, already in 1968, formalized by the committee on Medical Internal Radiation Dose (MIRD) of the Society of Nuclear Medicine

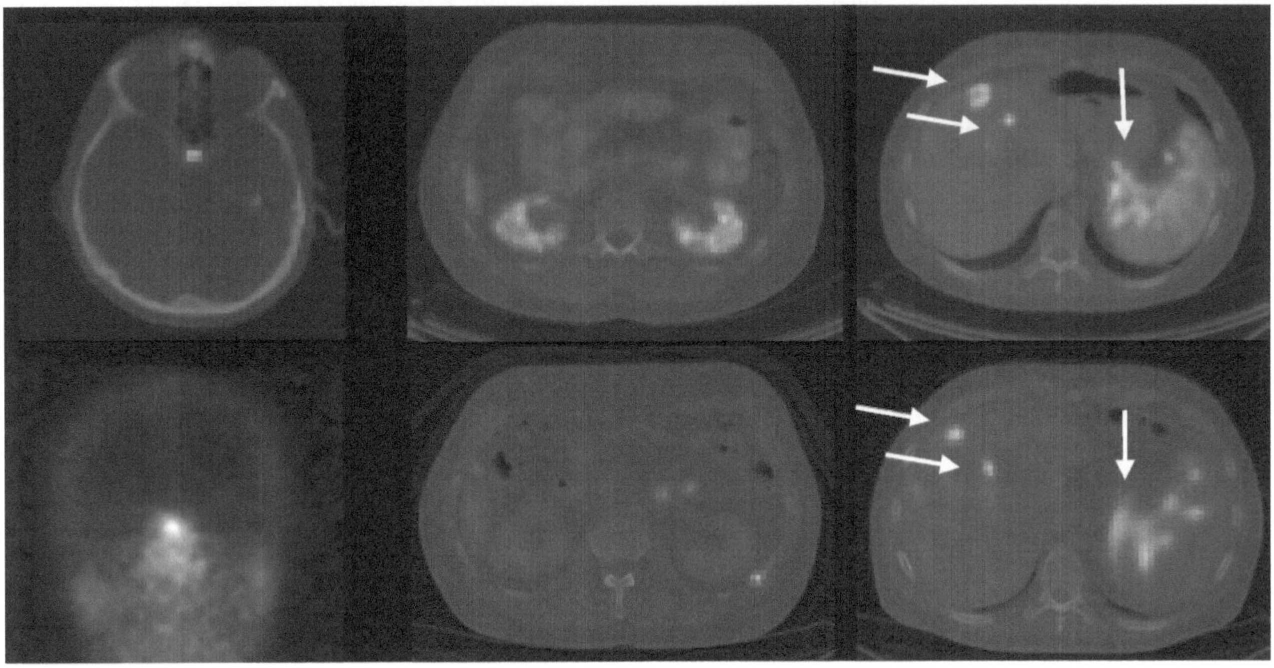

FIGURE 20.1 Pre-therapeutic ^{68}Ga-DOTA-TATE PET/CT (top row) and ^{177}Lu-DOTA-TATE post-therapeutic gamma-camera images (bottom), of the same patient treated for NET. Left: the pituitary gland in transverse PET/CT image, and in anterior planar gamma-camera scans for which images acquired d1, d4, and d7 p.i. have been co-registered and integrated over time. Middle: the kidneys, considered the principal organ at risk in PRRT. Right: tumours, indicated by arrows. Colour image available at www.routledge.com/9781138593312.

and Molecular Imaging [9]. Following the more recent *MIRD Pamphlet No. 21* [10] the mean absorbed dose $D(r_T)$ to a target region r_T is calculated from the time-integrated activity $\tilde{A}(r_s)$ in a source region r_s according to

$$D(r_T) = \sum_{r_s} \tilde{A}(r_s) \cdot S(r_T \leftarrow r_S). \tag{20.1}$$

The S coefficient $S(r_T \leftarrow r_S)$ represents the mean AD given to a target tissue r_T per unit of time-integrated activity in a source region r_S. This coefficient is specific for each radionuclide and source-target region combination. As indicated by the summation sign, all source regions that contribute to the AD in a given target region are taken into account. S coefficients are calculated by means of Monte Carlo calculations and are available from precompiled data sets for human-like reference geometries [11-14]. In order to make S coefficients patient-specific and applicable for individual patient dosimetry, a mass scaling needs to be applied. For charged-particle radiation this mass scaling is made by the ratio of the reference organ mass to the patient organ mass, following

$$S(r_T \leftarrow r_S)_{\text{pat}} = \frac{M_{\text{ref}}(r_T)}{M_{\text{pat}}(r_T)} \cdot S(r_T \leftarrow r_S)_{\text{ref}}. \tag{20.2}$$

Thus, the organ mass for the individual patient $M_{\text{pat}}(r_T)$ needs to be quantified based on anatomic imaging, such as CT. Other means to make the S coefficients patient specific are to perform Monte Carlo energy transport calculations directly on the voxel geometry obtained from quantitative SPECT/CT imaging [15, 16].

The factor $\tilde{A}(r_s)$ in Eqn. 1 is determined as the integral of the activity $A(r_s, t)$ in a source region over time. Activity estimates for dosimetry in NET are usually based on repeated SPECT/CT imaging. The number of image acquisitions is intimately related to the pharmacokinetics of the radiopharmaceutical and, in order to capture the entire washout phase, repeated measurements over several days are often required. Quantitative SPECT images, that is, three-dimensional images in which each voxel value represents the activity at a given anatomical location, are obtained by means of detailed, physics-based calculation models. A key element is the tomographic reconstruction, which needs to include corrections for photon attenuation, scatter, and preferably collimator-response modelling. Current tomographic reconstructions are based on iterative procedures and include the CT image to estimate the distribution of the attenuating properties of the tissues in the patient and, for some methods, also the scattering properties. The reconstructed voxel values are converted from counts to activity by the application of a calibration factor, determined prior to patient imaging. To analyse the voxels representing a particular organ or tissue (a source region, r_s) volumes-of-interest (VOIs) are delineated. These VOIs can be delineated manually or by some semi-automatic segmentation method, preferably using the CT as support for contouring [17, 18]. As an alternative to whole-organ segmentation, smaller VOIs have been proposed to determine the activity concentration [19]. An additional factor that needs to be considered for accurate activity quantification is the limited spatial resolution that modulates high-frequency components of the signal, thus producing blurry images. This limitation, termed the partial-volume effect, is usually taken into account by the application of post-reconstruction correction factors that are specific for the tomographic reconstruction and its settings, the volume and shape of the source region to be quantified.

Image-based dosimetry forms a complex measurement chain, and it was not until recently that studies of the combined uncertainty were published, approached by analytical or Monte Carlo–based propagation methods [20, 21]. The results of such uncertainty propagations are encouraging, as typically for regions with sufficiently large volumes, such as liver, kidneys, spleen and larger tumours, relative uncertainties of approximately 10 per cent can be obtained. The reliability of the absorbed-dose estimates forms a basis for their relevance in the clinical context.

20.4 DOSE-RESPONSE DATA

The pharmacokinetics of small molecules such as peptides is governed by one fast excretion component via the kidneys, and one circulating component that extravasates and is accumulated in tissues expressing SSTRs. For kidneys, there is also an active accumulation in the proximal tubule, which can be counteracted by co-infusion of an amino acid solution during radiopharmaceutical administration. The principal organs at risk in PRRT are the kidneys, bone marrow, and the pituitary gland.

The kidneys are a late-responding tissue, and effects of exposure to ionizing radiation can occur several years after treatment. When the linear-quadratic radiobiologic model is applied, a low value of the α / β ratio (between 2 and 3) is generally assumed for calculation of the biologically effective dose (BED) [22, 23]. For patients treated with ^{90}Y-DOTA-TOC the kidneys were observed to exhibit a dose-dependent change of renal function, measured as the annual loss in creatine clearance rate [24]. The authors highlighted the importance of individual mass estimation and calculation of the BED to take radiobiological effects from different absorbed-dose rates and fractionation schemes into account. For ^{177}Lu-DOTA-TATE the observed renal toxicity has so far been mild when administered according to the standard treatment schedule of 7.4 GBq in four cycles.

Both subacute and persistent hematologic effects have been observed in ^{177}Lu-DOTA-TATE treatment [25, 26]. Image-based bone-marrow dosimetry has shown an AD-related decrease of the platelet counts [27-29]. For ^{90}Y-DOTA-TOC, image-based dosimetry of selected vertebrae was performed and exhibited correlation to the decrease in platelet counts [30].

The pituitary gland was investigated for patients treated with ^{177}Lu-DOTA-TATE. The clinical trial was designed to tailor the number of treatment cycles to the renal AD, and patients thus received a variable number of cycles [31]. Both the number of cycles and the AD were shown to correlate with a decrease of IGF-1 levels during long-term follow-up, indicating late pituitary toxicity [32].

Although most dosimetry studies have so far been mainly focused on safety and organs at risk, tumour dosimetry reports are increasing in number. A dose-dependent tumour-volume reduction was reported in NETs treated with ^{90}Y-DOTA-TOC [30]. Similarly, an AD-dependent diameter reduction was observed for pancreatic NET treated with ^{177}Lu-DOTA-TATE [33]. For small intestinal NETs treated with ^{177}Lu-DOTA-TATE, a higher rate of response was associated with the totally administered activity over all cycles (which varied between patients) but not with AD [34]. The latter cohort of patients was later re-analysed with a longer follow-up time, and then exhibited an AD-dependent volume decrease [35]. Further support for a dose-response relationship in tumours was obtained in ^{90}Y microsphere treatment of NET metastases in the liver, where an AD cut-off of 191 Gy was associated with complete and partial response, as assessed by mRECIST [36].

20.5 PERSONALIZED THERAPY USING DOSIMETRY GUIDED PRRT

The pharmacokinetics of the internally administered radiopeptides is governed by physiologically and biochemically driven mechanisms, associated with various patient characteristics such as the plasma volume, the glomerular-filtration rate, the expression of SSTRs, and the tumour burden. As a result, the same administered activity will yield inter-patient variability in the ADs delivered. It is thus unlikely that a treatment protocol based on the administration of the same amount of activity to all patients is optimal in terms of balancing the likelihood of tumour effects against the risks of toxicity. In other forms of radiotherapy, such as external beam radiotherapy or brachytherapy, ADs are always planned and verified, with the intent of optimizing the treatment. This contrasts with radionuclide therapy and PRRT where fixed activity levels form the standard prescription posology. The standard protocol for ^{177}Lu-DOTA-TATE treatment, for example, is 4 cycles of 7,4 GBq given with an interval of 8-10 weeks. However, within clinical trials dosimetry-guided treatment has been explored.

Clinical trials have been designed to adjust the number of treatment cycles to each patient based on renal dosimetry [19, 31, 37]. Different renal AD or BED tolerances have been assumed for these studies, 23 Gy or 27 Gy, or 40 Gy for patients without risk factors. Largely, these tolerance doses are based on previous extrapolations from external beam radiotherapy [38]. While respecting these dose limits, the number of cycles, each of 7.4 GBq, has varied between 3 and 9 for individual patients, thus demonstrating the large inter-individual span in ADs given.

Another trial instead modified the administered activity per cycle, while keeping the number of cycles constant [29]. The administered activity for the first cycle was personalized based on the glomerular-filtration rate and body surface area, while the activities for the remaining three cycles was calculated from the estimated renal AD per administered activity for preceding cycles. A total AD limit of 23 Gy to the kidneys was applied, which was reduced in case of significant renal or hematological impairment. The authors found that on average the administered activity per cycle could be increased by approximately 25 per cent without surpassing the renal AD limit.

20.6 OTHER STRATEGIES OF PERSONALIZATION OF PRRT

In addition to using dosimetry to personalize PRRT, treatment can be further refined by improving patient selection with the help of so-called dual imaging (^{18}F-FDG-PET and ^{68}Ga-DOTA-PET). The prognostic value of ^{18}F-FDG-PET in NET has been thoroughly studied [39-41], and ^{18}F-FDG-PET-positivity has been shown to have a stronger correlation to overall survival than the current, Ki67-based grading system [41, 42]. This is not surprising considering that with PET-imaging the whole tumour burden is studied, in contrast to the microscopically small part of the tumour that is examined in a histological specimen.

The strong prognostic value of ^{18}F-FDG-PET-positivity raises the question of how the treatment can be adjusted to improve the outcome for these patients. One attractive alternative, given the low toxicity of ^{177}Lu-DOTA-TATE therapy, is to combine it with other antitumour drugs. Several retrospective and prospective clinical studies confirm the safety and efficacy of different combinations of chemotherapy and ^{177}Lu-DOTA-TATE. It has been studied in patients with a poor prognosis due to either high tumour load and/or ^{18}FDG-PET positive disease using combinations with either 5-fluorouracil/capecitabine or temozolomide and capecitabine [43–46]. One retrospective study of patients who had received a combination of 5-fluorouracil and ^{177}Lu-DOTA-TATE demonstrated an impressive mPFS of 48 months [45], mainly low-grade hematologic toxicity. Ongoing prospective clinical trials will add more data on the benefit of chemo-PRRT (e.g., NCT02358356, NCT02736500, NCT02736448).

Inhibitors of the Poly (ADP-ribose) polymerase-1 (PARP), an important enzyme in the detection and repair of DNA-damage, are another attractive candidate for combination treatment. The combination of a DNA-damaging drug, such as ^{177}Lu-DOTA-TATE, and a PARP-inhibitor increases the number of DNA double-strand breaks and thereby the tumouricidal effect of the treatment. This has been demonstrated in vivo and in vitro [47] and is now being tested in clinical trials (NCT04375267).

Personalized medicine in other areas of oncology is to a large extent based on molecular analyses of tumour tissue rather than imaging. This approach has gained little ground in the management of NETs, mainly due to a generally low level of mutations in the tumour DNA [48, 49]. A molecular test based on circulating RNA transcripts in peripheral blood, the NETest®, reflects the tumour gene expression at the time of the blood sampling. It has shown a high diagnostic, predictive, and prognostic accuracy [50], and has also been successfully applied to PRRT using a subset of the transcriptomic analytes reflecting growth-factor signalling and tumour metabolism. By combining this information with the Ki67-index of the tumour the PRRT Predictive Quotient (PPQ), the predictive accuracy for identifying responders versus non-responders reached 94 per cent [51].

20.7 NEW PRRT AGENTS

There are many lines of development of PRRT currently in progress. Developing a more individualized, tailored therapy with the help of dual imaging, combination therapies, and dosimetry-based treatment is one of them. Development of new, more effective PPRT agents is another, where PRRT with alpha emitters is an attractive approach thanks to the high energy and short range of the emitted radiation. Theoretically, this will translate into a high AD delivered to tumour lesions, while the surrounding normal tissue is spared. To date, the only published experience in humans is one series of seven patients who were refractory to ^{90}Y- and ^{177}Lu-DOTA-TOC, but all responded to ^{213}Bi-DOTATOC. There were a couple of cases of long-term hematologic and renal toxicity, while acute toxicity was low [52]. Another interesting line of development within PRRT is to use ^{177}Lu-labeled SSTR-antagonists (instead of agonists such as octreotide and –tate). These ligands have shown a higher uptake in tumour than the commonly used agonists, probably due to the fact that they can bind to both the active and inactive form of the SSTR and seem to have more binding sites [53]. In a first-in-human phase I trial the ^{177}Lu-labeled SSTR-antagonist satoreotide tetraxetan was given to 20 heavily pre-treated NET-patients. There was a high objective response rate (45 per cent) but also considerable hematological toxicity [54]. It is currently being further evaluated in a phase I/II trial (NCT02592707).

REFERENCES

[1] F. Inzani, G. Petrone, and G. Rindi, "The New World Health Organization Classification for Pancreatic Neuroendocrine Neoplasia," *Endocrinol Metab Clin North Am,* vol. 47, no. 3, pp. 463–70, 2018, doi: 10.1016/j.ecl.2018.04.008.

[2] J. C. Reubi and A. Schonbrunn, "Illuminating somatostatin analog action at neuroendocrine tumor receptors," *Trends Pharmacol Sci,* vol. 34, no. 12, pp. 676–88, 2013, doi: 10.1016/j.tips.2013.10.001.

[3] A. Rinke *et al.*, "Placebo-controlled, double-blind, prospective, randomized study on the effect of octreotide LAR in the control of tumor growth in patients with metastatic neuroendocrine midgut tumors: a report from the PROMID Study Group," *J Clin Oncol,* vol. 27, no. 28, pp. 4656–63, 2009, doi: 10.1200/JCO.2009.22.8510.

[4] M. E. Caplin *et al.*, "Lanreotide in metastatic enteropancreatic neuroendocrine tumors," *N Engl J Med,* vol. 371, no. 3, pp. 224–33, 2014, doi: 10.1056/NEJMoa1316158.

[5] A. N. Wymenga *et al.*, "Efficacy and safety of prolonged-release lanreotide in patients with gastrointestinal neuroendocrine tumors and hormone-related symptoms," *J Clin Oncol,* vol. 17, no. 4, p. 1111, 1999, doi: 10.1200/JCO.1999.17.4.1111.

[6] L. B. Anthony, E. A. Woltering, G. D. Espenan, M. D. Cronin, T. J. Maloney, and K. E. McCarthy, "Indium-111-pentetreotide prolongs survival in gastroenteropancreatic malignancies," *Semin Nucl Med,* vol. 32, no. 2, pp. 123–32, 2002, doi: 10.1053/snuc.2002.31769.

[7] J. Strosberg *et al.*, "Phase 3 trial of (177)Lu-Dotatate for midgut neuroendocrine tumors," *N Engl J Med,* vol. 376, no. 2, pp. 125–35, 2017, doi: 10.1056/NEJMoa1607427.

[8] R. Lhommel *et al.*, "Feasibility of 90Y TOF PET-based dosimetry in liver metastasis therapy using SIR-Spheres," *Eur J Nucl Med Mol Imaging,* vol. 37, no. 9, pp. 1654–62, 2010, doi: 10.1007/s00259-010-1470-9.

[9] R. Loevinger and M. Berman, "A schema for absorbed-dose calculations for biologically-distributed radionuclides," *J Nucl Med,* pp. Suppl 1:9–14, 1968.

[10] W. E. Bolch, K. F. Eckerman, G. Sgouros, and S. R. Thomas, "MIRD pamphlet No. 21: A generalized schema for radiopharmaceutical dosimetry – standardization of nomenclature," *J Nucl Med,* vol. 50, no. 3, pp. 477–84, 2009, doi: 10.2967/jnumed.108.056036.

[11] M. G. Stabin, R. B. Sparks, and E. Crowe, "OLINDA/EXM: the second-generation personal computer software for internal dose assessment in nuclear medicine," *J Nucl Med,* vol. 46, no. 6, pp. 1023–7, 2005.

[12] ICRP, "ICRP Publication 133. The ICRP computational framework for internal dose assessment for reference adults: specific absorbed fractions," *Annals of the ICRP,* vol. 45, no. 2, p. 74, 2016.

[13] M. Andersson, L. Johansson, D. Minarik, S. Mattsson, and S. Leide-Svegborn, "An internal radiation dosimetry computer program, IDAC 2.0, for estimation of patient doses from radiopharmaceuticals," *Radiat Prot Dosimetry,* vol. 162, no. 3, pp. 299–305, 2014, doi: 10.1093/rpd/nct337.

[14] M. Chauvin *et al.*, "OpenDose: Open-access resource for nuclear medicine dosimetry," *J Nucl Med,* vol. 61, no. 10, pp. 1514–19, 2020, doi: 10.2967/jnumed.119.240366.

[15] M. Ljungberg, E. Frey, K. Sjogreen, X. Liu, Y. Dewaraja, and S. E. Strand, "3D absorbed dose calculations based on SPECT: Evaluation for 111-In/90-Y therapy using Monte Carlo simulations," *Cancer Biother Radiopharm,* vol. 18, no. 1, pp. 99–107, 2003, doi: 10.1089/108497803321269377.

[16] M. Ljungberg, K. Sjogreen, X. Liu, E. Frey, Y. Dewaraja, and S. E. Strand, "A 3-dimensional absorbed dose calculation method based on quantitative SPECT for radionuclide therapy: Evaluation for (131)I using Monte Carlo simulation," *J Nucl Med,* vol. 43, no. 8, pp. 1101–9, 2002.

[17] M. Hatt *et al.*, "Classification and evaluation strategies of auto-segmentation approaches for PET: Report of AAPM task group No. 211," *Medical Physics,* vol. 44, no. 6, pp. e1-e42, 2017.

[18] J. Gustafsson, A. Sundlov, and K. Sjogreen Gleisner, "SPECT image segmentation for estimation of tumour volume and activity concentration in (177)Lu-Dotatate radionuclide therapy," *EJNMMI Res,* vol. 7, no. 1, p. 18, 2017, doi: 10.1186/s13550-017-0262-7.

[19] M. Sandstrom, U. Garske, D. Granberg, A. Sundin, and H. Lundqvist, "Individualized dosimetry in patients undergoing therapy with (177)Lu-DOTA-D-Phe (1)-Tyr (3)-octreotate," *Eur J Nucl Med Mol Imaging,* vol. 37, no. 2, pp. 212–25, 2010, doi: 10.1007/s00259-009-1216-8.

[20] J. I. Gear *et al.*, "EANM practical guidance on uncertainty analysis for molecular radiotherapy absorbed dose calculations," *Eur J Nucl Med Mol Imaging,* vol. 45, no. 13, pp. 2456–74, 2018, doi: 10.1007/s00259-018-4136-7.

[21] J. Gustafsson, G. Brolin, M. Cox, M. Ljungberg, L. Johansson, and K. S. Gleisner, "Uncertainty propagation for SPECT/CT-based renal dosimetry in (177)Lu peptide receptor radionuclide therapy," *Phys Med Biol,* vol. 60, no. 21, pp. 8329–46, 2015, doi: 10.1088/0031-9155/60/21/8329.

[22] B. W. Wessels *et al.*, "MIRD pamphlet No. 20: The effect of model assumptions on kidney dosimetry and response – implications for radionuclide therapy," *J Nucl Med,* vol. 49, no. 11, pp. 1884–99, 2008, doi: 10.2967/jnumed.108.053173.

[23] H. D. Thames, K. K. Ang, F. A. Stewart, and E. van der Schueren, "Does incomplete repair explain the apparent failure of the basic LQ model to predict spinal cord and kidney responses to low doses per fraction?" *Int J Radiat Biol,* vol. 54, no. 1, pp. 13–19, 1988, doi: 10.1080/09553008814551461.

[24] R. Barone *et al.*, "Patient-specific dosimetry in predicting renal toxicity with (90)Y-DOTATOC: relevance of kidney volume and dose rate in finding a dose-effect relationship," *J Nucl Med,* vol. 46 Suppl 1, pp. 99S–106S, 2005.

[25] H. Bergsma *et al.*, "Subacute haematotoxicity after PRRT with (177)Lu-DOTA-octreotate: prognostic factors, incidence and course," *Eur J Nucl Med Mol Imaging,* vol. 43, no. 3, pp. 453–63, 2016, doi: 10.1007/s00259-015-3193-4.

[26] H. Bergsma *et al.*, "Persistent hematologic dysfunction after peptide receptor radionuclide therapy with (177)Lu-Dotatate: Incidence, course, and predicting factors in patients with gastroenteropancreatic neuroendocrine tumors," *J Nucl Med,* vol. 59, no. 3, pp. 452–58, 2018, doi: 10.2967/jnumed.117.189712.

[27] L. Hagmarker *et al.*, "Bone marrow absorbed doses and correlations with hematologic response during (177)Lu-Dotatate treatments are influenced by image-based dosimetry method and presence of skeletal metastases," *J Nucl Med,* vol. 60, no. 10, pp. 1406–13, 2019, doi: 10.2967/jnumed.118.225235.

[28] M. Del Prete, F.-A. Buteau, and J.-M. Beauregard, "Personalized 177 Lu-octreotate peptide receptor radionuclide therapy of neuroendocrine tumours: A simulation study," *European Journal of Nuclear Medicine and Molecular Imaging,* vol. 44, no. 9, pp. 1490–1500, 2017.

[29] M. Del Prete *et al.*, "Personalized (177)Lu-octreotate peptide receptor radionuclide therapy of neuroendocrine tumours: Initial results from the P-PRRT trial," *Eur J Nucl Med Mol Imaging,* vol. 46, no. 3, pp. 728–42, 2019, doi: 10.1007/s00259-018-4209-7.

[30] S. Pauwels *et al.*, "Practical dosimetry of peptide receptor radionuclide therapy with (90)Y-labeled somatostatin analogs," *J Nucl Med,* vol. 46 Suppl 1, pp. 92S–98S, 2005.

[31] A. Sundlov *et al.*, "Individualised (177)Lu-Dotatate treatment of neuroendocrine tumours based on kidney dosimetry," *Eur J Nucl Med Mol Imaging,* vol. 44, no. 9, pp. 1480–89, 2017, doi: 10.1007/s00259-017-3678-4.

[32] A. Sundlov *et al.*, "Pituitary function after high-dose 177Lu-Dotatate therapy and long-term follow-up," *Neuroendocrinology,* vol. 111, no. 4, pp. 344–53, 2021, doi: 10.1159/000507761.

[33] E. Ilan *et al.*, "Dose response of pancreatic neuroendocrine tumors treated with peptide receptor radionuclide therapy using 177Lu-Dotatate," *J Nucl Med,* vol. 56, no. 2, pp. 177–82, 2015, doi: 10.2967/jnumed.114.148437.

[34] U. Jahn, E. Ilan, M. Sandstrom, U. Garske-Roman, M. Lubberink, and A. Sundin, "177Lu-Dotatate peptide receptor radionuclide therapy: Dose response in small intestinal neuroendocrine tumors," *Neuroendocrinology,* vol. 110, no. 7–8, pp. 662–70, 2020, doi: 10.1159/000504001.

[35] U. Jahn, E. Ilan, M. Sandstrom, M. Lubberink, U. Garske-Roman, and A. Sundin, "Peptide receptor radionuclide therapy (PRRT) with (177)Lu-Dotatate; Differences in tumor dosimetry, vascularity and lesion metrics in pancreatic and small intestinal neuroendocrine neoplasms," *Cancers (Basel),* vol. 13, no. 5, 2021, doi: 10.3390/cancers13050962.

[36] O. Chansanti *et al.*, "Tumor dose response in Yttrium-90 Resin Microsphere embolization for neuroendocrine liver metastases: A tumor-specific analysis with dose estimation using SPECT-CT," *J Vasc Interv Radiol,* vol. 28, no. 11, pp. 1528–35, 2017, doi: 10.1016/j.jvir.2017.07.008.

[37] U. Garske-Roman *et al.*, "Prospective observational study of (177)Lu-DOTA-octreotate therapy in 200 patients with advanced metastasized neuroendocrine tumours (NETs): Feasibility and impact of a dosimetry-guided study protocol on outcome and toxicity," *Eur J Nucl Med Mol Imaging,* vol. 45, no. 6, pp. 970–88, 2018, doi: 10.1007/s00259-018-3945-z.

[38] L. Bodei *et al.*, "Long-term evaluation of renal toxicity after peptide receptor radionuclide therapy with 90Y-DOTATOC and 177Lu-Dotatate: The role of associated risk factors," *Eur J Nucl Med Mol Imaging,* vol. 35, no. 10, pp. 1847–56, 2008, doi: 10.1007/s00259-008-0778-1.

[39] T. Binderup, U. Knigge, A. Loft, B. Federspiel, and A. Kjaer, "18F-fluorodeoxyglucose positron emission tomography predicts survival of patients with neuroendocrine tumors," *Clin Cancer Res,* vol. 16, no. 3, pp. 978–85, 2010, doi: 10.1158/1078-0432.CCR-09-1759.

[40] H. Bahri *et al.*, "High prognostic value of 18F-FDG PET for metastatic gastroenteropancreatic neuroendocrine tumors: A long-term evaluation," *J Nucl Med,* vol. 55, no. 11, pp. 1786–90, 2014, doi: 10.2967/jnumed.114.144386.

[41] D. L. Chan *et al.*, "Dual somatostatin Receptor/FDG PET/CT imaging in metastatic neuroendocrine tumours: Proposal for a novel grading scheme with prognostic significance," *Theranostics,* vol. 7, no. 5, pp. 1149–58, 2017, doi: 10.7150/thno.18068.

[42] I. Karfis *et al.*, "Prognostic value of a three-scale grading system based on combining molecular imaging with (68)Ga-Dotatate and (18)F-FDG PET/CT in patients with metastatic gastroenteropancreatic neuroendocrine neoplasias," *Oncotarget,* vol. 11, no. 6, pp. 589–99, 2020, doi: 10.18632/oncotarget.27460.

[43] P. G. Claringbold, P. A. Brayshaw, R. A. Price, and J. H. Turner, "Phase II study of radiopeptide 177Lu-octreotate and capecitabine therapy of progressive disseminated neuroendocrine tumours," *Eur J Nucl Med Mol Imaging,* vol. 38, no. 2, pp. 302–11, 2011, doi: 10.1007/s00259-010-1631-x.

[44] P. G. Claringbold and J. H. Turner, "Pancreatic neuroendocrine tumor control: Durable objective response to combination 177Lu-Octreotate-Capecitabine-Temozolomide radiopeptide chemotherapy," *Neuroendocrinology,* vol. 103, no. 5, pp. 432–39, 2016, doi: 10.1159/000434723.

[45] R. Kashyap *et al.*, "Favourable outcomes of (177)Lu-octreotate peptide receptor chemoradionuclide therapy in patients with FDG-avid neuroendocrine tumours," *Eur J Nucl Med Mol Imaging,* vol. 42, no. 2, pp. 176–85, 2015, doi: 10.1007/s00259-014-2906-4.

[46] T. W. Barber, M. S. Hofman, B. N. Thomson, and R. J. Hicks, "The potential for induction peptide receptor chemoradionuclide therapy to render inoperable pancreatic and duodenal neuroendocrine tumours resectable," *European Journal of Surgical Oncology: The journal of the European Society of Surgical Oncology and the British Association of Surgical Oncology,* vol. 38, no. 1, pp. 64–71, 2012, doi: 10.1016/j.ejso.2011.08.129.

[47] C. Cullinane *et al.*, "Enhancing the anti tumour activity of (177)Lu-DOTA-octreotate radionuclide therapy in somatostatin receptor-2 expressing tumour models by targeting PARP," *Sci Rep,* vol. 10, no. 1, p. 10196, 2020, doi: 10.1038/s41598-020-67199-9.

[48] A. Scarpa, "The landscape of molecular alterations in pancreatic and small intestinal neuroendocrine tumours," *Ann Endocrinol (Paris),* vol. 80, no. 3, pp. 153–58, 2019, doi: 10.1016/j.ando.2019.04.010.

[49] L. Bodei *et al.*, "Molecular profiling of neuroendocrine tumours to predict response and toxicity to peptide receptor radionuclide therapy," *Lancet Oncol,* vol. 21, no. 9, pp. e431–e43, 2020, doi: 10.1016/S1470-2045(20)30323-5.

[50] I. M. Modlin *et al.*, "The NETest: The clinical utility of multigene blood analysis in the diagnosis and management of neuroendocrine tumors," *Endocrinol Metab Clin North Am,* vol. 47, no. 3, pp. 485–504, 2018, doi: 10.1016/j.ecl.2018.05.002.

[51] L. Bodei *et al.*, "PRRT neuroendocrine tumor response monitored using circulating transcript analysis: The NETest," *Eur J Nucl Med Mol Imaging,* vol. 47, no. 4, pp. 895–906, 2020, doi: 10.1007/s00259-019-04601-3.

[52] C. Kratochwil *et al.*, "(2)(1)(3)Bi-DOTATOC receptor-targeted alpha-radionuclide therapy induces remission in neuroendocrine tumours refractory to beta radiation: a first-in-human experience," *Eur J Nucl Med Mol Imaging,* vol. 41, no. 11, pp. 2106–19, 2014, doi: 10.1007/s00259-014-2857-9.

[53] M. Ginj *et al.*, "Radiolabeled somatostatin receptor antagonists are preferable to agonists for in vivo peptide receptor targeting of tumors," *Proc Natl Acad Sci U S A,* vol. 103, no. 44, pp. 16436–41, 2006, doi: 10.1073/pnas.0607761103.

[54] D. Reidy-Lagunes *et al.*, "Phase I trial of well-differentiated neuroendocrine tumors (NETs) with radiolabeled somatostatin antagonist (177)Lu-Satoreotide Tetraxetan," *Clin Cancer Res,* vol. 25, no. 23, pp. 6939–47, 2019, doi: 10.1158/1078-0432.CCR-19-1026.

21 Lymphoscintigraphy

Rimma Axelsson, Maria Holstensson and Ulrika Estenberg

CONTENTS

21.1 THE LYMPHATIC SYSTEM

21.1.1 ANATOMY

The lymphatic system consists of lymphatic vessels, lymphatic nodes, and organized lymphoid tissue in different organs, such as tonsils, spleen, and bone marrow. Lymphatic vessels, lymphatic nodes, and lymphoid tissue are distributed throughout the whole body. The lymphatic system is more developed in organs that come into direct contact with the external environment, for example tonsils and skin.

The initial lymphatic vessels are composed of a porous basement membrane with a lining of loosely attached lymphatic endothelial cells. Such a structure permits easy drainage of fluid, macromolecules, colloids, cells, and cellular debris directly from extracellular space into the initial lymphatics. This fluid is called "lymph". In the extremities, the lymphatic system consists of a superficial system that collects lymph from the skin and a deeper system that drains muscle and bone. In the legs these two systems merge within the pelvis, and in the arms, they merge in the axilla.

The initial lymphatic vessels are composed of larger collecting ducts with valves preventing backflow of lymph. The lymph is therefore transported from the initial lymphatics to the collecting ducts and further on to the lymph nodes.

DOI: 10.1201/9780429489501-21

From there the lymph is transported in larger lymphatic vessels through one of two big lymphatic ducts before it enters the cardiovascular system.

21.1.2 Physiology

The lymphatic system has three main functions:

- It maintains the balance of fluid between the blood and tissues. The lymphatic system is sometimes called the "Third Circulation" [1] because of its capacity to return excess interstitial fluid and proteins that are too large to be returned through the blood vessels from the tissues to the cardiovascular system.
- It is a part of the body's immune system. Together with lymph, different foreign materials (for example bacteria or cancer cells) are delivered to the lymph nodes for assessment by immunocompetent cells. The lymphatic system produces white blood cells that are crucial for the body's defence against bacteria and other intruders.
- It plays a key role in absorption of fats and fat-soluble nutrients in the digestive system.

21.1.3 Pathological Conditions

The lymphatic system will be affected when lymph vessels, ducts, nodes, or lymph tissues become blocked, infected, inflamed, or invaded by cancer cells. Scintigraphic investigations of the lymphatic system are applied in patients with suspected lymphedema and for the detection of sentinel node (see section 21.2 "Sentinel Node") in cancer patients.

21.1.4 Lymphedema of the Extremities

The most common pathological condition with obstruction of lymphatic pathways is called lymphedema. When lymph remains in tissues for some time it will cause progressive and chronic swelling, fat deposition, scarring, and immunosuppression [2]. Lymphedema is divided into two categories – primary (as a congenital disorder with absence of or deficiency of lymphatic vessels [3]) and secondary. Secondary lymphedema can be caused by infections, inflammatory diseases, trauma, or cancer treatments. It is a widespread condition, and about 120 million people in developing countries have lymphedema due to a parasitic infection known as lymphatic filariasis [4], and 10 million people in the Western world have it as a complication after previous cancer treatments, such as surgery or radiation therapy. Lymphedema commonly affects one of the arms or legs but can also be bilateral. In rare cases it can be presented by swelling of the genitals or chest. Lymphedema is incurable, but the right treatment can help reduce the swelling and pain. It can be treated conservatively by Complex Decongestive Therapy (CDT) or, in limited cases, by surgery (transplantation of lymph nodes and vessels, by-pass procedure or liposuction). CDT is initially performed by specially trained lymphedema therapists who will train patients to perform the treatment themselves. CDT consists of four components: (1) Manual lymphatic drainage, or a special massage technique that aims to stimulate an increased flow in the remaining lymphatic pathways; (2) compression therapy in the form of multi-layered bandaging; (3) dermatological skin care to prevent skin infection, and (4) physical training, with light exercises aimed at encouraging movement of the lymph fluid out of the limb.

Early diagnosis is essential in preventing progressive lymphedema. However, the diagnosis of lymphedema is not easy and is often missed or confused with other conditions with swollen limbs, such as venous insufficiency or lipedema. Lymphedema diagnosis requires careful attention to the patient's history, risk factors, specific findings on physical examination, and use of different diagnostic techniques.

21.1.5 Diagnostic Radiological Methods

Today MRI, CT, ultrasonography, and lymphoscintigraphy are imaging modalities used for patients with swollen limbs. Radiological methods can demonstrate signs of lymphedema such as edema, skin thickening, fluid accumulation, and honeycomb pattern of the subcutaneous tissue, which are characteristic for the disease [5–7]. While all these are signs of consequences of lymphatic diseases, lymphoscintigraphic imaging provides direct visualization of the lymphatic vessels and physiological information of the lymph flow [8]. Lymphoscintigraphy is simple and is an easily reproducible imaging technique allowing assessment and visualization of the lymph distribution through the lymph vessels and nodes. Apart from the low radiation exposure associated with lymphoscintigraphy (further discussed in section 1.10), there are no known risks or side effects [9].

The principle of lymphoscintigraphy is based on one of the essential functions of the lymphatic system, that is, to transport large molecules from the interstitial space back to the cardiovascular system. Therefore, if a radioactively labelled large molecule or inert colloid particle is injected into the interstitial space, its transport through the initial lymphatics, the lymphatic vessels, and lymph nodes to the main lymphatic ducts and finally to the liver can be followed using imaging with a gamma camera.

Although this technique has been used worldwide since the 1990s, there is still a lack of international guidelines from organizations such as the European Association of Nuclear Medicine (EANM) and the North American Society of Nuclear Medicine. There are considerable variations in the protocols applied for lymphoscintigraphy at different nuclear medicine centres. Differences include the choice of radiotracer, the type and site of injection, the amount of injected activities, the imaging time-points and the interpretation of the images [1]. In a clinical situation, visual interpretation of lymphoscintigraphic imaging is used most frequently [10], even though studies have shown that quantification or application of scoring systems increase the diagnostic validity of this technique [1].

21.1.6 Indications

The main indication for lymphoscintigraphy is to distinguish lymphatic pathology from non-lymphatic causes of limb edema. It can also be used to identify lymphatic vessels and the level of obstruction before surgery.

21.1.7 Radiotracers

Two main types of radiotracers are used: Macromolecules and colloidal suspensions. Macromolecules are cleared faster than colloids and produce better images of the lymph vessels; on the other hand, colloids move more slowly than macromolecules from the injection site. Colloids are trapped more effectively in the lymph nodes which makes them more visible on the scan [11]. Macromolecules that have been used for lymphoscintigraphy are [1] ^{99m}Tc-HSA (human serum albumin), ^{99m}Tc-labelled dextran, and ^{99m}Tc-HIG (human immunoglobulin). The following colloids have been used for lymphoscintigraphy: ^{99m}Tc-antimony sulphide colloid, ^{99m}Tc-sulphur colloid, and ^{99m}Tc-albumin colloid (nanocolloid). The choice of radiotracer is mainly dependent on availability in different countries.

The size of the colloid particles will determine how well the radiotracer is transported from the injection site. Small particles (in the order of a few nanometers) will easily leak into the blood capillaries, whereas particles up to about 100 nm can enter the lymphatics and be transported further to the lymph nodes. Particles larger than 100 nm will be trapped in the interstitial compartment for a relatively long period with a long transit time, which makes them less suitable for clinical use [12]. The optimal colloidal size for lymphoscintigraphy is therefore approximately 50–70 nm and ^{99m}Tc-albumin nanocolloid with a particle size of < 80 nm (95% of particles) is widely used in Europe. It has also been argued that this is more "physiological" than some of the other colloids, as it is similar to the normal gel-like nature of the subcutaneous interstitial space [13].

The choice of radiotracer may also be affected by whether the lymphoscintigram is intended to be qualitative or quantitative in nature. As a majority of clinical studies are of descriptive, qualitative character aiming for visualization of the lymphatic vessels and lymph nodes, colloids will be the best choice. On the other hand, for quantitative studies, applied mostly for research purposes and aiming to measure kinetics of lymphatic flow, a macromolecule that is removed more rapidly from the injection site, may be the better option [14].

21.1.8 Injection

In the normal limb, transport of lymph through the deep system is slower and lesser in quantity compared to in the superficial system [1]. In lymphedema, the clinical pathology is mainly found in the subcutaneous tissues of the limb, making the subcutaneous injection with imaging of the drainage from this area of most interest. Intradermal or intramuscular injection of radiotracers can be used for investigations of the deep lymphatic system of the extremities. Two-compartment lymphoscintigraphy (subcutaneous and intradermal) can be applied in selected cases for the differentiation of various mechanisms of limb edema.

For lymphoscintigraphic investigation of the lower limbs, the subcutaneous injections of the radiotracer are made into the first and second interdigital spaces of the foot. For lymphoscintigraphic investigation of the upper limbs, the subcutaneous injections of the radiotracer are made into the first and second interdigital space of the hand [1].

The rate of uptake of the radiotracer may be influenced by the volume injected, as well as the concentration of the radiotracer. These factors may affect the local tissue hydrostatic and oncotic pressures at the site of injection. The most commonly recommended volume for subcutaneous injection is about 0.2 ml [11, 15].

21.1.9 Injected Activities

Recommended amounts of injected activities lie between ~20–50 MBq per limb and per compartment (intradermal and/or intramuscular) for ^{99m}Tc-nanocolloids [1, 15–17].

21.1.10 Dosimetry

Reported absorbed radiation doses in the literature for lymphoscintigraphic investigations of the limbs vary greatly. The absorbed radiation dose to the injection site and the rest of the body is strongly dependent on the rate of clearance of the radiotracer from the injection site and potential blockages of the lymph flow [18]. As described in section 21.1.7, the size of the colloid particles will determine how well the radiotracer is transported from the injection site. The injection sites receive the highest absorbed radiation doses, followed by the lymph nodes in which a small fraction of the injected radiotracer accumulates [18, 19]. A method to calculate the absorbed dose in lymphoscintigraphy has been described, for example, by Bergqvist *et al* [20].

21.1.11 Imaging Procedure

There are currently no EANM or international guideline recommendations regarding image acquisition protocols for extremity lymphatic scintigraphy. However, the majority of clinics use a dual-detector gamma camera equipped with low-energy high-resolution parallel-hole collimators in whole-body scanning mode. Typically, images are acquired using a 20 per cent energy window centred on the 140 keV photopeak of ^{99m}Tc, using a scan speed of 10 cm/minute. The data should be displayed with the upper windowing level set to display the small fraction of radiotracer that emigrates from the injection site to the lymph nodes (this setting usually causes substantial blooming of the image near the injection site but optimizes the likelihood of seeing the nodes).

Typically, the whole-body scan data are recorded within about 10 min of injection, at 1–2 h, and finally at 4–6 h after radiotracer administration. However, experience from our own clinic show that no additional clinically relevant information is to be gained from scans acquired later than three hours post injection. Figure 21.1 shows an anterior whole-body lymphoscintigraphy acquired 1.5 hours post injection.

Lymphoscintigraphic imaging with the patient lying on the investigating table will only reflect the passive lymphatic flow. Results could be enhanced by applying different factors provoking lymphatic flow – such as changes in temperature, physical exercise, or administration of a pharmacologic agent. When possible, patients with lower limb edema are asked to walk before imaging, and patients with upper limb edema are asked to squeeze a ball with their hands. A marked change in the appearance of lymph nodes or clearance of the radiotracer identifies a response to the intervention.

21.1.12 Interpretation / Evaluation

Qualitative and quantitative imaging procedures, as well as different scoring systems have been suggested for evaluation of the lymphoscintigraphic results during last two decades [7]. Qualitative lymphoscintigraphy shows the morphology of the lymphatic system and quantitative lymphoscintigraphy measures lymphatic flow [21]. The visual qualitative lymphoscintigraphy has been reported to have a very high specificity (about 100%) for diagnosis of dysfunction in the lymphatic system [22–24]. On the other hand, its sensitivity varies from 61 to 97 per cent [22, 25–27].

The reported sensitivity of different studies on quantitative lymphoscintigraphy also varies, from 55 to 100 per cent [19, 28]. These quantitative analyses with graphs showing how the activity varies over time may be more accurate in detection of incipient lymphedema [29]. However, this procedure is time consuming and not commonly used.

In a clinical situation, visual interpretation of lymphoscintigraphic imaging is more frequently used and is of a descriptive nature. Common analyses include visualization of the lymphatic vessels in the investigated limb (3–5 in the calf, 1–2 in the thigh and 1 in the arm); tracer uptake in regional lymphatic nodes (inguinal, pelvic for low limbs and axillary for upper limbs within 1 hour, para-aortic nodes by 2 hours); and the liver (indicating completion of the lymphatic circuit) by around 3 hours. The two most important findings to look out for and comment on are focal tracer accumulation outside vessels and dermal backflow, as these signs are pathognomonic for lymphedema.

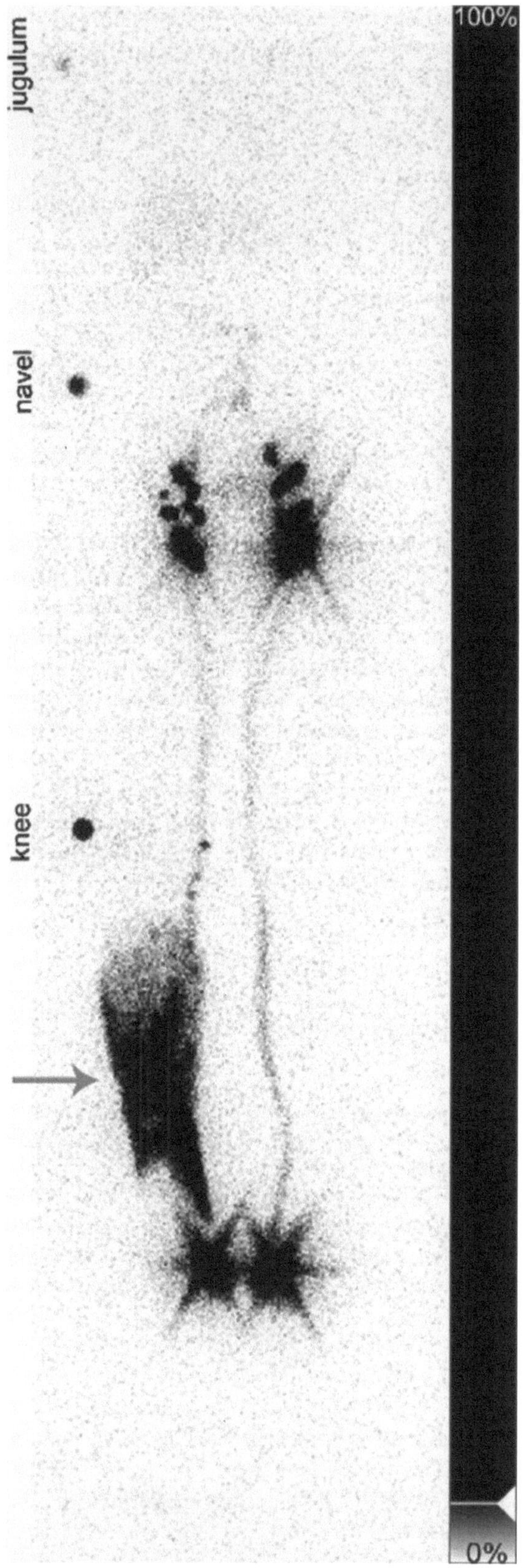

FIGURE 21.1 An anterior whole-body lymphoscintigraphy acquired 1.5 hours post injection in a patient with a swollen lower right limb. Radioactive anatomical markers are seen at the knee, navel and jugulum. The arrow indicates dermal backflow typical for lymphoedema. There is less uptake in the right pelvic lymph nodes as compared to the left side.

In order to increase the diagnostic validity of the visual interpretation of lymphoscintigrams, different scoring systems have been proposed [1, 30], however none have gained wide application in daily practice.

21.2 SENTINEL NODE

In patients with a cancer diagnosis, it is of utmost importance for treatment planning and prognosis to assess the possible spread to regional lymph nodes. Earlier, when conventional radiological methods could not answer this question, patients were often operated upon and as much as possible of the regional lymph nodes were removed. However, metastatic spread to these nodes was found in very few patients (5–40% depending on tumour type), which means that the majority of patients were overtreated. The extensive removal of regional lymph nodes is not only expensive, it is also associated with complications and side effects, such as persistent lymphedema.

The concept of sentinel node was introduced in the last century and gained frequent application for cancer types such as malignant melanoma, breast cancer, and gynecological cancers as well as cancers in the head and neck regions. A sentinel lymph node (SLN) is the first lymph node that drains lymph, and possible metastatic cancer cells, from the primary tumour. The absence of metastatic cancer cells in the SLN, according to this concept, means that other lymph nodes further along the lymphatic system also are free from cancer cells and extensive lymph node removal could be avoided.

There are three routine methods for detection of SLN: By injection of a coloured or fluorescent dye; by injection of a radioactive tracer followed by imaging (scintigraphy); or by injection of a radioactive tracer followed by per-operative identification with a hand-held gamma probe and/or a handheld gamma camera. For procedures with radioactive tracers, different guidelines (depending on tumour type) were published by EANM in cooperation with Society of Nuclear Medicine and Molecular Imaging (SNMMI) and European Society of Surgical Oncology (ESSO). Such guidelines provide general information about the SLN procedure, describe protocols in routine use, and offer assistance in optimizing the diagnostic procedure. The following guidelines are published and can be downloaded from the EANM website: Sentinel node in breast cancer procedural guidelines (2007); EANM-SENT Joint practice guidelines for radionuclide lymphoscintigraphy for sentinel node localization in oral/oropharyngeal squamous cell carcinoma (2009); the EANM and SNMMI practice guideline for lymphoscintigraphy and sentinel node localization in breast cancer (2013); EANM practice guidelines for lymphoscintigraphy and sentinel lymph node biopsy in melanoma (2015).

21.2.1 Radiotracers

There are no special requirements for patient preparation. The most widely used radiotracers are colloid particles: antimony trisulphide (Australia and Canada); ^{99m}Tc-nanocolloid albumin (Europe); sulphur colloid (United States). Another imaging agent recently introduced in clinical practice for imaging with gamma camera is ^{99m}Tc-tilmanocept (Lymphoseek®), which targets receptors on Tcell lymphocytes in lymph nodes.

21.2.2 Injection

The optimal injection technique varies depending on tumour type and local expertise. Common injection techniques include peritumoural, subdermal, intradermal, submucosal, periareolar, and subareolar (these last two are for breast cancer patients) as well as injections at a surgical scar (for patients with malignant melanoma or head and neck tumours). Data from different studies have confirmed that the method of injection does not significantly affect the identification rate of SLNs nor false-negative results.

21.2.3 Injected Activities

The amount of injected activity depends on the time between injection and imaging as well as the time to surgery. Activity levels stated in the guidelines for breast cancer vary from 5–30 MBq per injection site for injections performed on the same day and up to 150 MBq for injections performed the day before the surgery [31, 32]. The guidelines for sentinel node localization in oral/oropharyngeal squamos cell carcinoma and melanoma state that the injected activity should be adjusted for physical decay to ensure that the residual activity at the time of surgery is > 10 MBq [33, 34].

21.2.4 Dosimetry

When radiotracers are administered in an SLN procedure, both patients and staff are exposed to ionizing radiation. The absorbed dose at the injection site can be significant (for example 30–1500 mGy for an injected activity of 30 MBq [31]), however one should keep in mind in an SLN procedure that this region often will be removed during surgery. The effective dose to the patient depends on the radiotracer, the amount of injected activity, the site of injection, and the biodistribution of the radiotracer in the body. Estimates of effective doses to patients from radiotracers in SLN procedures reported in the literature vary from 0.002–0.03 mSv/MBq for breast cancer, and up to 0.002 mSv/MBq for melanoma. In patients who are breast feeding, nursing should be suspended for 24 h following radiotracer administration [31, 33].

The effective doses received by professionals from nuclear medicine, surgery and pathology involved in SLN procedures are below recommended limits for occupational exposures [31]. Studies in SLN procedures on breast and melanoma patients have estimated the maximum effective dose to the surgeon per operation to be $<2\mu Sv$ [34]. A study involving breast cancer SLN procedures reported mean equivalent finger doses per operation of $60\pm40\mu Sv$ for the surgeon performing the SLN biopsy and $120\pm230\mu Sv$ for the surgeon performing the tumour excision [35]. The radiation exposure to pathology staff is lower still, as the radioactivity in the tissue has decayed further by the time they receive it, and they spend a shorter time handling the radioactive tissue [33].

21.2.5 Imaging Procedure

Imaging is recommended before surgery for all types of cancers. Sentinel node detection could be performed as a one- or two-day procedure depending on the surgical facility's occupancy/schedule. SLNs are generally visualized within 1–2h after injection, and the patient should be operated upon within 2–30h after the injection of the tracer. Dynamic imaging is recommended for detection of SLNs in patients with malignant melanoma and in patients with tumours in the head and neck regions. Early (20 min after injection) and delayed (1–3 h after injection) static planar imaging as well as whole-body acquisition protocols are applied (whole-body imaging is essential for patients with malignant melanoma). However, neither planar imaging nor whole-body scanning provide the exact anatomical localization of SLNs. Therefore, in complex anatomical regions such as head and neck, tomographic gamma camera imaging fused with x-ray CT (SPECT/CT) is applied when the SLN is visible in the planar images in order to provide an anatomical road map for surgeons. The CT is usually acquired using a low-dose protocol without the use of a contrast agent. In patients with malignant melanoma or tumours in the head and neck regions, a higher overall SLN detection rate and better detection of SLNs located next to the injection site have been reported on SPECT/CT compared to on planar imaging [36, 37]. Utilizing SPECT/CT not only provides an anatomical localization of SLNs, but has also been shown to identify SLNs and clarify ambiguous SLNs on planar images. The region containing SLN should be marked on the skin using a permanent marker.

21.2.6 Tomographic Imaging

A dual-detector SPECT-system should be used when performing tomographic imaging for SLN detection. The collimators used should be either Low Energy High Resolution or Low Energy Ultra High Resolution. 120–128 projections over 360° should be acquired using a matrix size of 128×128 (with pixels of size 4–5 mm) and an acquisition time of 20–25 s/projection. A low-dose CT can be acquired and used for attenuation correction as well as for anatomical localization [31].

21.2.7 Interpretation

The rule of thumb is that the first visualized lymph node with the highest uptake of the radiotracer, located closest to the injection site, is defined as an SLN. In practice there could be more than one lymph node in a nodal basin. All lymph nodes in the same basin with tracer uptake are considered SLNs and are removed during surgery. This surgical procedure is called a Sentinel Lymph Node Biopsy (SLNB).

The results of imaging should be communicated to the surgeon before operation, and all these SLNs should be removed during surgery. Figure 21.2 shows transverse slices from a SPECT (top image), CT (middle image), and a fused image from a SPECT/CT (lower image) in a patient with tracer uptake in an SLN in the right axilla.

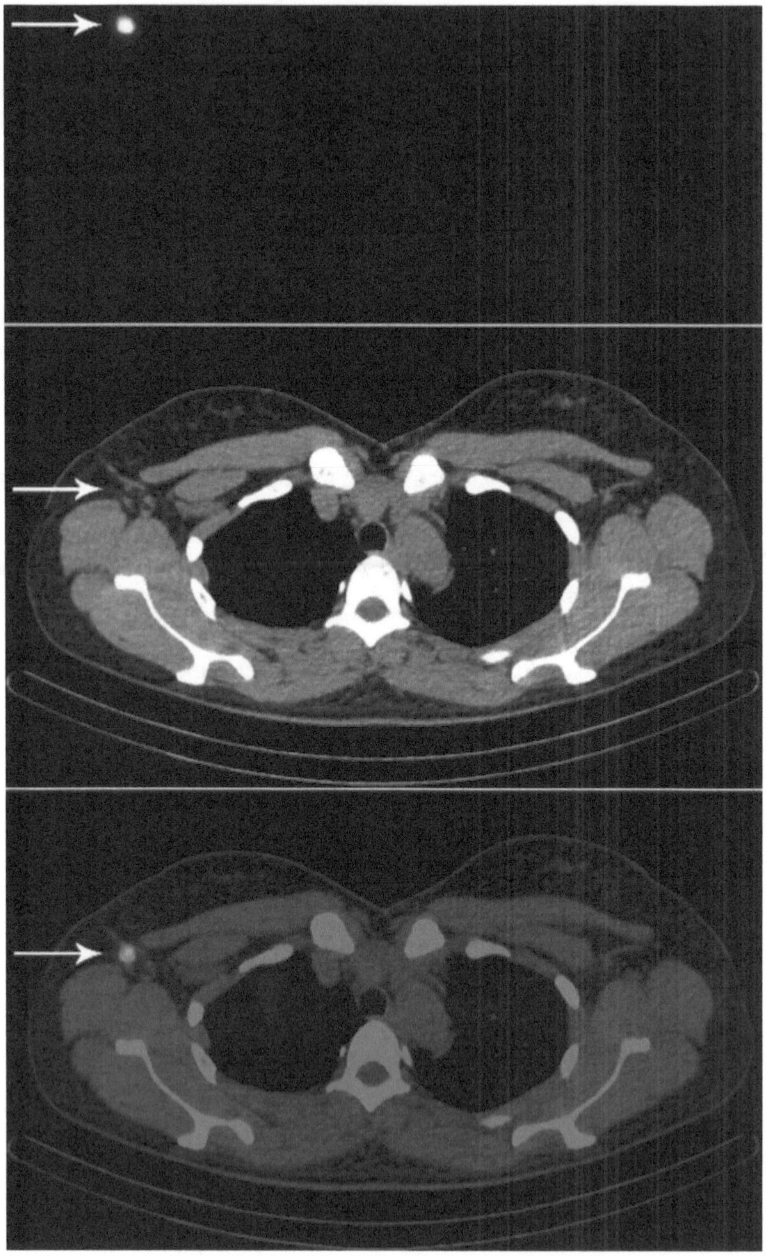

FIGURE 21.2 Transverse slices from SPECT (top image), CT (middle image) and a fused image from a SPECT/CT (lower image) in a patient with cancer in the right breast. Imaging performed 1 hour after tracer injection shows tracer uptake in a SN (see arrow) in the right axilla.

21.2.8 Gamma Probes and Hand-held Imaging Devices

Intra-operative gamma probes are used for the localization and detection of SLN during surgery. A range of probe systems are available on the market with different detector materials, detector sizes, and collimation [38]. As the SLN can be located at various depths within the body, there are certain performance demands on the gamma probes used; *high sensitivity and high signal-to-noise ratio* in order to enable detection of SNL that either have a low radiotracer uptake and/or are located deep within the body; *high spectral resolution* combined with a narrow energy window in order to discriminate scattered photons can enhance the localization of SLN near the injection site; *side shielding* can be added in order to absorb photons from the injection site whilst not decreasing the sensitive area of the probe; *confined angular sensitivity* in order to ensure precise localization of the SLN; *a user-friendly display of count rate proportional to the radioactivity level*, either by an audio signal and/or a display is important in order to more easily localize the SLN; *the ergonomic characteristics* of the gamma probe should be such that it is comfortable and easy to use [39].

The gamma probe procedure can be improved by the addition of a small imaging device that can be used during the surgery. As with conventional gamma cameras, there is a trade-off between *good spatial resolution* and *good sensitivity* [40]. There are a few prototypes of hand-held gamma cameras described in the literature [41]. *CrystalCam* is a commercially available semiconductor (CdZnTe) handheld gamma camera with a 40×40 mm^2 field of view and a pixel pitch of 2.46 mm. The physical size of the handheld unit is 60×60×160 mm^3, the weight is <800 g and the energy resolution is <7 per cent at 140keV [42]. *SurgeoSight* is another commercially available portable gamma camera mounted on a movable mechanical arm. It has a pixelated CsI(Na) scintillation crystal, a 42×42 mm^2 field of view, a pixel size of 1.2 mm and an energy resolution of 20 per cent at 140keV [43, 44].

21.2.9 Radioactive Waste and Personnel Education

During an SLN surgery procedure, certain equipment used in the handling of radioactive tissues and materials such as absorptive surgical sponges might present measurable contamination. The probability of material becoming contaminated is higher in the vicinity of the injection site. It is advisable to monitor surgery equipment and materials for contamination. Although the material presents a negligible contamination hazard, they constitute radioactive clinical waste and need to be handled as such as per institutional radiation safety procedures. The waste will also be a biohazard and should be handled accordingly. Personnel not accustomed to dealing with radioactive materials should undergo education as to safe handling and disposal of radioactive waste. Educating surgical and pathologist staff in radiation safety can also reassure concerned individuals and ensure appropriate processing of radioactive tissues [31, 33, 34].

21.3 SUMMARY

Successful application of the SLNB technique is dependent on good communication between the nuclear medicine department and the surgeons. SLNB has its known advantages (decreased morbidity and operational time/costs) as well as limitations such as non-visualization of SLN and 'skip metastases'. The majority of patients with preoperative lymphoscintigraphic SLN non-visualization will have at least one SLN detected intraoperatively, either by gamma probe alone or by gamma probe combined with blue dye. In approximately 1 to 2 per cent of patients, SLNs will not be detected preoperatively or intraoperatively, and the status of the axillary nodes cannot be determined by this technique. For example, for patients with breast cancer and old age, obesity, tumour location other than the upper outer quadrant, and non-visualization of SLNs on preoperative imaging were associated with failed SLN localization [45]. In these cases, extensive regional lymph nodes dissection will be performed.

Histopathological examination of SLNs is the "gold standard" procedure for the diagnosis of metastases. SLNs are assessed during ongoing surgery with imprint cytology, frozen sectioning, or both, and more thoroughly after the operation. Patients' further management is based on the results of SLNB.

21.4 FUTURE PERSPECTIVES

The standard technique of combined injection of ^{99m}Tc-labelled nanocolloid and blue dye has an SLN identification rate of 96 per cent reported in large meta-analysis of 8,000 patients [46], with a false-negative rate of 7.3 per cent [47]. The dual technique, nevertheless, has shortcomings, including issues with radiotracer availability, dependency on availability of nuclear medicine units, and allergic reactions to dyes. New techniques avoiding the use of radioactive tracers and risk for allergy have been developed to improve the clinical value of a sentinel lymph node biopsy with similar accuracy. Novel techniques studied in recent years include those using indocyanine green (ICG) fluorescence, superparamagnetic iron oxide (SPIO) nanoparticles, and contrast-enhanced ultrasound imaging using microbubbles (CEUS). Recently, a new combined medical device using super paramagnetic iron oxide (SPIO) particles (Sienna +) associated with a hand-held magnetometer (Sentimag) have been developed. This was first tested as an alternative to standard technique in the patients with breast cancer [47]. Among other approaches is a combination of indocyanine green fluorescence (ICG) and contrast-enhanced ultrasound (CEUS) [48]. Application of SLNB in other cancer types such as prostate cancer, gastric tumours, and pancreatic tumours are currently under evaluation.

REFERENCES

[1] A. Szuba, W. S. Shin, H. W. Strauss, and S. Rockson, "The third circulation: radionuclide lymphoscintigraphy in the evaluation of lymphedema," *J Nucl Med,* vol. 44, no. 1, pp. 43–57, 2003.

[2] Y. Saito, H. Nakagami, Y. Kaneda, and R. Morishita, "Lymphedema and therapeutic lymphangiogenesis," *Biomed Res Int,* vol. 2013, p. 804675, 2013, doi: 10.1155/2013/804675.

[3] B. Lee *et al.*, "Diagnosis and treatment of primary lymphedema. Consensus document of the International Union of Phlebology (IUP)-2009," *Int Angiol,* vol. 29, no. 5, pp. 454–70, 2010.

[4] M. R. Jensen, L. Simonsen, T. Karlsmark, and J. Bulow, "Lymphoedema of the lower extremities – background, pathophysiology and diagnostic considerations," *Clin Physiol Funct Imaging,* vol. 30, no. 6, pp. 389–98, 2010, doi: 10.1111/j.1475-097X.2010.00969.x.

[5] S. J. Simonian, C. L. Morgan, L. L. Tretbar, and B. Blondeau, "Differential diagnosis of lymphedema," in *Lymphedema.* London: Springer, 2008, pp. 12–20.

[6] K. Suehiro, N. Morikage, M. Murakami, O. Yamashita, M. Samura, and K. Hamano, "Significance of ultrasound examination of skin and subcutaneous tissue in secondary lower extremity lymphedema," *Ann Vasc Dis,* vol. 6, no. 2, pp. 180–8, 2013, doi: 10.3400/avd.oa.12.00102.

[7] C. L. Witte, "Quality of life," *Lymphology,* vol. 35, no. 2, pp. 44–45, 2002.

[8] F. Baulieu, G. Lorette, J. L. Baulieu, and L. Vaillant, "[Lymphoscintigraphic exploration in the limbs lymphatic disease]," *Presse Med,* vol. 39, no. 12, pp. 1292–304, 2010, doi: 10.1016/j.lpm.2009.11.023.

[9] R. J. Damstra, "Diagnostic and therapeutical aspects of lymphedema," Stichting Lymfologie Centrum Nederland (SLCN). Drachten, NL, 2009.

[10] E. S. Dylke *et al.*, "Reliability of a radiological grading system for dermal backflow in lymphoscintigraphy imaging," *Acad Radiol,* vol. 20, no. 6, pp. 758–63, 2013, doi: 10.1016/j.acra.2013.01.018.

[11] V. Keeley, "The use of lymphoscintigraphy in the management of chronic oedema," *Journal of Lymphoedema,* vol. 1, no. 1, pp. 42–57, 2006.

[12] S. M. Moghimi and B. Bonnemain, "Subcutaneous and intravenous delivery of diagnostic agents to the lymphatic system: applications in lymphoscintigraphy and indirect lymphography," *Adv Drug Deliv Rev,* vol. 37, no. 1–3, pp. 295–312, 1999, doi: 10.1016/s0169-409x(98)00099-4.

[13] P. Bourgeois, O. Leduc, J. P. Belgrado, and A. Leduc, "Scintigraphic investigations of the superficial lymphatic system: quantitative differences between intradermal and subcutaneous injections," *Nucl Med Communs,* vol. 30, no. 4, pp. 270–74, 2009, doi: 10.1097/MNM.0b013e328331bec4d.

[14] E. L. Kramer, "Lymphoscintigraphy: Defining a clinical role," *Lymphat Res Biol,* vol. 2, no. 1, pp. 32–37, 2004, doi: 10.1089/1539685041690454.

[15] G. Villa *et al.*, "Procedural recommendations for lymphoscintigraphy in the diagnosis of peripheral lymphedema: The Genoa Protocol," *Nucl Med Mol Imaging,* vol. 53, no. 1, pp. 47–56, 2019, doi: 10.1007/s13139-018-0565-2.

[16] J. R. Infante *et al.*, "Lymphoscintigraphy for differential diagnosis of peripheral edema: Diagnostic yield of different scintigraphic patterns," *Rev Esp Med Nucl Imagen Mol,* vol. 31, no. 5, pp. 237–42, 2012, doi: 10.1016/j.remn.2011.11.011.

[17] P. Brautigam, W. Vanscheidt, E. Foldi, T. Krause, and E. Moser, "The importance of the subfascial lymphatics in the diagnosis of lower limb edema: investigations with semiquantitative lymphoscintigraphy," *Angiology,* vol. 44, no. 6, pp. 464–70, 1993, doi: 10.1177/000331979304400606.

[18] M. J. Bronskill, "Radiation dose estimates for interstitial radiocolloid lymphoscintigraphy," *Semin Nucl Med,* vol. 13, no. 1, pp. 20–25, 1983, doi: 10.1016/s0001-2998(83)80032-4.

[19] H. Weissleder and R. Weissleder, "Lymphedema: Evaluation of qualitative and quantitative lymphoscintigraphy in 238 patients," *Radiology,* vol. 167, no. 3, pp. 729–35, 1988, doi: 10.1148/radiology.167.3.3363131.

[20] L. Bergqvist, S. E. Strand, B. Persson, L. Hafstrom, and P. E. Jonsson, "Dosimetry in lymphoscintigraphy of Tc-99m antimony sulfide colloid," *J Nucl Med,* vol. 23, no. 8, pp. 698–705, 1982.

[21] T. C. Kalawat, R. K. Chittoria, P. K. Reddy, B. Suneetha, R. Narayan, and P. Ravi, "Role of lymphoscintigraphy in diagnosis and management of patients with leg swelling of unclear etiology," *Indian J Nucl Med,* vol. 27, no. 4, pp. 226–30, 2012, doi: 10.4103/0972-3919.115392.

[22] J. Dabrowski, R. Merkert, and J. Kusmierek, "Optimized lymphoscintigraphy and diagnostics of lymphatic oedema of the lower extremities," *Nucl Med Rev Cent East Eur,* vol. 11, no. 1, pp. 26–29, 2008.

[23] P. Gloviczki *et al.*, "Noninvasive evaluation of the swollen extremity: experiences with 190 lymphoscintigraphic examinations," *J Vasc Surg,* vol. 9, no. 5, pp. 683–89; discussion 690, 1989, doi: 10.1067/mva.1989.vs0090683.

[24] Y. Ogawa and K. Hayashi, "[99mTc-DTPA-HSA lymphoscintigraphy in lymphedema of the lower extremities: diagnostic significance of dynamic study and muscular exercise]," *Kaku Igaku,* vol. 36, no. 1, pp. 31–36, 1999.

[25] G. Larcos and D. R. Foster, "Interpretation of lymphoscintigrams in suspected lymphoedema: contribution of delayed images," *Nucl Med Commun,* vol. 16, no. 8, pp. 683–86, 1995, doi: 10.1097/00006231-199508000-00010.

[26] G. Stewart, J. I. Gaunt, D. N. Croft, and N. L. Browse, "Isotope lymphography: A new method of investigating the role of the lymphatics in chronic limb oedema," *Br J Surg,* vol. 72, no. 11, pp. 906–9, 1985, doi: 10.1002/bjs.1800721120.

[27] S. E. Ter, A. Alavi, C. K. Kim, and G. Merli, "Lymphoscintigraphy. A reliable test for the diagnosis of lymphedema," *Clin Nucl Med,* vol. 18, no. 8, pp. 646–54, 1993, doi: 10.1097/00003072-199308000-00003.
[28] R. A. Cambria, P. Gloviczki, J. M. Naessens, and H. W. Wahner, "Noninvasive evaluation of the lymphatic system with lymphoscintigraphy: a prospective, semiquantitative analysis in 386 extremities," *J Vasc Surg,* vol. 18, no. 5, pp. 773–82, 1993, doi: 10.1067/mva.1993.50510.
[29] R. J. Damstra, M. A. van Steensel, J. H. Boomsma, P. Nelemans, and J. C. Veraart, "Erysipelas as a sign of subclinical primary lymphoedema: A prospective quantitative scintigraphic study of 40 patients with unilateral erysipelas of the leg," *Br J Dermatol,* vol. 158, no. 6, pp. 1210–15, 2008, doi: 10.1111/j.1365-2133.2008.08503.x.
[30] M. Ebrahim, I. Savitcheva, and R. Axelsson, "Reliability of a scoring system for qualitative evaluation of lymphoscintigraphy of the lower extremities," *J Nucl Med Technol,* vol. 45, no. 3, pp. 219–24, 2017, doi: 10.2967/jnmt.116.185710.
[31] F. Giammarile *et al.*, "The EANM and SNMMI practice guideline for lymphoscintigraphy and sentinel node localization in breast cancer," *Eur J Nucl Med Mol Imaging,* vol. 40, no. 12, pp. 1932–47, 2013, doi: 10.1007/s00259-013-2544-2.
[32] J. Buscombe *et al.*, "Sentinel node in breast cancer procedural guidelines," *Eur J Nucl Med Mol Imaging,* vol. 34, no. 12, pp. 2154–59, 2007, doi: 10.1007/s00259-007-0614-z.
[33] C. Bluemel *et al.*, "EANM practice guidelines for lymphoscintigraphy and sentinel lymph node biopsy in melanoma," *Eur J Nucl Med Mol Imaging,* vol. 42, no. 11, pp. 1750–66, 2015, doi: 10.1007/s00259-015-3135-1.
[34] L. W. Alkureishi *et al.*, "Joint practice guidelines for radionuclide lymphoscintigraphy for sentinel node localization in oral/oropharyngeal squamous cell carcinoma," *Eur J Nucl Med Mol Imaging,* vol. 36, no. 11, pp. 1915–36, 2009, doi: 10.1007/s00259-009-1248-0.
[35] W. A. Waddington, M. R. Keshtgar, I. Taylor, S. R. Lakhani, M. D. Short, and P. J. Ell, "Radiation safety of the sentinel lymph node technique in breast cancer," *Eur J Nucl Med,* vol. 27, no. 4, pp. 377–91, 2000, doi: 10.1007/s002590050520.
[36] T. Wagner, J. Buscombe, G. Gnanasegaran, and S. Navalkissoor, "SPECT/CT in sentinel node imaging," *Nucl Med Commun,* vol. 34, no. 3, pp. 191–202, 2013, doi: 10.1097/MNM.0b013e32835c5a24.
[37] A. Bilde *et al.*, "The role of SPECT-CT in the lymphoscintigraphic identification of sentinel nodes in patients with oral cancer," *Acta Otolaryngol,* vol. 126, no. 10, pp. 1096–103, 2006, doi: 10.1080/00016480600794453.
[38] A. J. Britten, "A method to evaluate intra-operative gamma probes for sentinel lymph node localisation," *Eur J Nucl Med,* vol. 26, no. 2, pp. 76–83, 1999, doi: 10.1007/s002590050362.
[39] T. Tiourina, B. Arends, D. Huysmans, H. Rutten, B. Lemaire, and S. Muller, "Evaluation of surgical gamma probes for radioguided sentinel node localisation," *Eur J Nucl Med*, vol. 25, no. 9, pp. 1224–31, 1998, doi: 10.1007/s002590050288.
[40] M. Georgiou, G. Loudos, D. Stratos, P. Papadimitroulas, P. Liakou, and P. Georgoulias, "Optimization of a gamma imaging probe for axillary sentinel lymph mapping," *J Instrum,* vol. 7, no. 09, pp. P09010, 2012, doi: 10.1088/1748-0221/7/09/p09010.
[41] C. Mathelin, S. Salvador, D. Huss, and J. L. Guyonnet, "Precise localization of sentinel lymph nodes and estimation of their depth using a prototype intraoperative mini gamma-camera in patients with breast cancer," *J Nucl Med,* vol. 48, no. 4, pp. 623–29, 2007, doi: 10.2967/jnumed.106.036574.
[42] "Handheld USB-Gamma Camera "CrystalCam"." https://crystal-photonics.com/enu/products/cam-crystalcam--enu.htm
[43] www.pnpmed.com/posts/SURGEOSIGHT
[44] M. Ay *et al.*, "SurgeoSight (TM): An Intraoprative Hand Held Gamma Camera for Precise Localization of Sentine Lymph Nodes," *Eur J Nucl Med Mol I,* vol. 39, pp. S385, 2012.
[45] G. Cheng, S. Kurita, D. A. Torigian, and A. Alavi, "Current status of sentinel lymph-node biopsy in patients with breast cancer," *Eur J Nucl Med Mol Imaging,* vol. 38, no. 3, pp. 562–75, 2011, doi: 10.1007/s00259-010-1577-z.
[46] T. Kim, A. E. Giuliano, and G. H. Lyman, "Lymphatic mapping and sentinel lymph node biopsy in early-stage breast carcinoma: A metaanalysis," *Cancer-Am Cancer Soc,* vol. 106, no. 1, pp. 4–16, 2006, doi: 10.1002/cncr.21568.
[47] A. Zada *et al.*, "Meta-analysis of sentinel lymph node biopsy in breast cancer using the magnetic technique," *Br J Surg,* vol. 103, no. 11, pp. 1409–19, 2016, doi: 10.1002/bjs.10283.
[48] Y. Wang *et al.*, "Variation of sentinel lymphatic channels (SLCs) and sentinel lymph nodes (SLNs) assessed by contrast-enhanced ultrasound (CEUS) in breast cancer patients," *World J Surg Oncol,* vol. 15, no. 1, p. 127, 2017, doi: 10.1186/s12957-017-1195-3.

22 Diagnostic Ultrasound

Tomas Jansson

CONTENTS

22.1 INTRODUCTION

Ultrasound is an established clinical tool, with its first use for diagnosis occurring in the beginning of the 1950s. Today's application areas are seen in more or less all clinical areas, where the more traditional are cardiology (assessment of cardiac function, such as ejection fraction), obstetrics (estimation of gestational age and fetal status), and radiology (status of the liver, spleen, kidney, and various other organs).

Ultrasound is a modality that is real-time, cost-efficient, and portable (nowadays even with a transducer connecting wirelessly to a smartphone or tablet). The real-time functionality makes the image available at the same instant as the transducer touches the patient, and it provides immediate feedback to the operator as to how the anatomy relates to the movement of the operator's hand. This is, however, also one of the drawbacks, the user dependence: How the image is acquired is in the hands of the operator. For this reason, images or image sequences are not stored to the same extent as for other modalities, such as computed tomography, and if so, for documentation purposes. Still, ultrasound is a very useful technique and technology development constantly pushes ultrasound to new applications, such as non-invasive sensing of tissue stiffness (elastography) and developing areas such as measurements of electromechanical activity in the heart, super-resolution imaging, nerve stimulation, photoacoustics, and other approaches for molecular imaging.

This chapter outlines the basic physical background for ultrasound imaging and some recent technical advances, together with some application areas that are relevant for medical physicists.

22.2 ULTRASOUND PHYSICS AND TECHNOLOGY

Ultrasound (i.e., sound at frequencies higher than what the human ear can perceive) is a distinction made by man. As a matter of fact, physically, ultrasound is no different from sound at audible frequencies, being vibrations transmitted in a medium. However, for practical reasons, when we speak of audible sound we mean sound with frequencies in the range of 20–20000 Hz, while lower-frequency sounds are referred to as infrasound, and those with higher, ultrasound.

The basic principle of ultrasound imaging is very simple, indeed. The fact is, that animals such as bats and dolphins have used this technique for millions of years. It is based on what is called the *pulse-echo principle*, whereby a short pulse (or shriek, if you will) is emitted, and the time until the return of an echo, is proportional to the distance to the structure from which the sound is reflected – given that the sound speed is known and constant in the medium.

First of all, one could ask why the animals use ultrasound and not sound in the audible range. The principle actually works just as fine for lower frequency sounds, but the answer lies in the need for the animal to locate the target of their

DOI: 10.1201/9780429489501-22

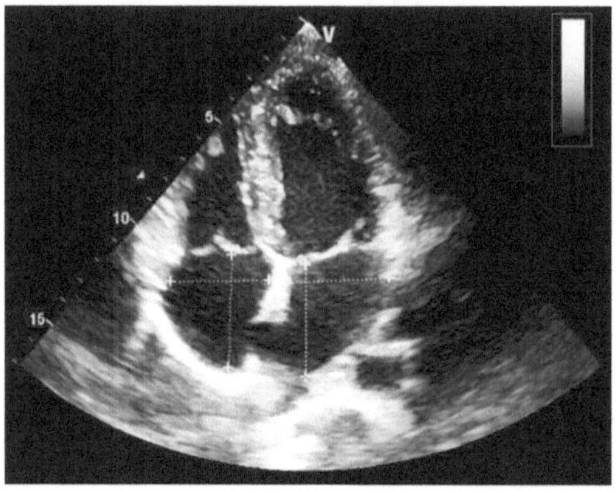

FIGURE 22.1 An ultrasound scan of an adult heart. The four chambers of the heart are clearly visible.

interest more precisely, for instance for food. We have all probably noted that bass speakers are quite insensitive in terms of their location, while tweeters are more directional. This is a result of the fact that with increasing frequency, the number of wavelengths that fit over the speaker membrane also increases. Conversely, if a sound has a low frequency, and thereby long wavelength, the speaker becomes relatively smaller and behaves more like a point source, so that the sound becomes omnidirectional. In other words, dolphins can make the sound beam narrower, to be more like a beam from a flashlight, by increasing the frequency. At the same time a shriek in itself can be made shorter (for a given number of cycles), which facilitates the separation of echoes originating from structures at nearly the same distance from the sound source. To sum up, the higher the frequency, the "sharper" the surrounding can be perceived, as a narrower beam allows better pinpointing from which direction a certain echo comes from and, with higher frequency, shorter pulses can be achieved that are better for distinguishing at what depth the target is. This is precisely the rationale to use ultrasound for diagnostic purposes – to increase the resolution – but here frequencies are even higher, in the MHz range. The practical factor as to why even higher frequencies are not used, is attenuation: Higher frequencies are attenuated more heavily, which sets the upper bounds for the selected frequency. Thereby is also the explanation for the large variety of probes used: For each application the resolution/penetration depth trade-off has been optimized.

This actually sums up very well the basic principles of ultrasound imaging, but there are further differences to ultrasound imaging as compared to how a dolphin perceives the world via its sonar. Consider the ultrasound image in Figure 22.1. The picture shows a sector scan of a heart. The sound is emitted and received at the top of the image, at the sector tip. The sound travels through numerous tissue types and produces echoes at each tissue interface. The dolphin or bat can normally anticipate the medium to have a homogeneous sound speed, but in the body the sound speed varies depending on tissue type. Fortunately, the sound speed does not vary too much, and an average value of 1540 m/s is normally assumed. The sound speed does not vary more than a couple of tens of m/s from this average among tissues. But this also means that the distance as indicated by the ultrasound scanner does have some error margin, and the ultrasound image in Figure 22.1 is actually a bit distorted – not more than a few percent, but it is difficult to deduce where the distortion arises, even though approaches have been proposed to overcome this.

Consider again Figure 22.1. The notion of echoes arising sequentially at numerous tissue interfaces are somewhat counter-intuitive. It would be similar to shouting at a large building, and instead of receiving a single echo (as expected), to also hear echoes from the rooms and corridors inside the building. In practice this is what happens in ultrasound imaging. So, what is actually the difference between these two cases? Let us examine in some more detail what happens when a sound wave is reflected.

To understand this, we start by introducing a model in which we can describe a medium from a relevant acoustical perspective. Any medium in which sound propagates can be modelled as a collection of small[1] volume elements. We think of each equal-sized element as having the mass m and being incompressible. Instead, each element can be thought to be connected to its neighbours by springs, much as in Figure 22.2. Thus, the medium is modelled as a collection of mass elements, described by the density of the medium (mass per unit volume of the medium), and the "springiness," elasticity, of the medium.

Return again to Figure 22.2 and imagine that one strikes a hammer on top of the "medium." The top masses will be displaced downward, compressing the springs beneath them, which in turn will compress their underlying springs, and

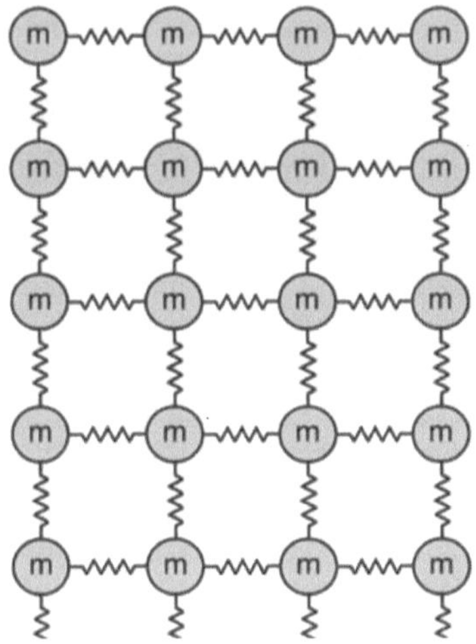

FIGURE 22.2 An acoustical model of a medium, comprising small incompressible masses connected by springs representing the "springiness" (elasticity) of the medium.

so forth. A mechanical disturbance will thus be transmitted in the medium and, after a rather lengthy derivation, it can be shown that the speed of transmission, c, is equal to

$$c = \sqrt{\frac{E}{\rho}} \tag{22.1}$$

where E is the elasticity (Young's modulus) of the medium, and ρ is its density.

The sound speed is thus a characteristic of the material in question, defined by the material's elasticity and density. Now, what happens at a reflection is that the wave propagating in a medium encounters a different medium with other acoustic properties. What determines the magnitude of the reflection is the difference in *acoustic impedance* between the two media. The acoustic impedance can be understood as the resistance for the volume elements in the model to move from their equilibrium position. It can be shown that acoustic impedance equals the sound speed times the density of a medium. Hence, this is also a material-specific parameter that also depends on density and elasticity, but in a different way than sound speed does.

In the body, in the same way as the sound speed varies only slightly between tissues, the same goes for acoustic impedance. As the difference in acoustic impedance is small, only a minute part of the incoming wave will be reflected, and the remaining energy will continue to propagate in the medium and *continue to produce echoes from more distal tissue interfaces*. This is why echoes can be produced from far inside the body (small differences in acoustic impedance), and why we only perceive a single echo from the outer wall of a building, and not its interior: The acoustic impedance of air is very low indeed, as opposed to any solid material, such as glass, concrete or stone. This also explains the need, when using diagnostic ultrasound in the clinic, for coupling gel – the necessary acoustic coupling between the transducer and the body. Air simply has too low an acoustic impedance to be able to bridge the sound from the transducer, that is, all sound energy is totally reflected at the transducer/air interface due to the large mismatch in acoustic impedance.

These are the two main physical principles needed for a basic understanding of ultrasound imaging. Of course, there are many more – for instance, refraction, scattering, and speckle formation – but for a basic understanding, these are the most fundamental. In this context it is of value to briefly discuss the emission and reception of ultrasound, as well as the formation of sound beams narrow enough to distinguish clinically relevant details.

The sound generated for imaging is of a much higher frequency than sound in the audible range, usually several MHz. For such high frequencies, the audio solution of a speaker and microphone does not work. Instead, certain materials are used, materials that possess what is called a piezo-electric property. This means that upon application of mechanical stress, charges inside the material are displaced such that a net electric field occurs and, conversely, the application of an electric field gives rise to a mechanical deformation. Thus, if an electrical impulse is applied over the faces of a disc of a piezoelectric material, the material will undergo a transient mechanical deformation. Depending on the geometry of the disc, the deformation will be larger at certain frequencies – frequencies that correspond to multiples of half wavelengths of the dimensions of the disc (i.e., resonance). For instance, a disc that is a half wavelength thick will vibrate at the frequency that has that wavelength for the piezo material in question. There are also other resonance frequencies depending on the overall dimensions of the piezo material, but an ultrasound transducer is normally designed to allow the thickness mode to produce the dominant frequency.

To produce a two-dimensional ultrasound image, one could transmit multiple beams covering an area of the object to be imaged and map the resulting echoes geometrically onto an image. In the early days of ultrasound imaging, the probe was physically translated, either by an articulated arm that could keep track of the probe position, or later, a probe that mechanically could oscillate back and forth, providing images of a sector.

This is, however, a somewhat inflexible solution, as the focus is fixed. Instead, imaging probes today are not single element probes, but a row of small transducer elements, called an array. Creating a beam relies on Huygen's principle in that an entire group of elements is excited and that the individual wavefronts generated at each element combine to make up a beam, so that just a few wavelengths from the array it is indistinguishable from a beam transmitted from a single transducer of the same size. The advantage is that each element can be controlled individually so that, for instance, the timing of transmission can allow focusing by first firing the edge elements and adjusting the delays on the elements towards the centre corresponding to the pathlength of a curved (focused) transducer. Beams may also be steered sideways if desired. A special case is the so-called phased array transducer in which all elements are used to create one beam (i.e., there are no element subgroups), steered from one side to the other like a fan. This makes all beams appear as if they originated from the centre of the transducer, and the footprint of the transducer can be kept small. This makes it especially useful for cardiac investigations, where sound needs to pass the small space between the ribs.

A real leap in ultrasound imaging came when electronics and computer power also allowed digital focusing on receive, in what is called dynamic focusing. As the echo data from all elements could be stored in memory, the optimum receive focus could be calculated for each depth separately. In this way an optimal focus can be achieved along the entire imaging depth on receive.

Nowadays, a common strategy is to insonify the whole image plane to be interrogated at once, collect echo data from all elements, and after the fact reconstruct image lines by focusing the received data as explained for dynamic focusing. Indeed, the image quality worsens, but on the other hand images can be acquired much faster as one transmission is sufficient for all image lines, as opposed to the conventional line-by-line transmission and detection. The advantage is that very fast events can be captured, such as electromechanical activity in the heart, or other events for which the technique was originally devised, tracking of *shear wave* propagation in tissue for estimation of elasticity (see below).

Three-dimensional (3D) imaging also relies on the same principle. Simply, the approach to transmit one image line at a time is not really feasible in 3D, since the number of lines needed to achieve a satisfactory spatial and temporal resolution will take too long to acquire. If the imaging depth needs to be 15 cm, then the round-trip time at 1540 m/s will be approximately 200 us. In other words, we can acquire 5,000 image lines in one second. If we choose to have 100 lines per image, we can achieve a frame rate of 50 Hz. If we then would like to have, say, 10 images to build up a volume, by translating the probe, for instance, the frame rate drops even further to 5 Hz.

The first 3D implementations were just a conventional transducer that produced a two-dimensional image, then translating that to build an image volume (Figure 22.3). This was done either by keeping track of the probe when the operator moved it, or mechanically angle the probe, and thereby the image plane, back and forth to cover a 3D volume. The mechanical solution is, however, not optimal, and it limits the frame rate. For it to be really effective, the transducer itself would need to consist of, not only a row of elements, but a whole two-dimensional plane of small ultrasound elements. In this way the beam can be steered in any desired angle of the plane. To reach the desired frame rate, a larger portion of the volume needs to be insonified, much like the plane wave imaging method mentioned above, but here image lines are reconstructed in many planes in the volume.

22.3 ADVANCES BEYOND ANATOMICAL IMAGING

In the beginning, ultrasound was used for *anatomical* investigations only, later developing into ultrasound imaging, as described above. Starting in the 1960s however, Doppler techniques developed that made it possible to measure blood flow, that is, a *functional* aspect of the body.

We likely all have experienced the Doppler effect from the change in pitch of the sirens of a passing ambulance. The same principle applies in diagnostic ultrasound, but with one fundamental difference: the sound is emitted from

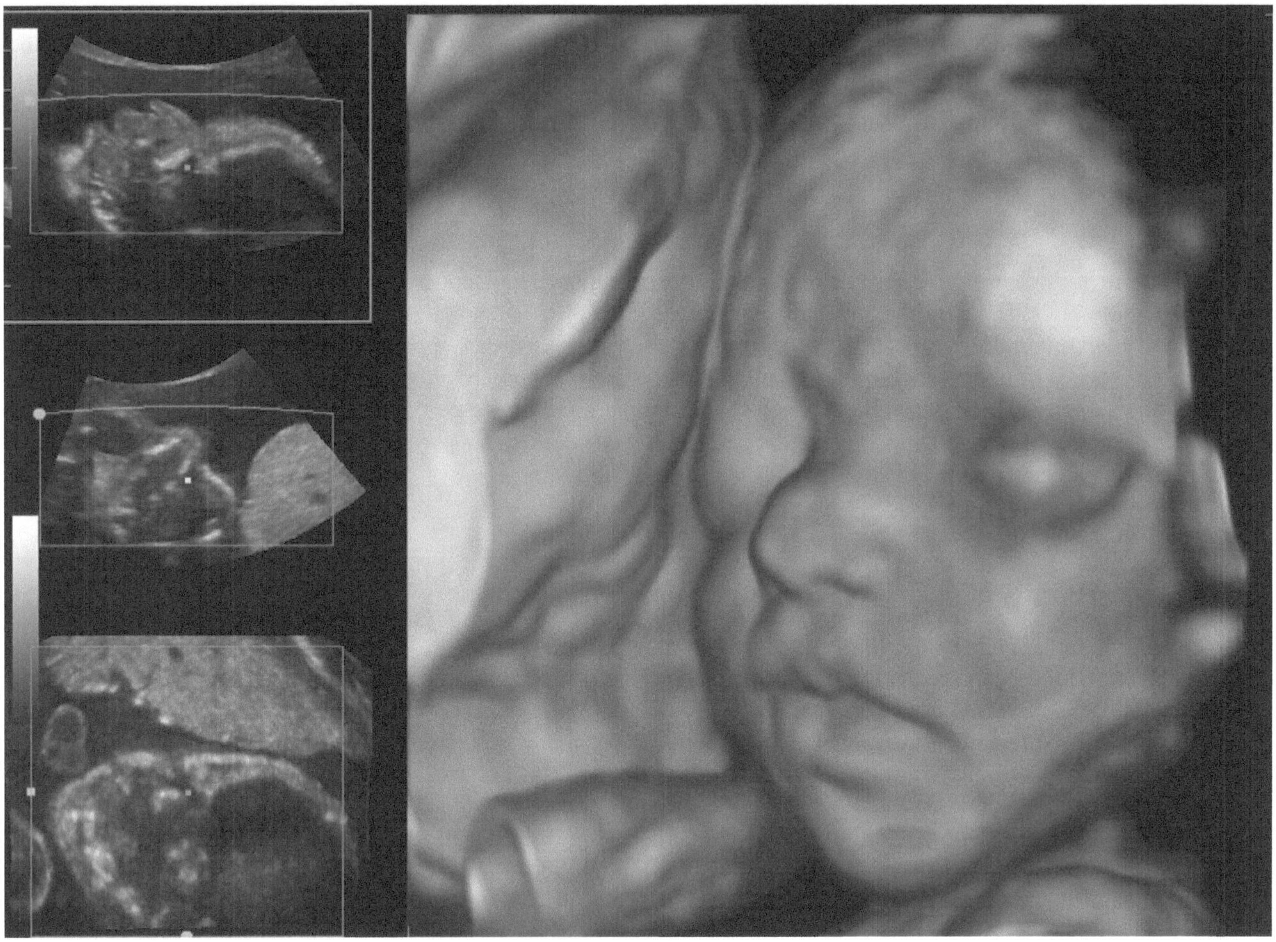

FIGURE 22.3 A three-dimensional ultrasound image of a foetus. To the left, three planes are displayed that are from three orthogonal directions in the 3D-image. Colour image available at www.routledge.com/9781138593312.

a transducer, reflected off a moving object, and received at a stationary receiver. In the case of the ambulance, the source is moving, and the observer is stationary. That has an implication in the detected frequency shift that soon will be noted. The full derivation is somewhat lengthy, and a bit out of scope for this book, so a more heuristic approach is as follows:

First, let us consider the basic nature of the Doppler shift, and again the case of an ambulance siren. As the ambulance moves forward, it will "catch up" with the sound waves it has just emitted and "pull away" from the sound waves emitted backwards. For the purpose of understanding, let us agree that it seems plausible that the compression and elongation of wavelengths is proportional to velocity of the ambulance. Further, let us agree that the ratio of the wavelength change, or conversely, frequency change, Δf, to the frequency of the stationary ambulance siren, f, will be the same as the ratio between the speed of the ambulance v, to the sound speed, c, like this.

$$\frac{\Delta f}{f} = \frac{v}{c} \tag{22.2}$$

In other words, we can express the frequency shift as the original frequency scaled with a factor that is the ratio of the ambulance speed to the sound speed. Of course, we realize that this relation breaks down if the ambulance speed would approach and exceed the sound speed, but it is approximately correct for small velocities. In the body, relevant blood velocities are on the order of m/s, which satisfies a condition for "small velocities."

For the case in ultrasound imaging there is a bit different situation: Both the transmitter and the receiver are stationary, and the sound bounces off an object in motion. The effect is then that the blood corpuscle in motion will perceive a sound that is already Doppler shifted by the amount indicated above. The reflected ("emitted" from the blood cell) sound, will again be Doppler shifted, which is why, in the relation for frequency shift in the diagnostic ultrasound case, a factor of two appears:

$$\Delta f = \frac{2vf}{c} \tag{22.3}$$

There is, however, one last parameter we need to consider in this relation, the angle dependence. From the initial reasoning above it is clear that when the ambulance is approaching us, sound waves are compressed, leading to an apparent increase in pitch. When the ambulance has passed, the pitch goes down by the same amount. The interesting limiting case is the moment precisely *when it is passing*. Then it is neither approaching nor receding, and hence the Doppler shift is zero. In light of this, it may be understood that the Doppler shift frequency describes a cosine function as a function of angle, that is, the angle between the sound path and the velocity vector, θ. The full expression for the Doppler shift, relevant for blood velocities in the body, can then be written as

$$\Delta f = \frac{2vf}{c} \cos \theta \tag{22.4}$$

What has been assumed in this derivation, is that we can speak of one single frequency, that is, that the transmitter emits a tone, where the pitch of the received echo changes with the velocity of the target. This requires one transmitting transducer and a separate receiving transducer. While this is an approach that is widely used for many diagnostic applications, is does carry the drawback that all Doppler shifts that occur within the intersection of the transmit and receive beams will be registered simultaneously. In other words, it is not possible to pinpoint exactly where in this intersection of beams, sample volume, a particular Doppler shift has occurred using this *continuous wave Doppler* approach.

The solution that immediately springs to mind would be to use a short pulse, as is done for imaging, and study the Doppler shift of the returning echo. While theoretically, each frequency component in the pulse undergoes a Doppler shift when the pulse is reflected off a moving target, other effects also occur that undermine this proposition. The most important is frequency-dependent attenuation. A pulse may be described as a sum of frequencies in terms of a Fourier transform. Generally, a shorter pulse means that a wider band of frequencies is needed to describe the time signal. Now, as attenuation is frequency-dependent, with higher frequencies being more attenuated, the pulse will undergo an apparent frequency shift towards a lower band. This shift could be interpreted as a Doppler shift and, in fact, in most cases the frequency shift due to attenuation is higher than that of Doppler shifts from moving blood.

So, interrogating the frequency content of the echo from a single pulse is not feasible since, although the Doppler shift is there, it is masked by other frequency-dependent effects. Instead, the *pulsed wave Doppler* was proposed to make use of a series of transmitted pulses. If, for instance, a reflector is moving towards the transducer, echoes from the reflector will appear closer and closer. By noting the phase of the returning echo, we can calculate how fast the target is moving. Actually, this analysis of how the phase changes gives mathematically the same result for the Doppler shift as the relation derived above for the continuous wave case, but with the frequency, *f*, substituted for the centre frequency of the transmitted pulse.

There is one important limitation with pulsed Doppler that is important to be aware of. Say that the velocity of the target is so high that the phase change is greater than 180 degrees. Then it becomes ambiguous whether the target has moved away or towards the transducer. This is a phenomenon known as aliasing, which occurs if the velocity exceeds a speed that would produce a Doppler shift exceeding half the pulse-repetition frequency, see Figure 22.4. This is completely analogous with the Nyquist frequency limit.

Pulsed Doppler may be applied so as to analyse the velocity distribution within the sample volume as illustrated in Figure 22.4 (note the aliasing effect, velocities exceeding the scale reappear as negative velocities). In Figure 22.4 can also be seen an example of colour Doppler imaging. This is basically also a pulsed Doppler system, but here the entire frequency content is not analysed, but only the average velocity at an image point is calculated and displayed in a colour code. As this is also based on pulsed Doppler measurements, colour Doppler also suffers from aliasing. Various other variants of colour Doppler exist, such as power Doppler, displaying Doppler signal energy. Elastography, even though implemented a bit differently, is also a modality that relies on motion estimation, but displaying tissue stiffness.

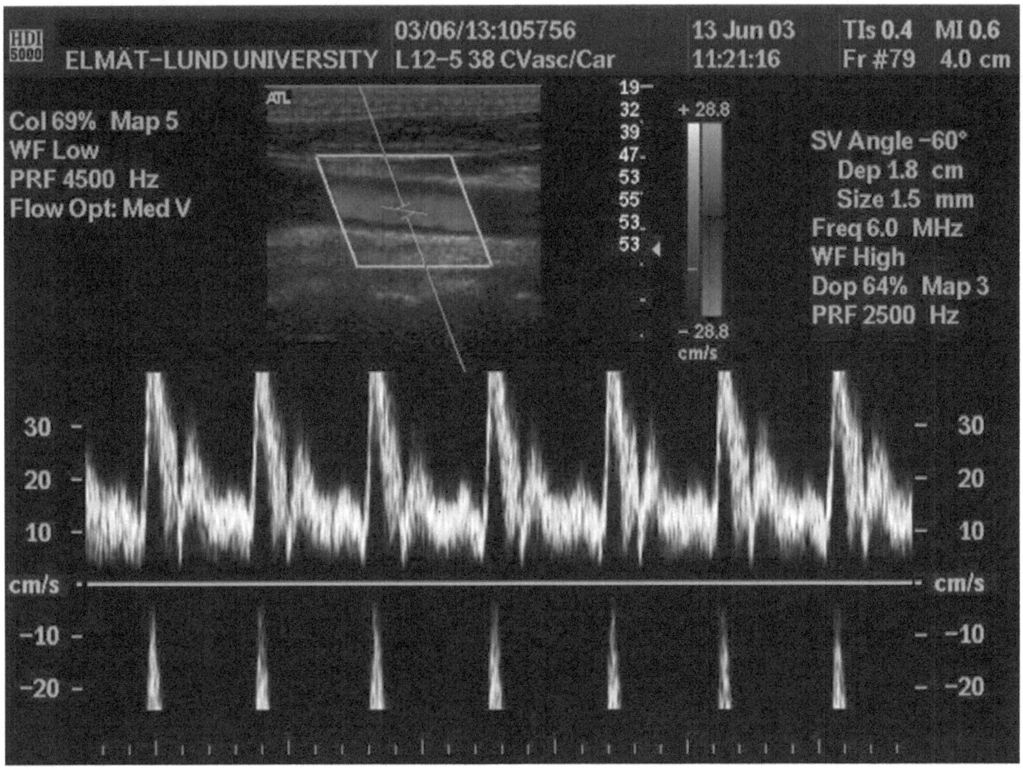

FIGURE 22.4 A screen shot from an ultrasound scanner in which both colour Doppler and pulsed Doppler modes have been activated. The top greyscale image with the red overlay shows a cross-section of a carotid artery. The red overlay is the colour Doppler representation of motion within the green box. The red colour can be compared to the colour bar to the right, where colour represents velocity. At the same time, a sample volume has been placed in the colour box in which velocities are analysed and displayed in the sonogram at the bottom. Here, time is on the horizontal axis and velocity on the vertical. Note how the peaks of the blood velocity waveform have been chopped off and reappear as negative velocities. This is a case of aliasing. Colour image available at www.routledge.com/9781138593312.

22.4 CONTRAST AGENTS

Contrast agents for ultrasound were developed to improve cardiac investigations, and the development took off in the 1990s, when intravenous injections were made possible. Cardiac applications remain a large application area with myocardial wall delineation and enhancement of Doppler signals, but also analysis of intracavitary thrombi or left ventricular aneurysms are examples of use. Liver lesions have become an important application field as well. Thanks to the dual blood supply to the liver, from the portal vein collecting blood from the gastrointestinal tract, and the hepatic artery, lesions show different dynamics in response to an intravenous bolus of contrast agent. The delay of the blood passing, for example, the intestines, means that lesions that feed mainly on arterial blood, as for instance, tumour tissue, show up in an early phase and can thereby be distinguished. Other application areas are, for instance, kidney or thyroid imaging.

Contrast agents in general are for increasing the imaging signal to identify the presence of blood or its absence, for example, to delineate vessel or heart walls. For ultrasound we thus need a substance that is *echogenic* (prone to produce echoes). Revisiting the first section, we remember that what caused a large reflection of sound energy was the difference in acoustic impedance between the tissue and the contrast agent. In that sense, heavy particles that exhibit a large density difference compared to tissue, could be possible candidates. These might, however, have other drawbacks such as clogging vessels. But gas bubbles, on the other hand, actually differ more in both density and sound speed (in fact the determining factor is the elasticity, or *compressibility*, which is a more proper term when speaking of fluids). The first contrast agents were neither stable nor small enough to be able to cross the lung capillary bed and therefore could only be used to identify, for example, shunts in the venous and arterial circulations.

The real enabler for the use of ultrasound contrast agents was the ability to stabilize bubbles and make them small enough to pass the pulmonary circulation. The stabilizing factor was the encapsulating of bubbles in a shell. Actually, the term "shell" is a bit misleading as it brings to mind something harder, like an eggshell: It is more proper to think of it as a film, like a soap bubble. What was exploited was also the fortunate coincidence that bubbles in a fluid can be seen as a mechanical system that has a resonance, and thereby a resonance frequency, f_0. This can, although a bit simplified, be written

$$f_0 = \frac{1}{2\pi r}\sqrt{\frac{3\gamma P_0}{\rho}}, \tag{22.5}$$

where r is the radius of the bubble, γ is the ratio of specific heats for the gas in question (being 1.4 for air), P_0 is the hydrostatic pressure, and ρ is the density of the surrounding fluid. Compare this expression with the expression for sound speed or, for that matter, a simple spring-mass system! Can you identify the parameter that corresponds to elasticity (spring constant)?

If we again look at the equation describing the resonance frequency of a bubble (sometimes referred to as the Minnaert frequency), we can think that for a given measurement situation where we imagine that we can change the radius of the bubble, all other parameters will be constant. In other words, the resonance frequency only depends on the size of the bubble. If we multiply both sides of the equation by r, we see that the product f_0 r, is equal to a constant number. For atmospheric conditions in water, this number turns out to be 3.26 [Hz m]. A bubble with a 1 m radius will thus have a resonance frequency of 3.26 Hz. In the same way a 1 μm radius bubble will have a resonance frequency of 3.26 MHz, which happens to fall right in the frequency range for diagnostic ultrasound. So, a bubble that is small enough to pass through the pulmonary circulation, thereby allowing intravenous injections, happens to be extra efficient as a sound scatterer at employed ultrasound frequencies – a most fortunate coincidence indeed.

Soon it was also realized that the non-linear behaviour of bubbles could be exploited for imaging purposes. In short, for a sufficiently large acoustic pressure, a bubble tends to become stiffer as it is compressed, thereby resisting further compression. On the other hand, it can readily expand and increase its radius by several times. Consequently, a large enough sinusoidal pressure wave will create an asymmetric compression and expansion of the bubble, a motion that translates to the scattered pressure wave from the bubble. An asymmetric waveform will mean that it contains harmonics, multiples of the fundamental frequency.

The first really contrast-specific imaging method that was employed was thus harmonic imaging. A pulse with a given frequency was transmitted, tuned to the expected resonance frequency of the bubbles, and the received signal was filtered to suppress all other frequencies but the harmonic. By this method it was for the first time possible to study myocardial perfusion.

Harmonic imaging, however, has the drawback that for efficient filtration of the harmonics, the signal needs to be relatively narrowband, that is, longer in the time domain. As discussed in the first section, this has a detrimental effect on the spatial resolution, because for a better spatial resolution, we need shorter pulses. The solution is as simple as it is elegant: A temporal cancellation technique. This means that first a pulse is sent out, and the echoes are stored in a memory. Then, a second pulse is transmitted, identical to the first, but with one important difference – it is inverted compared to the first. When adding the echoes from the two pulses, linear echoes (such as those arising from tissue) will cancel out, whereas non-linear or asymmetric echoes, such as those arising from bubbles, will add up. Effectively, this is a technique whereby odd harmonics are suppressed while even harmonics are added.

Contrast imaging today comprises variants on this theme with various pulse sequences that are used in different ways to enhance the contrast signal above the tissue background. Output energy is also tuned to maximize the non-linear effect, while not bursting bubbles.

As a concluding remark on bubbles, the disruption of bubbles by exposing them to high acoustic energy can be utilized to burst all bubbles in the image plane, and then, under the following seconds produce images with a low energy contrast specific modality to study the rate at which bubbles, and thereby blood, re-enter the image slice. The associated time constant and the relative number of bubbles in the slice, give a measure of perfusion. This is used to estimate myocardial or renal perfusion – for instance, a technique similar to what is employed in radiation-based methods (see Chapter 15).

22.5 SAFETY

Ultrasound is considered a safe imaging modality due to its non-ionizing nature. Still, it does rely on energy being transmitted into the body, and as such there are biophysical effects that one needs to be aware of. There are mainly two mechanisms by which ultrasound may affect tissue: First, as the sound wave propagates, energy dissipates, which leads to heat generation. Second, the high pressure excursions that are employed in diagnostic ultrasound may be of concern, and especially so, the large negative pressures. These may lead to cavitation, the formation of vapor-filled cavities in the liquid that, upon collapsing in the subsequent high pressure, may cause disruption of tissues.

There are two classes of indices that usually are displayed on the screen of the ultrasound equipment – the thermal index (TI), and the mechanical index (MI), both of which give an indication to the operator information on the risk-benefit-ratio for the current imaging settings for the two effects mentioned above.

The thermal index describes the likelihood of heating the tissue above a considered safe level in tissue. The heating is a result of internal friction losses in the medium, stemming from the oscillating motion of the propagating wave. The index is defined as the acoustic power at a point in tissue, divided by the power at the same point, which would raise the temperature by 1 degree. How much heat is generated is, however, a function of the tissue characteristics, and three clinically useful cases have been identified, defining one thermal index for each case. First, there is a case where only soft tissue is present, Tis. Second, where bone is at the ultrasound focus, Tib, which is especially important in foetal examinations and, third, when bone is near the transducer face, Tic, such as for cranial examinations.

The longer the pulses that are used, the more energy is transmitted. Normally, for imaging purposes, short pulses are used, not much exceeding a wavelength. For Doppler measurements, however, the pulses are usually longer (including more wavelengths) and, thereby, the risk of adverse heating is higher using Doppler. Also, more energy will be delivered to one point if using a scanning mode that is aimed at interrogating just one location in tissue, such as pulsed Doppler, for instance, while for imaging or colour Doppler energy is distributed spatially as the beam is translated over the interrogated region. This has led to further classes of the thermal index depending on scanning mode.

The answer to the question of how large value of the TI can be accepted is unfortunately not that clear, but more like, "it depends." The British Medical Ultrasound Society has formulated guidelines stating that for foetal examinations maximum exposure times can be 30 minutes at MI=1.0, while for MI=2.0 the total exposure time should be kept below 4 minutes. In other words, it is up to the operator to use these numbers as indications only and to be aware of potential risks.

The mechanical index, MI, is intended to indicate the risk for cavitation (or more precisely inertial cavitation). Basically, the risk increases with the amount of under-pressure applied, that is the peak negative pressure excursion of the ultrasound pulse. Also, the time during which this under-pressure acts also plays a role: Lower frequencies have longer period times, and the chance of cavitation is thus larger than for a higher frequency. Empirically, the following relation between peak negative pressure p_- (in MPa) and employed frequency f (in MHz), has been found to reflect the chance of cavitation, and thus serves as the definition of the mechanical index:

$$MI = \frac{p_-}{\sqrt{f}} \tag{22.6}$$

Ultrasound contrast agents, that consist of micro bubbles, may serve as cavitation nuclei, so when using contrast agents, the MI is more relevant. The MI is even more of an "it depends" index, but at least the index should at all times be kept below 1.9.

22.6 ARTEFACTS

As mentioned, ultrasound imaging relies on the assumption that the sound speed is constant in tissue. And as we have seen, it is not. In fact, the sound speed needs to differ, otherwise echoes would not be produced at tissue interfaces. This is however usually not that large of a problem, since sound speed differences are relatively small between tissue types. There are, however, other assumptions that are made when producing an ultrasound image – assumptions that are not always fulfilled in real life.

The first assumption is that an ultrasound beam is considered infinitely thin, propagating along a straight line. Of course, it is not, and it is the width of the beam that determines the smallest size an object can have in an ultrasound image (lateral resolution). The straight-line assumption has a number of consequences.

First, the beam may be deflected at tissue interfaces, such as bone, the diaphragm, or other gas pockets. In essence, the structure acts as a mirror and may erroneously put image objects at positions in which they are not. A variant of that theme is where the beam bounces off a strong reflector, back to the transducer, reflects there, and then produces a second echo that causes the travel time to be twice that of the first echo. Consequently, this produces a second image of the first reflector, but at precisely twice the distance. The way to spot these false echoes are that upon movement of the probe, false echoes move twice as fast due to the doubled path length.

Second, the beam in itself is not just one beam, but also contains side lobes. These are beams of sound energy that propagate in a different direction from the main beam, although with lower energy than the main beam. If a strong reflector appears in the sidelobe direction, it will give off an echo that will be interpreted as coming from the main direction. This may also be a source of noise, in the case of scanning a fluid-filled void, such as a cyst. Clear fluid does not produce echoes, but the sidelobes may pick up strong reflections from beside the cyst, and false echoes may appear within the cyst.

Third, certain structures like biopsy needles or gas bubbles, may produce the effect of a "comet tail" in which sound bounces back and forth within the structure, producing a constant stream of echoes imaged as if stretching out behind the target, hence the name comet tail.

Aside from the assumption of an infinitesimally thin, straight beam, the assumption is also made that the sound speed is constant at an agreed average value of 1540 m/s. This may give rise to distortions in the image, and also distort the focus of the beam, producing blurry images.

The user also sets the gain, and more specifically the gain with increasing distance. Thus, if a region where the attenuation is low, again as in the case of a fluid-filled cyst, the tissue more distal to the cyst may appear brighter, a phenomenon called enhancement. Likewise, bone, or other more compact structures, may cause shadowing.

22.7 DEVELOPING TECHNIQUES

This section deals with newer techniques, such as non-invasive sensing of tissue stiffness (elastography) and developing areas such as measurements of electromechanical activity in the heart, super-resolution imaging, nerve stimulation, photoacoustics, and other approaches for molecular imaging.

A common trait of emerging techniques for medical ultrasound is that even though the pulse echo technique still forms the basis, other means of stimulation are usually employed, and the following response is detected with ultrasound. This can be done in several ways but, for instance, approaches have been to push on tissue, shine light on it, or pull with a magnet.

The first approach, to push on tissue, was in the field of elastography where the objective is to estimate the stiffness of tissue. The rationale being that cancerous tissue normally is stiffer than normal tissue. By tracking echoes through a series of image frames as the probe is pressed against the skin, the displacement of underlying tissue structures is tracked. A hard or stiff region would tend to compress less than surrounding tissue, so a comparison of echoes before and after the compression can provide an estimate of elasticity.

Later a more elaborate way to push on tissue was developed. Here, the sound itself executes the push, by the effect known as acoustic radiation force. This is the result of the transfer of momentum from a sound wave in a medium, where the attenuation causes pressure and particle velocity to become out of phase. The result is a force in the propagation direction, which in a fluid is expressed as a flow (acoustic streaming), while in tissue it will be experienced as a displacement. Now, this displacement for a given sound intensity is dependent on the tissue elasticity. For example, the displacement in tumorous tissue would be smaller. Pulses used for imaging purposes do create a push as well, but it is too small to be used. Instead, a longer pulse, or even from a separate focused transducer is used.

Either the displacement itself can be measured, or the shear wave that naturally occurs as a result of the displacement. Consider the small tissue volume at the focus of the ultrasound beam being pushed in the sound propagation direction. When the wave has passed, the force is no longer present, and the tissue will recoil back to its original position. However, as tissue is cohesive with cells connecting to each other, tissue adjacent to the focus region will also be pulled along, but with a time delay. And tissue adjacent to that region, and so forth, followed by the recoil. This leads to a different wave type, shear wave, where particles displace *orthogonally* to the propagation direction, as opposed to the imaging pulse, which is *compressional*. It turns out that the elasticity is proportional to the square root of the shear wave speed, and by tracking the shear wave speed, a quantitative measure of elasticity may be obtained.

The shear wave has a significantly lower value than the average 1540 m/s for compressional waves – only a couple of m/s. Still, to be able to track the shear wave, a pulse-repetition frequency in the kHz range is necessary. The first implementations tracked the shear wave as it passed a single image line, but to cover the entire image plane, the plane wave imaging method was developed, as mentioned and described above. In this way, elasticity could be quantitatively mapped across the image.

The plane wave imaging method has found numerous other applications, as it allows tissue motions of a general nature, not only shear waves, to be imaged and measured. An interesting example is the tracking of electromechanical activity in the heart, the resulting tissue displacements, and minute contractions that result from electrical activation of the heart muscle. Development is under way to be able to diagnose conduction defects in the heart.

Optical irradiation of tissue is another developing field where ultrasound can be used to map out optical absorption. Conceptually, we might think of this as if we can "hear" light interacting with tissue. The absorption of a short (5–10 ns) laser light pulse in tissue causes some of the delivered energy to convert to heat, leading to a transient thermoelastic expansion. This sudden expansion causes a wide-band acoustic wave, which can be detected by an ultrasound transducer. In other words, we obtain strong ultrasound signals from strong absorbing targets, such as blood. Interestingly, the absorption differs between oxygenated blood and deoxygenated, so we can obtain oxygenation maps of tissue. Nanoparticles that serve as highly absorbing targets may be used as contrast agents. The method does have a limited penetration depth (few mms) due to the high absorption of light in tissue, but approaches to overcome this are investigated.

So, we have covered the pushing and shining light on tissue, followed by ultrasound detection. Another approach is to expose the tissue to a magnetic field and study the result. Tissue in itself is not magnetic, but a magnetic contrast agent, such as superparamagnetic iron oxide nanoparticles may be used. By exposing them to a magnetic field, they will move towards the magnet. However, nanoparticles are too small to produce any useful echo in themselves, but when endocytosed (taken up by cells) or otherwise bound in tissue, they will drag tissue along with them and create a motion signal that can be detected with ultrasound. A time-varying magnetic field will induce a vibration with the same frequency in tissue that is used a signature of where the particles are located.

As can be seen, future developments surely involve other combinations of physical phenomena to extract information of disease states, or even for therapy. Methodologies are also shared, as for instance the Nobel Prize–winning achievement of high-resolution microscopy, which is employed to obtain highly detailed flow maps using ultrasound contrast-agent tracking. Other multimodal combinations are bound to evolve, whereby modalities as we have learned to know them may become more and more integrated. A basic knowledge of other imaging methods is therefore a key to understanding the development in medical imaging, for which this text hopefully has served as a helpful introduction to diagnostic ultrasound.

NOTE

1 Small could mean arbitrarily small but, in any case, not larger than a tenth of the wavelength.

22.8 FURTHER READING

- Peter R. Hoskins, Kevin Martin, and Abigail Thrush, Eds. *Diagnostic Ultrasound- Physics and Equipment,* 3rd ed., CRC Press, 2019.
- Richard S. C. Cobbold, *Foundations of Biomedical Ultrasound.* Oxford University Press, 2007.
- Thomas L. Szabo, 2nd ed., *Diagnostic Ultrasound Imaging: Inside Out.* Academic Press, 2014.
- Jørgen Arendt Jensen, *Estimation of Blood Velocities Using Ultrasound: A Signal Processing Approach.* Cambridge University Press, 1996.

23 Clinical Trials

Purpose and Procedures

Anna Sundlöv

CONTENTS

23.1 WHY THE NEED FOR CLINICAL TRIALS?

Clinical trials are performed to obtain scientifically sound and robust data on the safety and efficacy of an intervention, while at the same time respecting the rights and guaranteeing the safety and well-being, of the trial subjects. The prospective nature of a clinical trial reduces the bias inherent to retrospective data collection and permits randomization to different interventions to be able to compare them to each other. Clinical trials are also the basis for obtaining marketing authorization for a new drug, that is, exclusive rights to commercialize a new drug within the geographic area of the relevant regulatory authority – the EMA (The European Medicines Agency. www.ema.europa.eu) for the European Union, the FDA (The U.S. Food & Drug Administration. www.fda.gov) for the United States, and so forth. In order to guarantee the scientific and ethical soundness of a clinical trial, as well as the quality of the data resulting from it, there is a whole system consisting of clinical trials offices, clinical research organizations, regulatory authorities, ethics committees, and others, dedicated to just that. In this chapter, you will get an overview of this system as well as some useful references for further reading.

23.2 WHAT IS A CLINICAL TRIAL?

In the European Clinical Trial Regulation (EU No 536/2014), the wider term "clinical study" is used to identify any research performed on human subjects with the intention to gain an understanding of the effect of a drug/intervention from a clinical, pharmacological, and/or pharmacodynamic perspective. The overall aim of any clinical study is to ascertain the safety and/or efficacy of a medicinal product. According to the same document, a clinical *trial* is a prospective clinical study where the treatment under study is one that is not considered normal clinical practice and the decision to include a subject in the study is directly related to the decision to offer them the investigational treatment. Furthermore, the safety and efficacy of the treatment is evaluated using procedures that are more elaborate than what would be considered standard clinical practice. The term "clinical trial" thus more specifically describes a prospectively conducted medical interventional experiment, while "clinical study" may include, for example, retrospective and observational approaches as well.

DOI: 10.1201/9780429489501-23

23.3 A SHORT HISTORY OF CLINICAL TRIALS AND RESEARCH ETHICS

When publishing results of clinical research in peer-reviewed journals, it is a requirement to declare that the research has been conducted in accordance with the Declaration of Helsinki and, in the case of clinical trials also according to ICH GCP (The International Council for Harmonisation of Technical Requirements for Pharmaceuticals for Human Use, Good Clinical Practice). This includes having the approval of an independent ethics committee. What does that mean? What is it we have to adhere to when we do research with human subjects? Before describing the current regulatory environment for clinical trials, let us take a look at the historical events that led to its development.

The first known written record of a clinical trial is from 1753, when the British surgeon James Lind published the results of a trial of different treatments for scurvy (vitamin C deficiency) [1]. He had isolated 12 sailors with scurvy on a ship and divided them into six pairs. Each pair received different treatments, ranging from "a quart of cyder a day," to "a course of sea water," "two spoonfuls of vinegar three times a day" or "two oranges and one lemon given them once a day." He reports the results in clear terms: "[T]he most sudden and visible good effects were perceived from the use of the oranges and lemons; one of those who had taken them, being at the end of six days fit for duty."

Clinical trial conduct was largely unregulated from the times of James Lind to the mid-twentieth century. The inhumane experiments conducted in the Nazi concentration camps during World War II were the immediate reason for the development of international ethical standards guiding future research in human subjects in what was called the Nüremburg Code (1947), which was replaced in 1964 by the more widely accepted Declaration of Helsinki. The latter has been revised on several occasions since its conception, until achieving its current form [2]. The Nazis were, however, far from alone in conducting what we would nowadays consider ethically unacceptable experiments, nor has medical research in human subjects been perfectly conducted since the conception of the Declaration of Helsinki. Recent history offers examples of clinical research with questionable ethics and/or scientific soundness. Some of the more well-known examples are the Tuskegee Syphilis Study, the Brooklyn Jewish Chronic Disease Hospital case, and the placebo-controlled HIV trials in developing countries in the 1980s [3].

The ethical standards set down by the Declaration of Helsinki are intended to guarantee the safety and well-being of trial subjects. This is further strengthened by the much more detailed and procedure-oriented document known as the ICH guideline for Good Clinical Practice. In its introduction, it is described as "an international ethical and scientific quality standard for designing, conducting, recording, and reporting trials that involve the participation of human subjects." It contains the instructions for the roles and responsibilities of those involved in doing clinical research, what the clinical trial protocol (CTP) and investigator brochure should contain and other necessary documentation to maintain before, during, and after a clinical trial. It forms the basis for many of the standard operating procedures in organizations specialized in conducting clinical trials and is thus a highly relevant document in everyday clinical research. To act as principal investigator in a clinical trial, sponsors and ethics committees require a certificate of recent participation in a course on GCP to ensure up-to-date knowledge on its content. The following list describes the 13 basic principles of GCP.

(1) Clinical trials should be conducted in accordance with the ethical principles that have their origin in the Declaration of Helsinki, and that are consistent with GCP and the applicable regulatory requirements.
(2) Before a trial is initiated, foreseeable risks and inconveniences should be weighed against the anticipated benefit for the individual trial subject and society. A trial should be initiated and continued only if the anticipated benefits justify the risks.
(3) The rights, safety and well-being of the trial subjects are the most important considerations and should prevail over interests of science and society.
(4) The available nonclinical and clinical information on an investigational product should be adequate to support the proposed clinical trial.
(5) Clinical trials should be scientifically sound, and described in a clear, detailed protocol.
(6) A trial should be conducted in compliance with the protocol that has received prior institutional review board (IRB)/independent ethics committee (IEC) approval/favourable opinion.
(7) The medical care given to, and medical decisions made on behalf of, subjects should always be the responsibility of a qualified physician or, when appropriate, of a qualified dentist.
(8) Each individual involved in conducting a trial should be qualified by education, training, and experience to perform his or her respective tasks.
(9) Freely given informed consent should be obtained from every subject prior to clinical trial participation.
(10) All clinical trial information should be recorded, handled, and stored in a way that allows its accurate reporting, interpretation and verification. This principle applies to all records referenced in this guideline, irrespective of the type of media used.

(11) The confidentiality of records that could identify subjects should be protected, respecting the privacy and confidentiality rules in accordance with the applicable regulatory requirements.
(12) Investigational products should be manufactured, handled, and stored in accordance with applicable good manufacturing practice (GMP). They should be used in accordance with the approved protocol.
(13) Systems with procedures that assure the quality of every aspect of the trial should be implemented. Aspects of the trial that are essential to ensure human subject protection and reliability of trial results should be the focus of such systems.

23.4 ROLES AND RESPONSIBILITIES IN A CLINICAL TRIAL

A considerable part of the GCP guidelines is dedicated to clarifying the roles and responsibilities of the key players in any clinical trial – the sponsor, the ethics committee, and the investigator. It is beyond the scope of this text to go into these aspects in greater detail, but some of the key points are the following:

The institutional review board (IRB)/independent ethics committee (IEC)

- Its core responsibility is to "safeguard the rights, safety, and well-being of all trial subjects. Special attention should be paid to trials that may include vulnerable subjects."
- Its composition, function, operations, and procedures are specified in the guidelines.
- In order to fulfil its role, the IRB has to review all clinical studies before their initiation. The review includes all the relevant trial documentation, written patient information and the investigator's qualifications.
- The full approval of the IRB is a *sine qua non* (an indispensable and essential action) for the initiation of any clinical study.

The investigator

- must have the necessary (and documented) qualifications, training and experience to assume the responsibility of the "proper conduct" of the clinical trial;
- must be able to demonstrate that he/she has the "adequate resources," including a sufficient number of patients, time and staff to conduct the trial;
- is responsible for supervising and training all personnel involved in the trial at the site;
- is responsible for all medical care and decisions related to the trial subjects at the site;
- is responsible for all communication with the Institutional Review Board/Independent Ethics Committee (IRB/IEC);
- has to ensure that the investigational product is used strictly according to the protocol;
- has to be familiar with, and comply with, the many aspects of how to correctly inform a patient about a study and ensure that the patient fully understands the information before consenting to participation;
- retains the responsibility for a function, even when that function has been delegated to another staff member involved in the trial (i.e. sub-investigators, research nurses, pharmacist, etc.).

The sponsor, defined as: "An individual, company, institution, or organization which takes responsibility for the initiation, management, and/or financing of a clinical trial," is responsible for

- trial design, data management, data analysis, results reporting and record keeping;
- quality management, assurance and control during all phases and aspects of the trial;
- data monitoring;
- medical expertise for the trial sites and investigators as needed;
- investigator selection;
- maintaining correct and updated information on the investigational product, including safety information and adverse drug reaction reporting,
- manufacturing, labelling, packaging, handling, and supply of the investigational product.

Clinical trials are thus either academic (the sponsor is a hospital or university) or industry-sponsored. For a more complete description, please refer the ICH GCP guidelines available at www.ich.org and www.ema.europa.eu/en/ich-e6-r2-good-clinical-practice [4].

23.5 THE CLINICAL PHASES OF DRUG DEVELOPMENT – PHASE I TO IV TRIALS

In order for a new drug to become available to a wider public, be included in reimbursement systems and be prescribed to patients, it has to have a marketing authorization from the relevant regulatory authority. To get to this point of approval, the pharmaceutical company has to present detailed data on both safety and efficacy in the disease for which the drug is to be used. This requires systematic development of the drug in several different phases (see Figure 23.1), each phase being designed to address different aspects of the drug and its effect on patients. Academic trials are often also categorized according to the same system, although the intention is usually not to seek regulatory approval but rather to prove or answer a scientific question.

Prior to testing any drug in humans, it has to have data from non-clinical testing confirming an acceptable safety profile and identifying a safe starting dose in the first trials in humans. Non-clinical testing, which is still mostly done in animals, is performed in order to get preliminary data on safety, toxicity, and pharmacokinetics (PK). Once the drug has successfully passed the stage of non-clinical testing, it can go into phase I trials in humans. Phase I trials have as the primary aim to gather early data on safety, toxicity and PK, and to identify the optimal dose for phase II. The optimal dose is found by treating a limited number of subjects (typically 3) on an initial, very low dose. If no serious toxicity is identified the next three subjects receive a slightly higher dose, and so on, until a dose with an acceptable balance between toxicity and likelihood of therapeutic effect is found. It usually requires a relatively small number of trial subjects (<40) to identify the optimal dose. In oncology, phase I trials are done in cancer patients while in other areas of medicine they are done in healthy volunteers.

Phase II trials are performed in the specific patient population for which the drug is being developed. The subjects receive the dose that was identified as optimal in phase I, and data is gathered on efficacy and safety and, often, also on more exploratory aspects such as biomarkers for prediction of efficacy or toxicity. The number of subjects required for reliable interpretation of the results in phase II is usually less than 100.

If a relevant level of efficacy is found in phase II, and no special toxicity concerns have been identified, the drug is ready for phase III – that is, a randomized trial comparing the new drug to placebo or standard treatment for the indication pursued. The gold standard for phase III trials is a randomized, double-blind, placebo-controlled and multicentre trial. Randomization means that the subjects are randomly assigned either to the experimental arm or to the standard/placebo arm. Double-blind means that neither the investigator/doctor nor the subject knows whether he or she is receiving the investigational drug or placebo/standard treatment. Placebo is an inactive substance that has been prepared in the

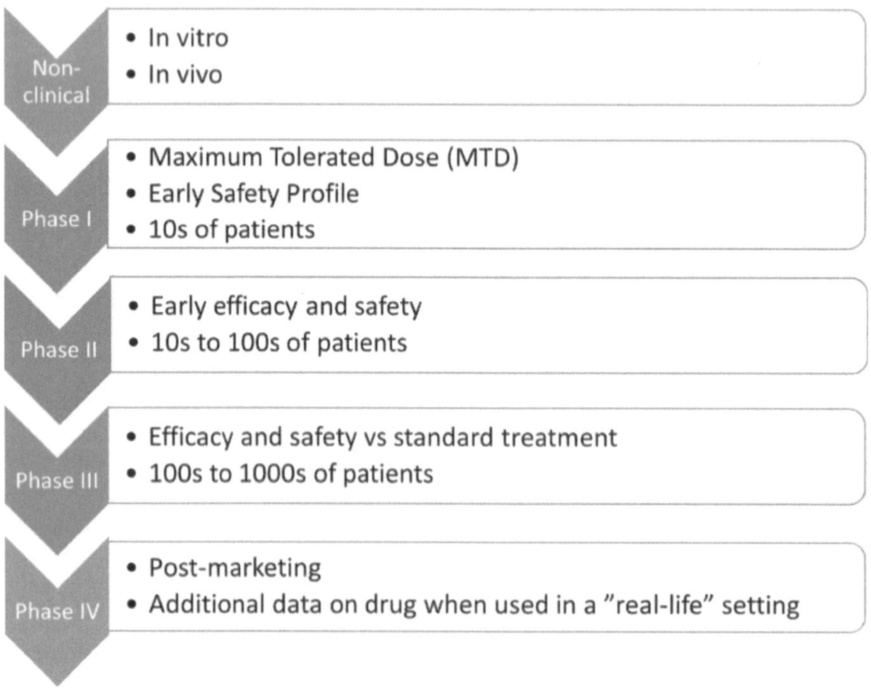

FIGURE 23.1 The different phases of drug development.

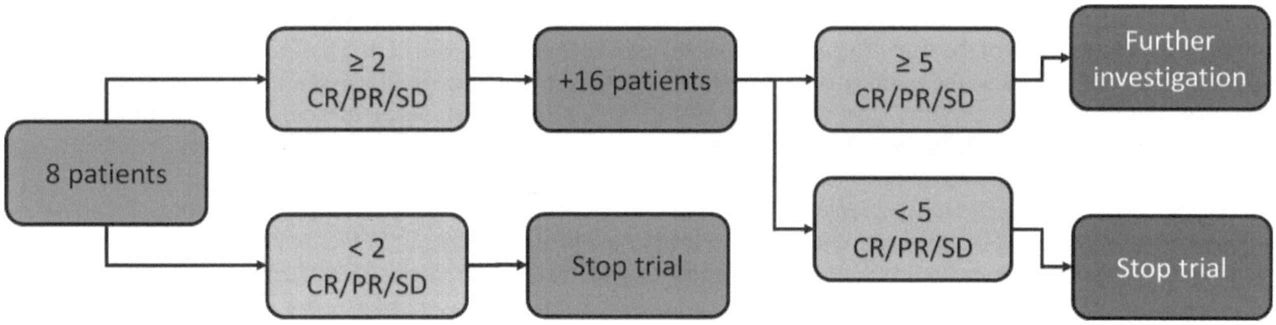

FIGURE 23.2 Example of a two-stage phase II design. In the first stage, eight subjects are included and treated with the experimental drug. After treatment efficacy has been evaluated, the number of subjects who have benefited from treatment is determined. In this example, demonstrated efficacy in at least two subjects was considered enough to justify including another 16 subjects, after which the final analysis was performed. If a total of at least 5 of 24 subjects had benefited from the investigational drug, phase III development was considered reasonable. CR: complete response; PR: partial response; SD: stable disease

same format as the investigational drug, and in such a way that it is impossible to know whether it contains the active substance or not. To include several centres in a trial is a way to minimize bias in patient selection, patient management, and so forth, that may otherwise differ between different sites/hospitals/countries in a way that may affect outcomes. The number of subjects required to prove a difference between the control arm and the experimental arm depends on how large the difference is between the two – the larger the difference the smaller the sample size required to prove it. Phase III trials generally include hundreds or even thousands of subjects.

There are often complementary aspects of a new drug that may be of interest to study even after receiving marketing authorization. This may include further safety data, quality of life, health economics, efficacy in a certain subpopulation of patients, and so forth. Clinical trials conducted within the approved indication but after commercialization, are called phase IV trials.

During the decades that have gone by since this process of drug development was formalized, several issues have been identified related primarily to time inefficiency, and the risk of treating too many subjects with a drug that is in the end identified to be ineffective. One way of improving the time efficiency in the drug development process is to combine two phases in one trial, that is, phase I–II or phase II–III in the same CTP. This requires more effort at the planning stage, but may save time and effort in the transition between the phases later on. To detect early signs of inefficacy, a two-stage phase II design (see Figure 23.2) can be used. With this design, a first analysis of efficacy is performed, for example, after one-third of the targeted number of subjects have been treated. The CTP will specify the minimum number of subjects with a positive treatment effect at that point, for the trial to continue. If that level of efficacy is not achieved, no more subjects are included, and the trial is discontinued.

Another special type of phase II trial is the so-called basket trial, where subjects with different diseases/indications are given the same drug in parallel arms. This is used when a drug is expected to be effective in more than one indication. With this type of trial design, inefficacy in one arm will lead to discontinuation of development in that indication, while the arms where some level of efficacy is identified can continue. The sample size in this type of trial is not dimensioned to detect significant differences between the arms, but only to detect relevant levels of efficacy in each arm.

23.6 THE PROCEDURE OF SETTING UP AND RUNNING A CLINICAL TRIAL

To design, set up and complete an academic clinical trial is long and hard work requiring experience, collaboration, funding, and time. It is often also richly rewarding as it gives the investigators access to a well-defined database from which new understanding and hypotheses can grow. Just how scientifically sound the data coming out of a clinical trial is will depend on how well designed the trial was, and how well it was carried out. It is therefore important to understand the intricacies of each and every part of the trial process. Many academic institutions have a clinical trials office that is specialized in assisting investigators in this process. It can offer advice on, for example, statistical, regulatory and data management aspects as well as guidance for the investigator team in the process of setting up the trial and conducting it in accordance with the GCP guidelines.

The first step in the creation of a new clinical trial is to clearly define the research question you want to answer, that is, *the primary objective* of the trial. The definition of the primary objective will be the basis for deciding the design of the trial and eventually also the number of subjects you will need to include, so it is essential not to lose sight of that objective despite all other questions that will inevitably attract your attention during this creative phase of trial design. Some of those questions may be included as *secondary objectives* of the trial to shed light on aspects of the investigational therapy that are not included in the primary objective. Once the primary objective has been defined, the next step is to decide what you have to measure in order to respond to the primary objective. That measure will be your *primary endpoint*. For example, if the primary objective is to study the efficacy of drug X in the treatment of metastatic breast cancer, the primary endpoint could be the proportion of patients that experience tumour shrinkage from the treatment (i.e. response rate).

The next step is to decide on how best to design the trial to meet your primary objective. For example, if your question is, "How efficacious is drug X in the treatment of disease Z?" you would formulate your primary objective as, "To study the efficacy of drug X in the treatment of disease Z," which would be the typical primary objective of a phase II trial. If, on the other hand the question is, "Is drug X better than drug Y in treating the disease Z?" you would state your primary objective as, "To compare the safety and efficacy of drug X and drug Y in the treatment of disease Z." That would lead to a randomized trial where some patients would be randomly assigned to receive drug X and others drug Y.

At this point, the *synopsis* of the trial is beginning to take shape. The synopsis is a short and concise summary, or skeleton, of the trial. Once completed, it will be the starting point for writing the full CTP (clinical trial protocol). A typical synopsis is a document stating the title of the trial, a short background, the objectives and endpoints, the main inclusion and exclusion criteria defining the study population, the overall design, a description of the study treatment, the basic statistical assumptions for sample-size calculations, and the estimated timelines for start-up and completion of the trial. It is often an essential tool for the involved investigators to agree on the fundamentals of the trial design and may also be the basis for applying for funding for the trial. It is at this early point of developing a clinical trial that the advice from an experienced clinical trials office should be sought. Input on statistical aspects, data management, a rough budget, and identification of regulatory challenges is better to receive early on, as it may affect the feasibility of the research project.

Once all the investigators involved in the trial agree on the content of the synopsis, the process of writing the full CTP can start. The CTP is one of the documents that will be sent to the regulatory authorities and ethics committees for approval before the trial can start, and its content is clearly defined in the ICH GCP guidelines. The CTP is also the document of reference for the investigators and their teams when conducting the trial in order to ensure that everyone follows the same procedures that guarantee patient safety and high-quality data. Other essential documents are the investigational medicinal product dossier, the written patient information, the informed consent form, and the investigator brochure. It may also be necessary to have an imaging guide, drug handling guidelines, and other documents that ensure the correct and consistent conduct of all aspects of the trial from beginning to end, especially in multicentre trials. The amount of documentation needed to start a clinical trial may be daunting to an investigator who may not have done it before and/or has other obligations to attend to. In some institutions and clinical research organizations (CROs) a medical writing service is offered. Medical writers are professionals experienced in creating all the documentation related to clinical research as mentioned above, and may also include grant applications, scientific manuscripts, and the final clinical study report (also a regulatory requirement). To get help from a medical writer may not only be time efficient, it may also improve the quality of the documentation, which in turn will influence the quality of the data and the results.

In parallel with the elaboration of the necessary documentation, it is wise to enlist help with what is generally referred to as "data management." Data management starts with the creation of the so-called case report form (CRF), which is the document in which all the data needed to analyze the endpoints is collected. The data that goes into the CRF comes from the patients' medical charts, radiology and pathology reports, laboratory assessments, and so forth. The CRF mirrors the CTP, and it is therefore a good idea to include the data manager in the CTP development process. From the CRF, which can be either on paper or in electronic format, the data is transferred to the trial's database for analysis according to the statistical analysis plan at the time points specified in the CTP. All data in the CRF and database are pseudonymized, meaning that each patient included in the trial is given an identity number that cannot be connected to the specific patient in any other way than through a code list wherein each identity number is coupled to the corresponding patient's name and birth date.

Once all the necessary documentation is in place, the approval process begins. The minimum for a clinical trial involving a medicinal product is approval by the "national competent authority," meaning the medicines agency in the country in which the trial is conducted, and the ethics committee(s). If it is an international multicentre trial, it will require approval in each country. In the European Union, a unified process is being implemented that will require only one application per trial for both medicines agency and ethics committee, independently of the number of participating member states. It is the intention that this unified and electronic process will reduce the time and effort needed to start a clinical trial, thereby improving the conditions for clinical research in Europe. Independently of the number of participating member states, any trial conducted within the EU has to be reported through the European Union Drug Regulating Authorities Clinical Trials (EudraCT) database system (https://eudract.ema.europa.eu/). Once registered, a unique trial identifier (EudraCT number) is given to the trial, which can then be sent to the regulatory authorities for approval.

The number and type of other approvals needed for start-up will depend on the type of trial and the site(s) where it will be conducted. Most trials will require approval for biobanking (collection of biological material), radiation safety (in the case of radiological or nuclear medicine procedures), and data protection, in addition to a varying number of local institutional research boards. There will also be a number of contracts to be signed where the type and content of the collaboration is defined, and any financial remuneration and timelines are specified.

When all the necessary approvals are in place, the trial can be formally initiated in each participating centre,. Patients can be enrolled, treated and followed up according to the specifications in the CTP. During the trial, the trial procedures and the data collected need to be monitored by a person with adequate training for that function. Monitoring is a GCP requirement, with the objective of ensuring that patients' rights and well-being are protected, the reported data is complete and accurate, and the conduct of the trial in accordance with the CTP. Once treatment and follow-up of the patients in the trial is completed, data is cleaned, and the database locked before proceeding to data analysis. The results of the trial are summarized in a final clinical study report (CSR), which is sent to the regulatory authorities for filing. The publication of the results in peer-reviewed medical journals does not replace the need for a CSR. All the essential documents and data records for a clinical trial have to be archived for a minimum of 15 years by sponsor and sites.

23.7 THE ROLE OF THE MEDICAL PHYSICIST IN CLINICAL TRIALS WITH RADIOPHARMACEUTICALS

The basic procedures and regulations are the same for radiopharmaceuticals as for any other investigational drug but has the added aspect of radiation exposure – wanted and unwanted, and to patients as well as health workers and caregivers/relatives. This makes the role of the medical physicist key in the planning, design, and conduct of the clinical trial. Specific guidance documents exist, intended to clarify the requirements of non-clinical data for radiopharmaceuticals in the EU [5] and the United States [6, 7]. In Europe, the requirements related to the use of radiopharmaceuticals in general, including within clinical trials, are specified by Euratom in the Council Directive 2013/59 published in the *Official Journal of the European Union* in December 2013. It delegates to each member state the duty to specify limits for exposure in different situations, and is clear in considering individualized dosimetry to be compulsory:

> For all medical exposure of patients for radiotherapeutic purposes, exposures of target volumes shall be individually planned and their delivery appropriately verified taking into account that doses to non-target volumes and tissues shall be as low as reasonably achievable and consistent with the intended radiotherapeutic purpose of the exposure.
>
> (Article 56)

The active collaboration of the medical physicist in clinical trials is seen as essential, and is therefore detailed specifically in the directive, in order to address the necessary concerns regarding planned and accidental exposure, procedure documentation, imaging, dosimetry, and so forth. The close collaboration between physician and physicist is also key in ensuring that "Medical exposure shall show a sufficient net benefit … against the individual detriment that the exposure might cause, taking into account the efficacy, benefits and risks of available alternative techniques having the same objective but involving no or less exposure to ionizing radiation" (Article 55).

23.8 SUMMARY

As we hope has been made clear in this chapter, the process of performing a clinical trial – including conception, design, essential documents, data management, statistics, data analysis, and reporting – is a very structured and regulated process. Before embarking on the often rather long journey of taking a research idea to a full-blown clinical trial, the importance from an early stage of contacting people with the right expertise cannot be stressed enough. It will make the product and the process better and smoother, which in the long run will result in better data quality and research. It is an arduous process, but well worth it once you get to the point of getting the answers to your original questions.

REFERENCES

[1] J. Lind. "The Treatise of the Scurvy." www.jameslindlibrary.org.

[2] "The Declaration of Helsinki." www.wma.net/what-we-do/medical-ethics/declaration-of-helsinki/doh-oct2008/.

[3] E. Emanuel, ed. *The Oxford Textbook on Clinical Research Ethics*. New York: Oxford University Press, 2008.

[4] *ICH: E6(R2) Guideline for Good Clinical Practice* by the International Council for Harmonisation of Technical Requirements for Pharmaceuticals for Human Use (ICH), 2017.

[5] European Medicines Agency. *EMA/CHMP/SWP/686140/2018, Guideline on the Non-clinical Requirements for Radiopharmaceuticals*. 2018. www.ema.europa.eu/en/non-clinical-requirements-radiopharmaceuticals.

[6] US Food and Drug Agency. *Nonclinical Evaluation of Late Radiation Toxicity of Therapeutic Radiopharmaceuticals*. 2011. www.fda.gov/media/72237/download.

[7] US Food and Drug Agency. *Oncology Therapeutic Radiopharmaceuticals: Nonclinical Studies and Labeling Recommendations*. 2019.

24 Introduction to Patient Safety and Improvement Knowledge

Tomas Kirkhorn

CONTENTS

24.1 INTRODUCTION

Why a chapter on patient safety? Is this not a self-evident ingredient in today's modern healthcare? The question might be especially relevant for those parts of healthcare which deal with radiation physics applications, an area where we traditionally are well aware of risks for both patients and personnel. However, the interventions and care of the individual patient depend on more than the safety aspects embedded in the technological methods and devices. The risk scenarios and the final outcome also depend on circumstances and conditions in the environment in which the methods or devices are being used.

During the last decades, the medical and technological quality in healthcare have developed in many ways. Our knowledge about giving care to patients, how to treat diseases, and give new types of support to patients with different kinds of disabilities have grown substantially. At the same time a considerable number of patients suffer from *medical harm* during their time in healthcare: it is commonly reported that approximately 10 per cent of all hospitalized patients experience harm, of which at least 50 per cent are preventable [1, 2].

The aim with this chapter is to highlight and increase the general awareness of the patient-safety area: An area closely connected to work environment and quality work in general. A healthy work environment contributes to creating positive conditions to achieve safer care, and continuously ongoing quality improvements will generally result in safer care. What we have learnt through the years is that there are no quick answers or solutions to be put in place. On the other hand, many examples from healthcare settings have shown that it is possible to improve safety, in big and small ways, when people and organizations take the time and show the effort to reflect on what we do, how we do it, and what improvements can be made.

A natural question related to the occurrence of medical harm is why it happens, and how we can improve the situation in order to prevent these events from reoccurring. However, it is just as important to reflect on why things go right, in spite of all the "threats" and challenges we recognize around us in our daily work, situations that easily could make things go wrong. A crucial factor to pay attention to is the patient-safety culture at our workplaces, and how it affects our behaviour and how we communicate with each other. The work towards safer care is not assigned a special profession or certain group of people, but is a job for everyone involved in healthcare: management on all levels, personnel working on "the floor," as well as the people we serve, the patients. The work for safer care can only be successful if we work together!

Before going any further, we will look at where we stand and, on some factors that brought us here: Let us take a look at the "patient-safety map."

DOI: 10.1201/9780429489501-24

24.2 DEFINITIONS AND MAGNITUDE

Let us start by considering the definition of "patient safety" and the magnitude of the problem. According to the World Health Organization (WHO) patient safety is "the absence of preventable harm to a patient during the process of healthcare." The word *preventable* indicates that the harms we discuss here do not include those that are a direct consequence of a known calculated risk associated with a certain therapeutic procedure. Neither is the preventable harm a direct consequence of a patient's severe condition as such. The fraction of preventable medical harm is estimated to be from ½ to 2/3 of all medical harm. For certain categories of harm studies show that this fraction is even higher. About half of all preventable harm is regarded to be mild and transitory, while the other half requires additional days of care at the hospital. Between 5 and 10 per cent is considered severe, causing permanent harm or disabilities, or death.

Medical harm includes not only physical injuries, but also suffering and psychological harm [1], which further indicates the importance of engagement for improved patient safety in a broader perspective.

As already mentioned, WHO states that 1 out of 10 patients in high-income countries are exposed to *preventable medical harm* at hospitals, causing suffering and loss for the patients and their families. Further, healthcare providers take a high financial toll. Available evidence suggests that 15 per cent of hospital expenditure and activity in the OECD countries can be attributed to treating safety failures [1].

Thus far, the terms *medical harm* and *preventable medical harm* have been used and explained. Another term is *adverse event*, which likewise refers to harm from medical care, rather than an underlying disease. Further, *preventable adverse event* refers to those that occurred due to error or failure to apply an accepted strategy for prevention of medical care rather than from an underlying disease.

We also talk about *near misses*, explained as unsafe situations that are indistinguishable from preventable harm or preventable adverse events except for their outcome. The patient harm was avoided by either early detection or by sheer luck. Alternative terms for near misses are potential adverse events and close calls.

24.3 SHORT ON RISK AND ACCIDENT MODELS

In the early days medical harm, or adverse events, were looked upon as unhappy, but at the same time, an unpreventable consequence of the on-going medical process. When something went wrong, it was also common to point out, and put the blame on, the person who was most closely involved in the situation. Not until the beginning of the 2000 did the discussion on medical harm and why it happens became more "public." One important milestone was the publication of the report from the Institute of Medicine, "To Err is Human - building a safer health system," which presented a very severe patient-safety situation [3]. However, in healthcare organizations in general, there were few established approaches and strategies to handle the problems of medical errors. Gradually, successful safety methods from other operations, like industry, were taken over and adapted to the healthcare area. Often, these methods were focused on component error, and on the law of cause and effect.

One of the earliest sequential accident models was described by Heinrich [4], the Domino model of accident causation. It describes how a chain of events, represented by dominos falling on each other, one after the other, finally results in an injury (Figure 24.1).

The different "components," or dominos, involved in this accident model, originally developed from human and physical failures, and later to versions including organizational and management issues, and further to the area of the safety culture. However, the focus on a "simple" cause and effect has survived the development of the different components involved in the process.

Another well-known model is the Swiss cheese model described by Reason [5], which is built up like a number of slices of Swiss cheese, each of which represents a barrier to preventing threats from reaching the patient (Figure 24.2). The barriers are created by, or represent, different "actors," all the way from legislation, organizational structures on different levels, technical resources, routines and protocols, and finally the healthcare personnel standing next to the patient. Like the slices of a Swiss cheese, the barriers have inherent holes at different locations. Some of the holes are well known, while others, referred to as *latent errors*, are unknown. Some holes are always present, while others show up only on certain occasions. The accident occurs when the holes in all the barriers happen to line up, allowing the hazard to slip through all slices, or barriers, and finally expose the patient. The last barrier, the medical personnel in "direct contact" with the patient, is often the rescuer, who manages to identify the potential risk of an error and prevents the results of that error from reaching the patient. However, as a natural part of being human, we sometimes fail, or, as the title says in in the earlier-mentioned report, "to err is human." For this reason, it is important to observe and secure the barriers located behind the personal, as being as safe, as free from holes as possible.

FIGURE 24.1 The Domino model of accident causation [4].

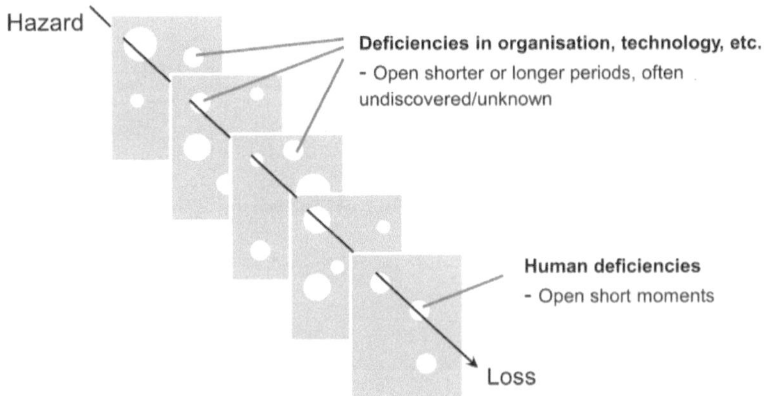

FIGURE 24.2 An illustration of the idea of the Swiss Cheese Model, described by Reason [5].

The Swiss cheese model in its fundamental outline is based on a one-directional, single process and does not take into consideration that today's healthcare happens in many steps and processes, in different locations, involving many different staff members, all with a high dependency on each other.

Another important model that helps in understanding why accidents happen is the Human-Technology-Organisation-model (HTO) [6]. It demonstrates how the three players – Human, Technology, and Organisation – are tightly connected and dependent on each other (Figure 24.3). Accidents most often happen as a result of deficiencies in all three parties and (as a standard rule) should not be blamed on only one. The introduction of this model has been important to changing a culture of blaming the personnel for being the major reason for medical harm – from a strict individual perspective, to a more balanced systemic perspective, taking many different factors and actors into account.

A commonly used tool to analyse and picture adverse events in healthcare today are different versions of Root Cause Analysis, which generally aims at finding the underlying root causes of why the event happened [7, 8]. The tool has its starting point from the system perspective, employing the Human–Technology–Organization model. From a detailed time sequence of separate events – of which some are identified as crucial to the adverse outcome – the goal is to search for root causes connected to these events. The root causes can be categorized as belonging to

Procedures, routines and protocols;
Communication and information;
Environment and organization;
Education and competence, and
Technology and devices.

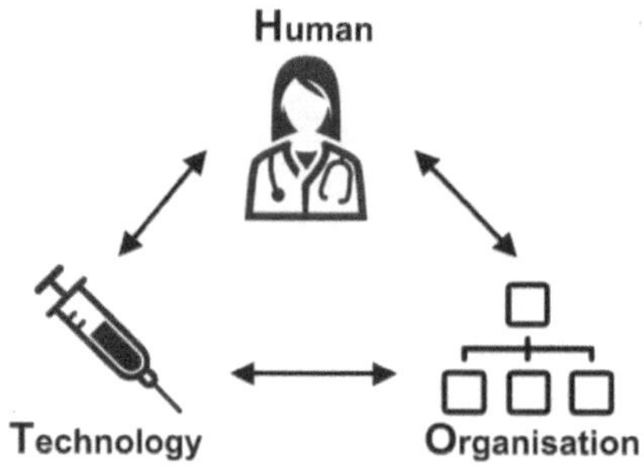

FIGURE 24.3 The HTO-model – Human Technology and Organization.

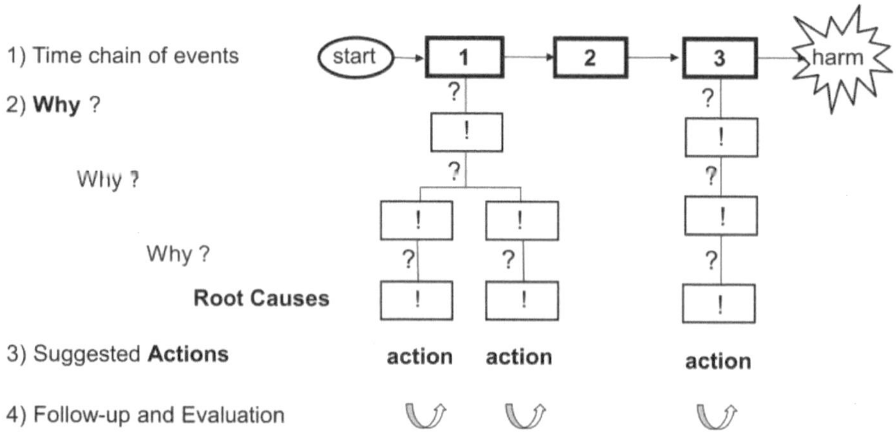

FIGURE 24.4 Root-cause analysis.

To reach down to the basic cause, or causes, the question "Why?" should be answered several times: For each answer, "Why?" is asked again, and so forth. When, finally, the root causes have been pointed out, actions are proposed to eliminate them, or if that is not possible, to limit or surveil each and every cause (Figure 24.4).

The limitation of using the root-cause analysis alone in investigations and in the follow up to adverse events, is that to a large extent it is based on the presumption of a direct cause-and-effect, which is not always the reality. For this reason, we need to complement our work with other methods, which take into consideration that the situations we find in healthcare are more complex, with additional factors that are not as controllable as the root cause model might suggest.

One reason as to why adverse events continue to show up in healthcare is the complexity with which modern healthcare is performed. The complex system does not build on parts that always follow rules such as cause and effect, but to a great extent builds on unpredictable entities with their own inner lives. Identical types of input at two different points of time will not necessarily give the same response. This is because the state of the entities changes from one point of time to another. Consequently, we will never be able to create a totally safe healthcare system based on only earlier knowledge about the system. In every moment, we need to study the current state, and adjust our input according to the present needs. Consequently, instead of relying too much on earlier preventive actions, taken when "things went wrong," it has been proposed that we should spend more time studying the situations when "things go right" and actively learn from those. When we take time to investigate why things go right, why we succeed in spite of difficulties and challenges we meet every day, we gain important understanding of how to approach and adapt our work for increased safety.

One model that describes both the traditional view, which takes off from situations when "things go wrong," and from those when "things go right," are the Safety I and Safety II models, suggested by Hollnagel [9, 10]. In the Safety I approach, work and activities are considered to be built on a bimodal view of the system (Figure 24.5). Activities can be performed along two different paths. In the first path activities go well, because the system functions correctly, and people perform as planned. This leads to successful outcomes, with no adverse events. The other path, which

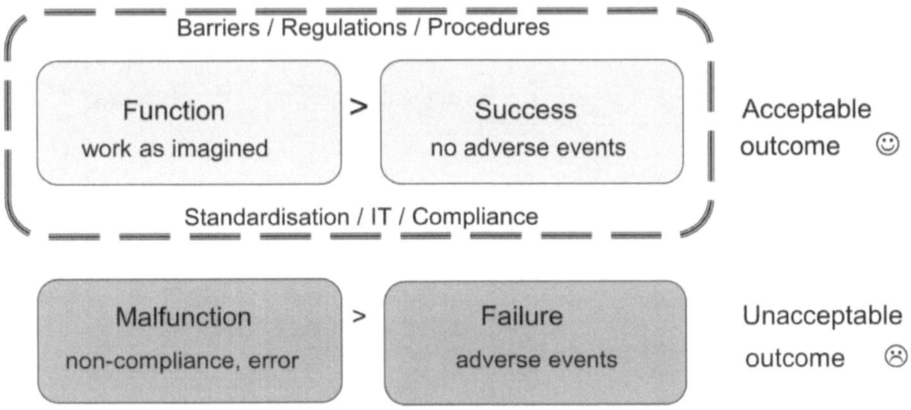

FIGURE 24.5 Safety I – a bimodal system approach [9]. Courtesy Erik Hollnagel.

exists parallel to the first, leads to unsuccessful outcomes due to malfunctioning systems, non-compliance, and persons who do not perform as planned. It ends in failure and accidents. The only way to avoid unacceptable outcomes is to stay on the first, successful, path. The means to stay on this successful path is to build a fence around it, by developing rules and regulations, and following them, and by building barriers to keep us on the safe side. Through analysing risks as well as adverse events (when these happen), we will try to find and fix deficiencies. Increased standardization, better supporting IT-systems, and high compliance to procedures will help us reach our goal – which is no adverse events.

A considerable part of patient-safety work has developed and been carried out on the bases of this approach, here named Safety I. A consequence of the efforts of remaining on the safe path will often be that problems tend to be solved separate from each other, one by one, and technological solutions will be chosen before socio-technological ones, which better take into account the interaction between humans and technology or technological processes. Further, legislation, rules, and guidelines on the subject are often based on the same preconditions. Of course, for limited activities in separate systems this approach can help us to reach a certain level of safety. However, many activities are developing towards bigger and complex systems, in which an increasing number of factors affect the overall outcome. Earlier, relatively limited dependency on IT-systems and technical support functions has developed into a large-scale dependency, as well as complexity. Also, a relatively low level of interaction between activities, which has been possible to monitor, is changing to a higher level of integration. However, the way we look upon, and work with, safety issues has not developed at the same pace. Do we need to complement the former, and often still present, approach characterized as Safety I with something else in order to meet the challenges ahead? Probably we all agree that the answer to this question is a yes.

In the Safety II approach, focus is turned from "when things that go wrong" towards "when things go right." The foundation in this approach is that those processes and actions that go right are performed in the same system, and by the same persons, as when things go wrong (Figure 24.6). The outcome of the process is instead dependent on variations of different kinds. Ingredients that need to be present to contribute to safety are flexibility, adjustment, and adaption to the present and variable situation. Thanks to the personnel who discover potential critical situations and make the proper adjustments, the adverse event will be avoided, or be of limited severity for the patient.

So, whatever happens, in the Safety II approach it happens by the one and the same path – everyday work. The outcome is dependent on variations in performance. Variability is found in different kinds of resources needed to solve the task, like time, information, skills, and competence, within both the system and the individuals at a certain point of time. The adjustments and adaption to the present variations are inevitable and essential – however, always approximate. Depending on to what level the adjustments and adaptions reach, the situation becomes either safe or unsafe. At a time when the combination of different variations, at a "functional resonance," exceeds the safety line, the accident is a fact (Figure 24.7).

A coarse way of defining safety work from the two approaches, Safety I and Safety II, could be, "How to avoid things going wrong? and "How to create conditions that make things go right?" However, depending on the rate of complexity and variability in the processes and organizations we study, the two approaches will be complementary, and used together in future patient-safety work.

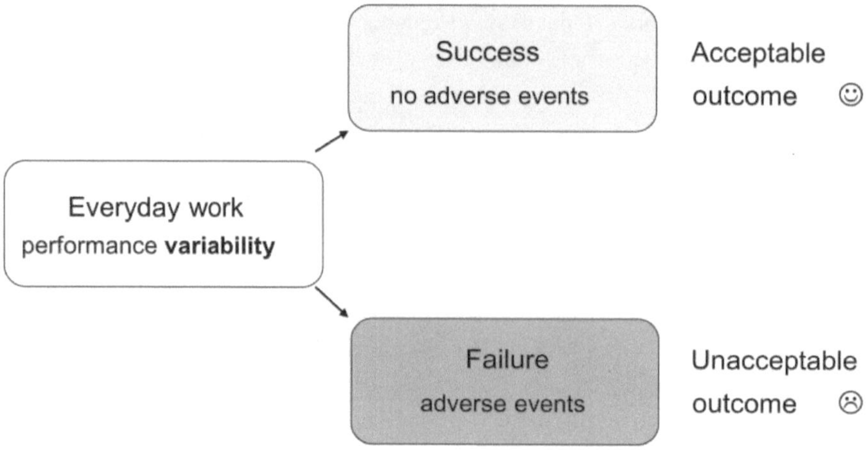

FIGURE 24.6 Safety II approach [9]. Courtesy Erik Hollnagel.

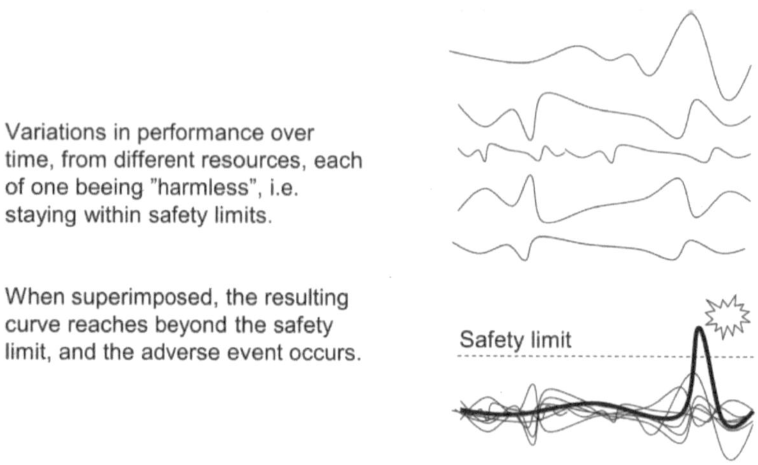

FIGURE 24.7 Functional resonance [9].

24.4 PATIENT-SAFETY CULTURE

No matter what risk models, tools for analyses, or preventions we use in our work towards improved patient safety, we need to be aware of in what "soil" we stand. The patient-safety culture that reigns in our workplace and organization, will define the level of success we will reach (Figure 24.8)

According to James Reason [11], a culture can be defined by

- shared values, about what is important,
- shared perception, about how things work, and

how these two interact with the structure and management of the organization. Together, these factors lead to standards for our behaviour: "How we do things."

In a patient-safety culture, the aim of our shared values and perception is the patient-safety situation in our organization. A well-functioning safety culture is suggested to be

- Informed: People in the organization are knowledgeable about risk and safety issues;
- Reporting: Cultivating an atmosphere where people have confidence to report safety concerns;
- Learning: The organization is able to learn from its mistakes and make changes;
- Just: No one in the organization will be blamed on false grounds; and
- Flexible: The organization and the people in it are capable of adapting effectively to changing demands [11].

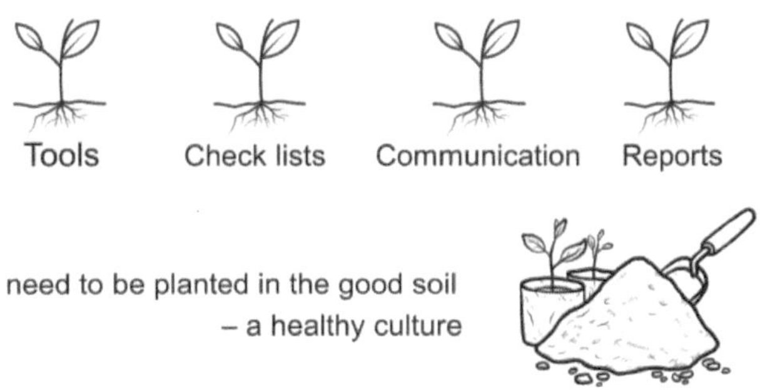

FIGURE 24.8 Safety culture – plants planted in the soil.

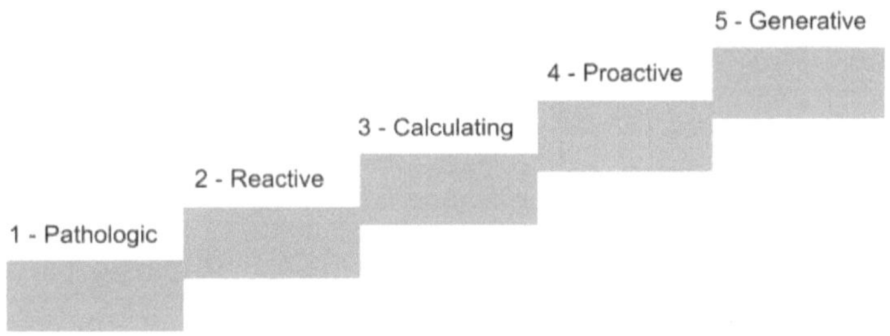

FIGURE 24.9 The patient-safety culture ladder [12].

The current level on which an organization stands in terms of safety culture has been illustrated by the 5-level safety culture ladder [12, 13] (Figure 24.9). It can be used within an organization as a dialogue tool to map and develop safety work. At the first level, known as the pathologic level, the general opinion is that safety work is waste of time: "We don't worry, as long as we don't get caught." At the second level, the reactive level, "Actions are taken when accidents happen." At the third level, the calculating level, "We have the systems in place to manage hazards." At the fourth level, known as the proactive level, "Safety leadership and values drive continuous improvement." At the fifth and highest level, named the generative or progressive level, safety is fully integrated in all activities: "Safety defines how we do business."

24.5 PROFOUND KNOWLEDGE OF IMPROVEMENT – TO SUCCEED IN IMPROVING SAFETY AND QUALITY

From analysing adverse events, but also our current status and possible risks, we gain a certain amount of understanding about where we stand in terms of safety and what factors or actions we should further develop in order to improve safety. Sometimes, these actions or adjustments seem to be quite straightforward and self-evident, and we may ask ourselves why they have not been implemented already! Filling this gap, from where we are to where we should be, is often more difficult than it seems at first sight. In order to accomplish successful change, we need to understand and practice the knowledges of improvement. This "Profound Knowledge of Improvement" described and developed by Deming [14] constitutes a valuable foundation for the necessary understanding. This improvement knowledge is separated from traditional professional knowledge, associated to the different professions we meet in the healthcare system as, for example, a physicist, engineer, physician, or a nurse. Batalden and Stoltz [15], have emphasized the necessity for all healthcare professionals to have knowledge not only in our own profession, but also in how to develop and improve our work practice and the system we work in. This can be illustrated by two parallel tracks of knowledge – the professional and the improvement knowledge - which combined lead to a development of practice which will benefit the individuals we serve – the patients (Figure 24.10).

While "professional knowledge" represents the traditional knowledge of the specific subject, personal skills and values, and ethics, the "improvement knowledge" comprises the knowledge of a system, the knowledge of variation, knowledge of psychology (especially psychology of change), and theory of knowledge (for learning-based improvement work).

Based on Deming's ideas on structured improvement work, a three-question model that is part of the Nolan improvement model [16] can be expressed as

(a) What are we trying to accomplish?
(b) How will we know that a change is an improvement?
(c) What changes can we make that will result in improvement?

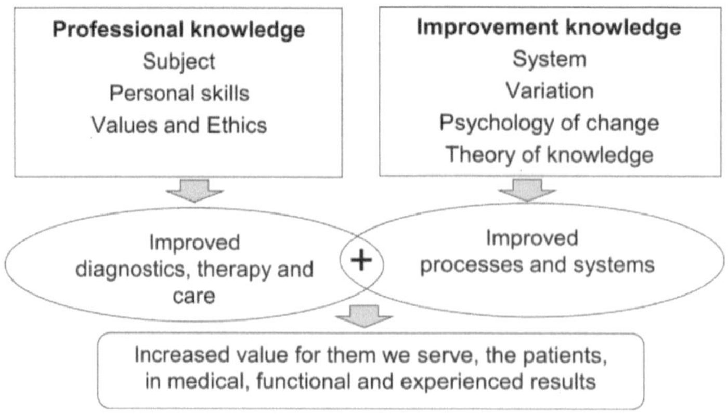

FIGURE 24.10 Professional knowledge and Improvement knowledge, to develop quality and safety in healthcare [15].

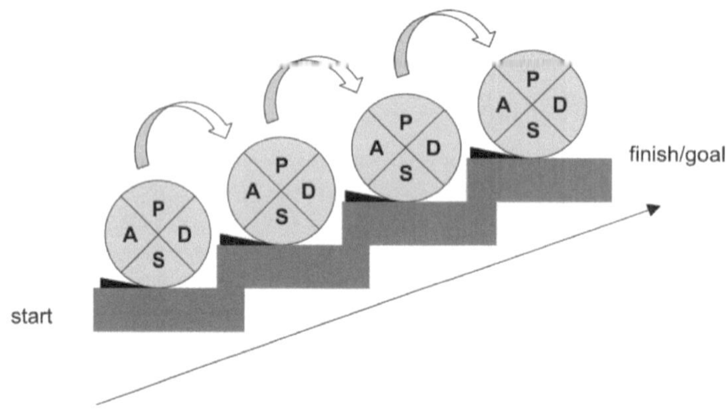

FIGURE 24.11 Securing the process towards the final goal by taking a number of smaller improvement steps.

These questions help us to define (a) the problem that we address, and the goal or objective of the improvement work; (b) means to measure and to get "evidence" that the procedure of change, results in an actual improvement in the end. The final answer to the third question, (c), about what we can do, is not always easy to get only by theoretical considerations and analyses, but needs to be developed step by step. We need to try out our solutions, see what happens, and adapt the next step according to the result. This systematic trial and adaption procedure can be utilized using the PDSA-cycle (Plan, Do, Study, Act). First, we plan what to do, when and how to do it, and by whom. Then we do it, while monitoring the procedure and gathering data on how the change affects the situation. After some time, we study and evaluate the outcome of our first trials. Depending on the results, we either have to alter some variables and try again, or proceed to the next level of implementation. Instead of taking "big" steps, aiming at the complete goal of improvement, smaller changes are preferable (Figure 24.11). By securing sub-targets of successful changes, the road to improvement will be well anchored at the workplace. Smaller steps in a wrong direction are also easier to change than bigger ones.

The number-one factor for a successful improvement work is to clearly identify the problem. From this recognized and acknowledged problem, we need to define the point where we are heading, and our goal, which should be concrete, limited, and measurable. If the goal is of a considerable measure, it needs to be divided into smaller parts that could be taken on, either one after the other, or by parallel tracks. In order to find and understand the problems and limitations in our work, we need to investigate our processes and the system we work in. A correct and clear picture of how we actually work and perform, or how the patient process looks, are crucial. Our actual work process, "work as done," does not seldom differ from the way we thought we worked, or were supposed to work, "work as imagined." To succeed in this mapping process, perspectives and experiences from all involved professions, as well as patients and their relatives, when applicable, need to be present. We should be aware of the inherent complexity of situations and processes that we work in. States and changes at one position may affect many other positions. Changes on different time scales need to be considered, and the relationship between cause and effect is not always clear. Further, a certain solution to one problem might be unfavourable to another.

When a problem, or a weak link, has been identified, the next step is to search for factors that contribute to this problem. One way to illuminate different contributions is to use a "fishbone-" or "Ishikawa-diagram," in which the problem, or objective, is marked at the head, and possible main causes, and to these are connected sub-causes, represented by the fishbones along the spine [17] (Figure 24.12). In order to help searching for different main causes, the fish bones can be allotted different categories of factors, such as people, environment, policies, procedures, and materials. When one or several critical factors have been identified, we know what to focus on when striving for solutions.

When our goal for the improvement work has been stated, and before we initiate the work of change, we need to find a measurable indicator to show the current state of matters and to be able to follow whether the changes we perform will give the expected improvement. These measurements can be obtained, for example, by pin statistics (when counting events), or through questionnaires to the staff or to patients (when focusing on experiences regarding the matter) during a defined period of time. By starting the measurements before any change is initiated, we get a baseline. After that, measurements are performed repeatedly throughout the improvement work.

When measuring over time we will discover a certain amount of variation, which needs to be taken into account and be understood (Figure 24.13). Some variations depend on natural causes, and cannot, or should not be controlled. Other variations are created, or systematic, and have a certain reason why they happen, some of which we may allow and adapt to, and others which we should affect and change. By measuring over time and studying the results we can gain understanding of the character of the variation. Generally speaking, as long as the variation is expected and manageable, it is considered okay, but when it goes beyond a certain limit, a situation can arise in the present procedure, jeopardizing our goal or safety.

When it comes to psychology of change, one basic prerequisite is that healthcare is performed by humans, not by institutions: It is still people who play the main roles. It is not the hospital that meets the person who seeks care, but the people who work there. As personnel, emotions and conceptions have a greater impact on how we function than we may think. And maybe we are not always as rational as we think we are.

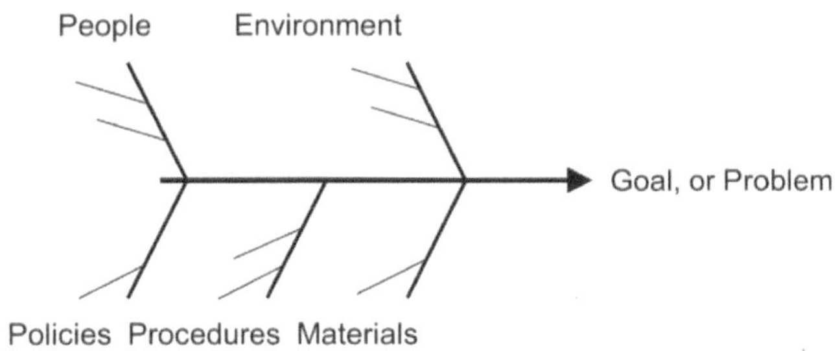

FIGURE 24.12 Fishbone or Ishikawa diagram, with factors (bones along the spine) contributing to either the goal or to the problem (the head).

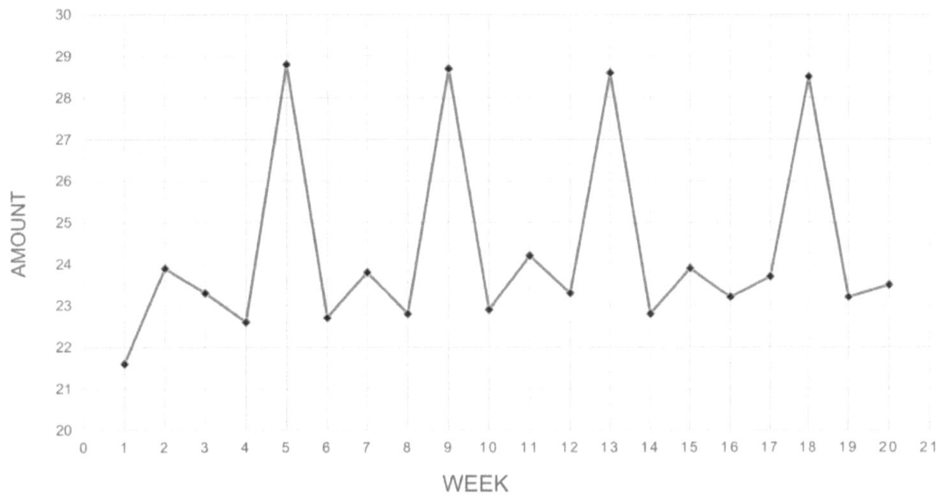

FIGURE 24.13 Natural and systematic variations

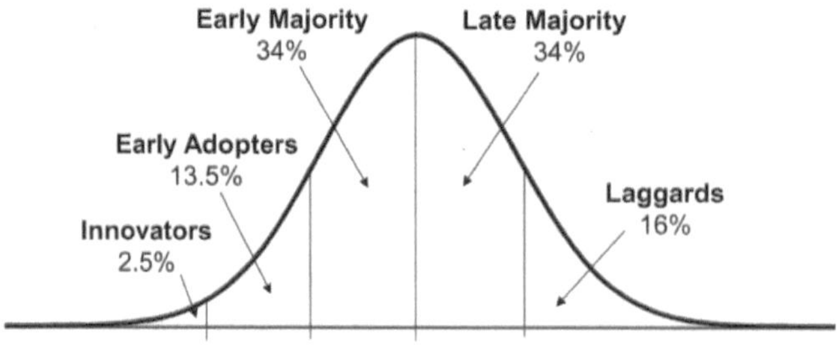

FIGURE 24.14 Diffusion of innovation [18]. Percentages represents the fraction of people who belong to each category.

When working with improvements, no single profession or staff member alone, can recognize and understand the whole picture. It is crucial to develop our ability to cooperate in multi-professional and multi-functional teams. The need for cooperation and collaborative work passing different kinds of borders is often mentioned, but seldom do we talk about how this will be accomplished. Borders between different organizational parties, medical specialties, and professions, are not bridged by directives, but through knowledge and understanding of one another's perspectives and psychological driving forces. We need to share perspectives and knowledge so that the gathered competence will become beneficial for the patient. Normally we apply psychological considerations in our relations with the patient. However, this is just as important in our daily relations with our colleagues, especially if we consider quality improvement and patient-safety work.

The philosopher Soren Kirkegaard said, "Everyone wants development, but no one wants change." However, improvements always mean change. To succeed with improvement work, we therefore always need to invest time and attention to the people the change will affect. Close collaboration with colleagues and a clear engagement from the management are two main factors we need to be aware of. As individuals we also act differently on changes, or suggestions of changes. Everett Rogers [18] illustrates in his theory Diffusion of Innovations, how innovations and technology are adopted by participants in a social system (Figure 24.14). The participants are divided into five categories of adopters: From the "Innovator," who got the idea (3%); early "adopters," who are ready to join (13%); the "early majority," who will follow in time (34%); to the "late majority," who will maintain a resistant approach for some time (34%); to finally the "laggards" or "over-my-dead-body" persons (16%) [18].

Any stronger reluctance in participating in a process of change needs to be considered. Naturally, reasons can vary. One set of three questions that can be useful in identifying hindrances is based on the three levels of information, emotions, and trust [19]. Questions related to the information level can be summed up with, "I don't understand"; to the emotional level "I don´t like it"; and to the trust level "I don´t trust you." Keeping an open dialogue and cultivating a permissive environment are essential components to reaching a level of understanding, acceptance, and trust for progress towards improved and safer work.

REFERENCES

[1] L. Slawomirski, A. Auraaen, and N. S. Klazinga, "The economics of patient safety," 2017, doi: 10.1787/5a9858cd-en.
[2] "Patient safety: Making health care safer," World Health Organization 2017, Geneva, 2017, https://apps.who.int/iris/handle/10665/255507
[3] *To Err is Human: Building a Safer Health System.* Washington, DC: Institute of Medicine (US) Committee on Quality of Health Care in America, National Academies Press, 2000.
[4] H. W. Heinrich, *Industrial Accident Prevention: A Scientific Approach.* New York: McGraw-Hill, 1931.
[5] J. Reason, "The contribution of latent human failures to the breakdown of complex systems," *Philos Trans R Soc Lond B Biol Sci,* vol. 327, no. 1241, pp. 475–84, 1990, doi: 10.1098/rstb.1990.0090.
[6] A. Karltun, J. Karltun, M. Berglund, and J. Eklund, "HTO – A complementary ergonomics approach," *Appl Ergon,* vol. 59, no. Pt A, pp. 182–90, 2017, doi: 10.1016/j.apergo.2016.08.024.
[7] (2021). *US Department of Veterans Affairs. VA National Center for Patient Safety: Root Cause Analysis.* www.patientsafety.va.gov/professionals/onthejob/rca.asp.
[8] M. F. Peerally, S. Carr, J. Waring, and M. Dixon-Woods, "The problem with root cause analysis," *BMJ Qual Saf,* vol. 26, no. 5, pp. 417–22, 2017, doi: 10.1136/bmjqs-2016-005511.
[9] E. Hollnagel, R. Wears, and J. Braithwaite, *From Safety-I to Safety-II: A White Paper.* 2015.

[10] J. Braithwaite, R. L. Wears, and E. Hollnagel, "Resilient health care: Turning patient safety on its head," *Int J Qual Health Care,* vol. 27, no. 5, pp. 418–20, 2015, doi: 10.1093/intqhc/mzv063.

[11] J. Reason, "Achieving a safe culture: Theory and practice," *Work & Stress,* vol. 12, no. 3, pp. 293–306, 1998, doi: 10.1080/02678379808256868.

[12] P. Hudson, "Implementing a safety culture in a major multi-national," *Safety Science,* vol. 45, pp. 697–722, 2007, doi: 10.1016/j.ssci.2007.04.005.

[13] D. Parker, M. Lawrie, and P. Hudson, "A framework for understanding the development of organisational safety culture," *Safety Science,* vol. 44, no. 6, pp. 551–62, 2006, doi: 10.1016/j.ssci.2005.10.004.

[14] W. E. Deming, *The New Economics for Industry, Government, Education*. Cambridge, MA: Massachusetts Institute of Technology, Center for Advanced Engineering Study, 1993.

[15] P. B. Batalden and P. K. Stoltz, "A framework for the continual improvement of health care: Building and applying professional and improvement knowledge to test changes in daily work," *Jt Comm J Qual Improv,* vol. 19, no. 10, pp. 424–47; discussion 448–52, 1993, doi: 10.1016/s1070-3241(16)30025-6.

[16] G. J. Langley, R. Moen, K. M. Nolan, T. W. Nolan, C. L. Norman, and L. P. Provost, "The Improvement Guide: A Practical Approach to Enhancing Organizational Performance," 2009.

[17] K. Ishikawa, *Guide to Quality Control* (Industrial engineering & technology). Tokyo: Asian Productivity Organization, 1976, pp. xiv, p. 226.

[18] E. M. Rogers, *Diffusion of Innovations*. New York: Free Press, 1995.

[19] R. Maurer, *Beyond the Wall of Resistance: Why 70% of All Changes Still Fail – and What You Can Do about It*. Rev. ed. Austin: Bard Press, 2010.

25 Closing Remarks

László Pávics

Since György von Hevessy first used radionuclides for monitoring biochemical processes in living beings, nuclear medicine has always adopted a multidisciplinary approach. Physicians, chemists, computer scientists, engineers, and physicists all work closely together, so we can do our best for our patients. But in the last few decades, there have been several changes in our discipline. The number of the therapeutical interventions is increasing and, besides beta-emitting radionuclides, alpha emitters have also been introduced. In the diagnostic field especially, the use of positron emission tomography has started to dominate. Nonetheless, Henry Wagner Jr's observation from 1991 remains just as insightful:

> If SPECT can do it, do it with SPECT, PET will always be able to do things that SPECT can't do. If SPECT can do tomorrow what PET is doing today, PET will go on to do other things.

There are growing challenges due to the combination of nuclear medicine imaging devices with other modalities, such as CT and MRI. Using new radionuclides in therapy and in diagnostics, in combination with radiological imaging devices, has produced an ever-increasing demand on dosimetry and radiation safety. More broadly, the public's view of nuclear plants and the possibility of nuclear accidents, and its perception of medical irradiation, have all moved in an unfavourable direction; Trust has been further undermined by less-informed medical personnel. The efforts to decrease unnecessary radiation exposures have never been more important.

We should do our best to increase the sensitivity of our detectors and to improve the image quality of our devices, within an acceptable cost. But we should not forget that the source of diagnostic information is basically the signal originating from the nucleus. If this primary signal is weak, we will lose the diagnostic information. The majority of unnecessary radiation exposure is caused by insufficient investigation. To find the balance between risk and benefit remains up to the individual nuclear medicine specialist's decision-making process. The use of morphological imaging techniques initiated the need to optimize devices to SPECT and PET and to learning to perform and evaluate radiological investigations. Nuclear medicine made several gestures towards collaboration with radiology but instead of a fruitful cooperation this resulted only in headaches for all parties involved. The conflict has grown to realizing several nuclear medicine departments being integrated into other imaging units. Often, professional training has not kept up with the growing number of different imaging modalities. Instead of competition, it would be wiser to be aiming for closer collaboration, not only within the imaging disciplines, but also between clinical medical fields. Radiology itself is more and more moved into specialization or is being overtaken by clinicians. Nuclear medicine should maintain its integrity. Armed with perfect nuclear physics, radiopharmacy, molecular biology, and medical knowledge in nuclear medicine with skills in radiology and clinical specialities will remain indispensable to our profession. In the end, the market should decide which technology (radiopharmaceutical, imaging device, therapeutical procedure, etc.) will become dominant.

The scientific life and education of nuclear medicine professionals are coordinated under the umbrella of national and international organizations. The two largest associations are the Society of Nuclear Medicine and the European Association of Nuclear Medicine. The Society of Nuclear Medicine is the nuclear medicine organization of the United States, but also has several members from other continents and countries. The EANM was founded in 1985 to integrate the former two European societies, the Nuclear Medicine Society Europe and European Nuclear Medicine Society. The Association brings together nuclear medicine professionals, not only from Europe, but also from other countries. Both organizations are in close cooperation with other societies and host joint meetings on various disorders, from pathology to pathophysiology, diagnostics and therapy, instrumentation and radiation safety. All physicists with interest and ambition from all corners of the word are welcome in our club. There is a flame that will shine all over the world – this is nuclear medicine. This will happen if you join us.

"When you reach turning point in life, it is best to turn." Henry Wagner, Jr.

DOI: 10.1201/9780429489501-25